Architectural Plans for Adding On or Remodeling

Architectural Plans for Adding On or Remodeling

Jerold L. Axelrod, Architect

TAB Books
Division of McGraw-Hill, Inc.
Blue Ridge Summit, PA 17294-0850

FIRST EDITION
THIRD PRINTING

Library of Congress Cataloging-in-Publication Data

Axelrod, Jerold L.
Architectural plans for adding on or remolding / by Jerold L. Axelrod.
p. cm.
Includes index.
ISBN 0-8306-3930-6 ISBN 0-8306-3929-2 (pbk.)
1. Architecture, Domestic—United States—Designs and plans. 2. Buildings—Additions—United States. 3. Buildings—United States—Repair and reconstruction. I. Title.
NA7205.A84 1992
643′.7—dc20 91-41375
CIP

Acquisitions Editor: Kimberly Tabor
Book Design: Jaclyn J. Boone
Book Editor: April D. Nolan
Director of Production: Katherine G. Brown WT1
Cover: Holberg Design, York, Pa. 3897

Contents

Acknowledgments

(Past and Future)

I would like to extend grateful appreciation to the National Association of Home Builders for providing access to their local home improvement councils, and to all those remodeling professionals throughout the country who provided me with plans and information on local remodeling practices. That information, together with my own personal research and travels, has enabled me to prepare this book of plans that should be relevant throughout North America.

I wish to further extend appreciation to those many clients our office has served, and whose remodeling projects helped inspire some of the plans herein, and whose experiences have afforded me the insight to be able to help others achieve their dreams. Sincere appreciation is extended to my dedicated office staff, for their marvelous efforts in putting this material together, and to Florence, Sharron, and Neil for allowing me to usurp the huge quantities of time that this undertaking required.

Finally, I want to encourage future acknowledgments. This book is a first-ever attempt to graphically quantify the most prevalent and common home improvement designs that make sense today. However, I know it is not finite; I am certain that there are likely many other commonly pursued plans that have not been shown. I encourage you to send these to me, either in plan or photograph, so that I may embark on the research for Volume II. And to those who eventually do build from these plans, I would love to see the completed photographs for a Volume III. My address is:

Jerold L. Axelrod, Architect
C.B. 9015
Commack, New York 11725

Introduction
What this book is and is not

This book is primarily a book of architectural plans. It is the first, to my knowledge, to ever attempt to provide to the homeowner contemplating remodeling his home with a veritable potpourri of hundreds of practical plans for that purpose. With over 800 illustrations and 240 different plans, it has something for everybody. There are beautiful room additions, exciting internal room remodelings, and even dramatic whole-house renovations; they range from a roof clearstory, which sheds new light to the interior, to a complete remake of a 4,000-square-foot colonial. There are porches, sunrooms, and vestibules, as well as kitchens, baths, bedrooms, and family rooms.

There are plans that expand up or expand out, and some that rearrange and even reduce rooms. There are contemporary additions and traditional ones. But most important, these are plans explicitly designed to remodel someone's home. The homes chosen are likely to include yours; there are one-stories, split-levels, one-and-one-half-stories, old bungalows and cottages, and recent two-story traditionals too. The plans presented are a product of extensive research; it is likely that there is a plan for remodeling that both suits your specific needs and matches your current home. Construction blueprints can be obtained for every plan shown, which will enable you to turn into reality any design—or combination of designs—you prefer.

This book is also an encyclopedia of ideas—both graphic and written. It is notable for its profusion of artwork, but the written words are not meant to be glossed over, either; much of my experience has been put down in the written material presented here. Although this is a book about remodeling your home, it should not be confused with the majority of books that deal with the subject; most are how-to books, written to assist you, once you know what you want to build and are ready to immerse yourself in the task of construction. This book is not that.

Many remodeling books attempt to teach the reader how to design his own project, and then how to draw the construction plans. For the majority of homeowners I believe this proves to be an extremely frustrating task to learn. Most books, unfortunately, gloss over the subject of design, presuming the reader will gain that elsewhere. But where? The major source of such ideas is the home journals, which do feature extensive examples of successful remodelings. But most such examples have little practical value to the homeowner frantically trying to plan a remodeling project. The majority of photographic features are, by nature, esoteric, one-of-a-kind essays on a custom renovation. In conversation with many clients, they tell me that they gain helpful ideas from the journals, but find it difficult to apply these to their own home.

Interestingly, the original inspiration for this book was provided to me by the editor of one of these journals more than ten years ago. He recognized the difficulty in the consumer's ability to find practical, readily usable, design ideas for their own home. I resisted the concept at first, primarily out of fear of the magnitude of the undertaking. Over the past few years, though, I did begin to research and collect ideas for common themes. However, the almost infinite number of variations that are possible for any one design did speak against trying to standardize them.

The answer to that came (or so I hoped) when we purchased our computer. After several years of frustration and several aborted efforts, we returned to using pencils, only finally to achieve success. All of the plans reproduced herein have been drawn on our CAD system, making it very easy to accommodate changes to effect individual nuances and variances in any home. This capability has enabled me to develop standardized, readily recognizable, existing home plans to work with as a base, with full knowledge that this base will grow over time, as will the remodeling projects themselves.

As to the organization of the book: resist the temptation to look at pictures first. There is a modest dose of helpful written material at the beginning to help you get started in the right direction.

The first three chapters of the book are designed to help you analyze your needs, evaluate your home for change, and develop a program. I do that with every custom client we have; it is a process you should not gloss over, as the satisfaction you ultimately receive from a finished remodeling project is directly related to the quality of the program you prepare.

The fourth chapter deals with what not to do. I debated frequently whether or not I should devote any valuable space to this subject. The problem was that to do it thoroughly could take a pictorial volume in itself. A brief discussion plus several pages of drawings was the compromise. If the project you are contemplating involves an addition, there could be some valuable information here.

A discussion ensues, in chapter 5, on how to best use the prototype plans that make up the balance of the book. There is a discussion on when you need an architect and on how to have these plans customized, if necessary, at moderate cost. I do believe that architectural help is important, but that is why I prepared this book. My relationship with you might lack personal contact, but the ideas—and plans—are the most important help that you can obtain.

The first chapter of drawings, chapter 6, presents 60 prototype shapes for successful additions. I call them *blockforms* as that is all they are: simple block shapes that you should be able to recognize easily and relate to your home. If properly followed, any of them should lead to an addition that meshes well with your current home.

The actual prototype plans are divided into four chapters. The first chapter, chapter 7, essentially presents room additions that go up or out in every direction, on an extensive variety of existing homes. The additions comprise the full gamut of needs ranging from apartments to family rooms and kitchens to bedrooms and baths. If your plans include an addition, it is likely you will find one here.

Following in chapter 8 is a voluminous collection of plans appropriately dubbed "Bumps, Bays, Extensions, and Interior Remodeling." Because the focus of the publishing world is usually on room additions, it is the identification and presentation of these smaller spatial undertakings that make this book most unique. The majority of home-remodeling projects usually include such smaller bump-outs or interior space rearrangements, yet they have never been analyzed and presented in a usable format. It is these that are readily prone to design error, more so than a self-contained addition. Furthermore, even most additions incorporate such smaller spatial changes to the home, and they can have a measurable affect on the ultimate success of the addition.

Chapter 9 presents a generous collection of stylish new front facade studies. Each portrays the old with the new so you can relate these drawings to your home. Exterior remodeling has become increasingly popular, and I have provided ideas for your consideration. Whether the exterior is being remodeled alone, or as a by-product of a space addition or alteration, this chapter is on target with today's trends.

In chapter 10, there is a wonderful collection of whole-house renovations. Included in this chapter are remakes of several old bungalows, cottages, and "cape cods." The remodeling of these lovely old homes is presented very clearly so even if yours is not exactly the same, you can learn from them.

A second group of whole-house undertakings is included, and these are derived from the prior chapters; this unique group of designs demonstrates how you can combine various plans shown in the book to create your own distinctively personal remodeling program. It is through the process demonstrated in this last group of homes that I hope you will learn how to take my designs and ideas and adapt them to your specific needs to achieve the personal satisfaction we all strive for in our remodeling projects.

I have said that this is not a nuts-and-bolts, or how-to-build book; well, chapter 11 does do a little of that. My motivation in this chapter is to provide several construction details that are design-driven, and which are not likely to appear in the how-to construction books you purchase.

Also included here is a handy chart to aid you in sizing girders so that you can open up rooms to each other—an important design tool. A detailed discussion and graphic presentation on staging a project is also provided. This frequently neglected aspect of remodeling is an essential part of the design process itself; whenever possible, I try to plan any intricate, involved project so that it can be developed in stages. The ability to do this can make the difference between a pleasurable and a miserable experience during construction.

Finally, chapter 12 takes a brief look at finishing your project—again, not with sandpaper and paint, but within the broad concepts of interior and exterior design.

So whether you're adding on, remodeling, or contemplating a little—or a lot—of both, let's get going. I want to see your finished snapshots.

1 CHAPTER

Why remodel?

The reasons to remodel are as individually different as are our fingerprints. Yet with all of our uniqueness, we tend to live in homes with a reasonable degree of sameness. These various homes eventually age the same, and although each of you might approach that problem differently, the decisions to remodel your particular design or style of home tend to fall into certain likely patterns. This book latches onto those commonalities in the designs presented. But before you choose any design it is first necessary to identify why you want—or need—to remodel.

A successful remodeling project has likely satisfied some predetermined need. It was successful because it began with a need that clearly surfaced and was properly identified—and a correct path was chosen to satisfy that need. As an architect, I will become involved with my clients in trying to help them identify needs so that the design I propose will be relevant. When someone complains about a disappointment in their completed remodeling, I usually suspect an irrelevant design, which did not mesh with that person's needs.

Put your needs to paper. Try to identify them, and don't skirt the process. Try to be as open as possible with yourself. Ask yourself what you expect to achieve in a contemplated remodeling. What changes have taken place, or will take place, that you should recognize? The answers to these questions are the beginning of a *program*, something we will delve into deeper in chapter 3. This program, which we will develop together, meshes your needs with the inadequacies and opportunities presented in your current home. But for now let us concentrate on needs. Hopefully you recognize that "needs" are personal requirements; we are not yet talking about the needs of your home.

Changing family needs

As families grow or decline or change, their needs change too. Changing family needs are the most common reason to remodel. As children come into the family, there is a need for bedroom space. As they grow older, there is a need for adequate play or study space. When they marry and leave, there is usually a surplus of space poorly serving the remaining occupants. These are the most often-cited reasons for adding or changing space—needs that frequently result in additions. They are *physical space* needs, as contrasted to *lifestyle* needs. The driving motivation is simply more space rather than the style or quality of the space.

A second marriage might demand some changes in lifestyle, but it also might bring physical space needs due to the changed complexion of the new family. In any changing family, try to carefully define the new spatial needs by functions needed—e.g., more sleeping space, play space, study space, private space, eating space, formal dining space, and informal and formal entertaining space. Don't forget changed storage requirements, too.

Requirements for a different amount of space

Many of these family changes result in a new requirement for the size or number of rooms. Determining the number of bedrooms or baths required might be easy, as is evaluating that you now need a family room. Determining that you have to increase the size of your kitchen or living room might prove a little more difficult because you will get trapped in the "how much" syndrome. Don't worry about size now; the goal is to determine needs. The quantification process can come later.

Try to define where you feel that your home least fulfills your spatial needs. Do you feel cramped in your kitchen? Do you enjoy a formal dining room? Would you prefer a kitchen eating area? If so, would you prefer a full breakfast table space, or is a snack counter with bar stools preferable?

What function will take place in the new family room and old living room? Do you need both? Would you prefer that they be interconnected for large scale entertaining, or should they be clearly separate? Remember to jot your needs down on paper.

Requirements for a different type of space

Many family changes also result in a need for new types of space that might not exist in your home. Some of these might also require that space be added, and as such, they also affect the amount of space that might be necessary. Many are reconciled by "rearranging underutilized space," which we will later quantify.

Perhaps you now work at home and need a home office. Is it simply a corner of your bedroom or living room, or is it a room? Do you enjoy adequate space for hobbies? Are you a garden enthusiast, and would you love a greenhouse?

Does your lifestyle now demand more space for adult entertaining? Is there any place for your new entertainment equipment? Do you no longer need distinct divisions between entertaining rooms?

How do you feel about an open kitchen? What about a great room where all (or just some) of the cooking, dining, and entertainment functions are combined, or opened to each other? Is a country kitchen to your liking? That's a great room of a less formal nature, where cooking, light meals, and casual relaxation are combined into one flowing space?

Do you feel the need for more space for bathing or dressing? And what about your closets? Do you have to benchpress to develop the strength to move your clothes? Would you prefer some extra space in your bedroom for reading?

The more your list of needs includes many of these contemporary living patterns, the more likely your current home will fall short, and the more likely you will need an extensive remodeling to satisfy them all. Therefore, unless you have no budget restrictions, which is unlikely, you should start to prioritize your list of needs—from those you absolutely must have to those that would be nice, but which you could live without. This is the time to dream, but let's try to make dreams come true.

Housing aging parents and newly married children

The nuclear family has stopped retreating. For many reasons, including emotional and economic ones, we are finding today that more and more extended families are living together again. Projects coming to my office often include the request for an apartment. If the apartment is for aging parents, it poses a major set of design restrictions as to where it can be located, particularly due to the potential need for handicapped accessibility. A first floor location is usually mandatory, but the need for a separate kitchen or entry might be unnecessary. But that is for you to define.

An apartment for a married child is a different need, with a very different set of requirements. A separate entrance and separate kitchen (building codes allowing) are usually required; however, the location of the apartment could be a second floor space. An apartment for income purposes could have the same requirements, but you should first check with your local municipality about legal requirements.

Psychological or emotional needs

As important as it is to identify your physical or spatial needs, it is equally necessary to clarify those very personal motivations that might be driving a decision to remodel your home. It is here that I might fail to connect with clients because of their inability to communicate a very private personal need. However, it is more important to be true to yourself. Now, I'm not talking about religion or mysticism. The issues that comprise this subject are real psychological or emotional needs that every remodeling project conjurs within the client or homeowner, and being honest with yourself will help ensure that successful remodeling I keep talking about.

Ask yourself, what are you looking to achieve beyond the addition or changes required? Do you expect to achieve a very personal gratification in your project? If you do, expect to be very involved in every decision. Are there sensual or aesthetic demands you would like to satisfy? If so, be honest in requiring that the proposed remodeling will satisfy those yearnings. It's too late once it is finished.

Perhaps it is a perception to others that must be addressed. There is nothing wrong with that as a motive for remodeling. So many people thrive on satisfying that need, but for some reason we hesitate to acknowledge it. If that is your motivation, it is clear that just adding walls and a roof will not suffice. Many of the plans presented in this book, particularly the new faces, acknowledge this need.

Possibly you must satisfy a need to express a newfound stature in life. This might result in the same requirements as the previous need, but it could be different if you are less concerned with perception to others than you are with achieving a personal gratification. Given this need, chances are you will be willing to be less conventional and more individual in your remodeling. So go ahead and be bold; if you don't, you could achieve less than satisfactory results.

The needs I am asking you to identify here are clearly very personal ones, and it is very possible that these could override purely spatial needs. In this vein would be a decision to remodel a clearly inadequate home, despite all the arguments to move, because of an extremely personal attachment to the home or to the community.

The identification of other personal requirements will likely help to clarify and define spatial needs. As an example, if you are a gourmet cook, you would probably prefer that your new kitchen be an expansive, beautiful space because you will spend much time there. It might include a large center preparation island, commercial ranges, and dual sinks. If, however, your favorite meals are prepared in your neighborhood restaurant, your focus for a new kitchen would likely be more practical. A love for plants will require sunny places to put them, whereas a penchant for horticulture might demand a plant room.

There is a clear danger in identifying such needs—our inability to ultimately satisfy them all in a remodeling project. So at this point it would be helpful to start asking yourself: are any of these needs compatible with others, or could they be combined? As an example, maybe a demand for a whirlpool spa could be merged into the same sunspace that houses the planting enthusiast. Or maybe that large-screen TV, and its associated equipment, doesn't need a separate room, but could find a happy place in a new country kitchen. Fortunately, many of the plans presented in the book have made some of these compromises for you.

Economic needs

Alas, it would be such a beautiful world if reality didn't present itself in the form of our checkbooks. However we may wish otherwise, in order to achieve that elusive, successful remodeling, I keep referring to, you must deal with the economic needs of a proposed remodeling.

Can you afford it?

Since the focus of this book is design, space does not allow me to delve into the myriad of issues concerning the financing of your project. I only touch upon these as they affect design decisions. There are numerous sources for you to consult for detailed information on this subject, including some of the how-to books. What I am concerned with is how you approach the subject as it affects the development of your design program. Once the scope of your proposed project is reasonably well defined, you might be able to get a handle on how much you have to budget by asking local contractors or neighbors who have completed similar projects. While you can't rely on these to provide you with your budget, they can help you establish a range so that you can determine whether or not you are reaching. Hopefully, the range is less than you expected, for you will hear it here, and everywhere, you have to allow at least ten percent more for those inevitable changes or unforseen problems.

The more precise method involves actual bidding from plans—and I am jumping ahead here, but purposely, so that when you reach this point you can have the proper response. Ready-made construction plans, such as those that can be obtained from this book, provide an inexpensive and potentially emotionally uninvolved way to obtain firm costs. Even if the plan isn't exactly what you want, it might be wise to price out what it would cost to build, before you go ahead and make plan changes, or have it redesigned. The emotional ramifications will be less straining, should the project come in way over budget, than if you have spent months laboring over your own plans. If that happens, it will be necessary to redefine your program and bring it down in scope. But better at this point than when you are halfway through construction.

Your project as an investment

I am sure that you have read that certain types of remodeling projects are very sound investments, in that they provide a return on the dollars spent. New kitchens, bathrooms, and extra bedrooms usually provide very close to an immediate return on what you have spent. New family rooms return somewhat less. However, this is a very subjective matter, and it does vary by community.

Exterior remodeling projects, such as windows, siding, etc., also tend to fare well, especially if they achieve some energy-saving benefits. It is my belief that any remodeling project that produces a significantly more attractive home will also bring a good return on the investment.

What you have to evaluate, though, is how important this all is to you. I frown upon a decision that dictates what to do, that hangs solely on a return on your investment; it tends to negate all the subjectively beautiful things you can do that can make your house a home. If you intend to remodel, and you expect to be there for a number of years yet, investment return should play a lesser role. A more important goal should be to better your living environment so that you can enjoy a more rewarding lifestyle.

Remodeling vs. moving

Although I tend to discuss the question of remodeling vs. moving under an economic heading, because the pivotal issues are often cost concerns, this is not at all to underestimate the very emotional, as well as the many spatial issues you must reconcile in order to arrive at a learned decision. Let us start with the emotional issue.

Are your emotional attachments such that, regardless of your answers to the other questions, you will find it extremely difficult to move? If, after careful analysis, the answer is yes to this question, it is likely to take precedence over all other issues. Hopefully, the program you develop can be satisfied, but if it cannot, you might have to make significant compromises on space needs, as well as on other aesthetic or personal wants.

Remember to carefully weigh whether your current neighborhood is likely to retain those attributes that give rise to your decision to stay. It is also necessary to acknowledge that living within a home undergoing a significant renovation is not fun. Be prepared to accept some emotionally upsetting times. While it might only last a few months, it could seem like an eternity.

Will remodeling satisfy your changing needs? The answer is usually yes. How can I say that so unequivocally? Because of the inherent flexibility of most homes and the infinite number of solutions that I can develop for almost any given home. Not finding a solution to your particular needs and your particular home in this book is not an indication that it can't be solved. On the contrary, it is likely that one or more solutions can be developed for you. You'll find a discussion on the subject of custom designs in chapter 5.

There are, however, certain circumstances where your current home would not satisfy your needs. An example is a home on a very small lot where there is no room to expand out, but your requirements demand new ground floor area—such as a new kitchen, a new garage, or a new handicap-accessible apartment—and there is nothing available from existing internal spaces. Another example could be a very old home with historic relevance, which might demand that you move if your special needs cannot be compromised with the special requirements of that home. But these are exceptions.

Now to the economic issues. Will the remodeling be an over-improvement? This is usually the only economic issue that might argue for moving. It is distinctly possible that your needs result in a remodeling project that could cost so much that, when added to the home's current value, it outstrips the values in the neighborhood by an amount that is economically not sound. How will you know? Obtain some ready-made plans, estimate their cost, add it to your home's current value, and talk to real estate brokers and bankers. If the answer is affirmative, reduce the scale of your project, or move. Remember, you don't have to equal the average resale price of your neighborhood; you could exceed it by 20%, but talk to your local real estate professionals.

Can you afford to move? The answer to this question involves a myriad of issues such as financing options, the cost of new money, the true total cost of moving, and the cost of new houses. The availability of equity financing and the additional costs associated with moving—such as landscaping, furniture, financing costs, brokers fees, etc.—usually speak favorably toward remodeling. However, it is something that you should carefully evaluate before finalizing your decision.

Summing up

We have looked at identifying your physical needs, emotional needs, and economic needs; together they form the basis for why you want to remodel. Although it might prove difficult at first to try to think your way through these issues, it is an essential first step before embarking on the writing of a successful remodeling program.

2
CHAPTER

An overview of remodeling projects

Once you have identified your needs, you are ready to start developing a detailed program. However, before you plunge directly into another heady task, it might be helpful to take a broad-brush look at the myriad of remodeling possibilities. Even if you think you know what you want, and you don't believe that writing a program will be meaningful, this could be a valuable exercise.

Specific vs. general projects

Specific needs are easier to find solutions for. For example, an identified need "for an extra bedroom" is fairly straightforward, whereas a need that concluded that "your house is too crowded" is too broad a general statement; it will require a more detailed analysis before you get much further, to see what types of solutions best satisfy you and your existing house.

Exploring all options

Whether your project is specific or general in nature, there are likely to be a number of solutions that will satisfy the requirements. It is important for you to look at all the solutions. You might find an alternative plan that runs contrary to a preconceived notion you had, but which, in actuality, works much better. Be open with yourself at this point, and avoid preconceived answers.

Let's try an example. Assume your goal is an enlarged dining room. The existing rear facing room is hemmed by the kitchen at one side and the garage on the other. The likely solution, you say, is to expand into the rear yard. Well, maybe—but what if you also recognize that the kitchen is due for major remodeling, too? Maybe the dining room should expand into the old kitchen space, and you should build a brand new kitchen out the rear. The results of this might be far more exciting than expanding the dining room to the rear and remodeling the kitchen in place. In addition, it is easier to stage this project, and it will avoid much of the mess.

The above remodeling happens to be pictured in the book, as plan KD006 on page 130. One of my persistent themes is exploring alternatives. You will find a number of plans that provide different solutions to the same problem or need, that I have done with bedrooms, bathrooms, family rooms, kitchens, and even whole-house renovations.

Types of projects

There are four types of remodeling projects. Let us look at each.

Projects that add a room or rooms

Whether the room is added up, out, or down, the following are possible room additions:

Bedroom
Kitchen
Country Kitchen
Family Room
Dining Room
Living Room
Great Room
Office
Master Suite
Bedroom Wing
Media Room
Library/Study
Playroom
Bath
Breakfast Room
Laundry Room
Porch
Sunroom
Exercise Room
Apartment
Professional Suite
Barrier-Free Apartment
Foyer
Garage
Walk-in Closet
Storage Room

Making the possibilities more intricate, any of these can be combined into a single project. For example:

Family Room, Bedroom, Bath
Kitchen, Breakfast Room, Family Room
Porch, Office, Garage
Kitchen, Family Room, Master Suite

When you consider that any of these can be combined, depending on your needs and your existing home, and that the solutions can be any mix of one- or two-story plans, again depending on your home, you do realize that there can be an infinite variety of types of remodeling projects. This book presents an extensive variety of such combined programs. It also goes beyond by showing you how to combine various ideas on your own to create your own personal program. An important aspect of a room addition is the possibility of relocating other features of your home—something you might not have considered originally.

Projects that add space to or remodel existing rooms

The same list of rooms applies, but the actual remodeling projects are very different. In this type of project, the existing room most often retains its original use, but the space has been redesigned to better service your needs. The potential variety of such projects is even more infinite than the prior list because you can alter virtually any space within your home.

Such a project can be as diverse as the following (all one project): remodeling your kitchen in place, bumping out a bay in your breakfast room, popping a clearstory for light in your living room, combining two bedrooms to create a new master suite, creating a garden bath, opening up a loft in the attic over your living room, creating an extra powder room from one bath, opening up walls to create a great room, popping a greenhouse on your dining room, and bumping out a sitting alcove.

The list is potentially endless. Whatever your program eventually tells you, if it concerns adding or remodeling space (as contrasted to rooms), it is likely you can accomplish it.

Projects that update your home

Included amongst these are: Projects that simply update fixtures and equipment in bathrooms and kitchens; energy-related projects, such as new windows, heating systems, insulation, etc.; projects that update the exterior of your home by creating a new facade.

Because this is a design book, I have dealt only with the last one. Updating your home's exterior can be a very important part of the design program you develop. You can readily combine a new facade with any of the other types of projects so that it becomes an integral part of your overall project.

Projects that remodel an entire home

Sometimes the scope of what you want to achieve is so all-encompassing that it involves virtually the entire house. While this is more likely to occur in an older home, it is not necessarily a rule. The same could apply to a home only 15 years old, if the program you develop suggests wholesale changes, to bring all parts of the home up to contemporary standards.

Many homes that are more than 50 years old might require a complete remodeling, especially if very little has been done to them over the years. Some of these homes have unredeeming floor plans, which so inhibit modern living patterns that you have no choice but to redo them stem to stern. Frequently someone buys such an old home because it is charming to look at and is situated on a lovely parcel of land. Hopefully you bought it "right," leaving yourself enough to completely remodel it.

Such concludes my brief review of alternative types of remodeling projects. Keep in mind that many programs lead to a project that can simultaneously involve several of these types. It is less common to find a project that is tidy and neat and simple to define. Let us now begin to develop your program.

3 CHAPTER

Developing your own program

I have mentioned this "program" before—what is it that I am talking about? A *program* is your guide to a successful remodeling project. It will lay out, in written form, the main objectives you are seeking to achieve, which rooms are to be remodeled, what is to be added, where it is to be added, what features you expect in each room, what exterior concerns you have, your stylistic preference, and any other personal inclinations that are important to you, such as relationships between various rooms and where privacy is essential. It is a wish list. You will need to prioritize these, for if your list is long, it is likely that you will have to make certain compromises, and you should know which items are not adjustable and which are. Don't worry about format.

Developing a program

The program begins by writing down the personal needs you identified in chapter 1. Anything you write down from there on should be viewed against these needs. Do these now.

The second step will be an analysis of your home. The purpose of this step is to identify problem areas within your home—which areas fall short, which need updating, which are OK.

The third step will be an evaluation of your home for change. The purpose of this step is to identify where you can make changes or where you can add. There is a tendency to short-circuit the process by combining steps two and three; do not fall to this temptation. You will achieve better results if you first identify problems, without simultaneously recording potential solutions.

The fourth and final step is to assimilate this information and develop the remodeling program as previously defined. Once the program is written there is another step, which cannot be overlooked, that concerns retaining flexibility.

It would be helpful, after you finish your program and prioritize its components, if you develop an alternate list that deals with issues such as: what if your program is way over budget? Should you just strike something, or is there an alternative course to pursue?

A simple example: your program listed in priority order 1.) a slightly expanded, remodeled kitchen; 2.) a new breakfast room; and 3.) a new sunporch. You are over budget. You could just strike the sunporch, but maybe you could enlarge the breakfast room some and add enough windows and skylights that it simulates a sunporch and leaves enough space for a small sitting area beside the table.

Also remember that there are many potential solutions to the same problem. You can achieve that new larger kitchen by: building out a new room; reworking interior space, with or without a small addition; or going up with something else and thereby freeing first floor area. So try to avoid preconceived solutions and give a fair review to all possibilities. The plans shown in this book should help you do that.

Analyzing your current home

I will now detail a systematic approach for you to follow in analyzing your home. But before we start, let's talk about obtaining a plan to work with. You will find it helpful to the overall process if you have a floor plan(s) of your current home. You might have been given a set of plans when you purchased your home, or maybe you have a builder's brochure (actually, that's the best, because it is simple to read). Unless your home is very old, you should also be able to obtain a set of plans from your local building department. If all else fails, it will be necessary to measure your home and prepare a rough, sketchy layout. If you are unsure how to do this, talk to someone who has done it; it is really not that complicated. Once you have a plan in hand, you will analyze your home by visually walking through and examining the rooms, one at a time, to identify possible problems.

Circulation problems

Circulation problems are most common in older homes and tract homes, but they can be prevalent in any home. The paths you and your family use in your daily travels through and around your home are the circulation paths you should observe. Do these paths cut through certain rooms? If they do, do they affect your ability to use or furnish these rooms? Such problems are common to living rooms, dining rooms, family rooms, and kitchens. Sometimes just the simple relocation of an opening or a doorway can solve the problem. More often, though, it requires more serious remodeling. But that's the next step; let's avoid looking at solutions now.

Circulation problems frequently result from the fact that the home was never designed with ample circulation routes to begin with. Rooms were expected to provide part of their area to serve as connection to the next room, and to the next, and so on. I am a fervent advocate of defined circulation routes. I believe they add to the functionality, furnishability, and livability of any home. I am not dogmatic about circulation routes necessarily being confined to designated halls, but if a room is to serve a circulation pattern, it must be studied as it concerns your ability to furnish it, and, if necessary, the room should be enlarged or combined with another.

Congested places are another possible symptom of circulation problems. Although congested places are frequently a symptom of inadequate space (and we will deal with that shortly), it is also possible that a horrendous circulation problem is the cause. Study your habits. It is possible that the family has unknowingly adapted its living patterns to solve the problems of the home. A small example: if everyone must take off their shoes before entering, if it is not by social custom, it is likely that your circulation patterns are not in order.

Wherever you identify a circulation problem, write it down on your program notes and make a note on the floor plan.

Underutilized areas

Every home has some underutilized space; the questions are how much and where it is and whether it has some other potential use. The whole subject of space utilization has some disputable aspects. For example, your bathroom and kitchen probably get more use than your dining room or your living room. Should your bathroom be larger than either of these? Well, that is happening today, especially if the bathroom is going to be used by more than one person at a time. This discussion was improbable for the architect designing a bungalow in 1925.

The decision of where to allocate new space should include the possible redeployment of underutilized space. However, the decision to declare space surplus is a personal one. Can you truly live without that formal dining room? Could you do away with your living room? Only you can answer these questions.

Sometimes underutilization might be only a part of a room's problem. It might not be that you don't need it, but that some other problem, such as poor circulation, has rendered it underutilized. Look at each room carefully, and try to identify the actual problem.

Finally, don't confuse this phase with finding space. That is part of the next step. What you are seeking to clarify here is whether or not you have any excess existing space not serving you well. If so, write it down on your list and plan.

Congested spaces

I have spoken of crowded bathrooms; well, that's almost endemic to the species. If that's so, then write it down. What else is congested? Is your kitchen too crowded all the time? What about your family room, if you have one, or your living room? Is it crowded because there are too many activities that must take place there (i.e., eating, cooking, watching T.V., games, etc.) or is it that there is a poor circulation pattern that forces everyone through a tight space? These are questions that might take some careful analysis to resolve.

One of the best ways to answer the question, as it concerns activity rooms, is to list your family's activities, where they are currently being carried out, and where you would prefer they be centered. This will help you later decide how many different rooms are necessary, whether or not they should be connected, and how big they should be.

Some of the problem areas you identify as being congested might be very subjective. What appears too cramped for one person might be more than adequate for the next. That is why each remodeling project develops its own program—and ultimately its own personality. So don't necessarily be guided by norms for room sizes; make your decision based on what will satisfy you. For this reason you will see in this book, for example, a new family room 11 ft. × 16 ft., another 20 ft. × 30 ft., and a third plan without a family room, but with a great room instead.

Inadequate places

The discussion of inadequate places might overlap with congested spaces, but there is a reason for me to discuss them separately. You might ultimately deal with these as one subject, but let us see the distinction. While your master bedroom might appear congested because it is too small, that might be so because what you seek is a sitting room or alcove for reading. Is it not, then, that you are short a place or your home is inadequate in this regard—not congested?

This is a subtlety, but it is important to distinguish whether you need larger spaces to satisfy congestion, or whether you need new places to satisfy missing functions. An easy example is a shortage of the number of bedrooms or bathrooms; that is a problem that will be solved quantitatively by increasing the number of rooms, not by increasing the size of a room.

So however you decide to review these potential shortcomings in your home, try to distinguish whether the space deficiency is one that is best resolved by increasing the size of rooms, or by adding new rooms or spaces or alcoves.

Rooms in need of repair or modernization

One of your tasks in going through each room is to try to list areas in obvious need of repair or modernization. If the repairs are purely cosmetic (something a paint job will take care of) and the room is otherwise a well-functioning part of your home, you would not likely look to focus any remodeling here. If, however, the room is in desperate need of remodeling or updating (an old kitchen, for example) then there is a greater likelihood that other efforts might be directed here simultaneously. Skylights, new windows, or an extension might be readily accomplished at the same time.

Whatever it is, put the item on your list. It could have some bearing on the overall scope of your project. In addition, the ability to "package" miscellaneous repairs into one project has cost-saving potential, as well as financing opportunities that might not otherwise exist.

Exterior repairs or modernization

Do your roof, siding, windows, gutters, or trim need repair or replacement? If so add the items to your program list. As discussed, there could be benefits in doing it all together with space additions or changes. In fact, it might well be that this is one of your motivations—or even your prime concern—in undertaking a remodeling of your home. If so, state it in your program. Chapter 9 deals with this subject in somewhat greater detail.

One of the major side benefits of any large remodeling project is the ability to purposefully redesign the exterior of the home to afford it some enhanced aesthetic appeal. I encourage this approach because I believe that to undertake a costly project and not to provide any consideration toward improving the appearance of the home is an opportunity lost.

Landscaping, walks, patios, driveways

There are several areas you should consider here. First, do any of the exterior site improvements to your home need any repair? If so, add them to your program list; the same logic prevails about doing them simultaneously with other improvements.

But there is a second area of concern. If your program will lead to room additions or space enlargements it is very likely that some disturbance will take place in landscaping, walks, or patios, and possibly even your driveway.

You should consider these disturbances very carefully within your overall program. You might have to move some extremely ornate shrubs which could affect the timing of your project. More often than not, however, this is an opportunity you should seize. As with the opportunity to redesign the exterior of your home, look upon this as a chance to improve the landscaping and the patios. Add any of your ideas to your program notes.

Before you think I am callous with your money, remember there is a distinction between planning the design of new landscaping and patios and actually doing the work. You should certainly seize the opportunity to replan your yard, but you could defer the work. We'll talk more about finishing your project in chapter 12.

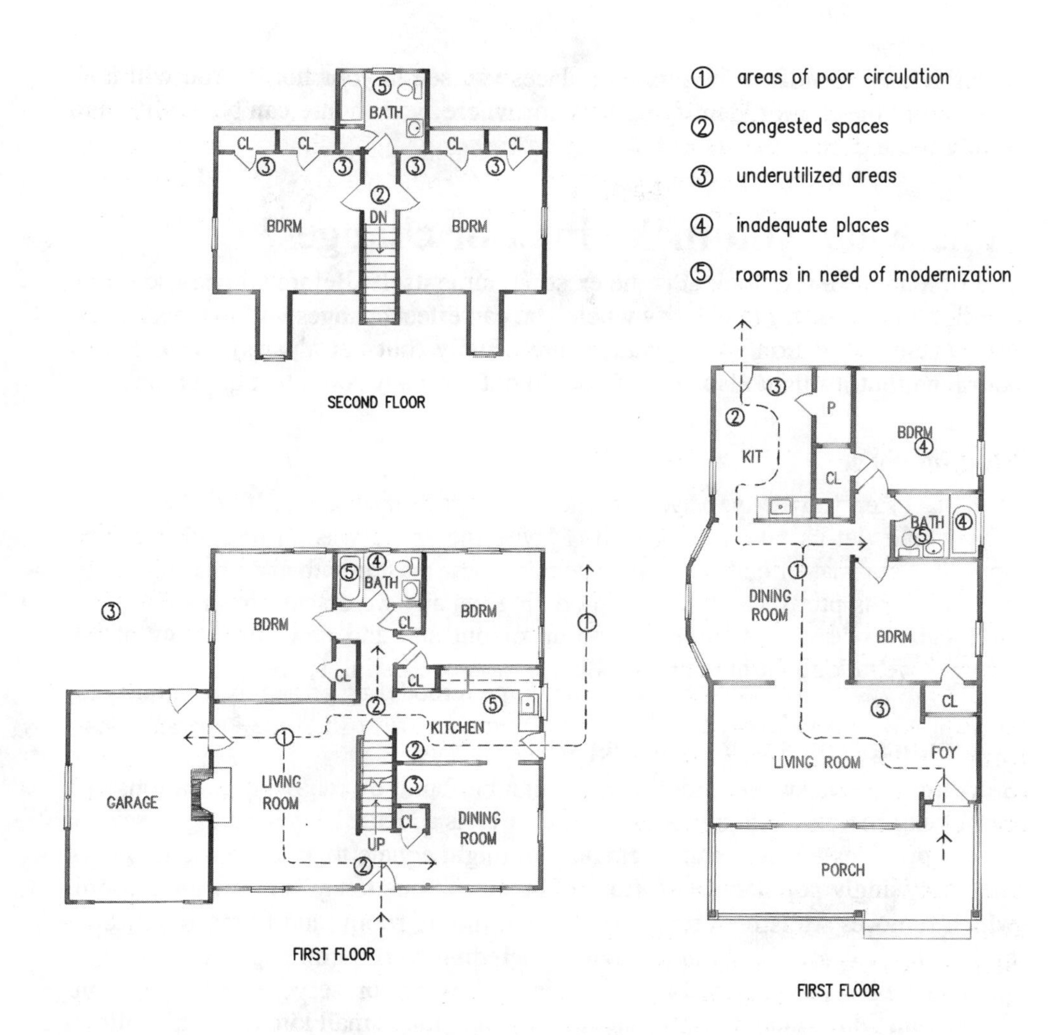

ANALYZING YOUR CURRENT HOME

Determining current market value

One of the unrelated items you might consider doing as you are analyzing your home is to try to determine its current value. This could be an opportune time since you will be focusing intently on your home, its room sizes, and its pluses and minuses.

The only reason to do it, though, is if there is any serious concern about the potential for overimproving. I tend to minimize this issue, as I said before, but if you are worried, then this is an opportune time to deal with the subject.

As an aid to help visualize the potential deficiencies we have discussed so far, several floor plans are presented above, with their problem areas identified.

You have identified the problem places you see in your home. You will now move on to the task of identifying how, or where, your home can be modified to satisfy these problems.

Where can you make interior changes?

The first area we will look at is the existing home itself. Before you consider any exterior additions, try to identify where you can effect changes within the shape of the current foundation. Such changes are usually (but not always) cheaper than additions, but it might also be that you do not have any room to expand out.

Finding space

In the last step you analyzed your home for underutilized space. If you have identified some, let us take a look at it. Maybe the space was an unused bedroom opposite your master bedroom; your goal is a new master bath and dressing room. The answer is pretty clear. This remodeling, as any we discuss in this part, is a reallocation of space. Unless you go up or out and add new space, any of the changes we talk about here are a redistribution of existing space.

Combining rooms and removing walls

Sometimes the answer to finding space can be found by removing partitions and combining rooms. Although no new floor area is added, the combining of several spaces provides a visual enhancement that might equate to an expansion. One of the increasingly popular applications of this technique today is in the great room, which removes walls between your kitchen, dining room, and living room. Several such plans are shown in this book, including KDL03 on page 223.

Maybe it's only a partial removal of a wall, which preserves some separation. A low wall with a counter above can do the same for a small kitchen, while allowing it to visually be part of a larger space.

Sometimes you can achieve most of the same effect by creating larger openings in existing walls. This might be necessary if the walls are structurally bearing, which we will discuss later. It also could be beneficial where you feel the need to maintain some greater definition of a room because of wallpaper or some other decorating concern. An excellent example of this is the opening up of a small dining room to the adjacent space, whether it be a foyer, a family room, or living room. Similar spatial enhancement occurs, but short walls at each side and a header maintain the definition of the room. You will also see plans in this book that use interior columns in lieu of walls—another very stylish technique that is frequently used today.

Moving partitions

On occasion, it might be advantageous to move an interior partition a few feet. This could prove beneficial, say in a kitchen, where two feet more might allow a return counter, making an inadequate kitchen workable. Of course the two feet would have to be such that the room it is taken from can still function.

Another common example might be several feet taken from a secondary bedroom to create a new modern bathroom. If an up-to-date bathroom is high on your program and there is no other place to gain the area to achieve it, someone in the family will learn to adjust to a smaller bedroom. Plan number B0006 on page 179 is an example. Maybe that bedroom can be enhanced by larger windows, or by raising its ceiling to provide volume or a loft. Compromises are going to be necessary every step of the way, and there might be some way to compensate the losing part of the compromise.

Creating halls and foyers

In certain circumstances, you might want to reduce the size of a room to create a hall or a foyer. When? Only when you have such a serious circulation problem that nothing else will work. We can sometimes satisfy circulation problems by a change in flooring, by using columns or low walls, by moving an opening, or even by using furnishings to define the traffic flow. But there might be occasions where the best solution is a new wall. Plan CAP04 on page 275 is such an example.

Using volume

One of the best methods available to visually enhance a cramped space is the utilization of volume. Volume is that space above your head, usually covered up in an attic. Usually all it takes to uncover volume is the removal of ceiling beams. If your house has a trussed roof, you would be limited to a smaller area, but it can still be achieved. Skylights are a common addition to a vaulted or cathedral ceiling. With the addition of extra light, an otherwise small room is further enhanced.

Another place to find living space is from unused volume in the attic; space under the eaves is potentially available for storage needs, or to expand second floor rooms or create new rooms. It will take a little investigation on your part, but if you find attic area that you can stand in for a width of at least four or five feet, it is space worth pursuing. There could be space for a bath lurking there (see plan B0010 on page 182). If it is 8 or 10 feet in width, you might have a whole playroom (see plan F0013 on page 201). Another potential use for unused attic space is the development of finished loft space that overlooks or serves as an adjunct to first floor rooms.

Going up

Finding some space in an attic is great, but if your needs can be met only by a large infusion of new square footage, you should look up. Unless there are zoning restrictions that prohibit your use of the air and sky overhead, adding up is a very potent solution for cramped spaces. It becomes the only solution if you are prevented from expanding out. There are a number of different styles of such expansions, depending on what type of home you have and on how much space you need. Because this book is full of many such expansions, too numerous to list here, you should peruse the plans first. Let's talk about five types of additions that go up.

Partial second floor If your current home is a one-and-one-half-story design with a partial second floor, you could add the balance of the area over the first floor to create a full two-story home, enabling you to free much first floor space for living or entertainment needs. You also might be able to create a luxurious first floor master bedroom. Such an addition requires care in developing a new facade. Because your home is now a two-story design, new relationships are created between first and second floor forms, windows, and rooflines. Many disasters have been created by an unthinking remodeler adding a full second floor. Follow the examples shown in this book, and you won't go wrong.

Split level If your current home is a split level, the area over the living room wing has ready potential for adding space. Likely uses include a spacious new master suite or an apartment. Stairs are easily located. Extreme care must be exercised in the forms added here. A setback from the front wall is important and rooflines must echo the existing house.

One story If yours is a one-story home, there's a great deal of possible space obtainable on a second floor. You could add more bedrooms, a whole bedroom wing, or a private master suite—possibly even a whole apartment. As with the one-and-one-half-story home, the relocation of some bedrooms here offers the potential freeing of first floor area to enhance other uses.

There are, however, two areas of significant concern when adding up on a one-story home. The first involves finding a location for a stair. If you have a basement, the likely place to look is over the basement stair, but that could be poorly located. If you have to create a new stair, the 3-ft.-×-10-ft. space required might be an unpalatable loss that forces you to abandon the idea that very much depends on your exact floor plan and what you hope to achieve on the second floor.

The second concern with adding up on a one-story home involves the aesthetic balance of the home. A small dormer or clearstory will always look fine on a one-story roof, but once the space reaches the outer walls, it must be studied very carefully. If the element on the second floor is too small, it could be out of proportion with the home. The size of the addition, plus the styling of its roofline, are best picked from these ready-made plans or left to a professional.

Adding space above the garage The fourth type of plan, that utilizes space overhead, is one that adds above a garage. The space over a garage is ideal for a playroom, an apartment or a bedroom. On a two-story design it can also be used to expand the second floor. Caution is the word again, as concerns exterior forms.

Adding dormers The fifth is actually the simplest. It involves the addition of bumps for small second-floor additions to a one-and-one-half-story design. These are commonly called dormers.

New windows and the use of glass

What do windows and glass have to do with making interior changes? A whole lot! Much of our efforts are aimed at trying to find space to enhance rooms. The addition of large windows, maybe even some very stylish shapes, or the use of large unbroken areas of fixed glass or glass block can help to visually expand interior space. By doing so, the outdoor spaces you face can help enhance tight indoor spaces. With the addition of garden lighting, privacy walls, and landscaping, this enhancement is available day and night. It is a tool you should consider using.

Remodeling kitchens and baths in place

Just a few words to those who choose to remodel kitchens and baths by merely replacing fixtures and finishes. If your decision is purely economic, I will not dissuade you. There is absolutely no question that the cheapest solution is to remodel in place. If, however, you choose this route because you truly believe that no space exists to enhance them, look again. I assure you there is some way to achieve those few feet of space so essential to solving the problems of many older kitchens and baths. Chapter 8 presents a number of such plans.

All of the subjects we have just discussed are graphically shown in the drawing on the next page. In that drawing you will see how I have proposed to resolve the problems we had identified in the plans shown on page 17.

Potential problems of interior remodeling

There are several problems common to all remodeling projects that stay inside the walls. The first is identifying bearing partitions that can limit your freedom in removing or relocating walls. If you have obtained a set of plans on your home, these will be labeled. If you are not sure, ask a contractor with experience. All of the plans shown in this book have addressed that concern. If you discover an unexpected bearing wall during construction, the chart in chapter 11 should help you get a quick answer on necessary girders.

① finding space

② moving partitions

③ combining rooms

④ creating halls and foyers

⑤ creating volume

⑥ going up

⑦ adding out

⑧ bearing partitions

⑨ remodeling baths or kitchens in place

- - - - new circulation

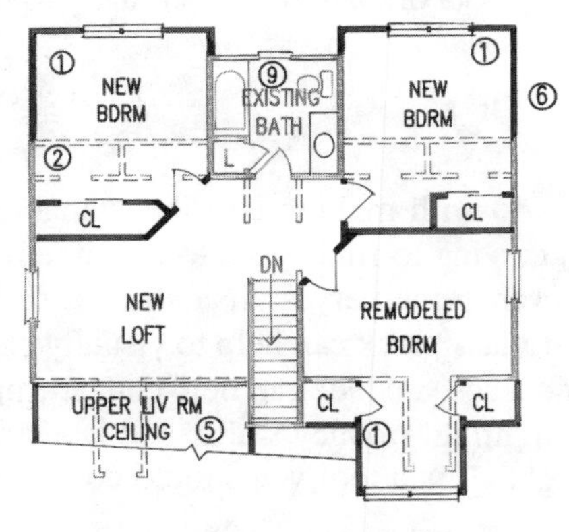

REMODELED SECOND FLOOR ⑥

UPPER LIV RM CEILING ⑤

ALL NEW SECOND FLOOR ⑥

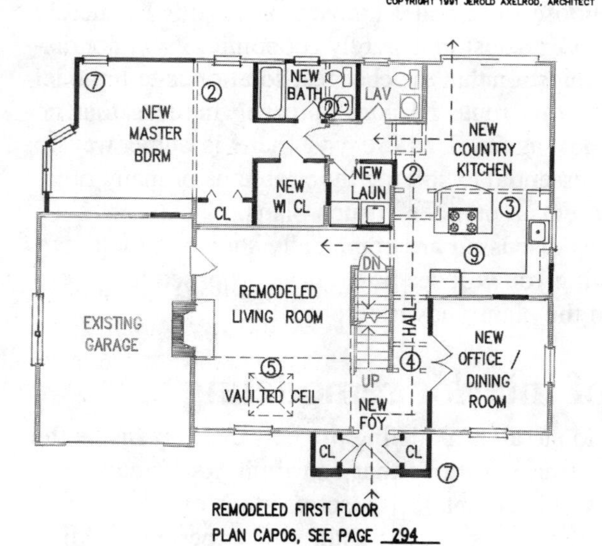

REMODELED FIRST FLOOR
PLAN CAP06, SEE PAGE 294

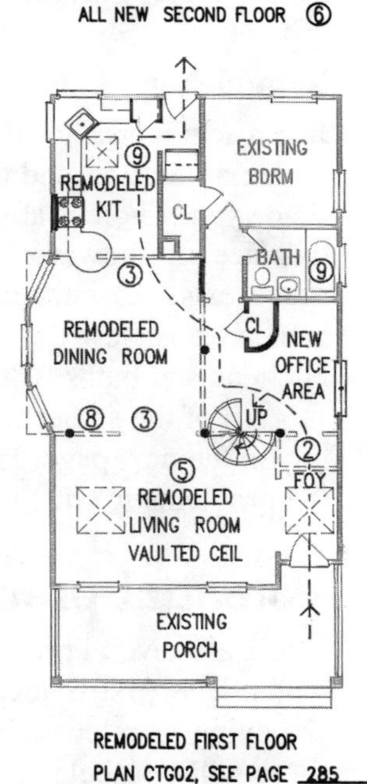

REMODELED FIRST FLOOR
PLAN CTG02, SEE PAGE 285

IDENTIFYING WHERE YOU CAN MAKE CHANGES

Homes with masonry exterior bearing walls might inhibit your ability to create large new openings, not necessarily because of the loads on girders, but because of the extensive labor required to make the openings. If yours is a masonry home, you might have to reduce your expectations—or increase your budget.

Interior renovations could obviously cause disruptions in your heating, electrical, and plumbing systems. While cost factors are one concern, the larger, unanticipated one is the disruption factor that is impossible to prepare for. No matter how well you do plan for these and prepare yourself for being without them, it will seem like an eternity. The same holds true for the mess in general, but it is an unavoidable factor that you must try to plan for. One avenue of help is to plan for staging the construction. A detailed discussion, plus some examples, are presented in chapter 11.

Where can you add to your home?

In the previous pages, we have reviewed where you can effect changes within your home, including going up, but not going out. Additions bring a different set of issues and concerns that I will now review.

Finding space

Finding space for additions presents entirely different search criteria. The space you are looking for to create an addition is exterior space—part of your front, rear, or side yards. The space could also be in the form of a recess or a courtyard. Other places to look at are underutilized parts of your lot, like the area behind a garage, or the area in front of a garage on an L-shaped home. Wherever you find this space, there are a number of significant considerations and constraints you will have to identify before you can consider using the space.

Zoning considerations

Foremost amongst the possible constraints are zoning considerations. Most communities have established criteria for minimum front, rear, and side yards. There are also restrictions on the percentage of your lot that can be covered by structures and sometimes there are even total floor-area restrictions and height restrictions. All of these could inhibit your use of what you thought was found space.

The first step necessary to evaluate these is a survey of your home. You likely received one at closing, when you purchased the home. If you can't find it, call your attorney or your bank; they should have it. The survey is to solving exterior space needs as the floor plan was to solving interior problems. From a survey, you can sketch a site plan that sets out all the constraints and benefits that will control your ability to build out. From the survey, you will find out what your current yards are. (Don't rely on what you think they are; verify with the survey.) You can then contact your municipality and inquire about the zoning restrictions. You will usually find that there is plenty of room to expand to the rear, and very little, maybe even no room at all, to expand to the front and sides. However, if you believe that your home would be best served by adding within one of these required yards, you can usually pursue an appeal process called a variance.

Variances Depending upon your municipality, the process of filing for a variance could either be a routine matter you could handle yourself or a legal proceeding that demands you hire an attorney. It is usually necessary to show a hardship to convince the board that your request should be granted, as well as some proof (or a statement) that granting your variance will not result in reduced property values.

An example for discussion: say your kitchen is in the front; it is dark, dingy, and awfully cramped, but the front of your home is right on the setback line. Well, you could easily argue a hardship; you should not be prevented from enlarging your kitchen, and to relocate it elsewhere is a total impracticality. The way to sell it, though, is to bring a sketch of the new front (like one of the pictures in this book) and demonstrate that you will be upgrading the streetscape. The same two-part approach should be used for any variance. Sometimes certain types of structures like porches, vestibules, and terraces are permitted to encroach into a required yard, so check your ordinance carefully.

Natural site constraints and easements

Sometimes it's not zoning that controls where you can place an addition, but a natural condition of your site. A steep drop-off or up-slope are such restraints, as are rock outcroppings—unless you are prepared to remove the rock. A modest drop-off in grade or a modest up-slope, however, can actually be advantageous in creating an attractive setting for an addition. If it drops away, a walk-out basement could be a bonus. On an up-slope a small retaining wall could add interest to the view. Such a wall is also technically a necessity, so that when the addition is finished, your lot will still maintain a positive flow of drainage away from your home.

A drainage ditch or swale could be an impediment; so could a buried sewer main. Both of these are probably located within easements, which are areas of your lot that restrict building. These would be shown on your survey.

Other buried items to look for are water mains, utility lines, septic systems, and oil tanks. None of these would prevent you from building, but you would have to plan to relocate them elsewhere—and that will cost you money.

Finally, you might have planted shrubs, and patios and walks that are in the way of a potential addition. It isn't likely that such things would prevent you from adding on, but keep them in mind in terms of developing a thorough program that lists everything to be done.

Orientation

Where the sun rises and falls and how it affects your lot are design issues you should examine. All too often, the orientation of your existing home doesn't allow you to take full advantage of —or protect against—the effects of the sun. However, when you are planning an addition, you might find some opportunities to exploit. If your goal is to take advantage of natural light, you could try to locate the addition where it would receive more hours of sun. If that were not practical, you could place windows in the appropriate walls, and skylights in the appropriately facing roof plane. If your goal is to avoid sun, you could readily make sure that windows facing south would be protected by overhangs, and make certain not to locate windows on a westerly wall.

Choosing the size and shape of your addition

The determination of size (assuming there are no zoning or physical constraints) is a very subjective matter, which only you can decide. The size of your current home should not be a factor; the purpose of the addition is to provide you the space you need, so don't limit it for some abstract reason like that. Size could be an aesthetic concern but I discuss that under "problems," coming up next.

As concerns shape, an addition is a wonderful opportunity to break with the rigid shapes of your current home. This requires a little daring and a good design; you provide the former and I provide the latter. You will see good examples of some exciting shapes in the chapters on plans.

All of the factors that affect the design of an addition are shown in the sample site plan pictured on page 26.

Potential problems of additions

As with the remodeling of interior space, you should be aware of the potential problems associated with all additions. The first is the potential loss of light to the room(s) that adjoin the addition. It is usually necessary to compensate for this loss with the use of skylights, if possible, and the use of recessed lighting. However, another method is to design the new addition—particularly if it is a family room, kitchen, dining room, etc.—to be fully open to the old room it adjoins. Of course, this requires careful placement of the addition.

A second possible problem is a significant design concern: the problem of roof design and the attachment of the addition so that it appears as an integral part of the home. It is easy to make a mistake and end up with one of those additions that clearly looks like an addition, and which is at odds with the original home. The only certain way to avoid such a problem is to follow the predesigned plans shown in the book, or to hire a professional. This is an area so prone to error that I have even pursued it further by developing a special chapter, known as Blockforms,

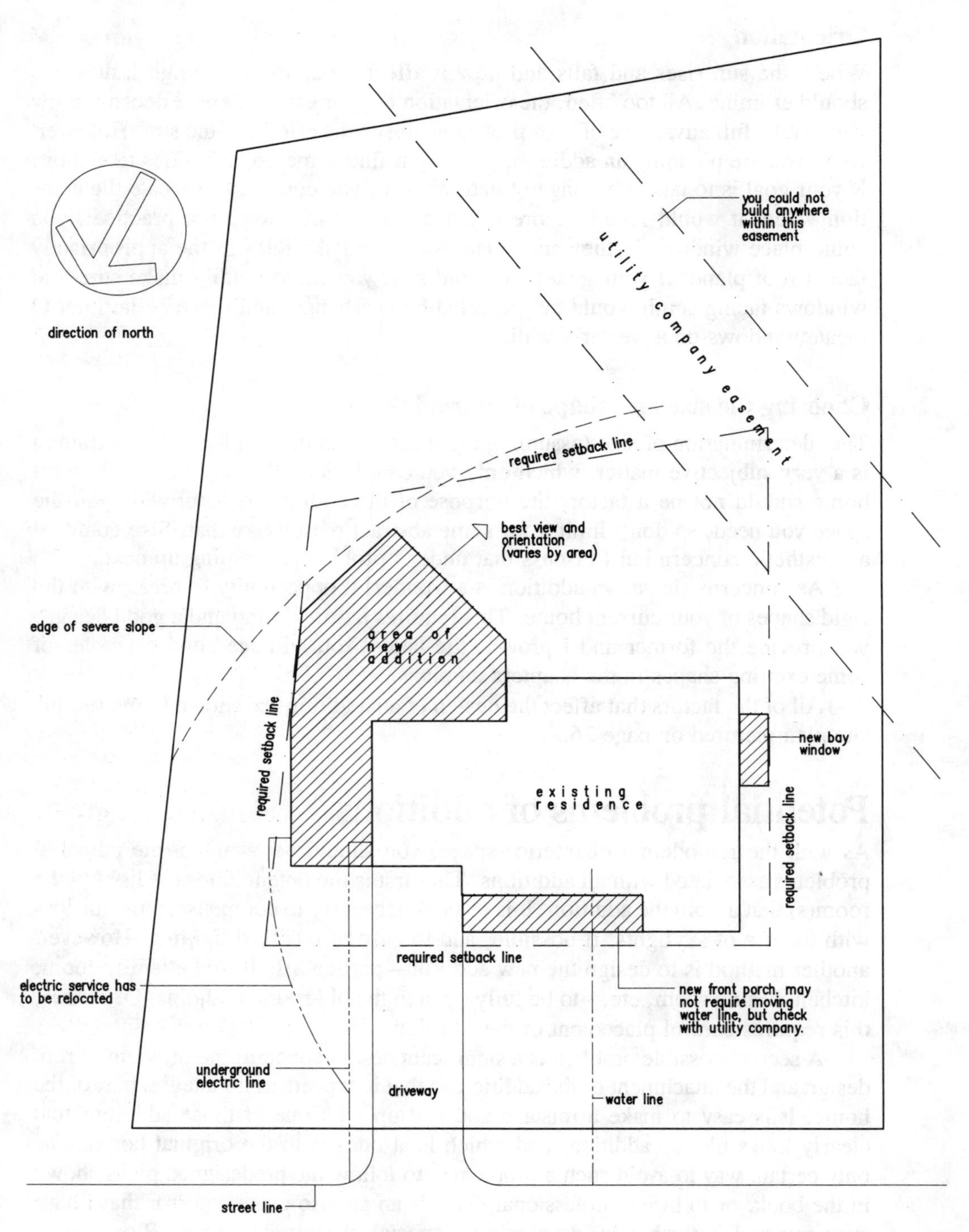

SAMPLE SITE PLAN

which are 60 schematic forms showing attachments that work. There is also another group, located in the very next chapter, that shows you what *not* to do.

Additions also have a tendency to expand the scope of exterior work. When you add something new it is likely the sidings won't match. Chances are you will also use lots of windows in the addition and that you might prefer windows of a different style and design than the original house. What do you do? Do you replace all the windows in the house and redo the siding? Very often the answer is yes, not just for the sake of exterior harmony, but also because these items were due for replacement anyway. If your house does not warrant this wholesale change—or your budget prescribes against it—you will have to carefully choose new sidings and windows that blend with the existing, and locate the addition so that there is no flush match line. The best match line to avoid this problem is a 90-degree intersection.

Finalizing your program

Unless the scope of your project was narrow and well-defined to start, the likelihood is that a final program will involve a mix of some interior remodeling of space and some exterior addition. The addition might only be a bay or a few feet, but it is also very common to see large new additions combined with significant renovations to interior space. Many of the plans shown in chapter 7 do just that.

Your program might not be very tidy. It is probably several pages of some scribbled notes you have made in trying to evaluate your needs and your home. You might have many uncertainties, and be unsure, for example, whether to build out, or add up, or do both. But do not fret; such a dilemma is not uncommon. To try to gain a firmer picture of what might ultimately satisfy your needs, now is the time to rewrite your notes and spend a reasonable amount of time to prioritize this wish list. It will then be necessary to prepare the list of alternates as discussed in the introduction to this chapter. Once that has been completed, you have a program that can be used as a guide in talking to a professional you hire, or in looking through the plans that comprise the bulk of this book.

Remember, compromises will be necessary. You will not likely find a plan that meshes 100% with your program, but do not back off on features that you strongly desire. The ability to make changes to any plan is a given fact, and you can customize any of the plans shown in this book. We will talk more about that in chapter 5.

4
CHAPTER

What not to do

I could probably prepare a complete volume, including hundreds of plans, of what you should not do. Twenty years of private practice and thousands of miles of travel have offered me the perspective to do that, but I do not believe it is a productive avenue to follow. Rather than perusing the endless possibilities for failure, this book is devoted to providing positive ideas and sound examples for you to follow.

One of the clearest traps to avoid is not being true to yourself and your needs. Somehow when it comes to our homes, every friend, relative, or contractor has their own helpful idea. Listen to them all, but try to analyze their thoughts against your needs and your heretofore prepared program. Your contemplated remodeling is one that should bring you years of enjoyment, but it is only you who can evaluate what will achieve that goal.

This is not to suggest being smug. Acknowledge where you are uncertain and where you need assistance. Foremost, unless you possess the talent to be your own designer and you can visualize three dimensionally, use the plans in this book or hire the skills of a professional. It is possible that you or your family possess the talent to build an addition or to undertake the remodeling, but learning to design is not the same as learning to use a circular saw.

Although I encourage you to pursue the creation of home improvements that clearly suit yourself, I caution against producing a remodeled home that is so out of place that it loses marketability. Inputting your personality into a project is usually a positive; the only exception is when the remodeling becomes so unique that it is bizarre. If you are not certain, that is when it is important to bounce your ideas off others. An underground addition to a tract split level might preserve your lawn and be an energy-efficient improvement, but not only will it excite few others, it could detract from your home and reduce its marketability. As I indicated earlier, not every improvement has to be measured in terms of its return on investment. A return for personal gratification is perfectly acceptable. However, an improvement should not decrease the value of your home.

The potential for planning errors exists whether the remodeling is an interior renovation or an exterior addition. An interior renovation, however, is not likely to be fatal—an exterior addition can be. For that reason I have prepared a few pages of sketches to show some examples of what not to do; these deal primarily with the issue of attachment. Although certainly not all-inclusive, they are suggestive of potential error. The first chapter of drawings presents a whole series of sketches that show what is correct.

Presented on the next two pages are common examples of additions that attach very poorly to their homes, or are awkward shapes that relate poorly to the existing home. The addition is indicated by shading. All should be avoided.

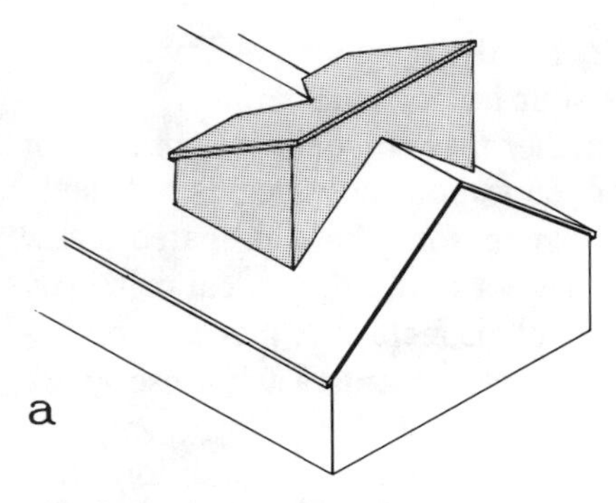

The roof structure of the new addition is not harmonious with the simple gable roof below.

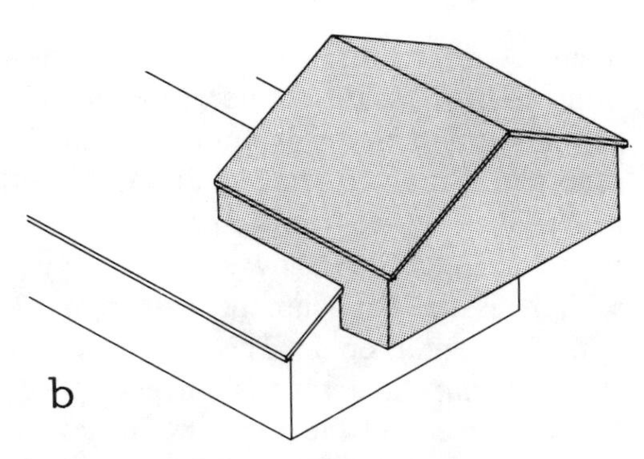

The second floor addition looks like it was just stuck on—and it might fall off. Don't cantilever a gable wall.

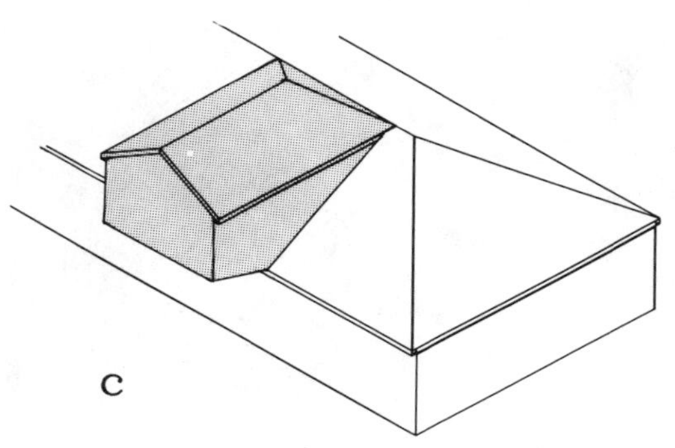

The new dormer element looks like it wants to fly away. Keep small second floor elements inside the roofline.

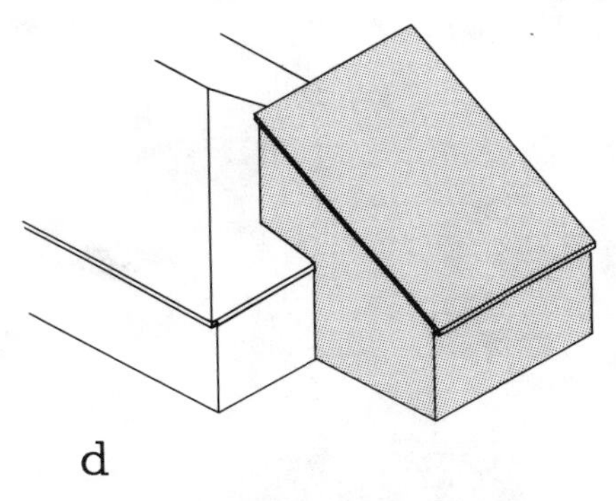

The shed roof side addition is fighting the hip roof structure it is attached to. Use another hip roof.

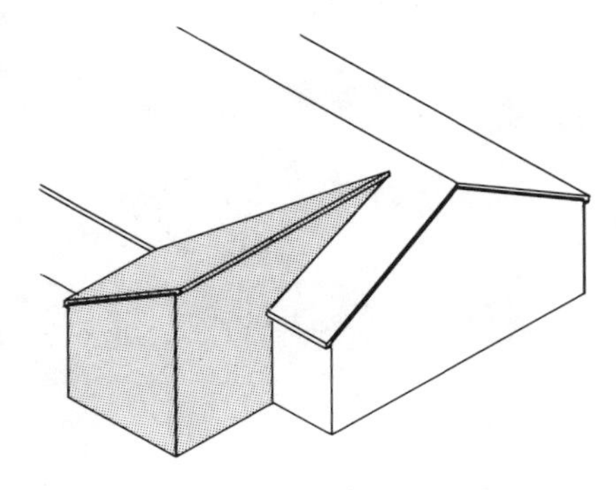

A side-to-side shed roof on the gable doesn't work without lots of study. Stay away from it. Use a reverse gable.

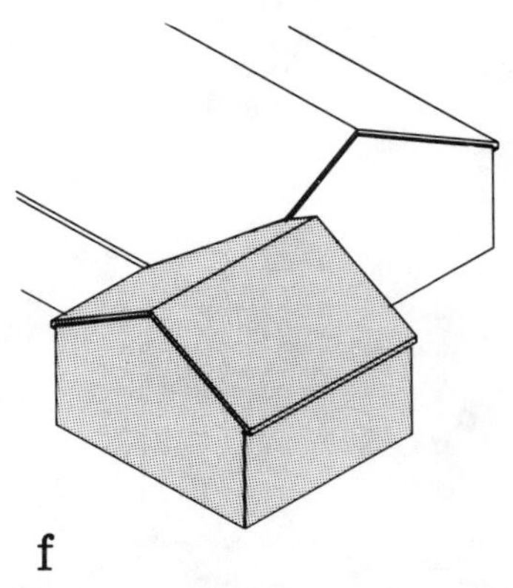

Attaching at a corner is dangerous. Would be better if the addition also extended along the side wall.

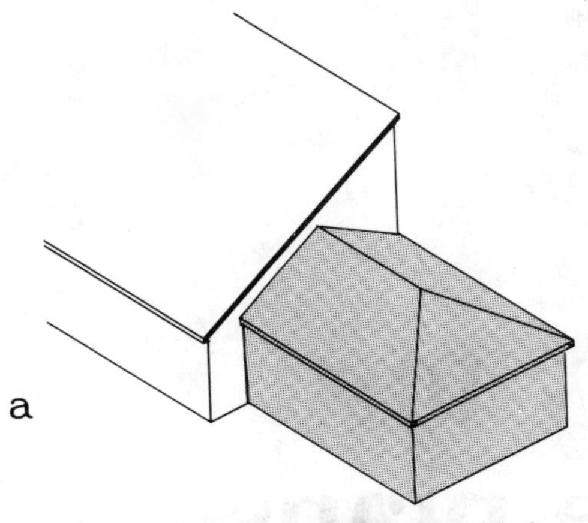

The hip roof addition is discordant with the existing shed. Use another shed roof or a flat roof.

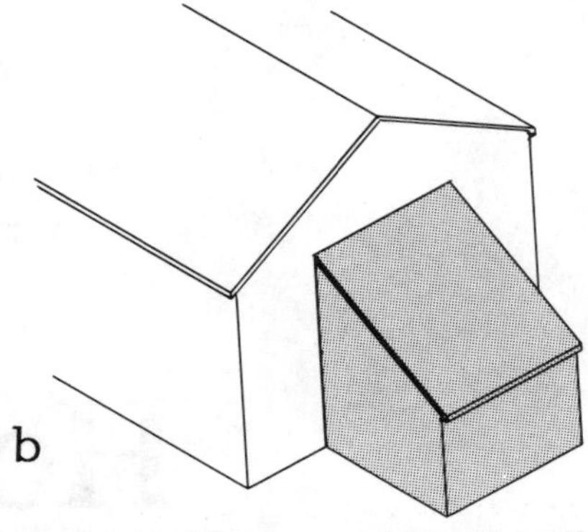

A shed roof addition on a side gable wall is an unappealing form, and it looks like an addition.

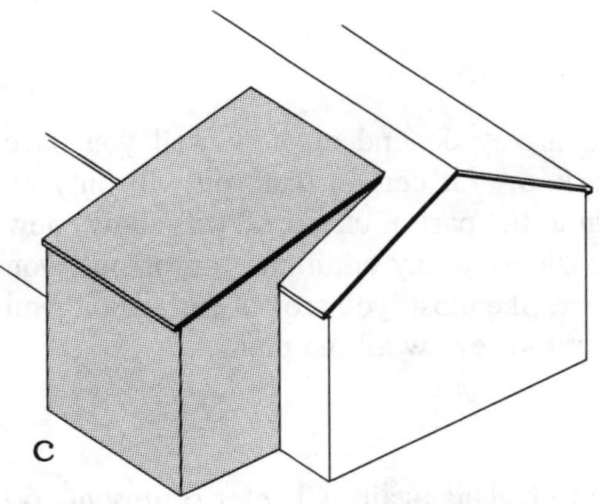

The two-story shed roof form is simple to build but unattractive. A reverse gable is much nicer.

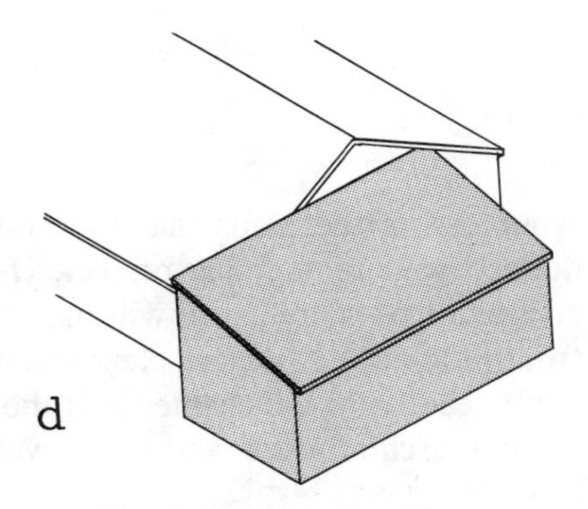

The side addition makes no effort to harmonize with the existing form. Needs a reverse gable at the front.

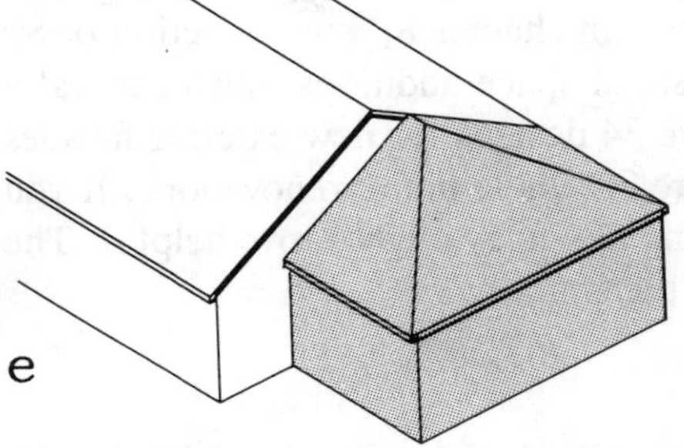

The shape is fine but the new roofline doesn't belong. Should be a matching gable.

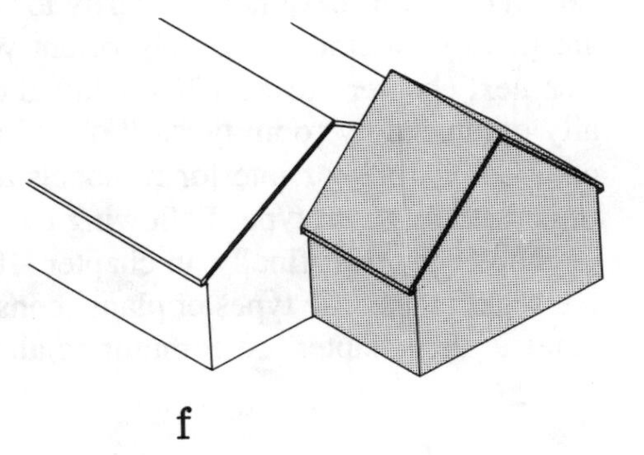

The pitch of the new addition does not match the existing roofline and should be flush or lower than it.

5
CHAPTER

How to use the plans

Your use of the plans that follow will largely depend on how well you have defined your remodeling project. If you know for certain that you will only be adding a front porch, you will likely turn to the part of chapter 8 that shows new front porches. If you are clearly not considering any additions, up or out, you could readily bypass chapter 6. If, however, like most, you are not certain and you are in search of ideas, you might well want to review all the plans.

Reviewing the plans

I will briefly summarize the organization of plans again. Chapter 6 presents 60 sketch diagrams designed primarily for the person considering an addition. These are merely blockforms to help orient you on shapes that might suit your home. The next chapter presents 104 detailed designs for additions of every sort, generally organized by room types. Following that, in chapter 8, is a collection of 80 detailed designs for interior remodels and small space additions—also generally organized by room type. Following these are 34 designs for new exterior facades in chapter 9. And, finally, in chapter 10, are 22 whole house renovations. If you are seeking specific types of plans, consulting the index might prove helpful. The notes in this chapter are common to all the plans.

Text

It is important to read the text accompanying each plan. It not only highlights the salient components of each plan, but also provides helpful tips and hints for better design. In some cases, the text also provides the critical overall dimensions of an addition, which indicates whether or not the size would require adjustment to suit your lot. It could also direct you to other plans for consideration, especially if there are alternates for your review. Finally, the text might provide tips on staging, cost considerations, or other facts for review.

Combining, reversing, or modifying plans

Keep in mind that all plans can be reversed; if your current home is a mirror image—or you would prefer a mirror image of an addition—any of the plans could be constructed in reverse.

Keep the door open for changes. Remember: these are generally very flexible plans that can be readily modified to suit your specific requirements, so don't look for a precise fit. Custom tailoring is expected. Also remember that plans can be combined. Many of the plans shown remodel two or more rooms, and all of those in chapter 10 remodel the whole house. The potential for other combinations is almost infinite.

Rules for viewing plans

The general orientation of each plan presumes that the street is at the bottom of the page. Although that might not be the case for your particular home, it is the general rule. As such, a front addition is generally indicated below the existing house, and a rear addition is indicated above the existing rooms. Even on additions to a side or on internal renovations, the plan has been drawn the way it is most frequently found. Where the perspective view shows an exterior addition, the view could be from the side, rear, or front. The direction is not indicated, but the addition is shaded, so you should be able to tell from which direction the view is taken.

When rooms are dimensioned, the first number is always the horizontal dimension, the second number is the vertical dimension. This holds true regardless of the shape of the room.

Study all the options

One of my persistent themes is exploring alternatives. You will find throughout these chapters on plans various solutions to the same problem. In certain instances where I invite direct comparison, I will point out which plans to compare. However, the majority of options are dependent on your requirements, so unless your program is extremely well-defined, and you have very few alternatives listed, it is probably worthwhile to review all the plans that deal with your needs.

As an example, assume your home was a typical three-bedroom split-level, and your goal (or one of them) was a new master suite. You could consider the addition of only a new private master bath, or the rear or side addition of an entire new master suite, or going up internally over the living room. These are three distinctly different solutions to the same problem. All are pictured in the book, but it will be up to you to perceive them as viable options for your consideration.

The extras or bonuses of each plan

You will find, occasionally, in the text, a reference to an "extra" or a "bonus." These are not intended in the same way a builder of new homes uses these terms. Very often there is a secondary benefit of a plan, which was not the main focus of the plan but which results as a by-product. It is not that this is an accident, but in the design process, as I see it occurring, these other elements are studied and are perfected in the final design along with the main subject. If they are not studied, the secondary effects of most remodeling projects could prove to be an unwelcome afterthought. These I have seen all too often, as a result of an unprofessionally planned design. By being aware, in advance, of all the potential impacts of a particular improvement, the designer can study them and make them a beneficial aspect of the project.

As an example, take the split level previously mentioned, and say you opted for a side addition. Do you build it on posts? Never, unless you absolutely had to. An extra garage becomes a likely bonus to the new master suite, but it could be more. If you didn't need an extra garage, you could increase the potential living area by expanding into the old garage. You might even be able to develop an apartment. These then are all studied, preplanned bonuses to the plan. Many are more subtle and significantly smaller than this example.

Obtaining construction blueprints

You can obtain construction blueprints for any of the plans presented in this book. These plans can serve several different purposes. The first is the obvious function of providing the technical information necessary to build the remodeling as pictured. Do not attempt to build from the pictures in the book. Unless you are planning a facade remodeling or the rehabilitation of an existing bathroom or kitchen in place, construction plans are necessary.

Because these are ready-made plans intended for multiple use, they are priced very reasonably. All of the blueprints include a list of materials to aid you in obtaining a cost for major material components—a big help, whether you are considering doing the labor yourself or are hiring a contractor—and come in multiple set packages for construction needs. Orders for several different designs will be accorded pricing credits. For further information on availability and pricing, write to me at the following address:

Jerold L. Axelrod, Architect
C.B. 9015
Commack, NY 11725

Construction blueprints can also serve two other important functions. If you are unsure what direction to pursue because of an uncertainty about costs, obtaining estimates is a significant benefit of predesigned plans. The blueprints for these plans provide an inexpensive, less-involved method of verifying your budget. Even if a plan isn't the precise design you are seeking, it might be wise to price out the construction cost before becoming emotionally and financially involved in making changes or having custom plans prepared for you. Remember that although you can generally obtain a budget bid based on design drawings, such as those in this book, a firm price requires estimating from construction blueprints.

Blueprints also can serve as a guide for you to review the project in detail before finalizing decisions about making changes. You and your contractor will have an opportunity to better plan those custom features and changes that will satisfy you and your remodeling program. Usually only one set of plans is required for this purpose.

Making changes to blueprints

As I have stated before, it is not likely that any two remodeling projects will look precisely the same, even if they are built from the same blueprints. Every project will involve your decisions about materials, colors, and possibly many other choices. These choices will personalize the project for you.

It is also very likely that you will make changes from the construction blueprints. Specific site conditions at your lot or caused by your home might require certain adaptations or changes to the blueprints. This is very common. Should you change them yourself or have the plans professionally changed? That depends on the type of change and on your (or your contractor's) skills in adapting the change. If you are uncertain, write to my office, and we will provide an answer.

Because all of the plans in this book are on our CAD system, it is feasible for us to make changes, and, when we do, our system automatically updates the material list. We should be able to accommodate your needs, even if your requirements cause very extensive changes.

Finally, if you do not see precisely what you require, write us. It is distinctly possible that we have already prepared a plan similar to what you need.

Hiring a local architect

As I indicated, even if you have very special requirements, we might be able to provide you with the necessary plans. For some people, though, this is too impersonal. Mail, telephone, and fax communications are, by nature, impersonal. It is not likely that I can physically see your lot, or walk through your home to see how you live, or sit down and have a face-to-face, personal discussion about your needs and habits so that we may develop a personalized program for you. For those reasons, you might well consider engaging the services of a local architect.

Also, if your existing home is very different, and your program requirements demand a very unique design, you might need to have local services. The names of such professionals are usually provided by reference from friends, contractors or your local professional associations. Fees will vary, depending on the scope of services requested and the extent of the project.

6
CHAPTER

Blockforms: An introduction to additions

In chapter 4, I presented a few sketches showing common errors encountered in do-it-yourself additions undertaken without any professional guidance. Only a handful were shown. The current chapter provides 60 examples of building shapes that will lead to successful additions. You might say, "Why waste time? They are only simple blocks; I don't see windows and doors or other details." While perusing these, however, you might be able to recognize and then match the shape of your existing home to a potential addition. (The addition is always pictured shaded.)

The process of designing an addition involves this somewhat abstract concept of creating a "shape" or "form" that will complement your home. If planned and executed properly, this new shape could do more than just match your home. It could also make it exciting to look at, thereby enhancing its market appeal, too. Therefore, I would recommend that if you are planning an addition and are uncertain about shapes, the process for you to follow would involve first finding those blockforms that fit, and then reviewing the next chapter for floor plans that would match these shapes.

The text under each form might also list other possible roof shapes than the one pictured. One of the important decision factors that this section deals with is the shape of the roof on the addition. The decision about which roof shape to use is a complex one that involves the relationships created between the existing and the new. To make matters more complicated, there are frequently several combinations that are suited to any one situation. As a simple example, a rectangular addition to the front or rear of a simple rectangular home can be roofed as a shed, a reverse gable, a hip, or even a flat roof. Because all could be correct, the decision then is one of personal preference as to the style of the exterior preferred or the quality of the interior space (sloped ceilings, high ceiling, skylights, etc.).

In many circumstances, however, there might be only one suitable roofline—that usually occurs when the existing home is a distinct roof shape in itself. But it is also possible that the shape of certain additions can dictate a particular roof design, which we will address again in the next chapter.

One-story additions to the gable end of one-story homes

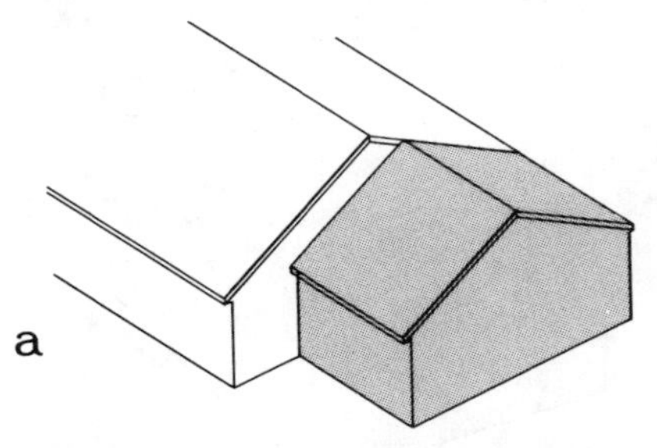

The setback form is attractive and could save you from matching siding. On a hip roof, use a hip-roof addition.

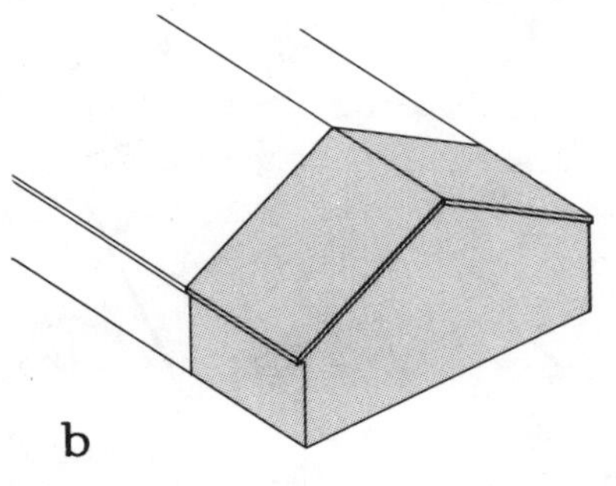

The flush match is OK, as long as it is not a garage. Will require the perfect match of roofing and siding.

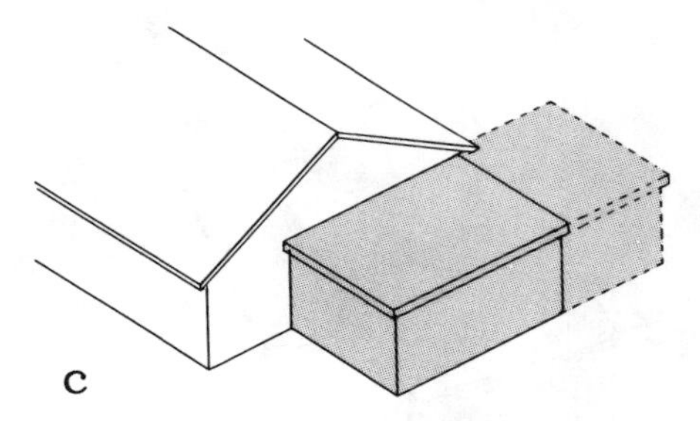

Flat roof is acceptable on the side, even though it won't help to broaden the front facade.

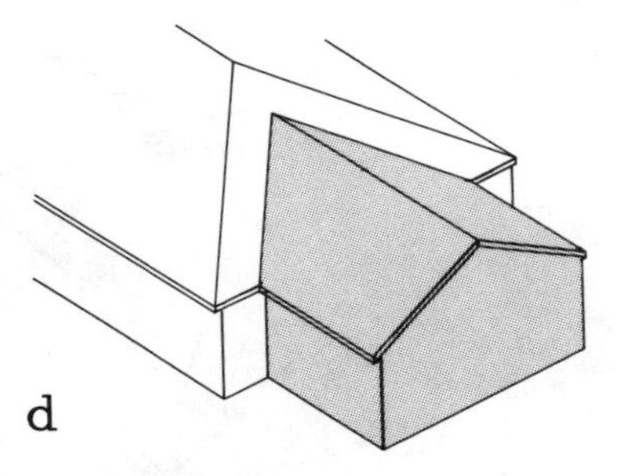

A gable intersecting a hip is a very stylish treatment. Make sure the gable is smaller than the hip.

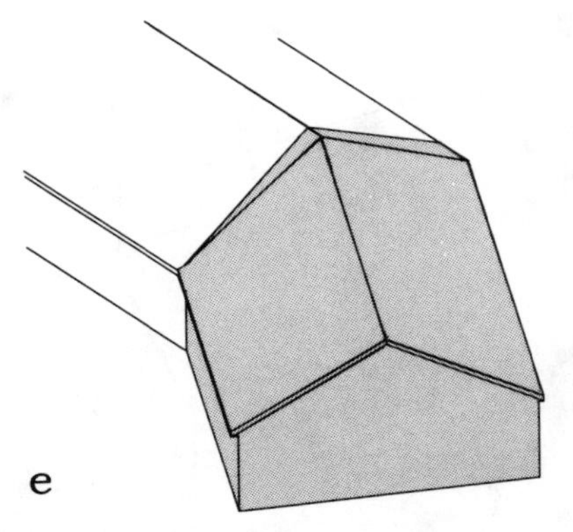

Attaching at an angle can be attractive. Gets a little tricky at the intersection. Match pitches.

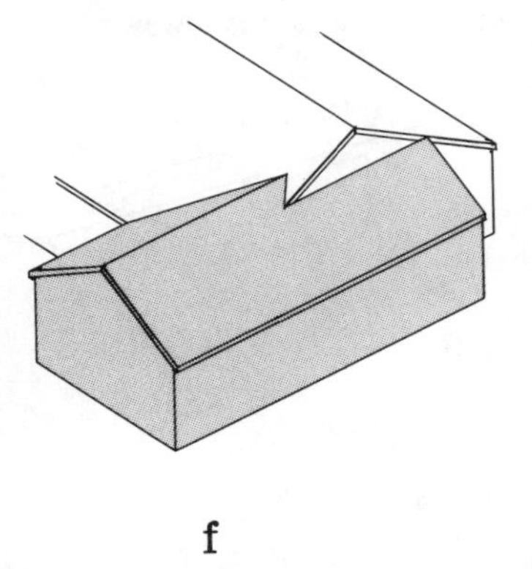

Attaching at a corner with a reverse gable that becomes a shed is a very appealing form.

One-story additions to front or rear of one-story homes

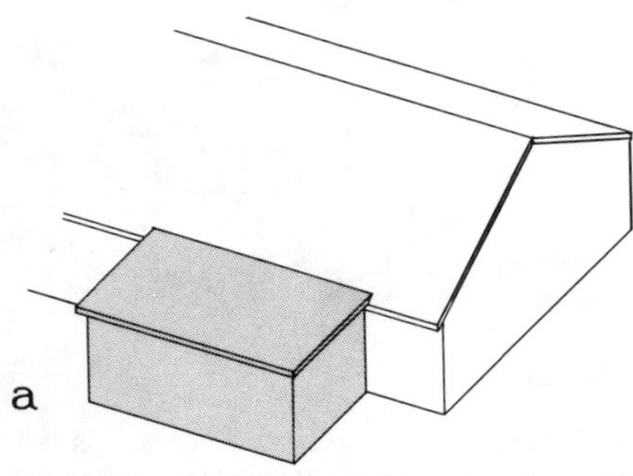

A flat roof addition at the eave of a gable roof addition is fine for a porch or vestibule.

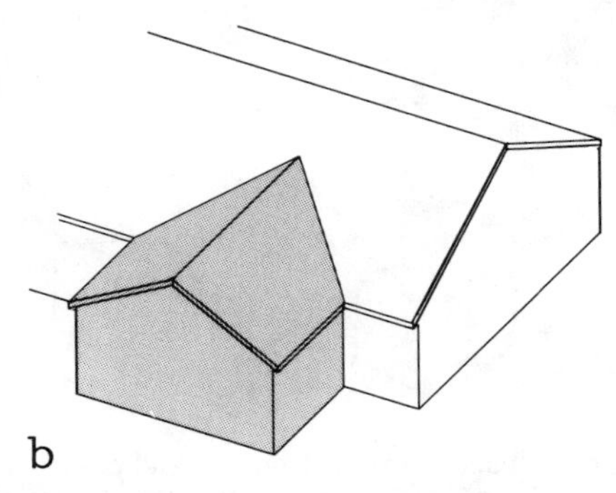

A reverse gable intersecting another gable is probably the best form. A hip would be acceptable, too.

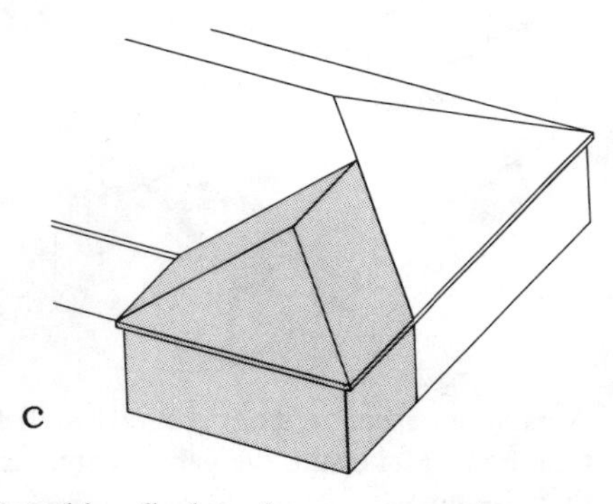

Attaching flush to the corner requires care. The hip roof matches the existing hip perfectly as long as it is lower.

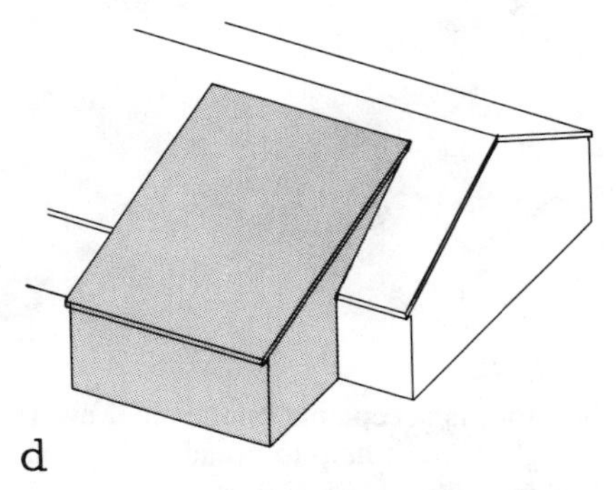

The one-story shed is acceptable as long as the width of its front wall is twice its height.

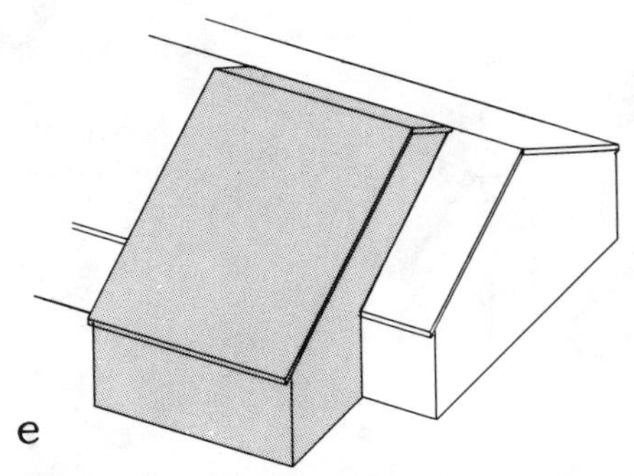

A better way to use a shed by matching the existing pitch of the roof. Requires matching roof shingles at top.

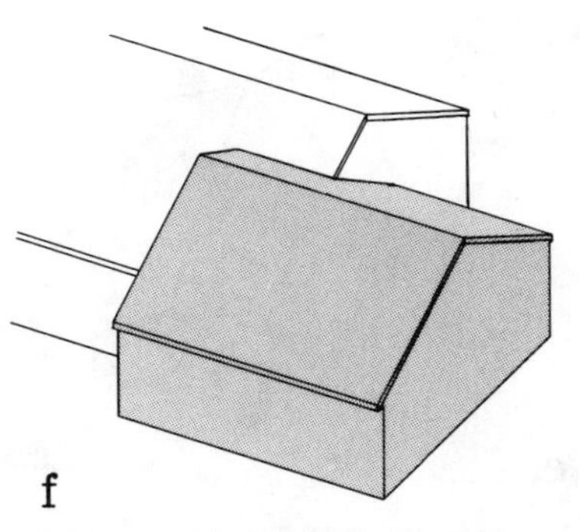

Another acceptable corner attachment. Match the pitch of the existing roof. A cricket might be required.

One-story additions to one-story shed or flat-roof homes

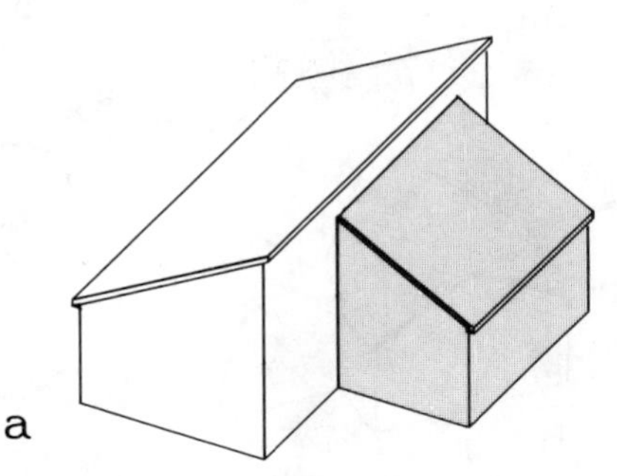

Use an opposing pitch to attach a narrow shed roof addition to the tall side of an existing shed form.

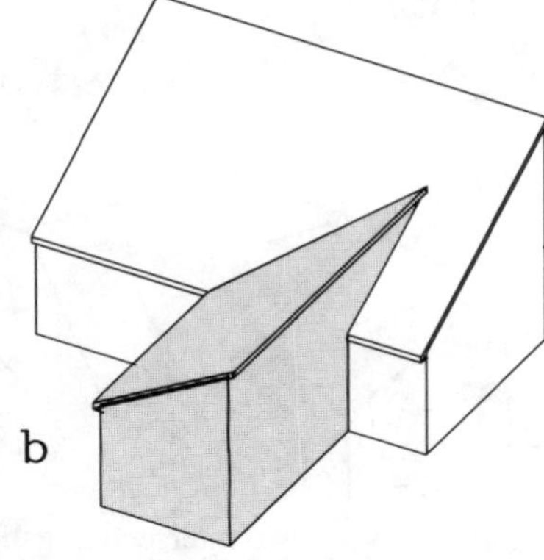

An intersecting shed works, but be careful. Keep the addition smaller than the existing form.

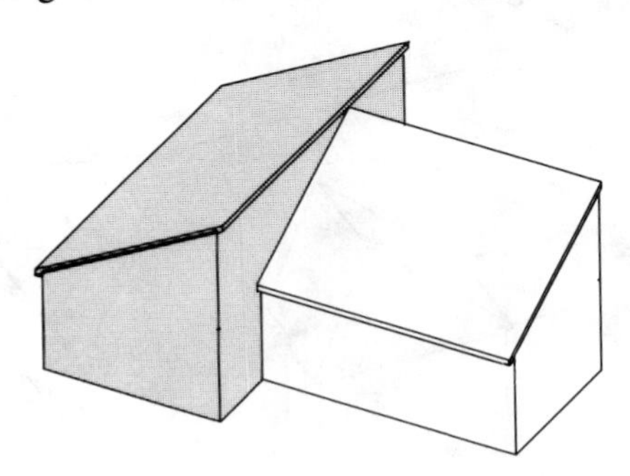

The attractive opposing shed is turned 90 degrees to the existing. Keep the heights different, even if it's inches.

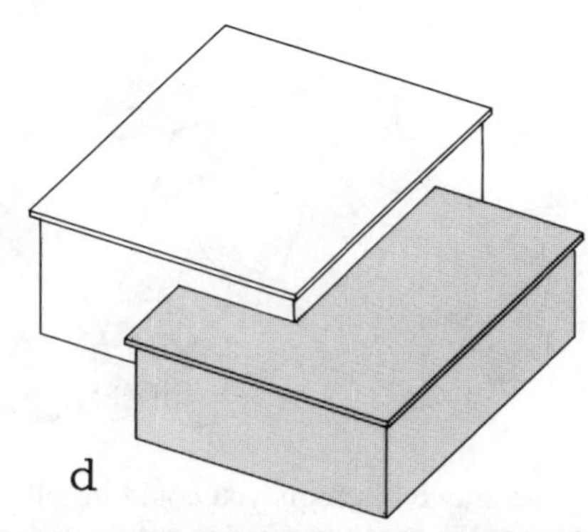

A flat roof addition that is lower than the existing looks just fine—better if it wraps as shown.

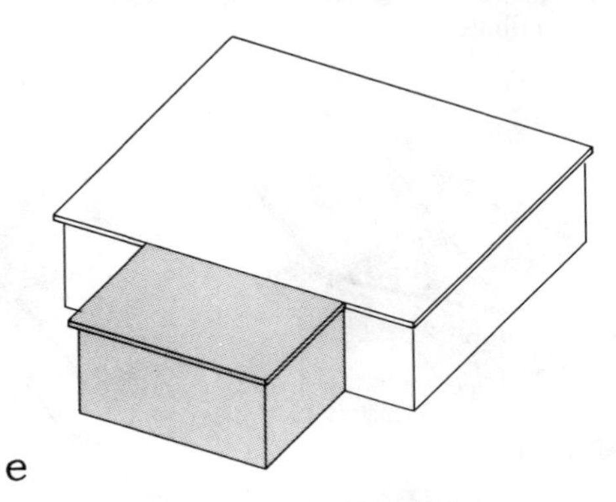

When adding to a large flat roof you can also keep the addition flush with the existing.

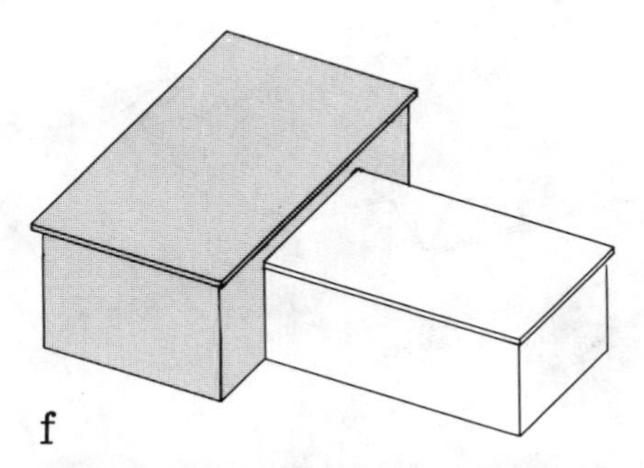

A flat roof addition can also be higher than the existing. The two contrasting forms look good.

One-story additions to one-story complex roof homes

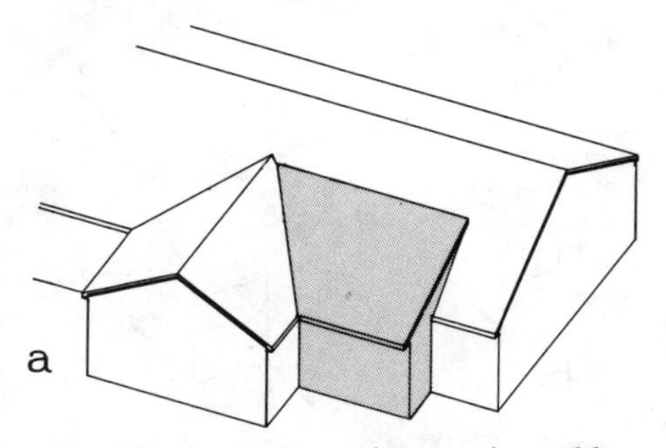

Attaching between two intersecting gables is tricky. The shed is an excellent solution as long as it's not too big.

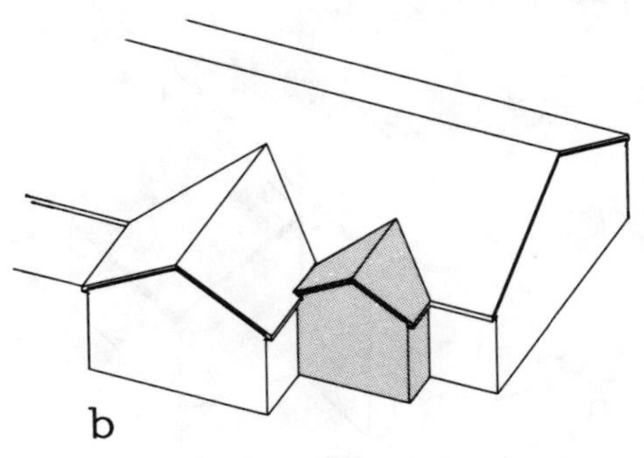

A matching reverse gable can be a stylish solution to the same addition. A cricket might be required.

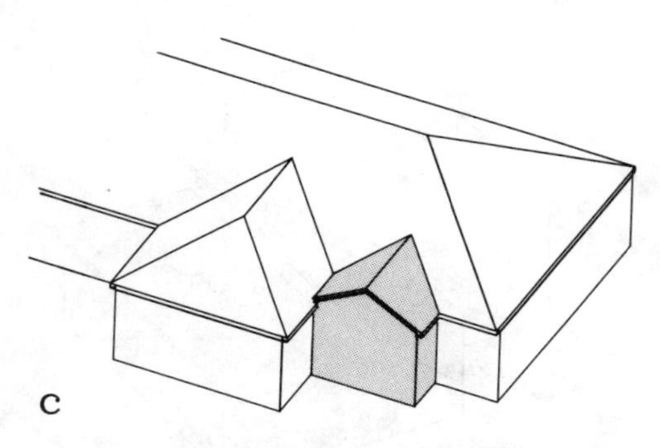

Between two hip roofs you could install another hip roof for certain, but the reverse gable might be more exciting.

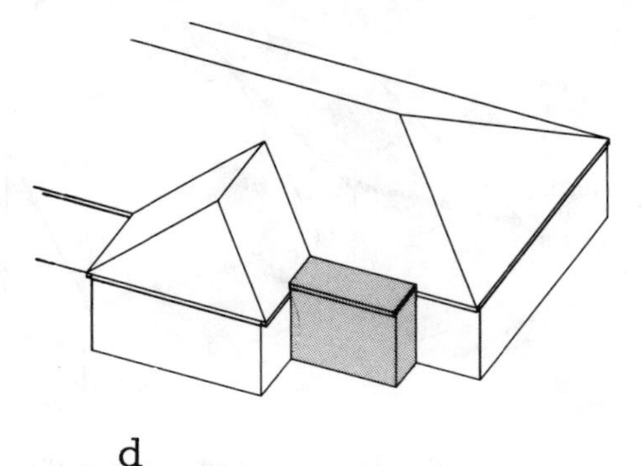

The flat roof connection between the two hips—or two gables—is OK for a porch or vestibule.

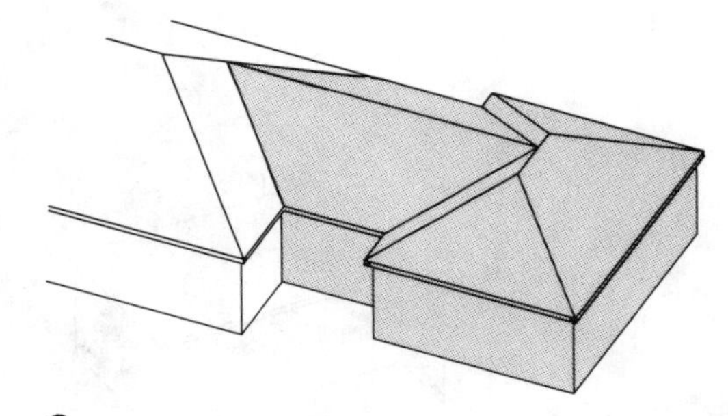

Multiple roof forms are very common in better-looking additions. Roofs blend or intersect at similar pitches.

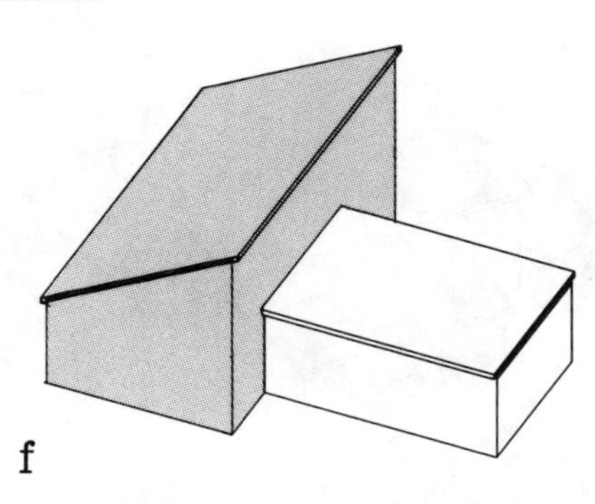

A shed and a flat roof can be successfully combined. Keep the forms contrasting or opposing each other.

One-story additions to the gable end of two-story homes

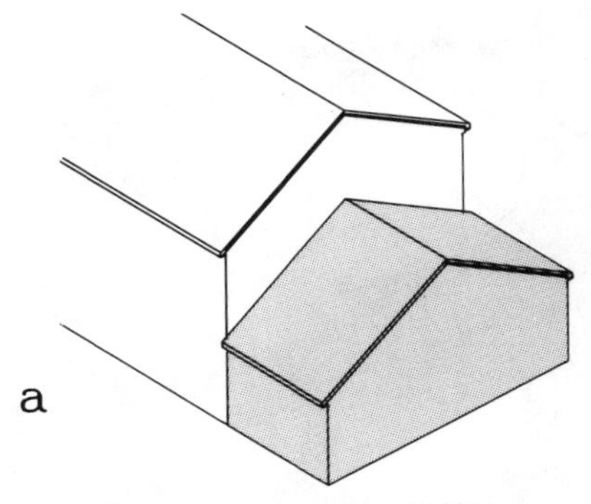

Flush ends mean matching siding. If the gable is a side wall, the scale must be proportional to the existing front.

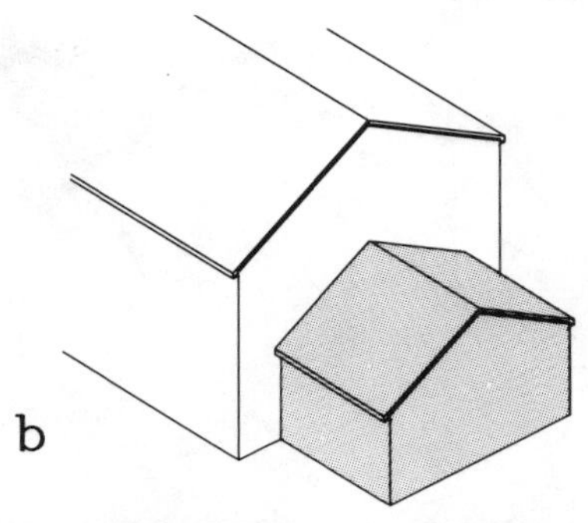

Setback form is fine, especially if gable is a front or rear wall. As a side wall, the size is crucial.

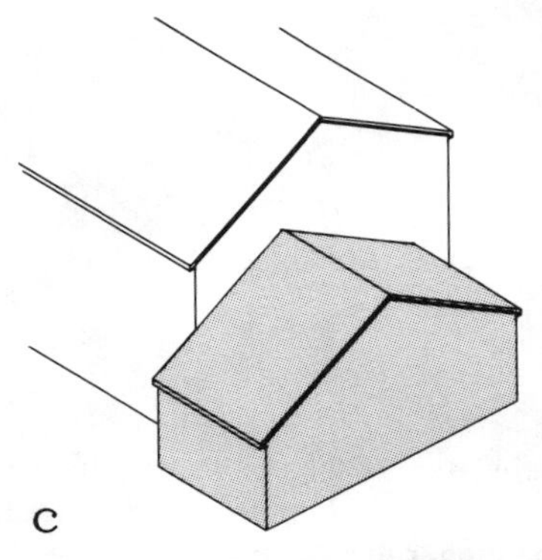

Looks excellent if attachment is a side wall. Width of front wall should be a minimum of 1/3 of existing front.

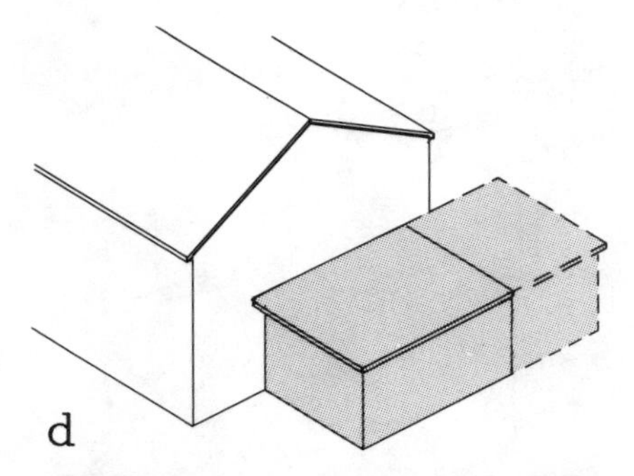

Acceptable for a porch or carport whether gable end is a front or side wall.

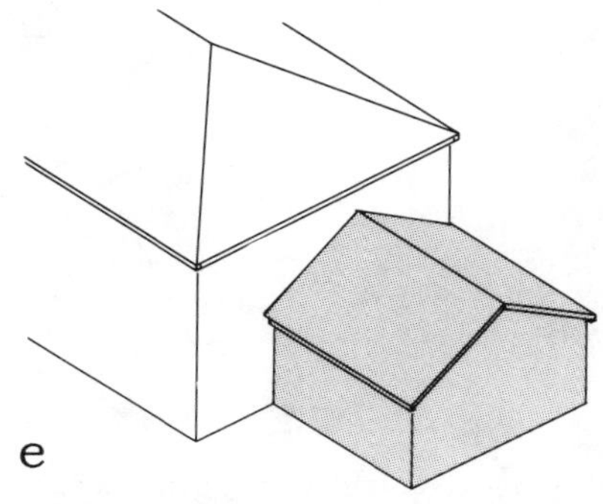

A hip on the new addition would be best if it is large. If it is small use gable as shown.

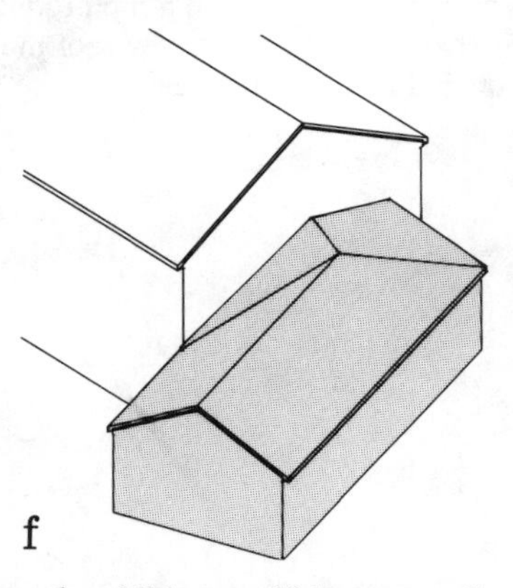

The angle addition could be an excellent solution. Pitches do not have to match. Be careful along common wall.

One-story additions to the eave wall of two-story homes

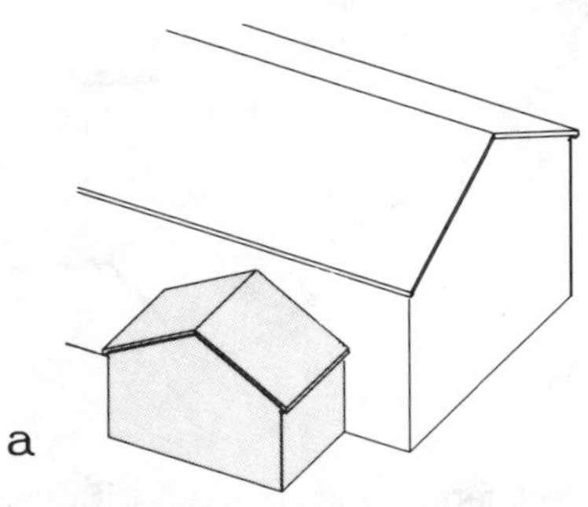

Reverse gable is a perfect form for almost any size addition. Pitch determined by aesthetics of new element.

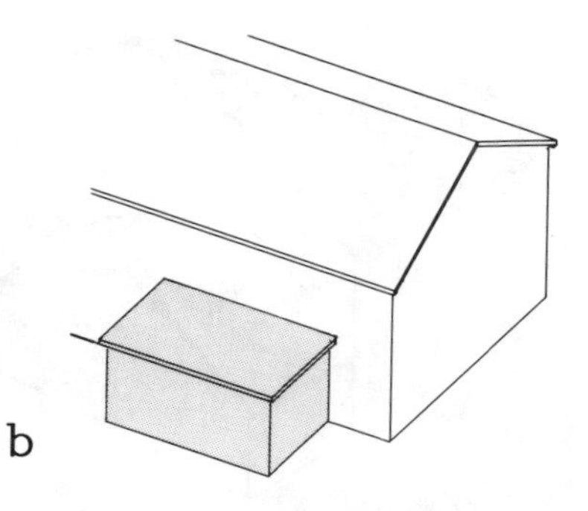

Flat roof is just fine for many additions. Perfect for a vestibule, porch or carport.

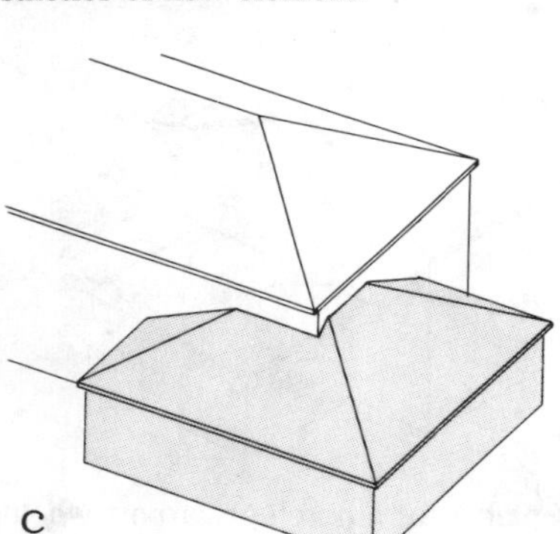

Wrapping the corner with a hip produces attractive forms. Pitch of new roof must match and be lower than eave.

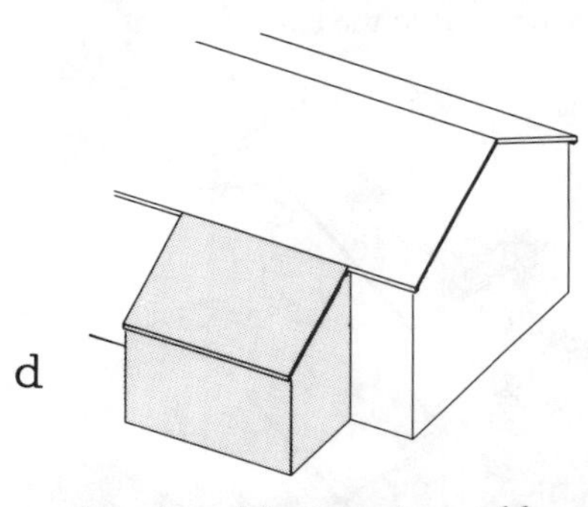

A continuation of the two-story gable roof creates a unified addition. New end wall may be higher than standard.

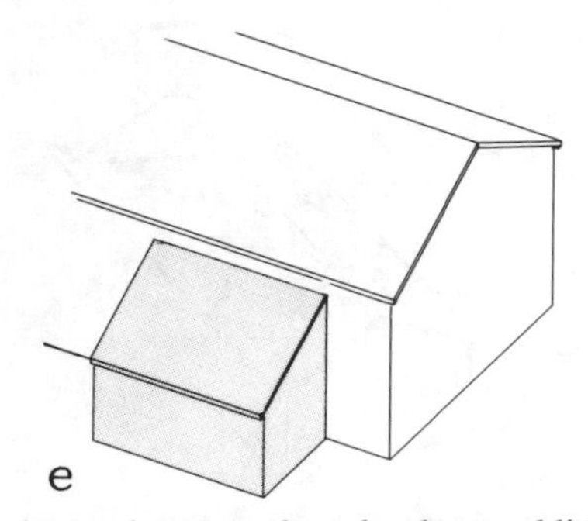

Almost the same, but clearly an addition. The separate shed roof does not have to match existing roof pitch, but could.

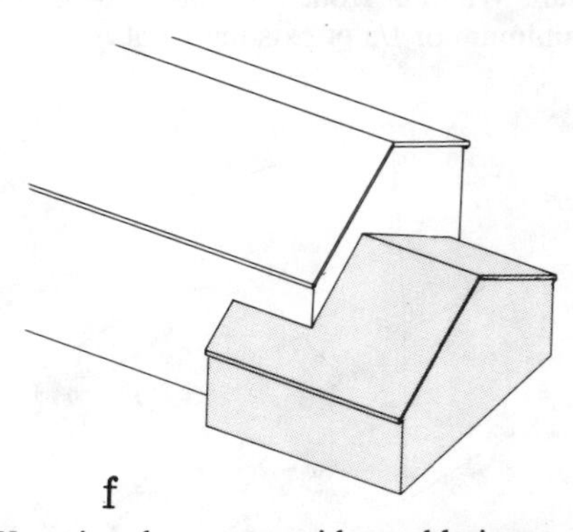

Wrapping the corner with a gable is an excellent marriage of new to old. Maintain the same roof pitches.

Two-story additions to one-story homes

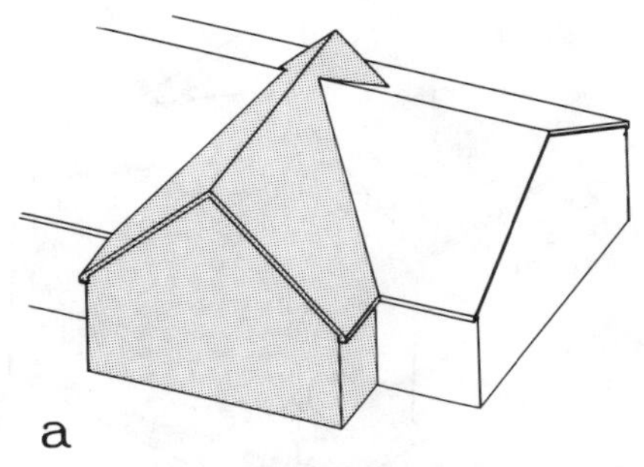

a

The steep one-and-one-half-story reverse gable provides a small second floor room. It is an excellent attachment.

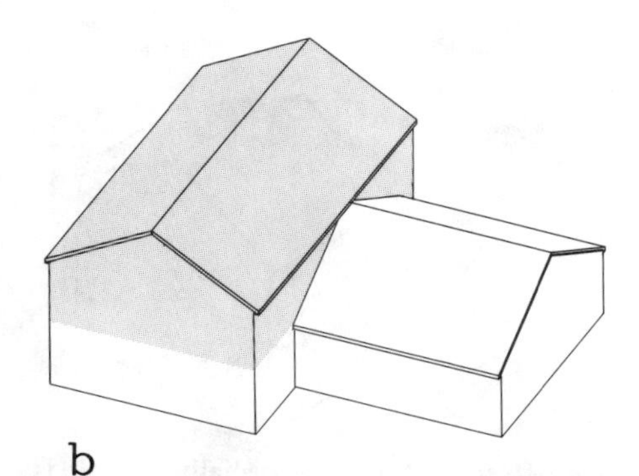

b

A second floor over one wing of an L-shaped one-story is a good-looking solution to adding lots of space.

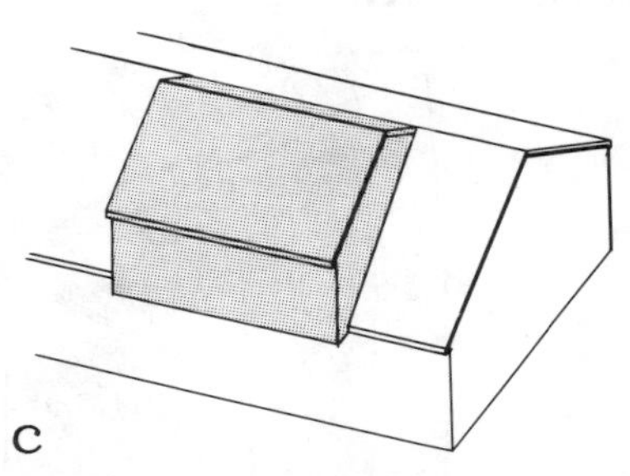

c

Extending the ridge to create second floor space at one side is an excellent tool. Keep overhangs to a minimum.

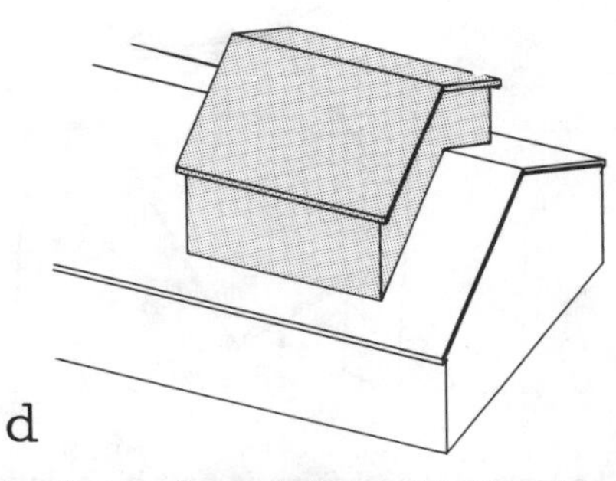

d

Popping up through the roof is fine, but dangerous. Proportions must be carefully analyzed to the existing bulk.

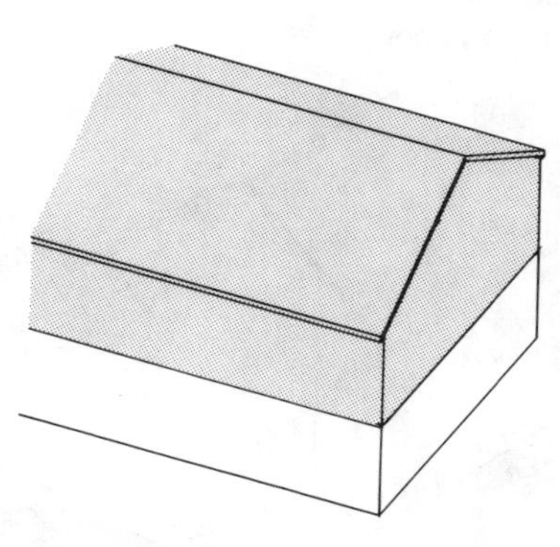

e

Adding a full second floor is also fine, and dangerous. First floor windows must relate to new second floor openings.

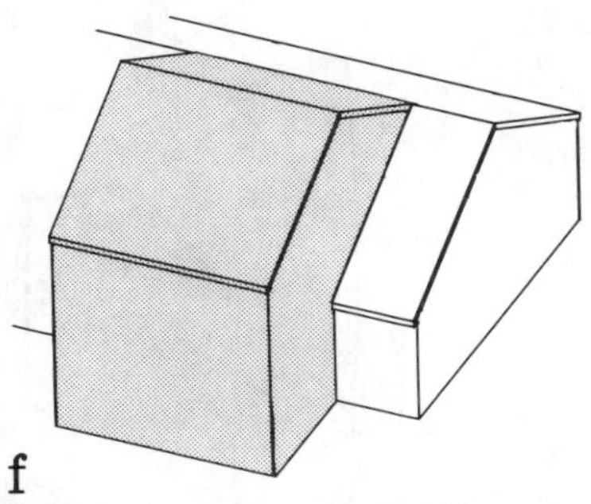

f

Extending the ridge to create a full two-story addition produces a thoroughly integrated solution. Match pitches.

Two-story additions to the gable end of two-story homes

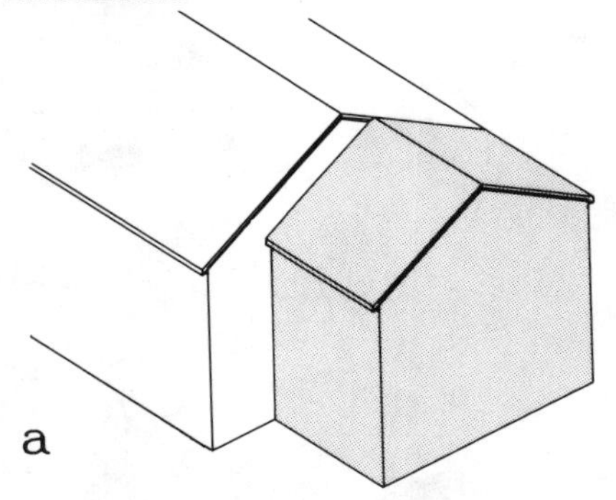

The offset is attractive especially if gable is a front or rear wall. If it is a side wall, watch the scale.

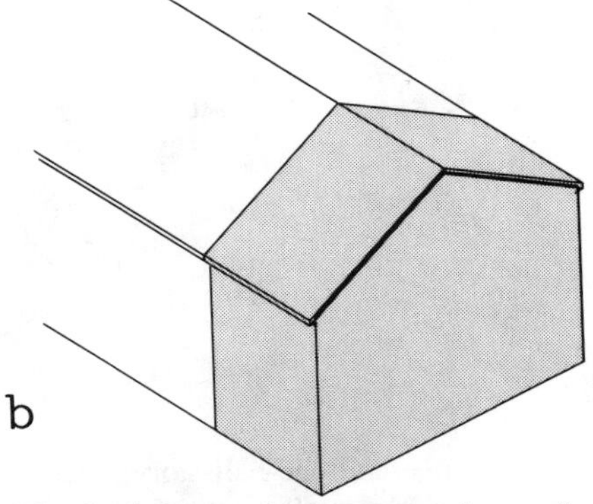

The flush form achieves a seamless end result. It is best for a small addition, but requires perfect matching.

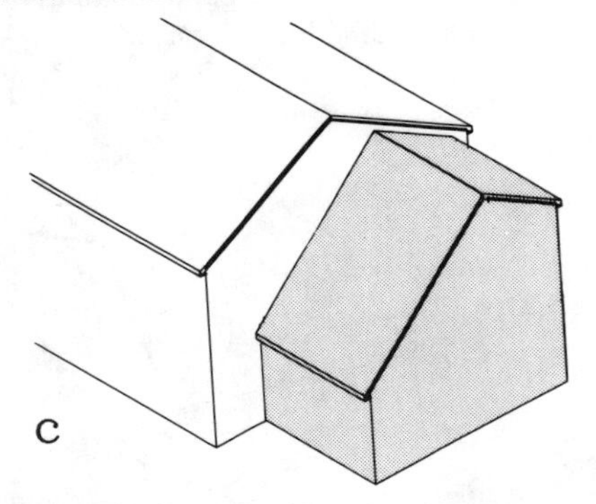

The asymmetrical form is best if gable is an end wall, then the addition looks like a one-story from front.

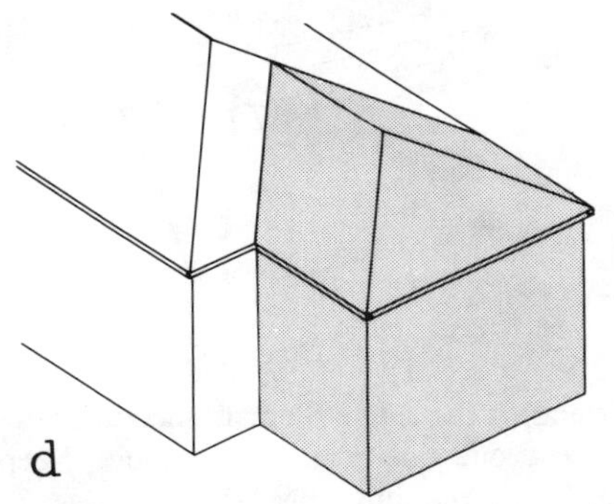

Match a two-story hip with a two-story hip addition. Same rules for flush or setback walls as above.

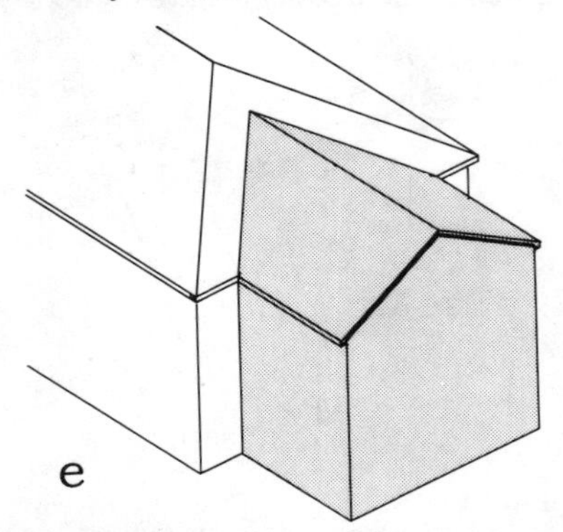

Attaching a two-story gable to a hip can produce a good look if the gable end is made a feature wall.

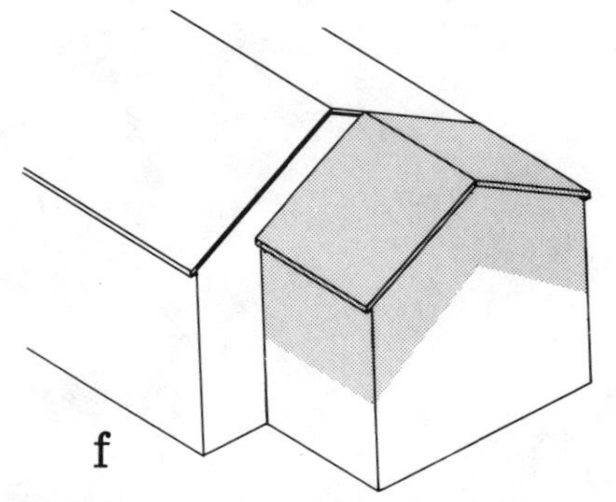

Adding a second floor over a one-story element is fine but requires coordination of first and second floor openings.

Two-story additions to the eave side of two-story homes

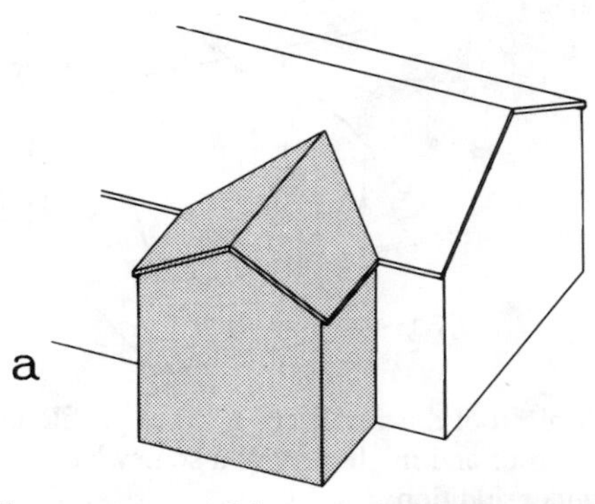

The ever-successful reverse gable produces best results on vertical additions. Depth of the addition is unlimited.

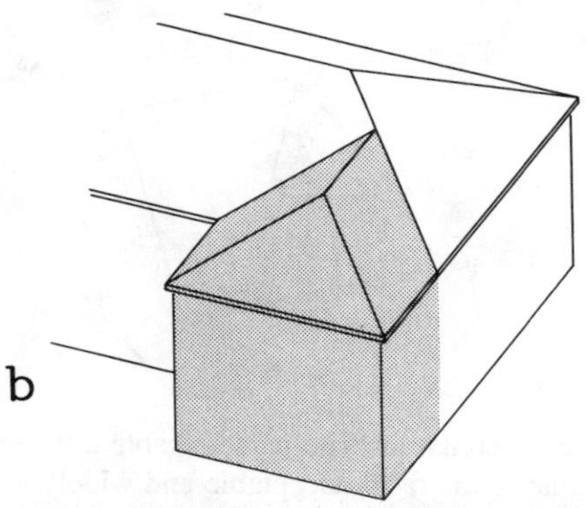

The hip roof addition is best for an existing hip, and especially so at a corner.

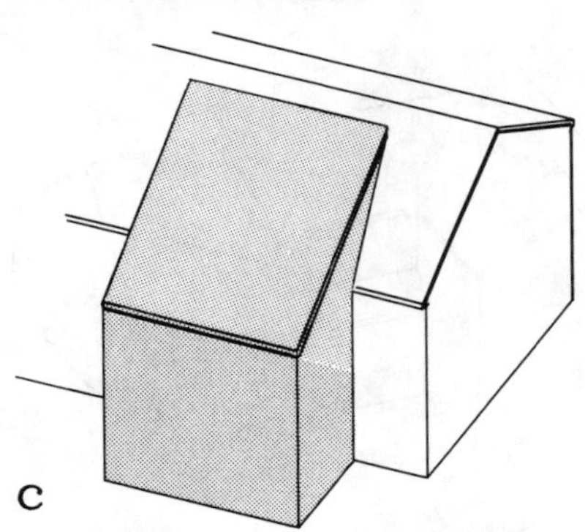

The shed roof form is best for wide but shallow additions, where a reverse gable would be too high.

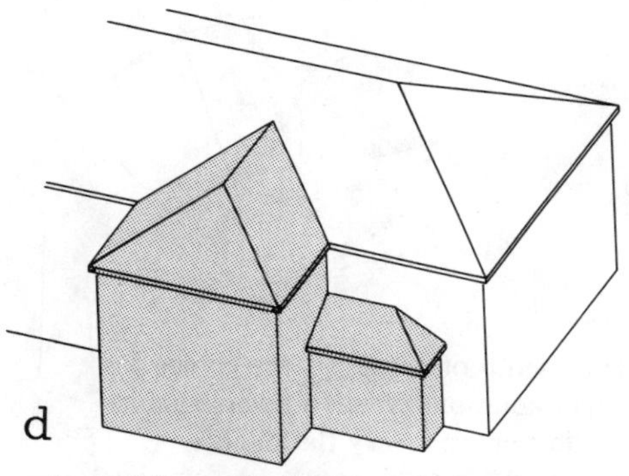

A complex roof of one- and two-story additions requires you to follow rules for both. Keep roof types similar.

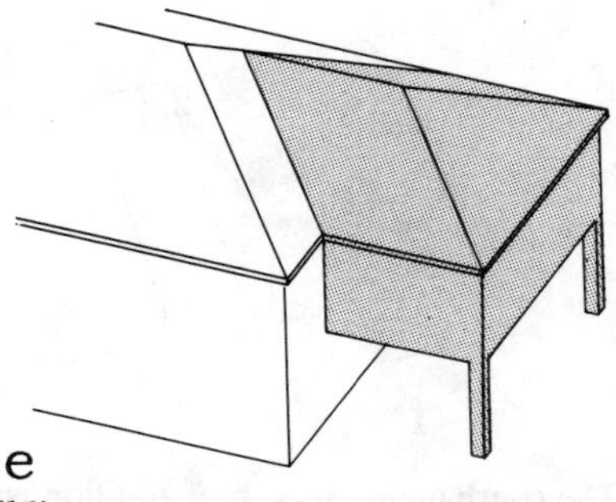

Building up over a driveway requires you to follow rules for two-story forms, but proportions may be different.

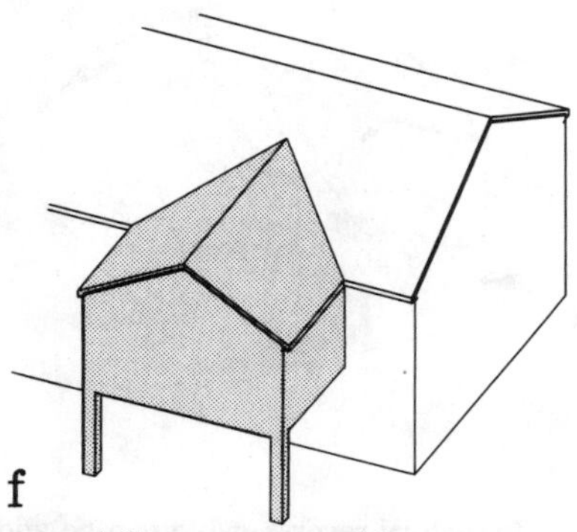

A second floor reverse gable on posts requires you to understand that the mass of the addition is less than a full two-story.

Second-floor additions to one-and-one-half-story homes

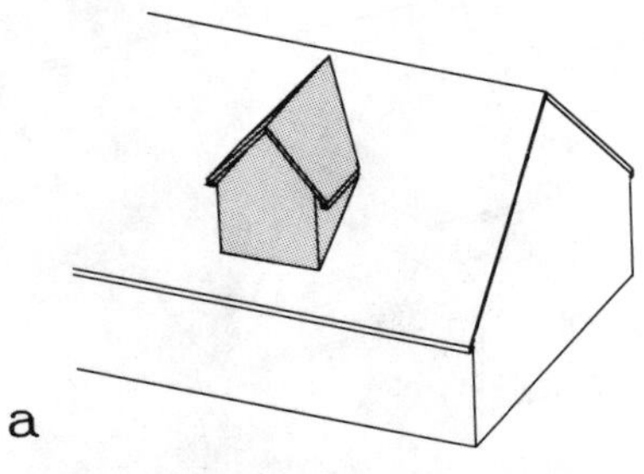

a

The perennial favorite reverse gable dormer is one of the most acceptable and widely used forms.

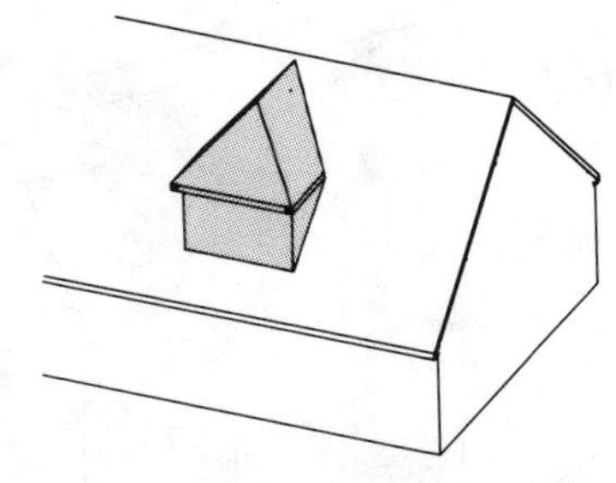

b

A hip roof dormer keeps a lower profile on the roof and might permit a somewhat wider addition.

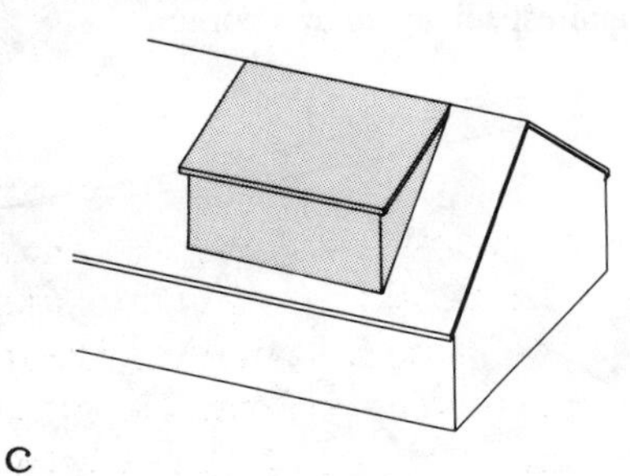

c

The shed roof dormer is the choice for larger additions where a reverse gable would require a very flat pitch.

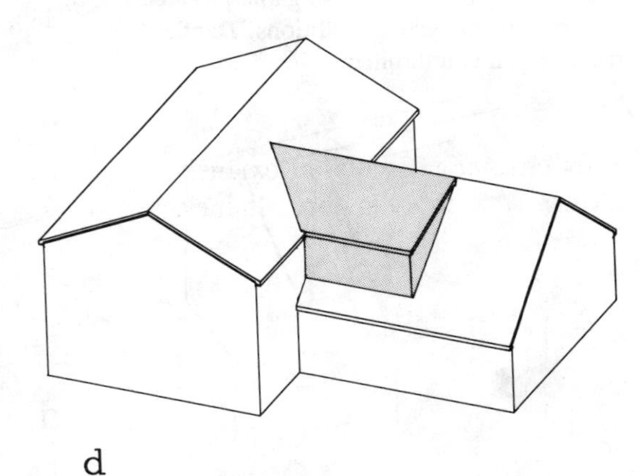

d

Adding at a complex intersection requires roofs to marry to both forms, as this shed on the lower roof.

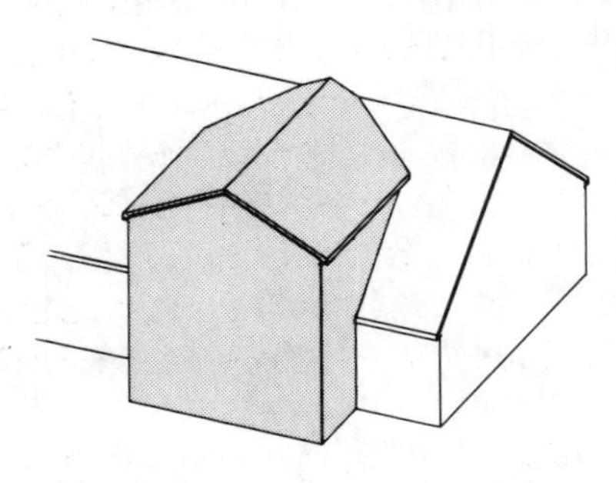

e

A full two-story reverse gable can be added for a tall vertical addition. Use a shed for a wider addition.

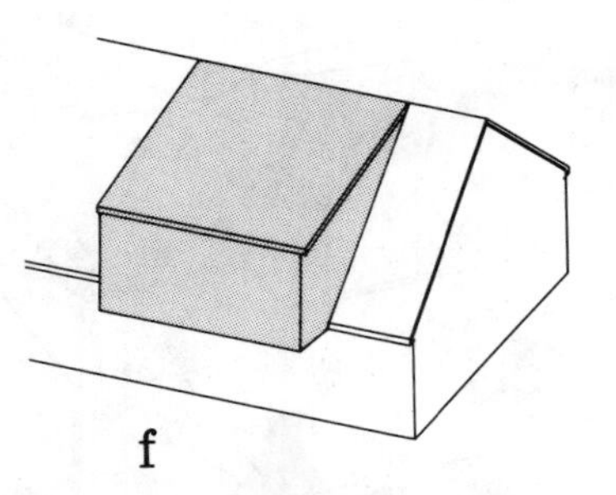

f

The overhanging shed roof addition is no longer a dormer since it breaks the roofline. It adds more space.

7 CHAPTER

Room size additions

This chapter presents an extensive collection of architectural designs for room additions of every nature. It includes one-room bath or kitchen additions, and five-room additions involving a kitchen, family room, master bedroom, dining room, and laundry room. It presents room additions that go up and out. The plans are generally grouped by room type, but that is not universally possible. In a complex plan, like the five-room addition mentioned, the plan might be grouped with other kitchen plans, so the bedroom or laundry room would not be with other similar rooms. Therefore, depending on your precise needs, you might have to peruse all the designs.

The plan viewing rules discussed in chapter 5 prevail as to front orientation and dimensions. There are, however, other viewing instructions. All dark-colored (black) walls are new walls. All light-colored (gray) walls are existing walls. Dotted gray lines represent walls to be removed or fixtures being removed. When clarity is required, I have prepared a separate existing plan, representing the way it is assumed that the home currently exists. Numbers on the plans (i.e., 1, 2, 3, 4, etc.) refer to explanatory notes listed on the page. A sample page entitled "Viewing the Plans" follows this introduction and explains all the viewing conventions.

With each plan there is also a perspective view. For the majority of these additions, the view is an exterior of the addition, showing its likely attachment to your home. The area of the addition is shaded so you should be able to tell the direction the view is taken; views could be from the front, rear, or side, depending on which seemed more appropriate. Materials and window details are intentionally removed to keep the views simple to read. Materials should obviously complement (or on occasion, contrast) with your existing home. See chapter 12 for further discussion on materials.

Under each exterior view is a notation about "rooflines available." Remember the discussion in the prior chapter about the various possibilities for rooflines that often exist for a given addition? Well, it is here that I indicate all the rooflines possible for the addition pictured. If only one is preferred, it is so indicated. The note also indicates which roofline is shown in the view.

On occasion, rather than an exterior view, I have chosen to show an interior view where it provides more information on the addition. Shading has not been used on these interiors, as it would make the drawings very difficult to read, but these are only utilized for very simple additions, and you should be able to discern the area of addition from the floor plan itself.

A few words about the specificity of the existing home pictured. The majority of additions are shown attached to a home of some fairly defined shape or character. That is not to say that the same addition could not be attached to a very different house. This is where the blockforms in chapter 6 should help provide you with the confidence that the two (addition plus existing home) will match well. Also keep in mind that one-story additions can usually attach to most one-, one-and-one-half-, and two-story homes. Furthermore, where a two-story addition has been pictured, it is possible to take either floor and build it as a one-story addition. Our CAD system could readily accommodate that for you.

There are also a number of generic types of additions shown where the existing home is only vaguely outlined. These additions are planned to permit the greatest flexibility of attachment to any home. Only the rooflines have to be considered, and these are usually pictured.

Finally, many designs also show some renovation to interior space as well. As I have indicated, this is a very common practice; the addition leads to some reason to alter the interior space as an integral part of the remodeling project. You might decide to forego this part of the renovation, or to modify it to suit your specific home. I fully expect that any remodeling project undertaken from these designs will produce many changes necessary to suit your needs, lifestyle, or existing home, so don't feel inhibited by change.

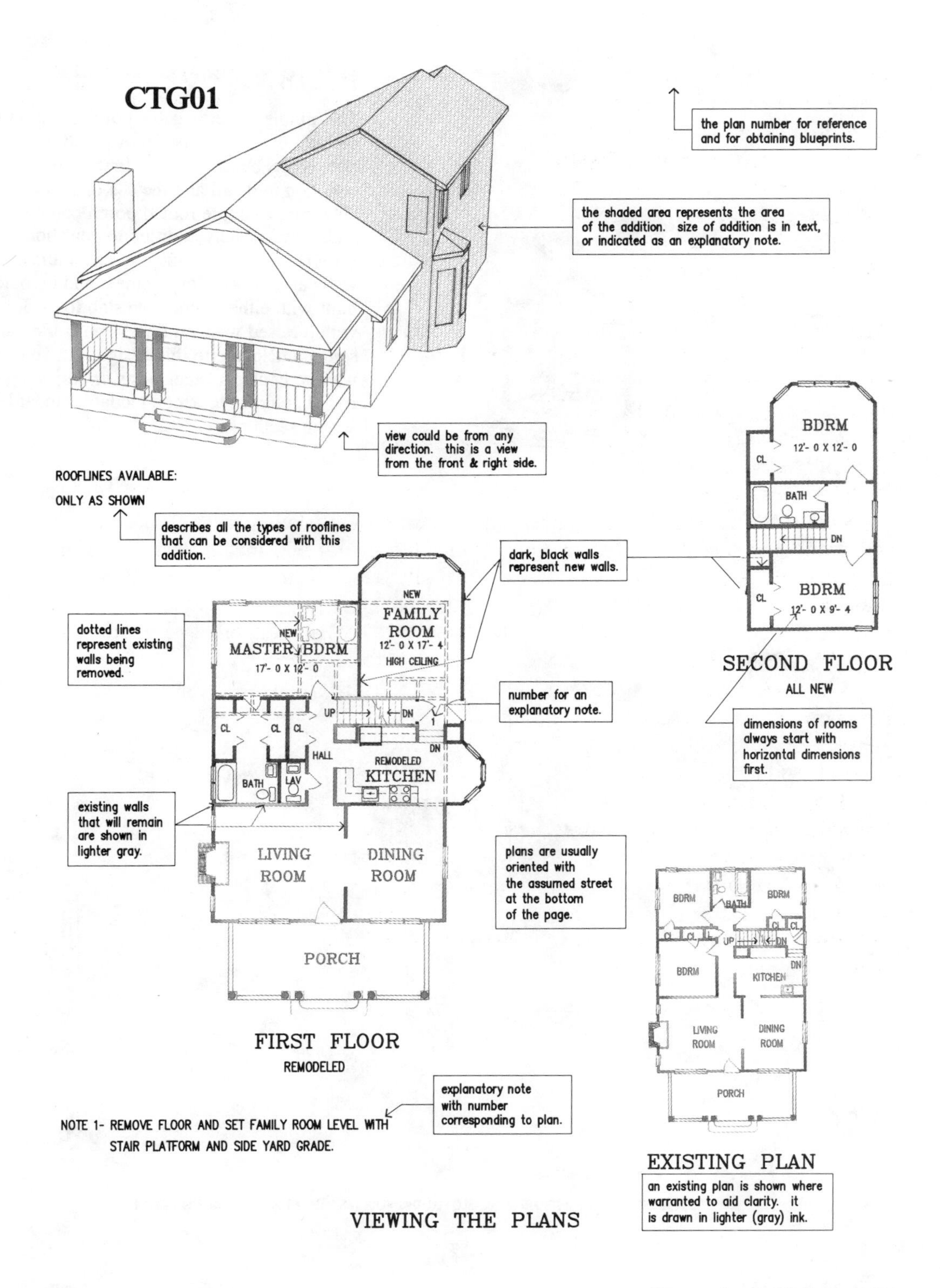

VIEWING THE PLANS

P0018

The simple, reverse-gable porch addition is second only to the shed roof porch in popularity, but it could be first in looks. With attention to detail and the use of round columns, a simple roofed porch could become a classical, stately addition to your home. The form as pictured is suited to end walls or front walls and to two-story homes, and it could be built with either a concrete slab for a floor, or with a raised wooden floor. You could adapt a screen enclosure to this design, but you might want to consider other posts or respacing the round columns to accommodate standard screen sections.

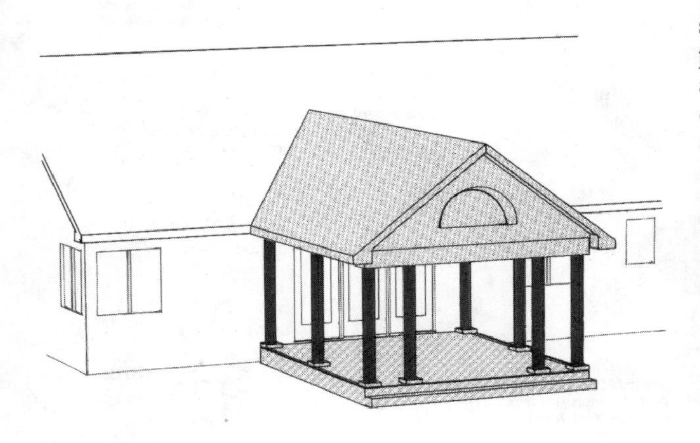

ROOFLINES AVAILABLE:
GABLE (SHOWN)

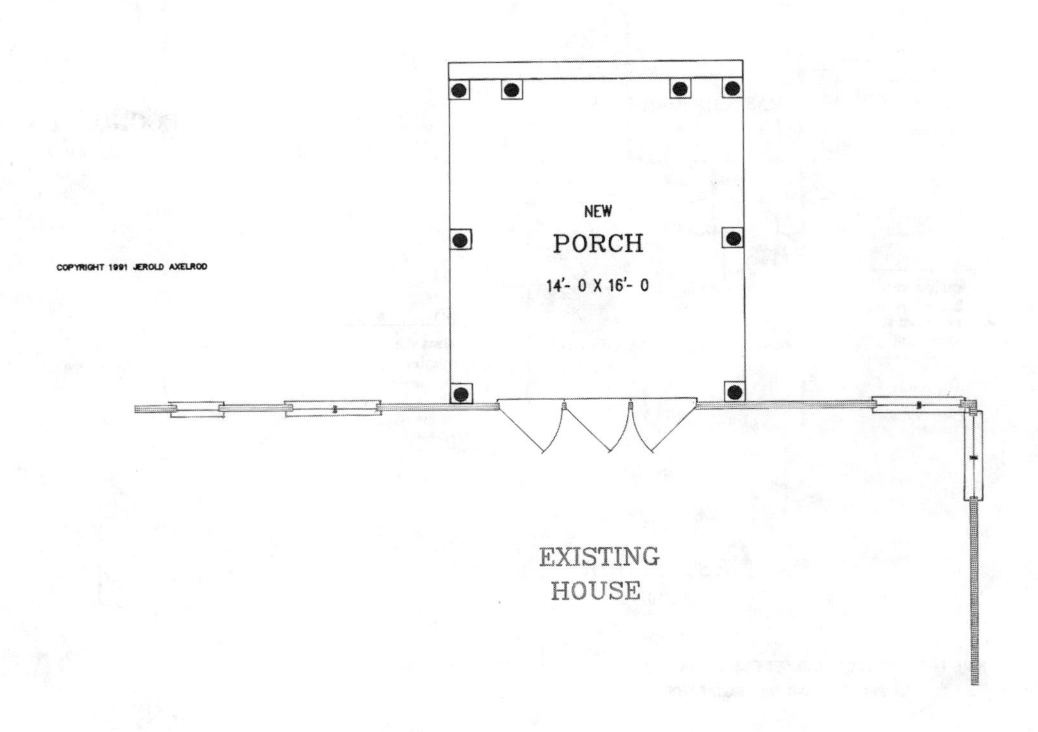

CONSTRUCTION BLUEPRINTS ARE AVAILABLE FOR THIS PLAN (AS WELL AS ALL OTHERS) AND INCLUDE A MATERIAL LIST

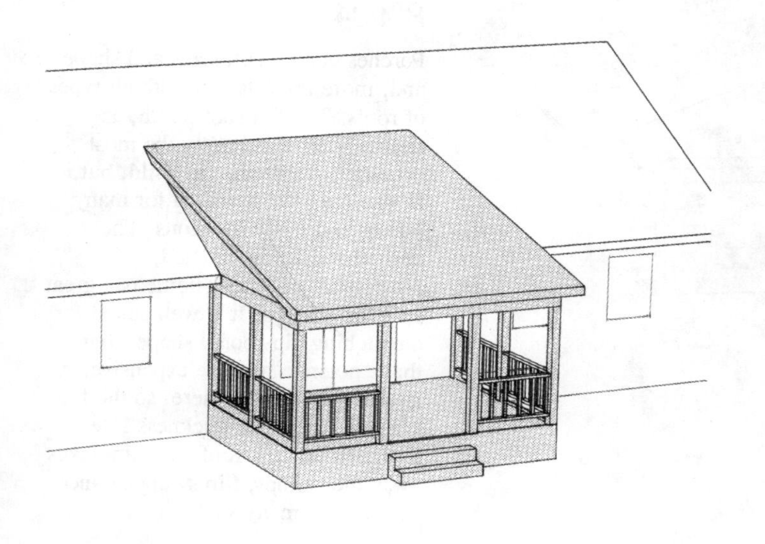

ROOFLINES AVAILABLE:
SHED (SHOWN)

P0017

The shed roofed porch is probably the most common type added to the rear of homes. It is also the most common addition of all, but with a little flair it could have some character, too. A fairly regular spacing of posts, even on the nonbearing side walls, and a railing help provide the porch pictured with a warm, inviting character. The porch could be built on a concrete slab or with a raised wooden floor, depending on your wishes and on the style of house. The same design is also suited to the rear of a two-story home, but it would not look good on the side. A reverse gable, such as in P0018 on page 52, is more appropriate to a side gable wall.

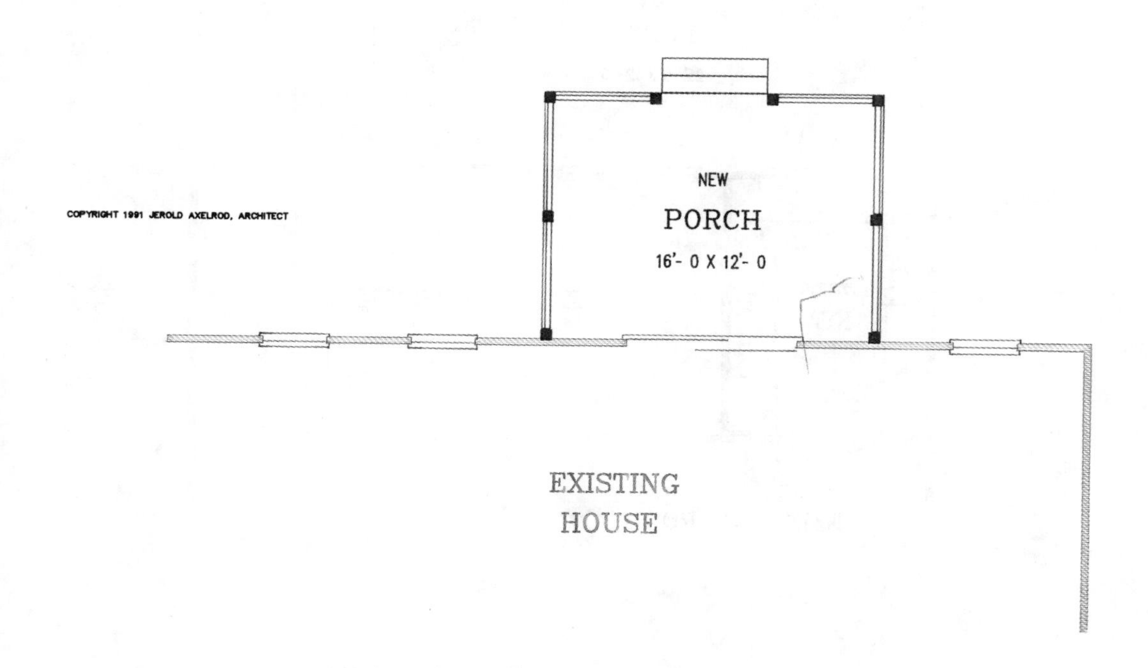

NOTE: CONSTRUCTION BLUEPRINTS CAN BE READILY MODIFIED TO ACCOMMODATE YOUR SPECIFIC HOUSE OR ROOM SIZES.

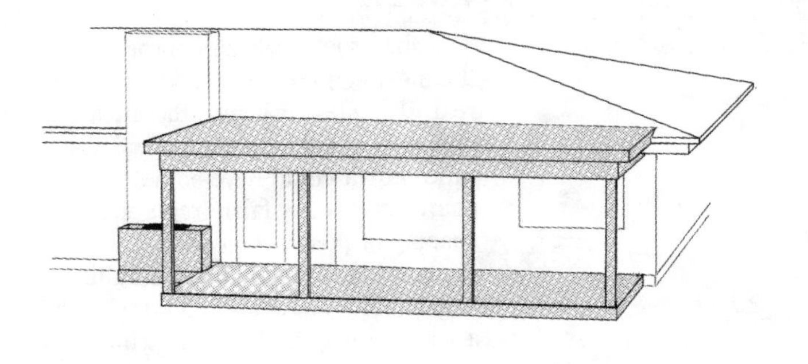

ROOFLINES AVAILABLE:
FLAT (SHOWN), HIP, SHED, GABLE

P0004

Porches come in all sizes and shapes and, more importantly, with all types of roofs. The flat roof porch, as shown here, is certainly the most economical structure to build, but it is also the most practical for many homes and many locations. The corner of this low-pitched, hip-roofed, one-story home is a great spot for a porch. It is well suited for a matching hip-roofed shape—but that's potentially more expensive. A gable roof won't go here, so the flat is ideal. The extra thickness I've designed into the roof structure takes away the skimpy, flimsy appearance common to many such porches.

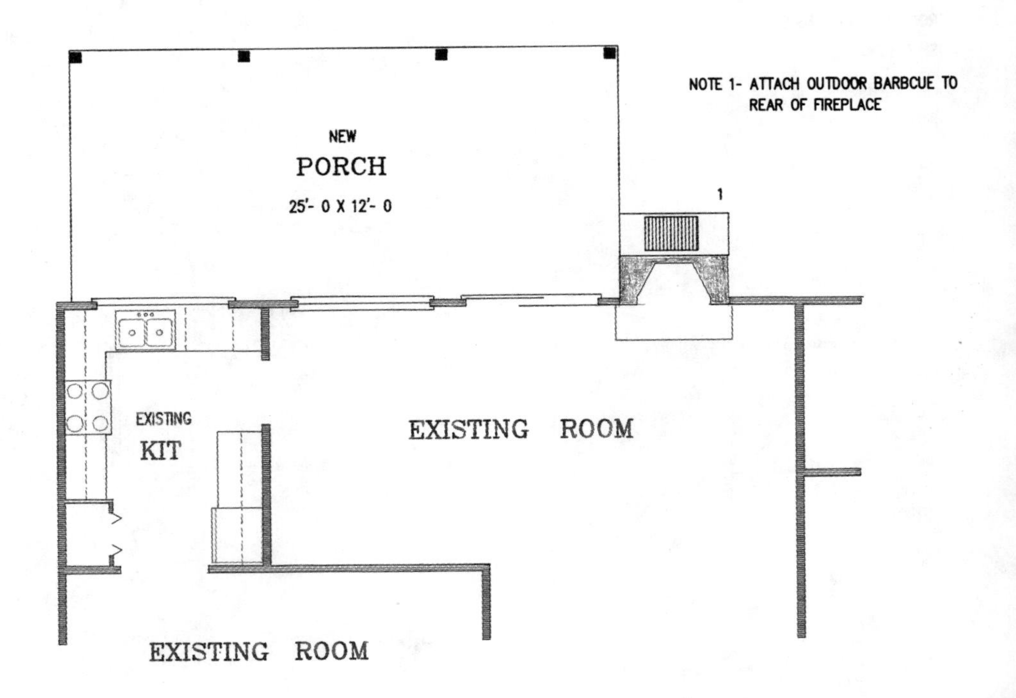

NOTE: CONSTRUCTION BLUEPRINTS CAN BE READILY MODIFIED TO ACCOMMODATE YOUR SPECIFIC HOUSE OR ROOM SIZES.

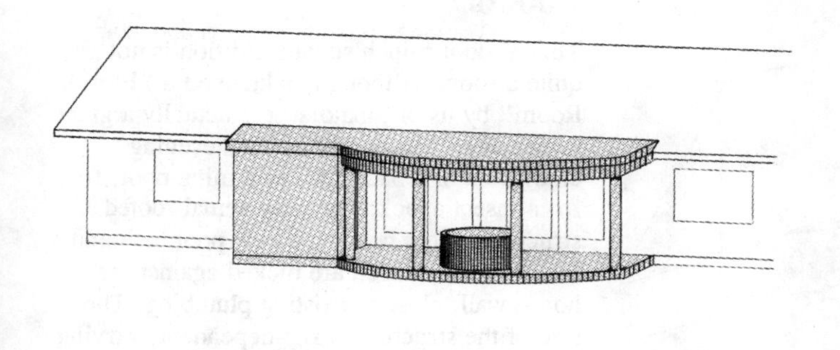

ROOFLINES AVAILABLE:
FLAT (SHOWN), GABLE, HIP, SHED

P0005

There's a rear porch in your program, and a pool cabana and bath . . . well, why not combine them? This classy-looking addition artfully combines a dressing room and shower-bath with a semicircular flat-roofed porch. Round columns provide a nice detail, as does a dropped header frieze following the circular shape. The bath is located on the rear wall and might well be in the vicinity of existing plumbing as indicated. This porch could be a great place to locate a spa as shown.

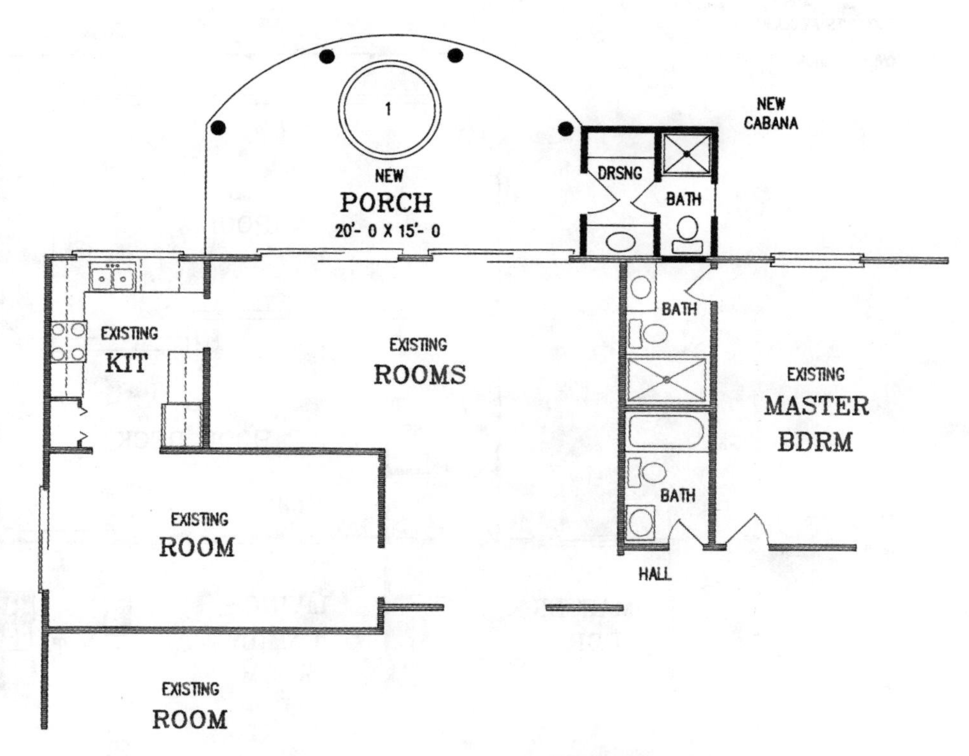

CONSTRUCTION BLUEPRINTS ARE AVAILABLE FOR THIS PLAN (AS WELL AS ALL OTHERS) AND INCLUDE A MATERIAL LIST.

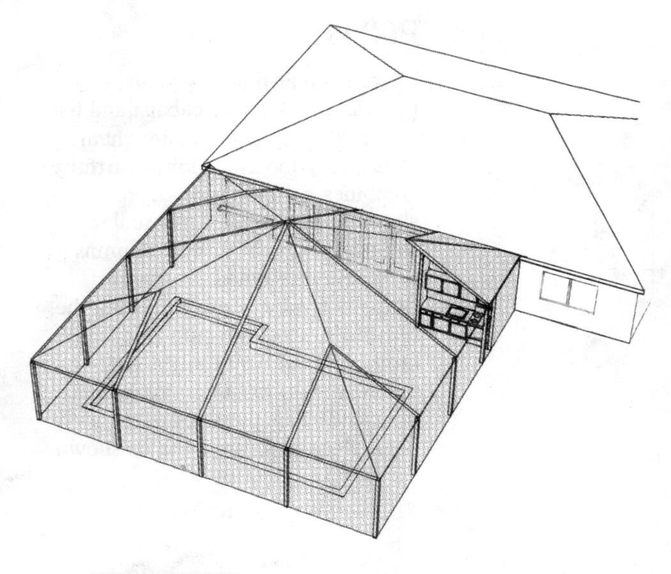

ROOFLINES AVAILABLE:

ONLY AS SHOWN

P0006

This 37-foot-6-inch square addition is not quite a room. Although nicknamed a "Florida Room" by its originators, it is actually a giant screened enclosure that permits evening enjoyment of a backyard swimming pool, free from insect attack. The only actual roofed structure is that housing a new pool bath and a mini-kitchen, which are tucked against the house wall, close to existing plumbing. The size of the structure is site-dependent, varying by virtue of the size of the pool and the pool deck. A glass roof is an option to consider for colder climates. However, before doing anything, check on local codes, zoning requirements, and taxing rules. Also try to locate a local manufacturer or distributor of such roof systems.

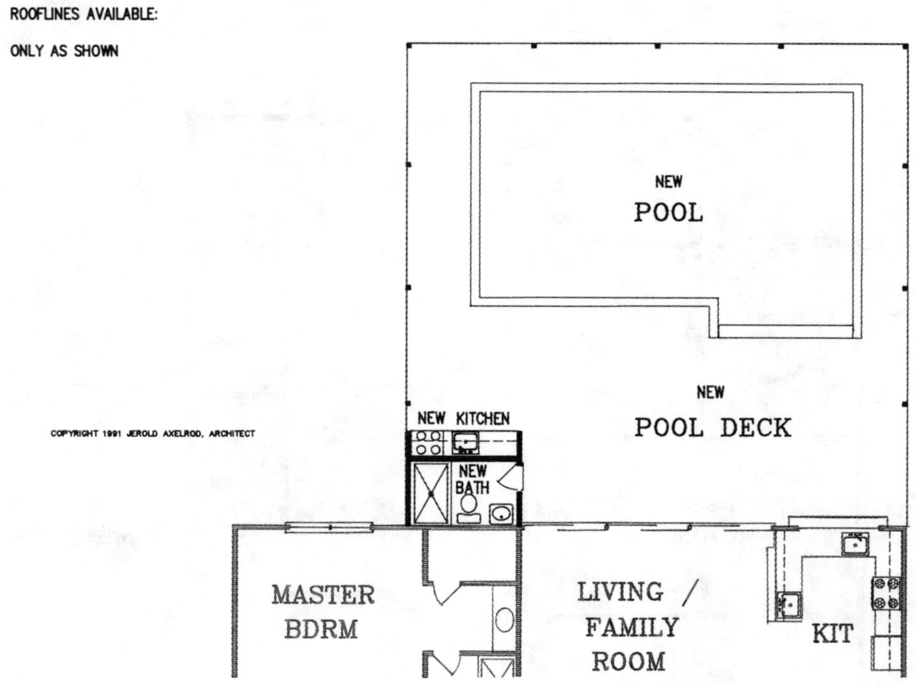

CONSTRUCTION BLUEPRINTS ARE AVAILABLE FOR THIS PLAN (AS WELL AS ALL OTHERS) AND INCLUDE DETAILS ON THE POOL BATH & KITCHEN PLUS SUGGESTIONS ONLY FOR THE POOL ENCLOSURE.

P0002

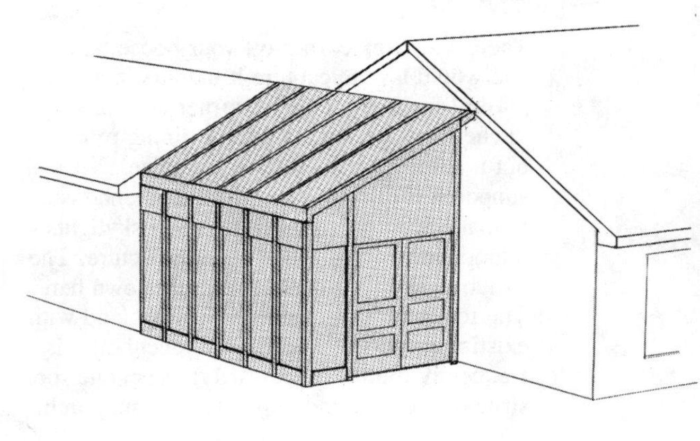

While most new rooms are for people, and sometimes for cars, this room is for plants. It is a solar greenhouse structure available from a number of manufacturers. There are other greenhouses in this book designed as living spaces, but this one is truly a plant room for the garden enthusiast. It is shown tucked into a corner—it would obviously be a sunny corner of your home, where the plants would receive ample sunlight. The room is shown accessed from the garage and viewed from the kitchen. The kitchen windows should be left to open to the new greenhouse. The greenhouse structure requires that it not be flush with adjacent walls or roofs, but be somewhat lower for proper flashing.

ROOFLINES AVAILABLE:

ONLY AS SHOWN

CONSTRUCTION BLUEPRINTS ARE AVAILABLE FOR THIS PLAN (AS WELL AS ALL OTHERS) AND INCLUDE GREENHOUSE MANUFACTURER'S DETAILS AND SPECIFICATIONS.

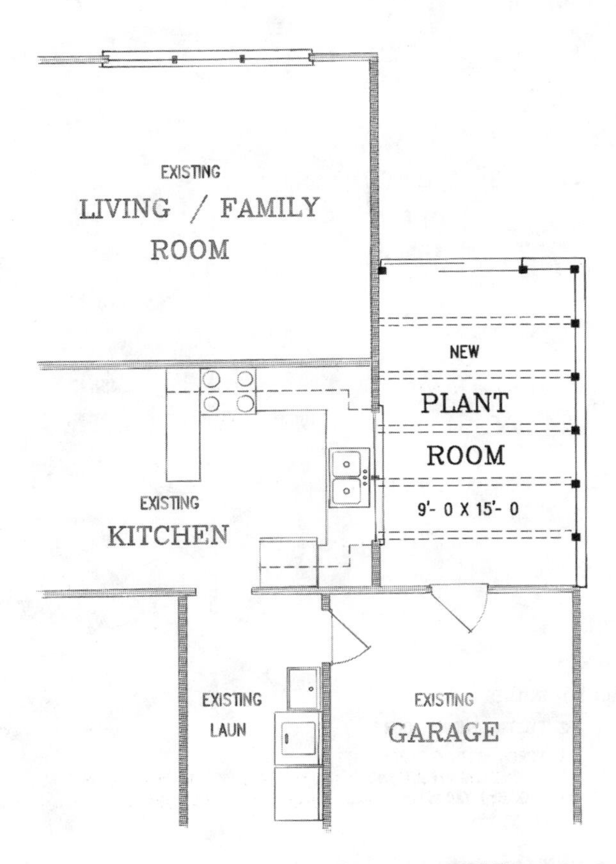

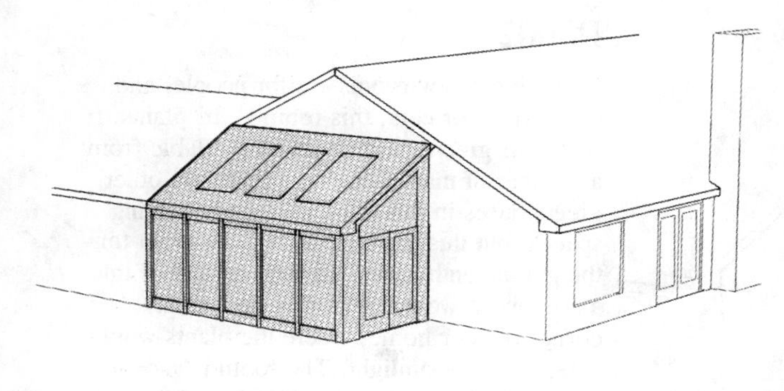

ROOFLINES AVAILABLE:
SHED (SHOWN), FLAT, GABLE

P0003

There's a sunny corner off your house, and you would love to capture it indoors in a glassed-in sun room. The corner shown is a kitchen, garage, and family or dining room, but it could be any set of rooms. The sunporch is distinguished from a greenhouse in that it incorporates windows and skylights into a conventional wood frame structure. The windows could be wall-to-wall as shown here. The roofline of the sunporch has to blend with existing rooflines, whereas the greenhouse is frequently (but not necessarily) a separate roof structure. Costs tend to be less in a sunporch.

CONSTRUCTION BLUEPRINTS ARE AVAILABLE FOR THIS PLAN (AS WELL AS ALL OTHERS) AND INCLUDE A LIST OF CONSTRUCTION MATERIALS REQUIRED.

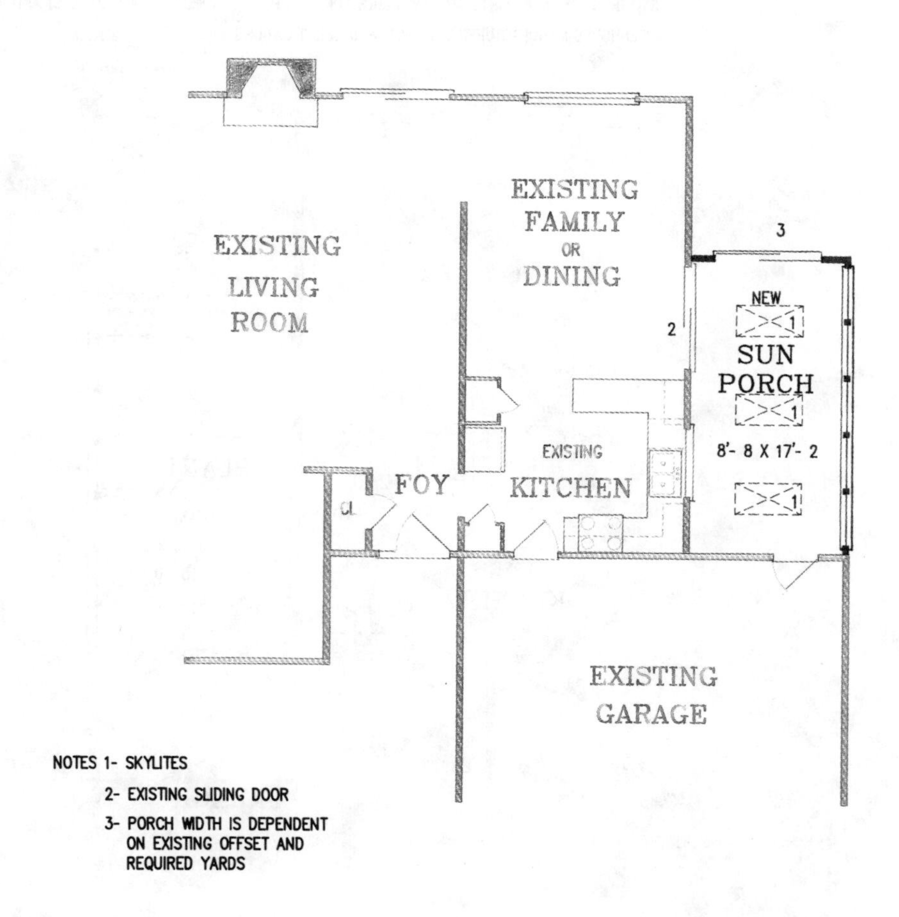

NOTES 1- SKYLITES
2- EXISTING SLIDING DOOR
3- PORCH WIDTH IS DEPENDENT ON EXISTING OFFSET AND REQUIRED YARDS

P0001

This classy-looking square-shaped addition adds a sparkling, bright, new sun porch to your home. It features windows wrapping its three exposed sides, an expansive vaulted ceiling, and four large skylights, thereby creating an open feeling similar to that of a prefabricated metal-and-glass structure. You might find that a conventionally built structure such as this is more appropriate to your home than is a greenhouse structure—and it could be less expensive. For even more glass than shown, all the windows could extend to the floor and become sliding-glass doors. To prevent bowing walls, do not attempt to build a room such as this (or any other, for that matter) without proper construction plans.

ROOFLINES AVAILABLE:
GABLE (SHOWN)

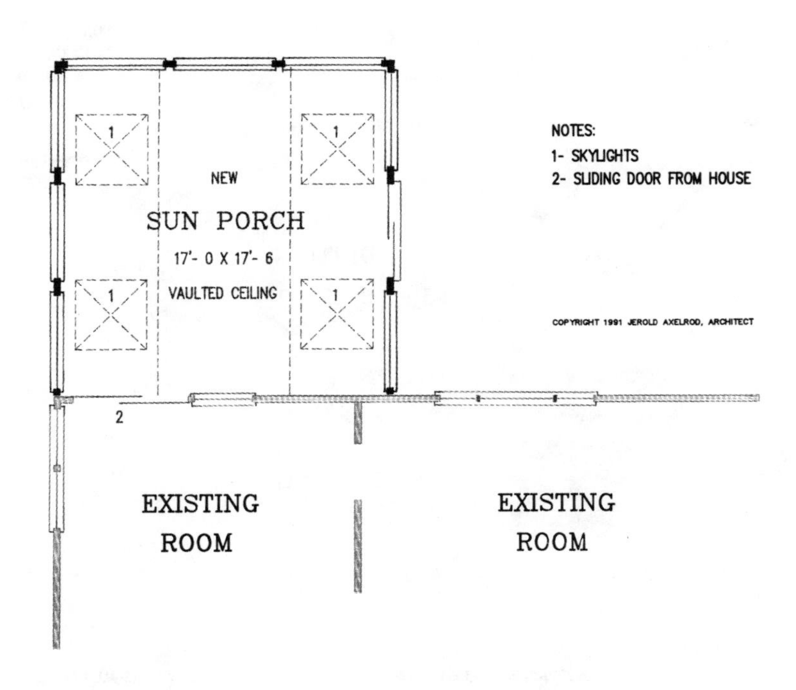

CONSTRUCTION BLUEPRINTS ARE AVAILABLE FOR THIS PLAN (AS WELL AS ALL OTHERS) AND INCLUDE A MATERIALS LIST AND INFORMATION ON HOW TO MAKE CHANGES TO PLANS

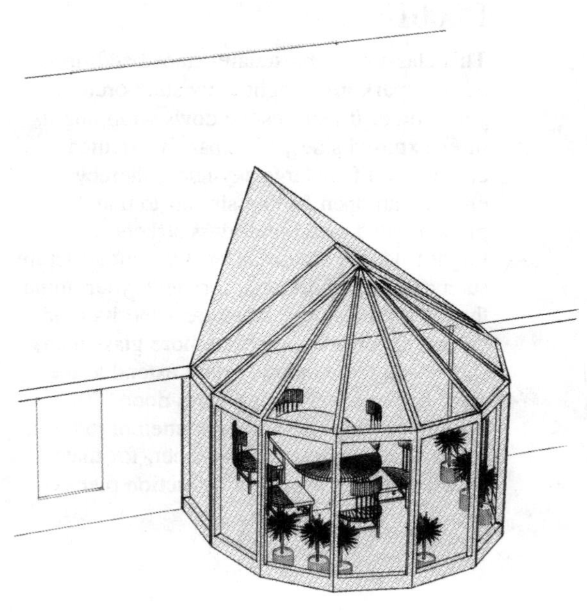

P0020

Additions don't have to be enclosed by stereotypical walls and roofs. If you have a penchant for a glass enclosed space—whether to use for dining under the stars, or to use as a sunny family room addition—this could be the plan for you. There are a number of other glass-roofed porch additions that could also serve your needs, but this one is unusual for its shape. Its eight sides are glazed, as are its eight sections of roof, and it will likely require custom fabrication. Note the new reverse-gabled section of conventional roof beyond, which is necessary to enable you to attach such a structure to a one-story home.

ROOFLINES AVAILABLE:

ONLY AS SHOWN

NOTES:

1- ADDITION MAY ALSO BE ADDED TO SIDE OR REAR OF TWO STORY

2- REMOVE EXISTING REAR OR SIDE WALL

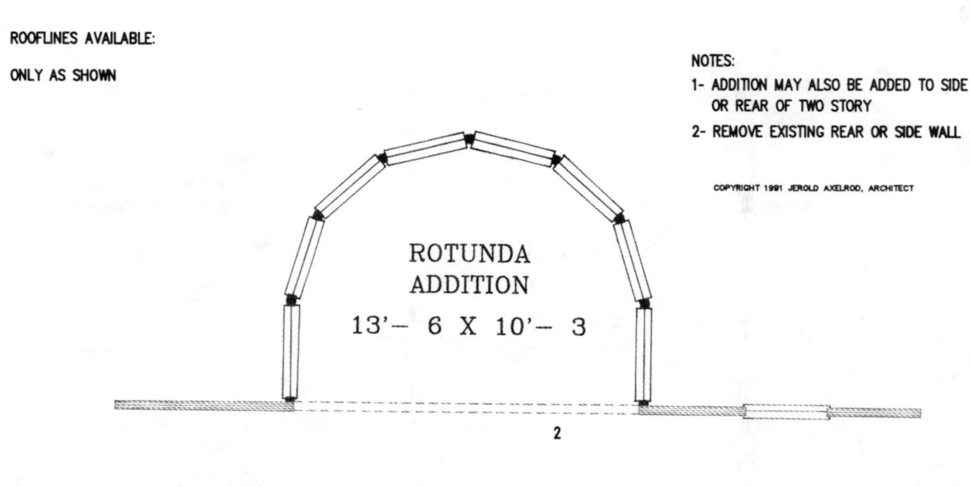

CONSTRUCTION BLUEPRINTS ARE AVAILABLE FOR THIS PLAN (AS WELL AS ALL OTHERS) AND INCLUDE GREENHOUSE MANUFACTURERS INFORMATION AND STRUCTURAL DATA FOR REMOVING REAR WALL

P0022

Another in the series of beautiful enclosed sunspaces presented in the book, this is a generous size room—large enough for a delightful, built-in whirlpool spa and a wood stove to heat the room. The rear wall is designed as a masonry heat storage (or *trombe* wall) to capture and store the heat from the sun and the wood stove. If you do install a whirlpool, make sure to provide an oversized ventilation system. This type of structure is available from many manufacturers and can be purchased as a kit or fully installed. Note the piece of new conventional roof framing required.

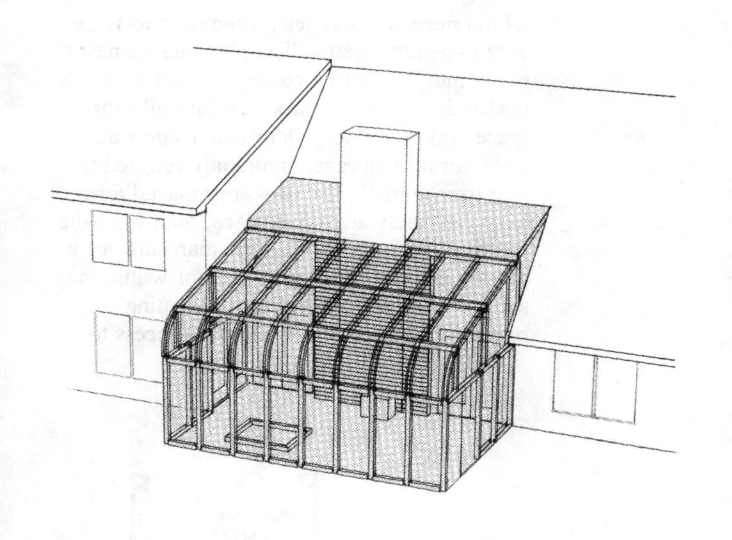

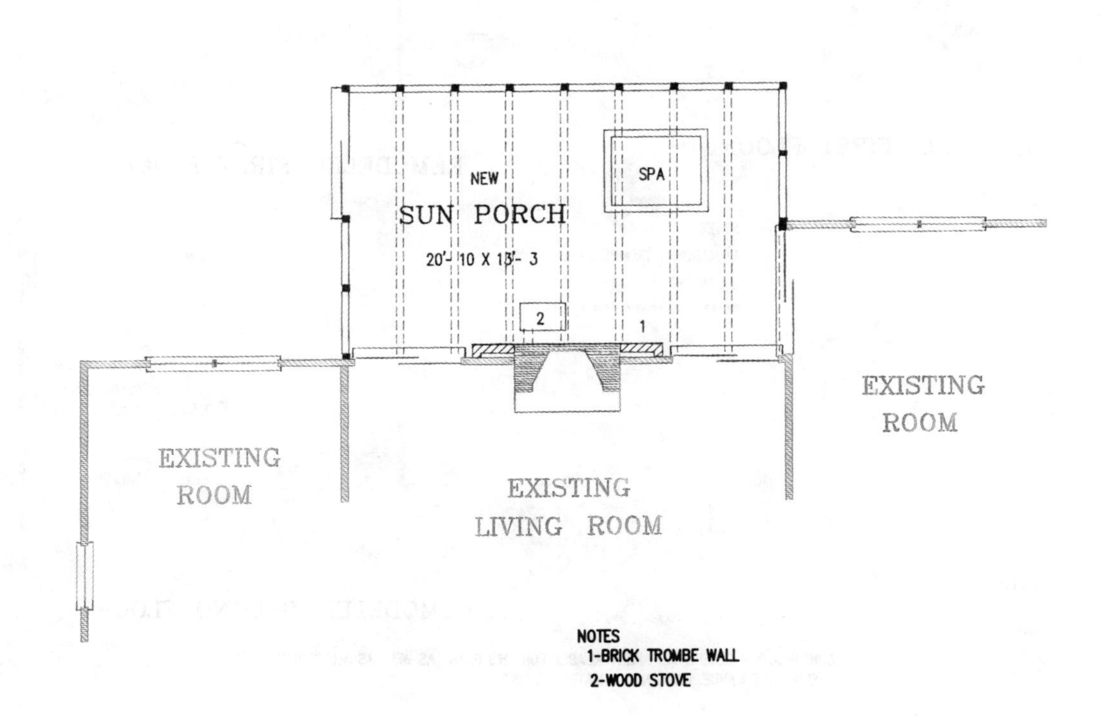

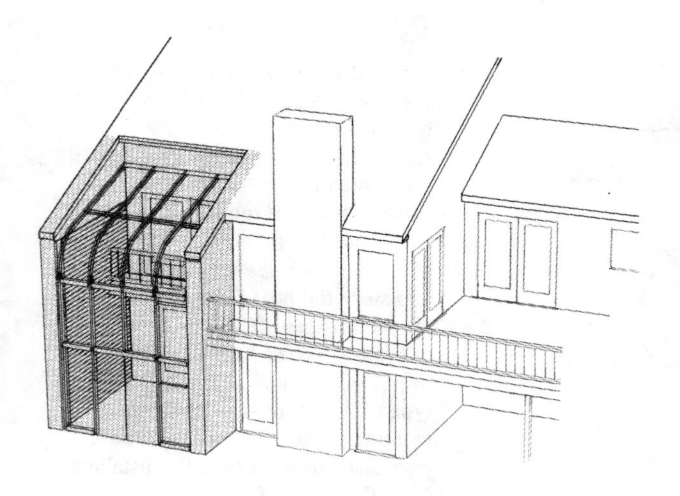

P0023

Of the many sunporches presented, this is the only two-story design. The plan tucks a new porch into an existing corner, on two floors. It is also designed as a passive solar collector space with a masonry floor and a side wall of brick for heat storage, commonly referred to as a *trombe wall*. The two-story glazed form is an extremely attractive space, both from the exterior and from within. The plan calls for a narrow balcony at the second floor within the sunspace, thereby providing a charming retreat from the rooms it provides access to.

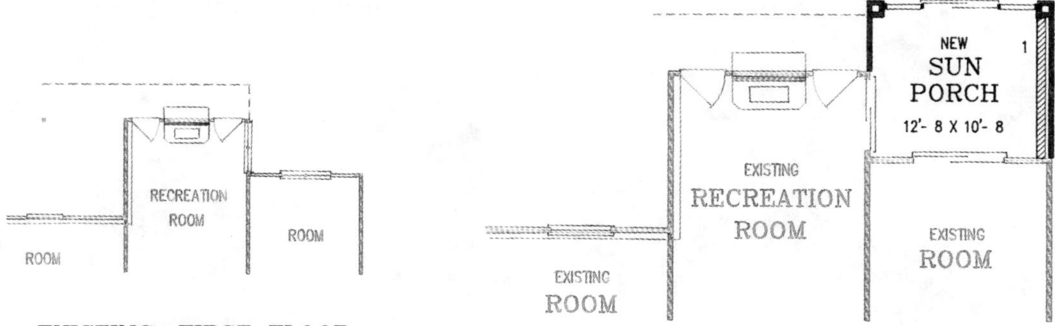

EXISTING FIRST FLOOR

REMODELED FIRST FLOOR

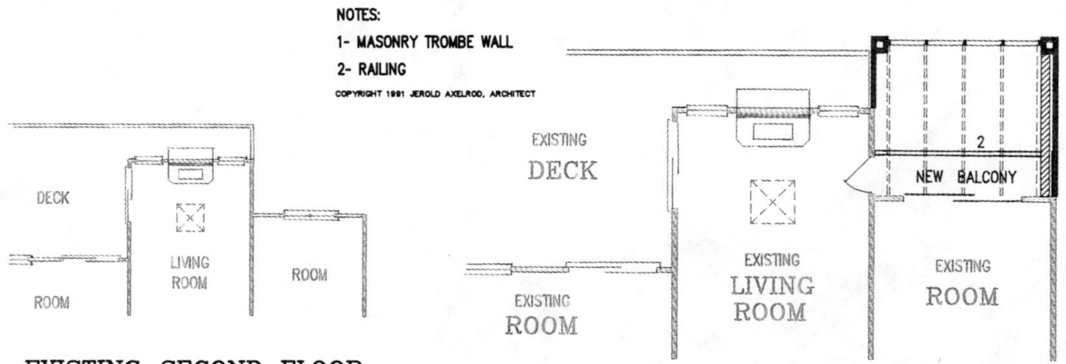

EXISTING SECOND FLOOR

REMODELED SECOND FLOOR

CONSTRUCTION BLUEPRINTS ARE AVAILABLE FOR THIS PLAN (AS WELL AS ALL OTHERS) AND INCLUDE GREENHOUSE MANUF. DATA AND A MATERIALS LIST

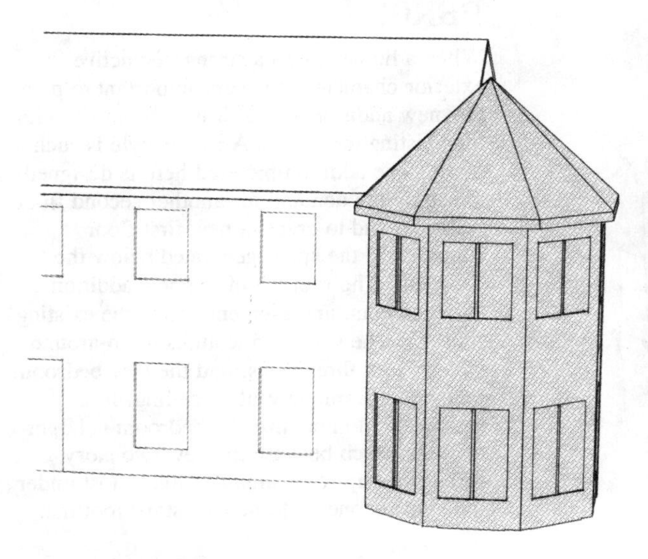

PBR04

This two-story tower, with its conical shaped roof, might be just the addition for you if you are searching for the unusual. The plan is a picturesque octagon, 12 feet square, enveloped with windows on all its sides except for one—that being the side that provides connection to the house. The same plan is duplicated on each floor; just the use may vary. In the design shown, where it is attached to the corner of a two-story home, the first floor is a likely sunroom or porch, and the second floor could be an engaging new sitting room connected to your master bedroom. Whatever its use, it could become a striking addition to your home.

ROOFLINES AVAILABLE:

ONLY AS SHOWN

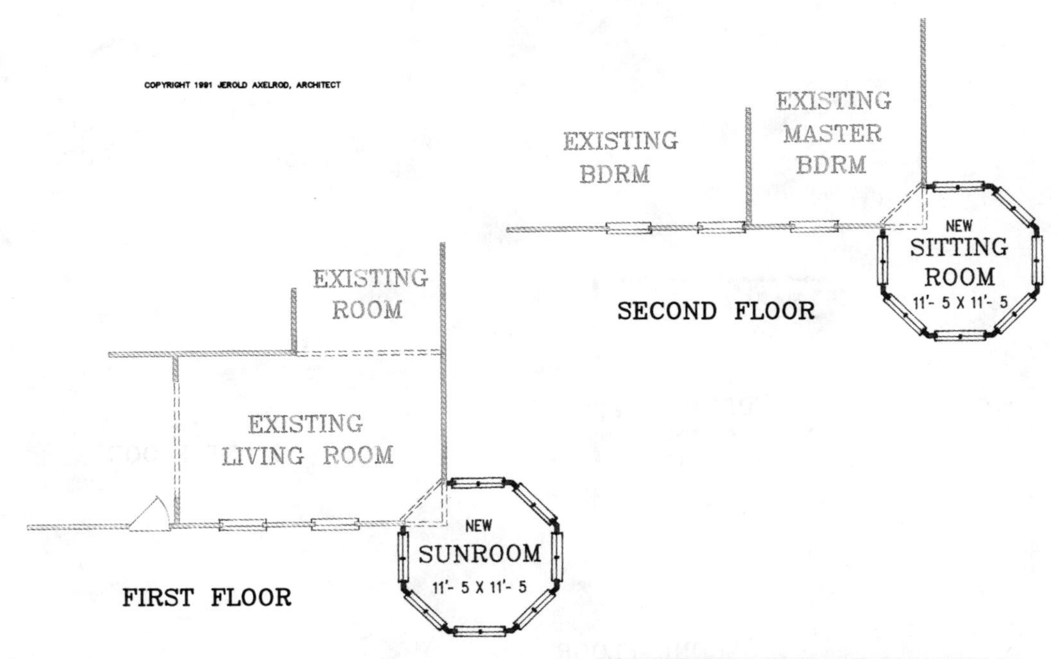

NOTE: CONSTRUCTION BLUEPRINTS CAN BE READILY MODIFIED TO ACCOMMODATE YOUR SPECIFIC HOUSE OR ROOM SIZES.

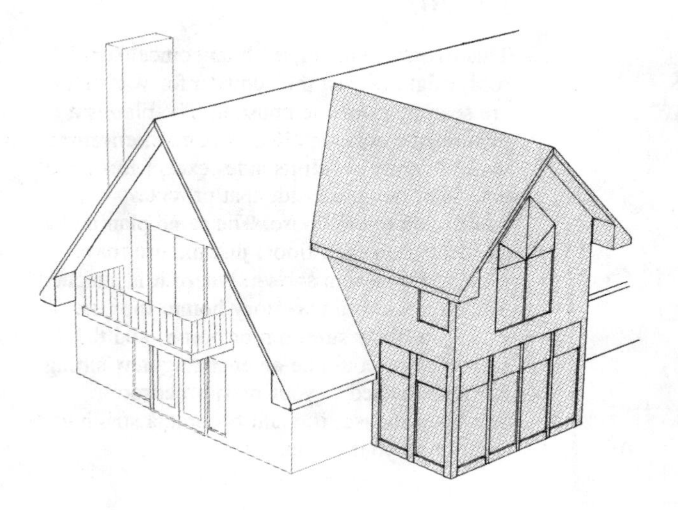

PBR01

When a home enjoys a strong distinctive exterior character, it is very important to plan any new addition so that it is in harmony with the existing forms. An A-frame style is such a home. The addition pictured here is designed to satisfy the demand for another second floor bedroom and to create a new first floor sunporch in the space generated below the bedroom. The roofline of the new addition complements, and even enhances, the existing roofline. The sunporch features wrap-around windows on three sides, and the new bedroom includes a dramatic vaulted ceiling. It is worthwhile to note that the bedroom is larger than the porch because the new two-story form picks up some area heretofore lost under the existing one-and-one-half-story roofline.

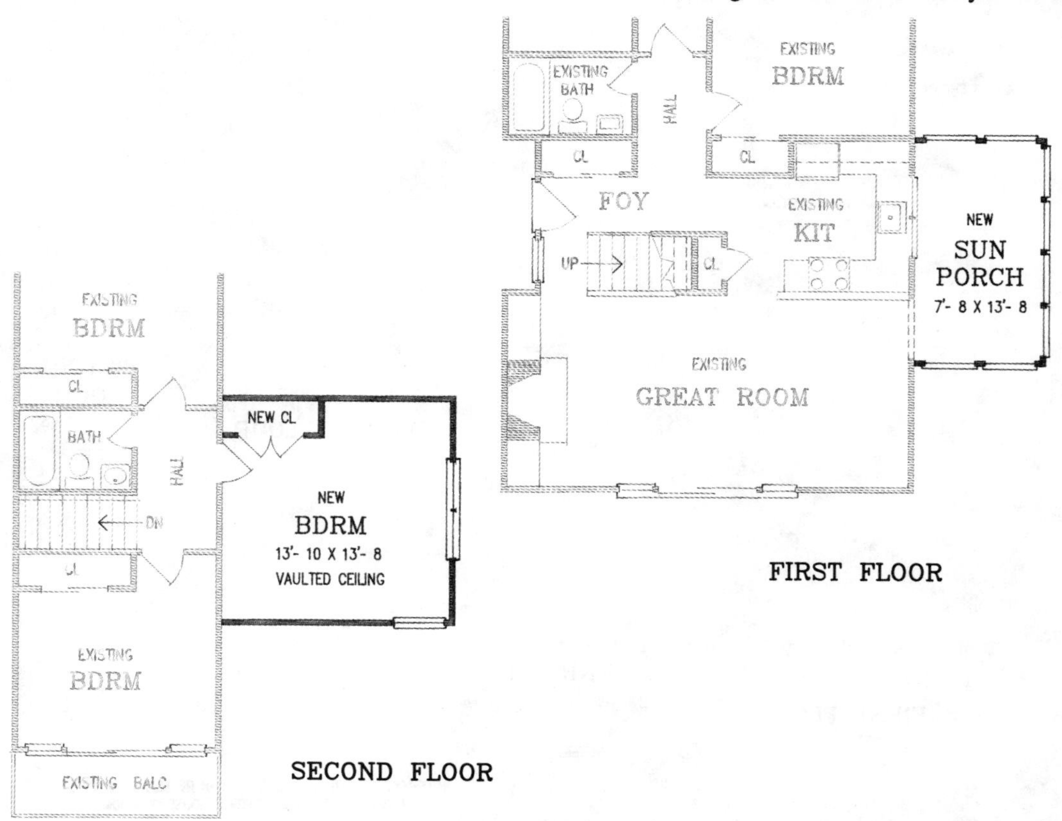

CONSTRUCTION BLUEPRINTS ARE AVAILABLE FOR THIS PLAN (AS WELL AS ALL OTHERS) AND INCLUDE A MATERIALS LIST AND WINDOW MANUFACTURER INFORMATION.

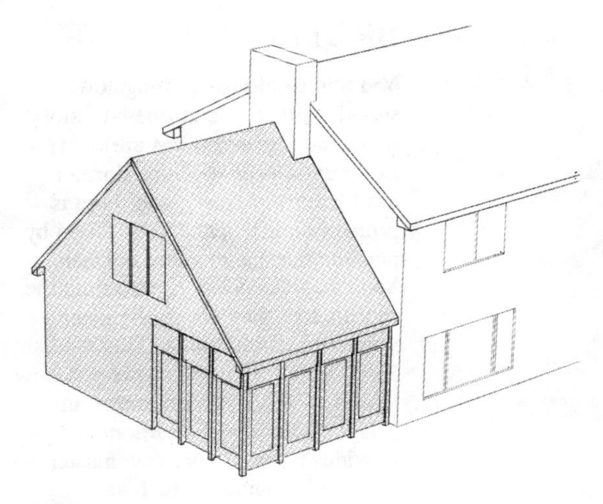

ROOFLINES AVAILABLE:
GABLE (SHOWN), HIP

PBR02

Here's an example where the "bonus" or "extra" (see chapter 3 for discussion) could be as important as the main purpose of the remodeling. While the major goal of this plan is to add a first floor home office and enclosed porch (or sunroom), the second floor dressing room or sitting room is not an afterthought. In plan LD001, on page 136, we see a one-story addition to the side of a two-story home. Its aesthetic proportions could be a problem, especially on a wider home, so I recommend the steeper, one-and-one-half-story roofline to achieve better balance and to gain the second floor space. If you want to consider going all-out, there are a number of plans in chapter 8 that show you how to remodel the master bathroom, too.

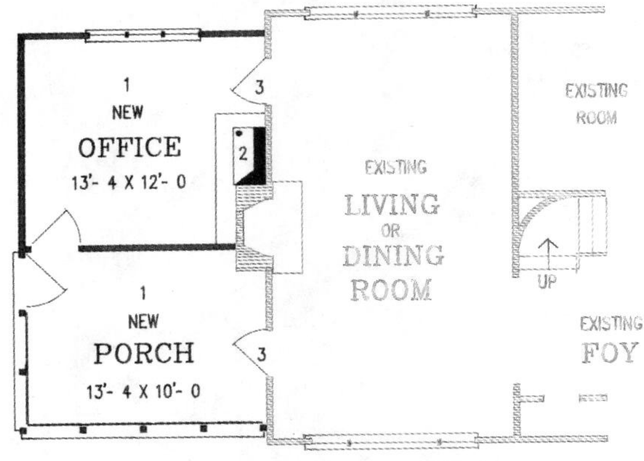

FIRST FLOOR

NOTES
1- PORCH AND OFFICE COULD BE REVERSED IF YOU WANT
2- OPTIONAL NEW FIREPLACE
3- CUT EXISTING WINDOWS TO NEW DOORS
4- CLOSET / BATH AREA COULD BE REMODELED

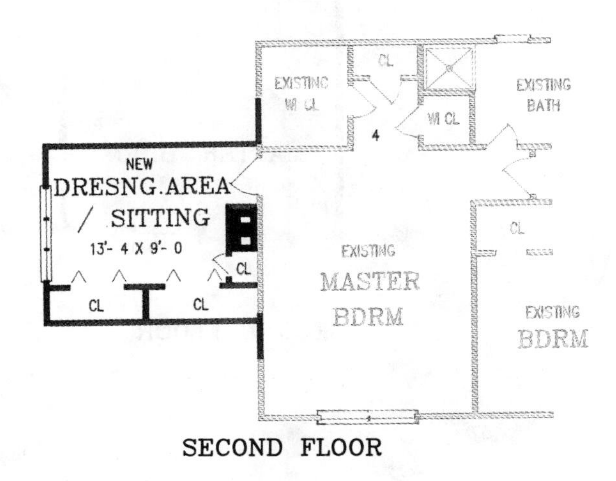

SECOND FLOOR

NOTE: CONSTRUCTION BLUEPRINTS CAN BE READILY MODIFIED TO ACCOMMODATE YOUR SPECIFIC HOUSE OR ROOM SIZES.

ROOFLINES AVAILABLE; ONLY AS SHOWN

BRB17

Yours is an old cape, bungalow, or shingle style one-and-one-half-story home and the bedrooms are tight? Look up at your roof; the dormers are likely very small, and there is probably much space to be found by pushing out one or more dormers. Your major concern is extending the rooflines so they are in character with the existing home. This modish 27′0″×7′0″ addition shows you how to do it with class. It removes an existing small front bedroom and provides a sensational new master suite with a fabulous bath and spacious walk-in closet. Most importantly, the new rooflines echo the existing lines of the house, and the addition looks smashing.

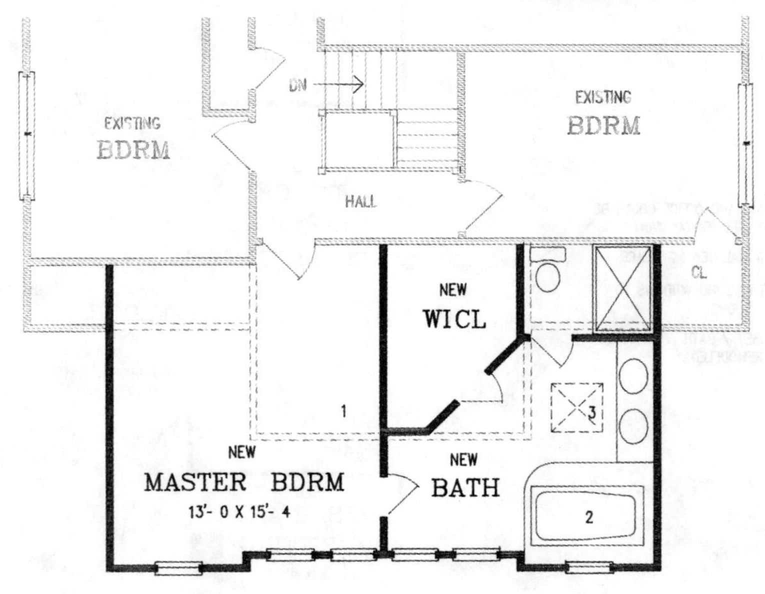

SECOND FLOOR

NOTES 1- REMOVE EXISTING FRONT DORMER
2- PLATFORM WHIRLPOOL TUB
3- SKYLITE

BRB15

A splendid master bedroom suite is created by this 9′6″×21′8″ second-floor addition over an existing first floor sun porch. By combining an existing bedroom with the newfound area over the porch, more than ample footage is available to create a delicious, up-to-date master suite. Two small closets are combined and enlarged into a spacious walk-in closet. The new master bath includes a compartmented water closet, a platform whirlpool tub, a separate corner-style stall shower, and a double vanity. Bedroom windows complement the porch below, creating a bright, cheerful bedroom. An alternate plan for this space might call for the extension of the hall and the creation of a smaller extra bedroom located only within the addition, similar to plan BRB08 on page 75.

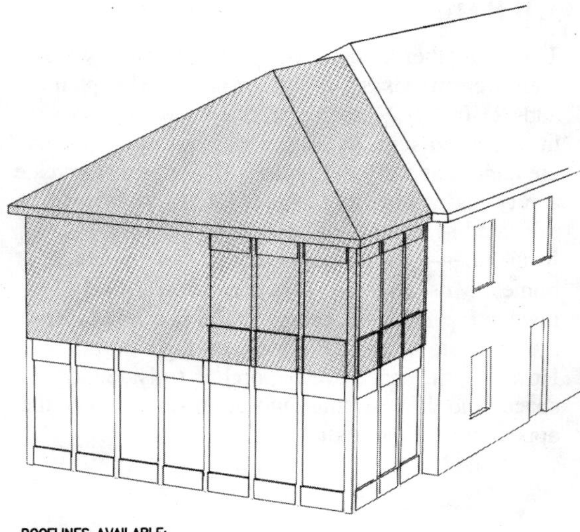

ROOFLINES AVAILABLE:
HIP (SHOWN), GABLE, SHED, FLAT

NOTE 1- SKYLIGHTS

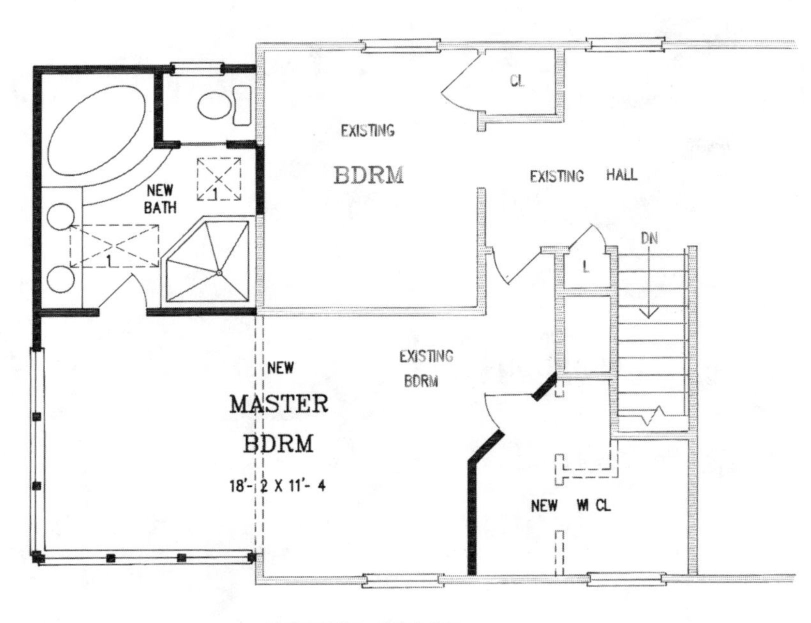

SECOND FLOOR

CONSTRUCTION BLUEPRINTS ARE AVAILABLE FOR THIS PLAN (AS WELL AS ALL OTHERS) AND INCLUDE THE DESIGN OF:
SOLUTIONS TO PREVENT FREEZE-UP OF PLUMBING OVER THE PORCH

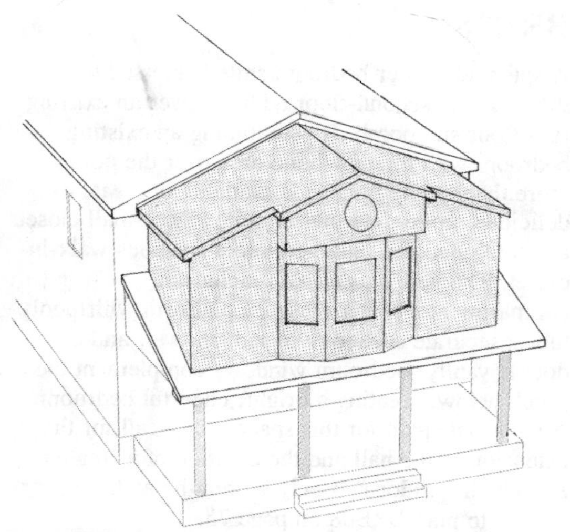

ROOFLINES AVAILABLE:
GABLE (SHOWN), HIP, SHED, FLAT

BRB16

This is another second floor bedroom expansion plan over an existing first floor porch. This plan adds 21′0″×8′0″ to enlarge one bedroom into a master-sized suite. A private bath and walk-in closet are added, as well as additional floor area, to create a very inviting bedroom suite. Closet doors are relocated to provide ample wall space for furnishing—a factor often ignored in many older homes. Most additions over a porch will likely involve replacing the ceiling of the porch because the structure is usually inadequate for a floor. However, the space over a porch is truly found space, and the resulting renovation can enhance the appearance of the home.

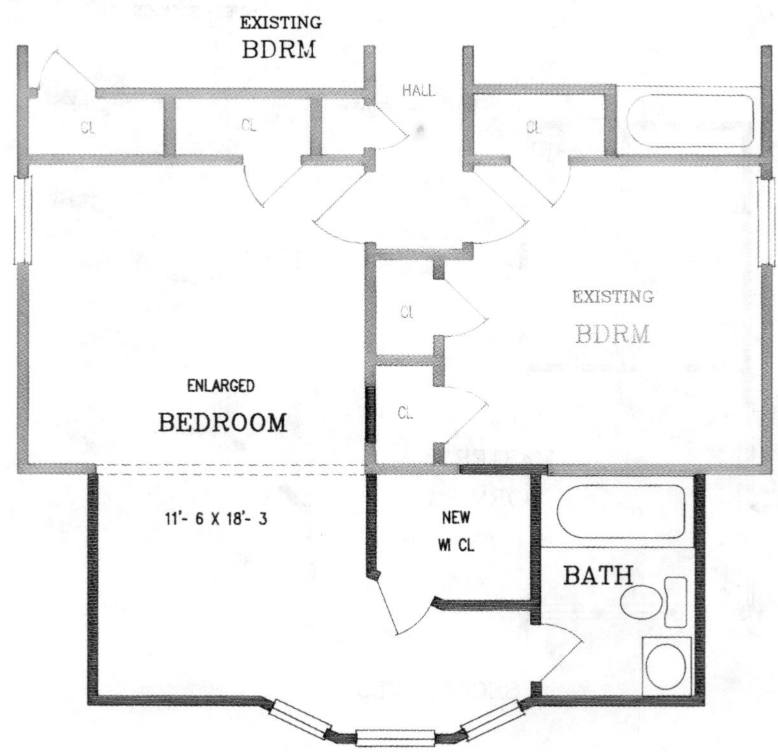

SECOND FLOOR

BRB05

Your home has a spacious four-bedroom layout, but the master bedroom is just not up to today's standards. Your dream is a sensational new suite with a sparkling new bath, plenty of closets, and some office space to boot. The roof over your garage and laundry room is waiting. Pictured here is an architecturally delightful addition that satisfies those wishes. It provides all the space you asked for and does it with flair. It even redesigns the front of your home to give it a stylish new character. Vaulted ceilings, skylights, and built-in space are some extra bonuses of the plan.

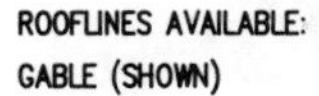

ROOFLINES AVAILABLE:
GABLE (SHOWN)

NOTES:
1- 27'- 8" DEEP OVER FIRST FLOOR
2- RELOCATE CLOSET AND HALL
3- THREE ROD WALK-IN CLOSET
4- BUILT-INS
5- SKYLITE OVER WHIRLPOOL TUB
6- RELOCATE WINDOWS FOR NEW REVERSE GABLE ROOF BELOW

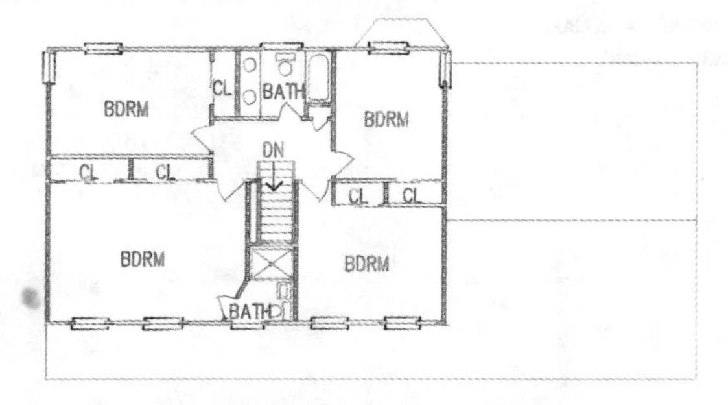

EXISTING SECOND FLOOR

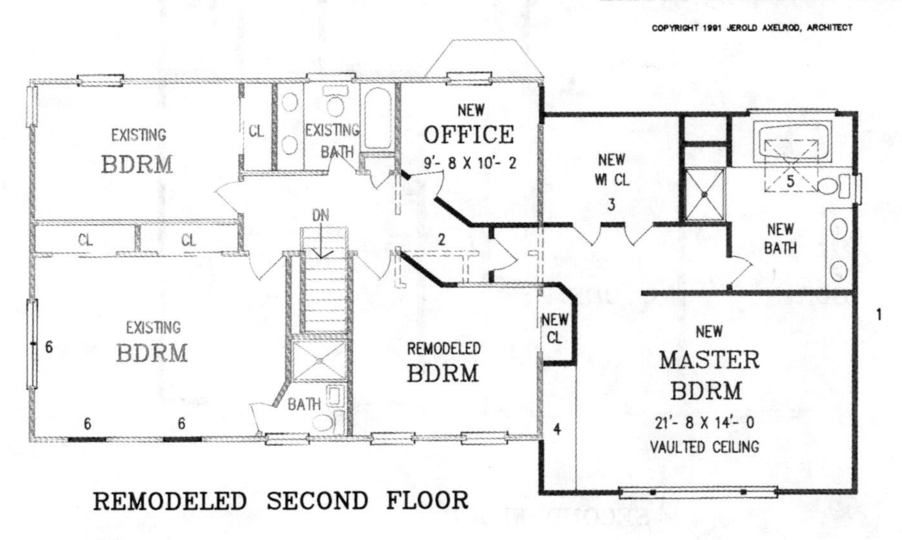

REMODELED SECOND FLOOR

CONSTRUCTION BLUEPRINTS ARE AVAILABLE FOR THIS PLAN (AS WELL AS ALL OTHERS) AND INCLUDE A MATERIALS LIST AND INFORMATION ON CUSTOMIZING THE PLANS

BRB20

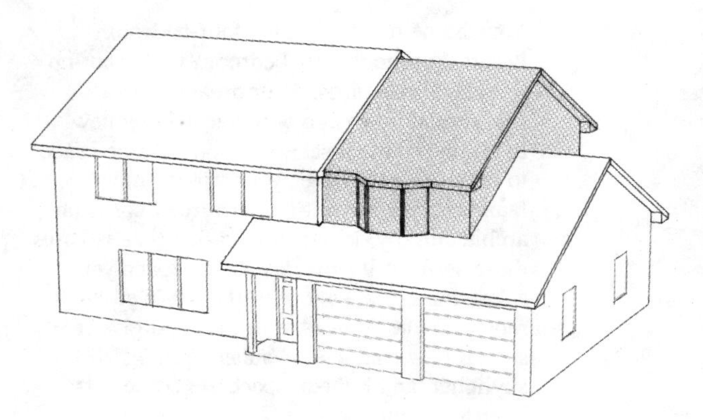

ROOFLINES AVAILABLE:
GABLE (SHOWN)

Yours is a three-bedroom side-hall two-story with an attached two-car garage and nothing but sky above. Well, if you would like to add a fourth bedroom over the garage, this plan shows how to do it. The plan, as drawn, calls for an entire new master suite, complete with its own shower-bath, dressing alcove, and walk-in closet. It is predicated on a one-story wing below that is 29 or 30 feet deep, which is fairly common if there is a laundry, utility room, or kitchen behind the garage. The room could be made wider or shallower, as necessary, to accommodate the existing dimensions of your house.

NOTE: CONSTRUCTION BLUEPRINTS CAN BE READILY MODIFIED TO ACCOMMODATE YOUR SPECIFIC HOUSE OR ROOM SIZES.

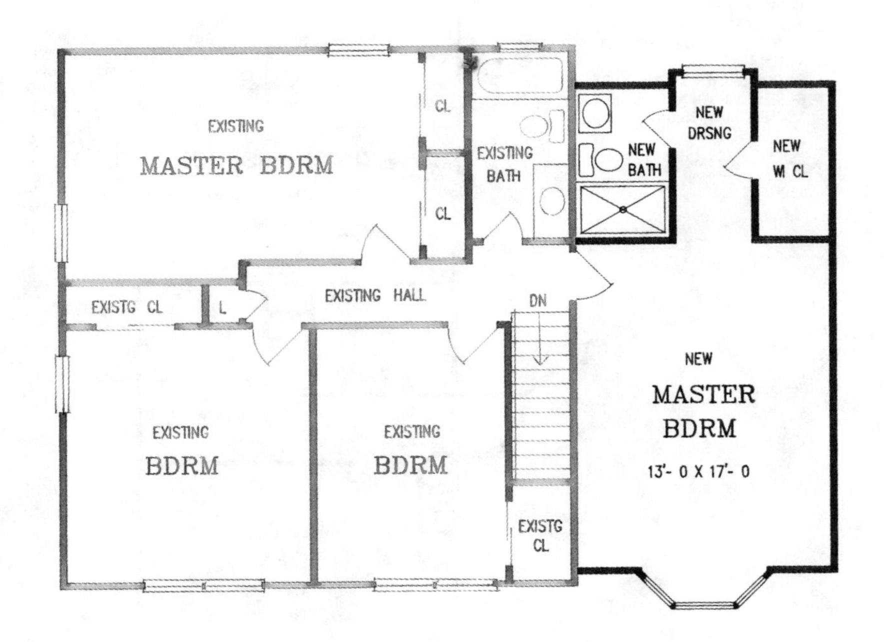

SECOND FLOOR

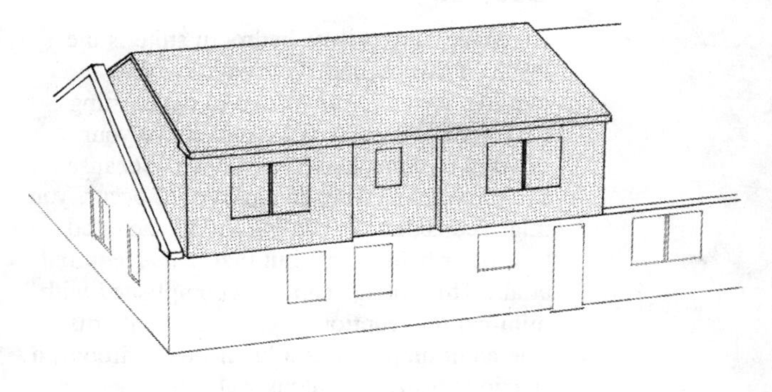

BRB01

Given the rear corner of a one-story home, you need at least one new bedroom and you need to explore the options of going out or going up. This plan raises the roof and places a new stair to the second floor over the basement stair, a new bath over the existing hall bath, and two bedrooms flanking on either side. The bedrooms are cantilevered, which adds some square footage, while improving the aesthetics of the partial second floor. Although the plan is similar to BRB12 on pg. 83, the roof form is more specific to a flatter pitched one-story home.

ROOFLINES AVAILABLE:

ONLY AS SHOWN

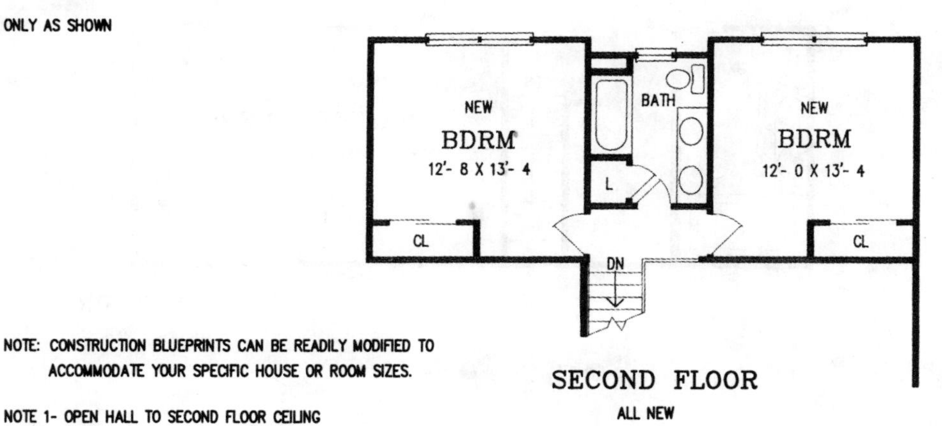

NOTE: CONSTRUCTION BLUEPRINTS CAN BE READILY MODIFIED TO ACCOMMODATE YOUR SPECIFIC HOUSE OR ROOM SIZES.

NOTE 1- OPEN HALL TO SECOND FLOOR CEILING

SECOND FLOOR

ALL NEW

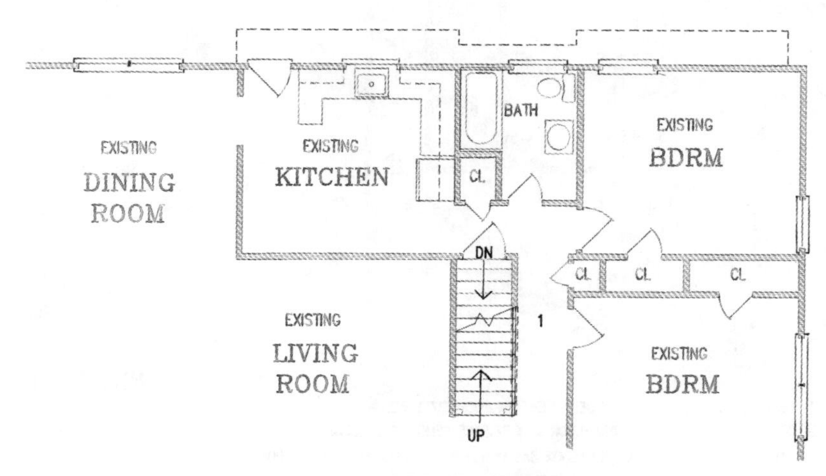

FIRST FLOOR

BRB03

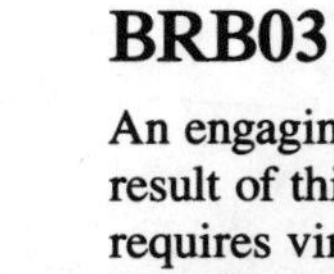

An engagingly private bedroom suite is the result of this second floor addition that requires virtually no change to the existing first floor. The stair is located above your existing basement stair, as shown. Because some modest ceiling damage could occur, you might consider remodeling the kitchen and first floor bathrooms, but that is not required at all. This change can be accomplished with minimal interruption to your daily patterns. The addition provides a luxurious bedroom, a spacious bath, a fabulous walk-in closet, and a lovely sitting area in front of the bay window. The bath includes a whirlpool tub, separate stall shower, double vanity, and a skylight.

ROOFLINES AVAILABLE:

ONLY AS SHOWN

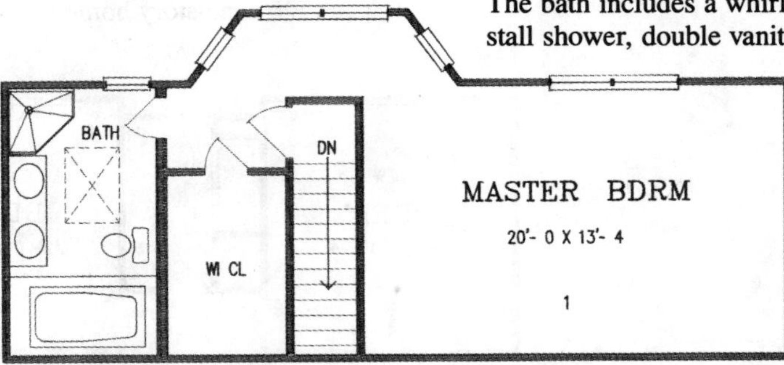

SECOND FLOOR

ALL NEW

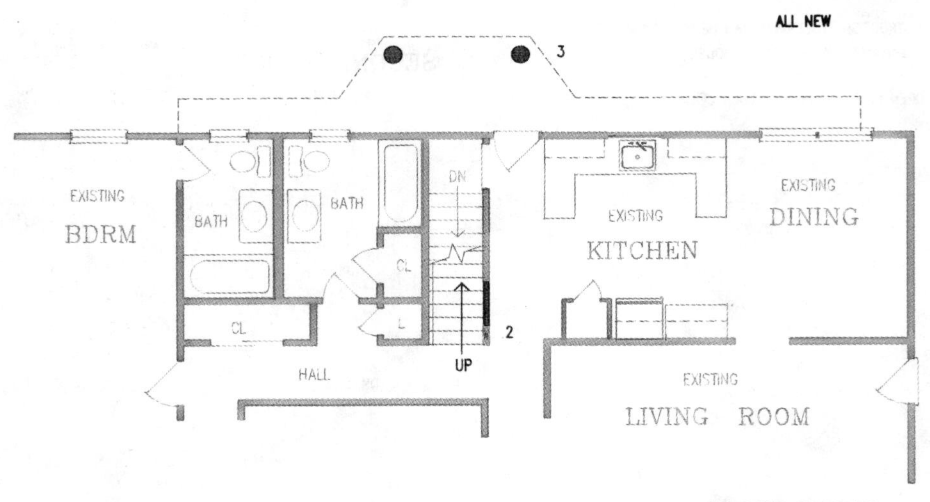

FIRST FLOOR

NOTES 1- BEDROOM SIZE IS A VARIABLE DEPENDENT ON ROOM SIZES BELOW
2- OPENING UP STAIR TO SECOND FLOOR MAY REQUIRE REMOVING A CLOSET
3- COLUMNS TO SUPPORT SECOND FLOOR BAY COULD BE INCORPORATED INTO A PORCH

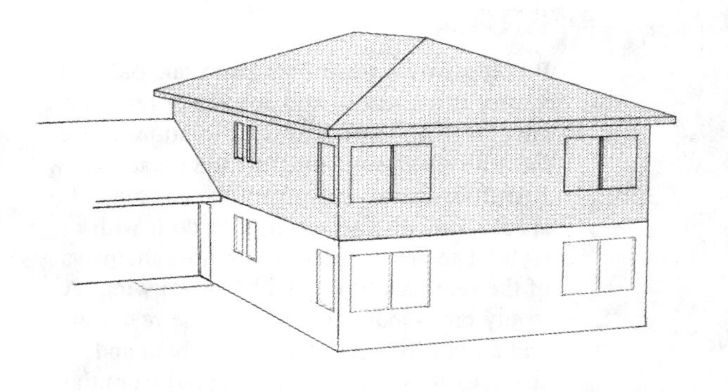

BRB02

Cramped, small bedrooms are eliminated when this modest L-shaped ranch becomes a stunning two-story in this remodel. The roof is raised over the bedroom wing, and three bedrooms and a bath are added on the second floor. There's now loads of closet space and even a dramatic balcony overlooking a new raised ceiling area of the living room. The first floor bedroom wing is remodeled, and a fabulous, private master suite is carved out. It includes two walk-in closets and a great bath with a whirlpool tub, separate shower, and curved vanity. Such an extensive project will likely require major decisions on new windows, siding, roofing, and even the heating and air conditioning system. (See chapters 2 and 12 for further discussion.)

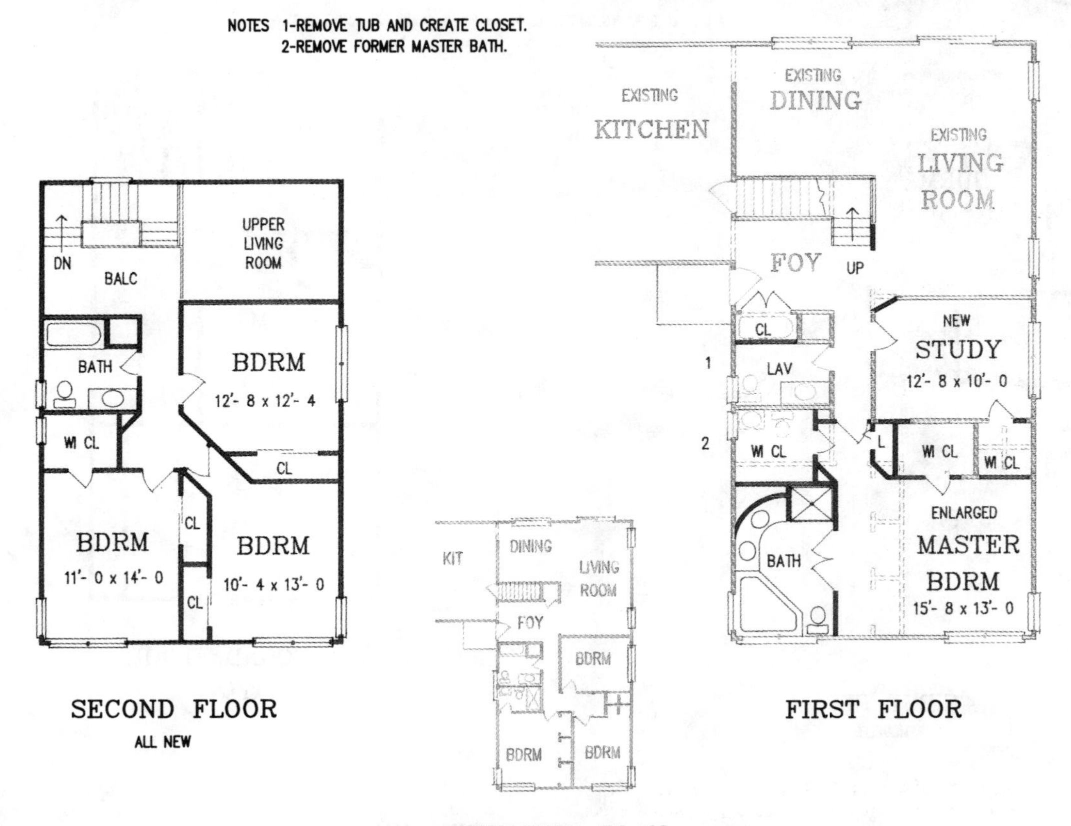

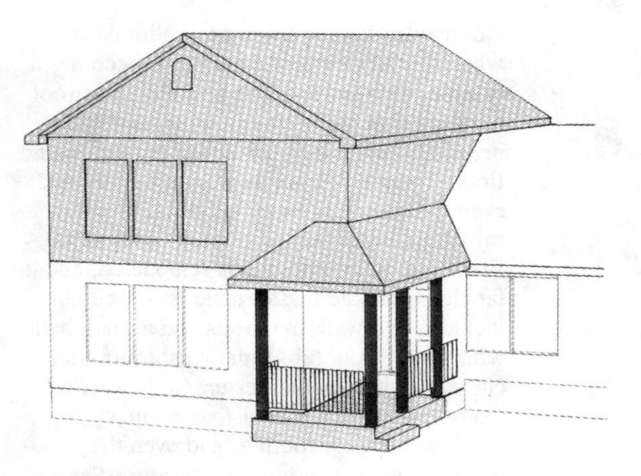

BRB04

Problem: three bedrooms share one bath, the master is too small, and you really don't have enough room to expand out. Solution: move the entire bedroom wing up and create a new first floor master suite from the existing space. This plan shows how to do it with style. The new master suite enjoys the privacy of the rear part of the old bedroom wing. A lovely bay window is added to the rear wall, and a luxurious compartmented bath and spacious walk-in closet are carved from the old middle bedroom. The front bedroom remains intact and serves as an ideal guest or parent's bedroom, or as an office. The covered first floor porch is a smart by-product of the design.

ROOFLINES AVAILABLE:
GABLE (SHOWN), HIP, SHED, FLAT

CONSTRUCTION BLUEPRINTS ARE AVAILABLE FOR THIS PLAN (AS WELL AS ALL OTHERS) AND INCLUDE A LIST OF BUILDING MATERIALS REQUIRED TO COMPLETE CONSTRUCTION.

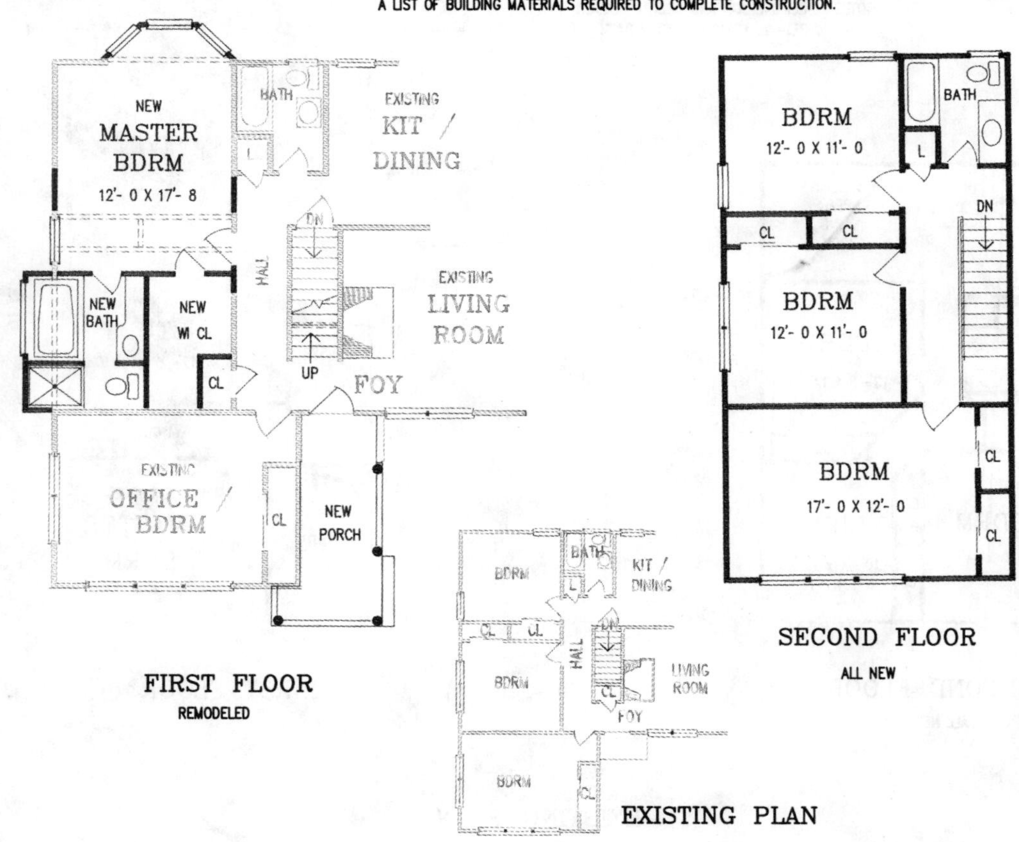

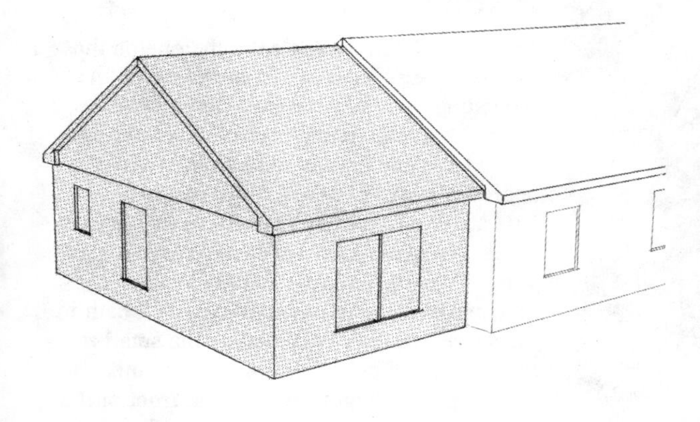

ROOFLINES AVAILABLE:
GABLE (SHOWN), HIP, SHED, FLAT

BRB08

This 13′10″×25′4″ one-story addition provides a comfortable new master suite, comprising a lovely bedroom, a private tub bath, and a spacious walk-in closet. The attachment shown is to the side or rear of a very commonly found bedroom wing. The plan shows you how to extend the bedroom hall by removing and relocating three existing closets. A remodeling of the existing hall bath is also indicated. Don't be tempted to gain every inch of space by aligning the addition flush on both the front and rear (or both sides). Doing so would likely require reroofing, residing, and a hard look at matching windows, too. The break shown avoids those concerns.

NOTES 1- REMOVE CLOSETS AND EXTEND HALL
2- REMODEL EXISTING BATH

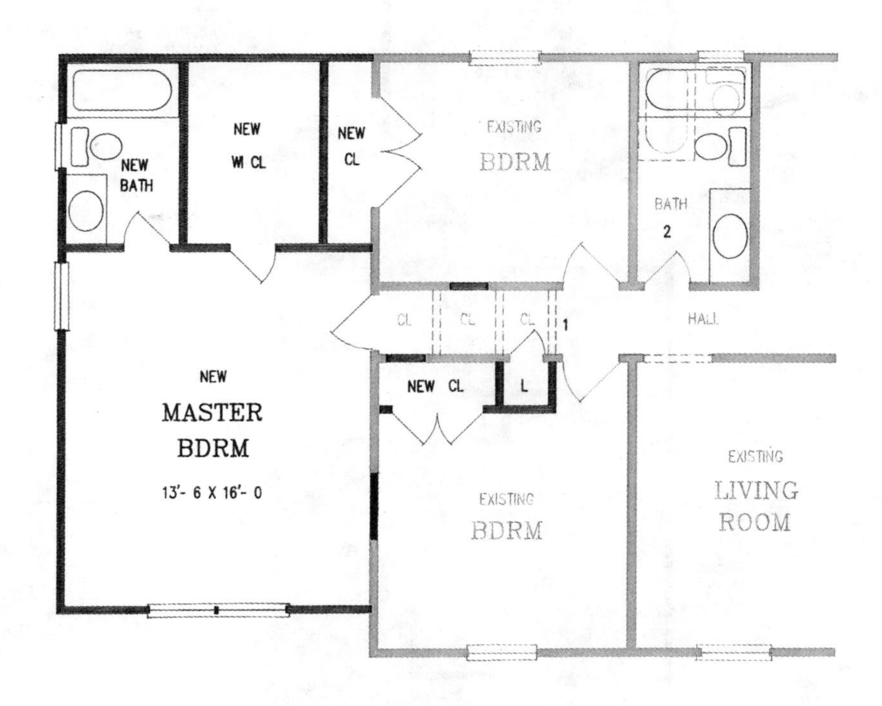

NOTE: CONSTRUCTION BLUEPRINTS CAN BE READILY MODIFIED TO ACCOMMODATE YOUR SPECIFIC HOUSE OR ROOM SIZES.

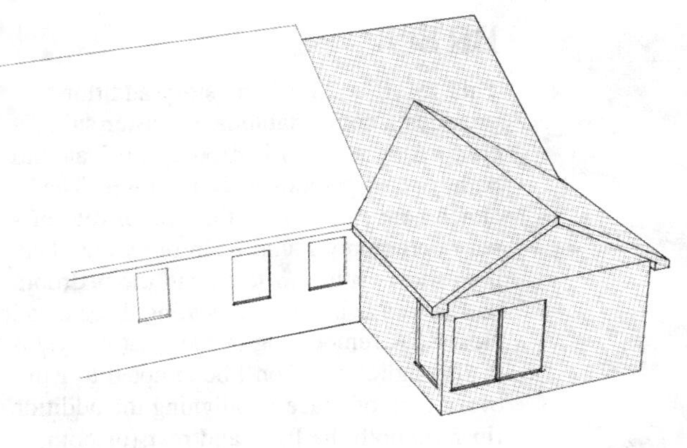
ROOFLINES AVAILABLE:
GABLE (SHOWN), HIP, SHED, FLAT

BRB09

This plan, when compared to the one on the previous page, shows the importance of evaluating options when deciding how to approach a remodeling project. This remodel also adds a new master suite to the same bedroom wing; however, it does so in a 13′4″×37′8″ addition that places the new master bedroom in a rear-facing reverse-gable wing. Other differences include locating the new master bath and a brand new hall bath in the space formerly occupied by the small rear bedroom. The new addition replaces this bedroom, in a larger form, in the front of the new wing. Other nuances: a tray ceiling in the master bedroom, corner windows, and the old hall bath converted to a spacious laundry room.

NOTES 1- REMOVE CLOSETS AND EXTEND HALL
2- EXISTING BEDROOM CONVERTED TO TWO NEW BATHS
3- EXISTING BATH CONVERTED TO LAUNDRY ROOM

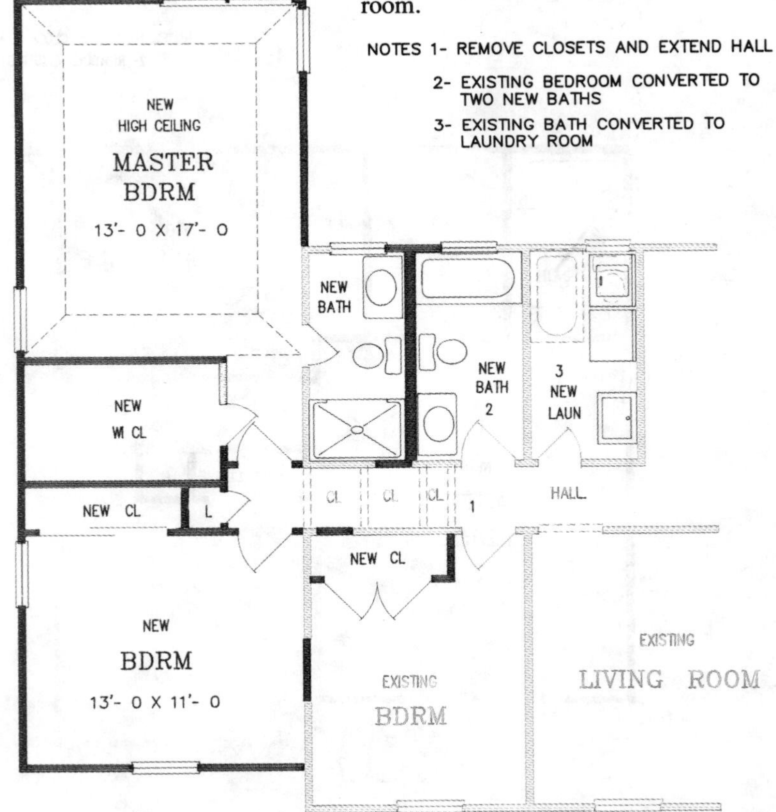

BRB10

Given: the same existing bedroom wing and the same program as the two prior plans, namely the need for a new master bedroom and bath. The difference here: the entire extension (18′6″×21′2″) is to the rear, not to the side. That decision, whether by choice or set-back restrictions, results in a very different solution. Two new tub baths are constructed in the new wing, which allows a very easy staging process. The old bath would be the last to be done; removing it is necessary to provide the hall access to the new wing while also providing laundry space. The new side door is an interesting aspect of this design; it could be the access to a new family room such as that shown in plan F0003 on page 106.

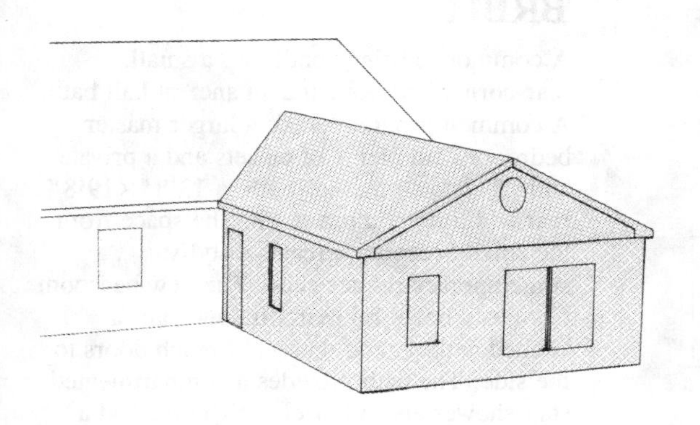

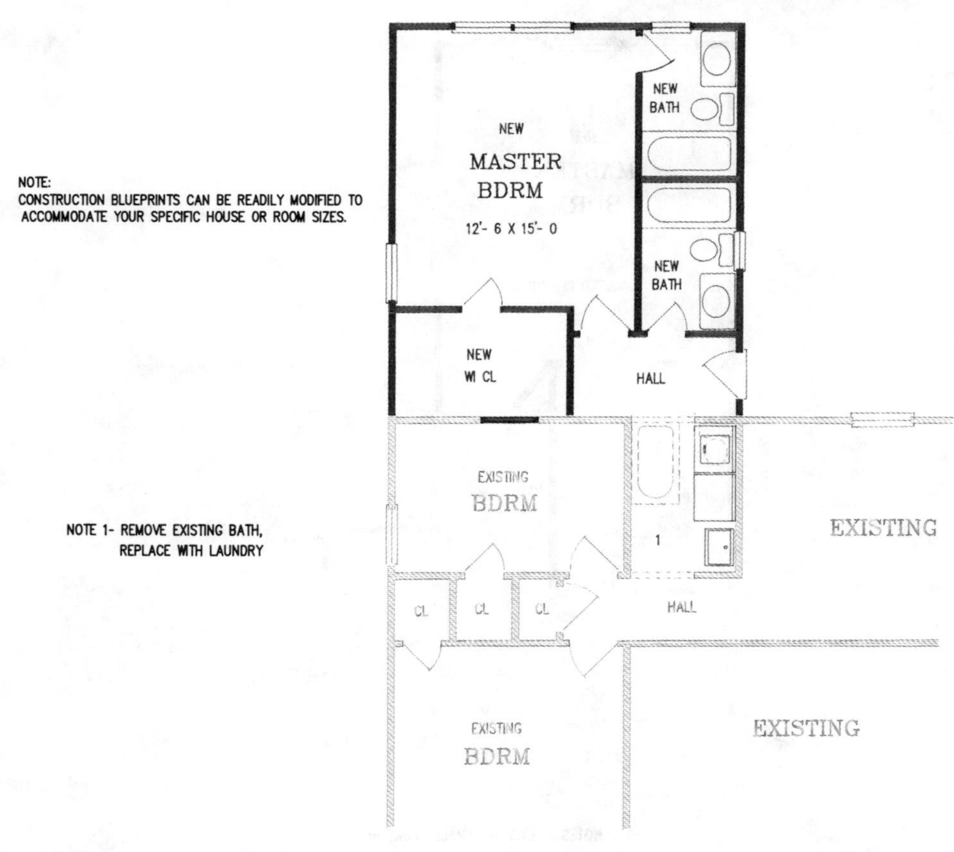

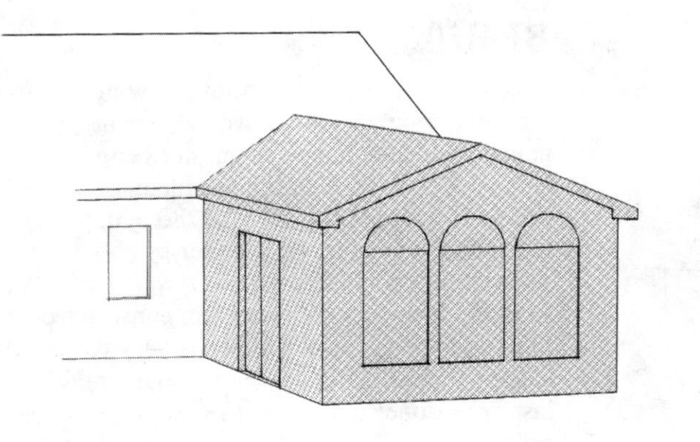

BRB11

A common existing condition: a small, rear-corner bedroom and an ancient hall bath. A common program need: a larger master bedroom with plenty of closets and a private bath. This plan meshes both; a 13′8″×19′8″ rear addition, combined with the space from the small bedroom, creates a stylish contemporary master suite. The new bedroom features a beautiful rear window wall, a vaulted ceiling, and optional French doors to the side. The bath provides a compartmented stall shower and water closet alcove, and a lovely bay window at the tub area. The old hall bath can be simply remodeled or divided into a powder room and a separate laundry alcove, depending on your needs.

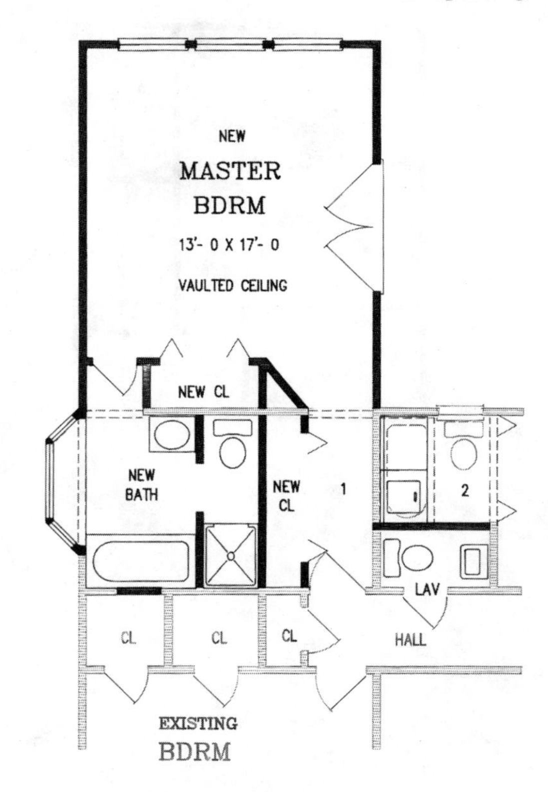

NOTES 1- EXISTING SMALL BEDROOM
2- REPLACE EXISTING HALL BATH WITH LAUNDRY AND POWDER ROOM

BRB06

Problem: your three bedrooms share a hall bath; you don't need an extra bedroom, but you want a grand new master suite. Furthermore, the only place to expand out is in the rear, which requires removing the old path. Actually, the removal of the old hall bath becomes a benefit, as the solution shown creates a new bath that can be finished off early in the renovation process, allowing you to comfortably stage the alteration. The new master suite boasts a wrap-around bay window, a vaulted ceiling, and a private bath. The bath includes an oversized platform whirlpool tub, a large separate stall shower, and a walk-in closet.

ROOFLINES AVAILABLE:
GABLE (SHOWN), HIP, SHED, FLAT

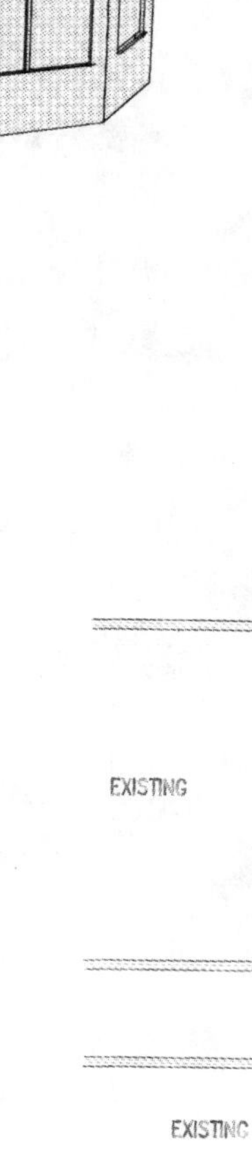

EXISTING PLAN

REMODELED PLAN

PBR03

The new first floor master suite pictured here is a stunning example of the use of found space. The old breezeway, blank side of the dining room, and rear of the garage are likely to be underutilized areas of your lot. What a perfect spot for a private master retreat! Although a one-story home is pictured, the same plan works for an existing split-level or two-story home. The old breezeway is now transformed into a beautiful sunporch, which not only enhances the existing living room, but also serves as a lovely entrance and adjunct to the new master bedroom.

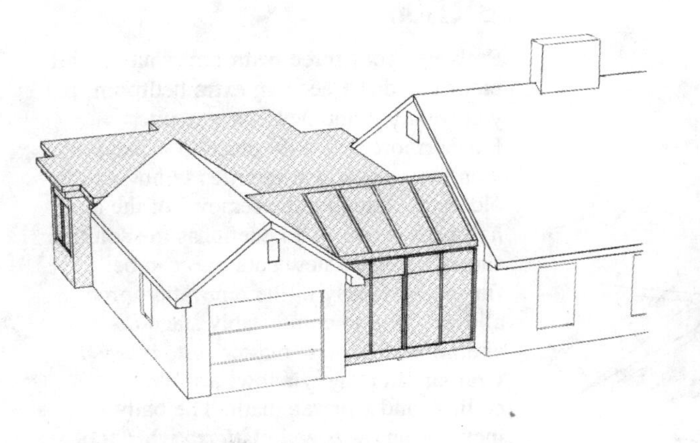

ROOFLINES AVAILABLE:
GABLE & FLAT (SHOWN)

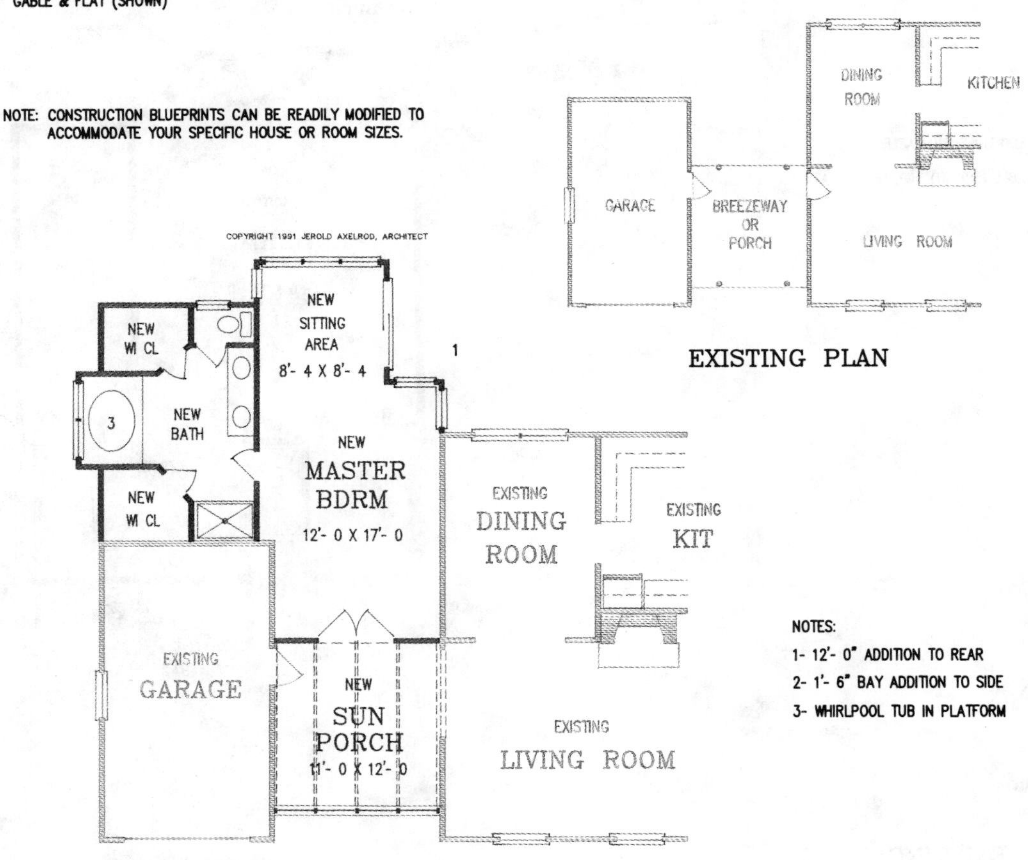

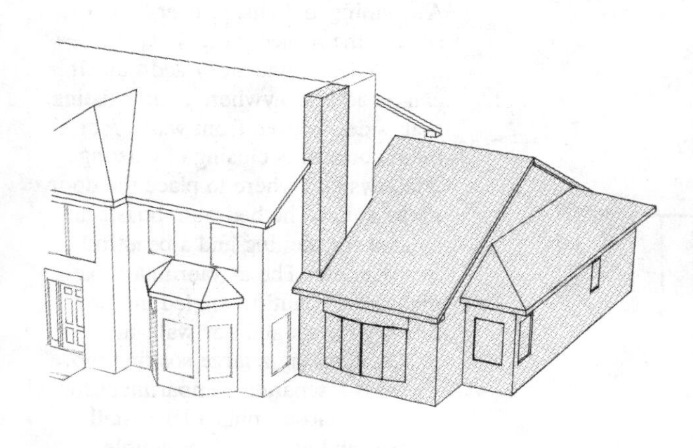

ROOFLINES AVAILABLE:
GABLE & HIP (SHOWN), FLAT

BRB24

A winning new master suite is the subject of this large one-story addition. Although specifically planned to attach to the living room side of a two-story home, this splashy design could be connected to many different homes. The plan provides for a large, tray-ceilinged bedroom with an adjoining sitting area, and a luxurious bathroom, dressing, and closet suite. A large bow window provides an abundance of light to the sitting area, where there is also a fireplace. The spacious dressing area includes a dramatic double vanity that wraps around a corner, a large whirlpool tub, and two walk-in closets—all designed into a stylish plan dominated by striking angles.

NOTES:
1- 6'- 0" WHIRLPOOL TUB
2- NEW FIREPLACE
3- BUILT-IN
4- 18'-0 X 34'-4 ADDITION

NOTE: CONSTRUCTION BLUEPRINTS CAN BE READILY MODIFIED TO ACCOMMODATE YOUR SPECIFIC HOUSE OR ROOM SIZES.

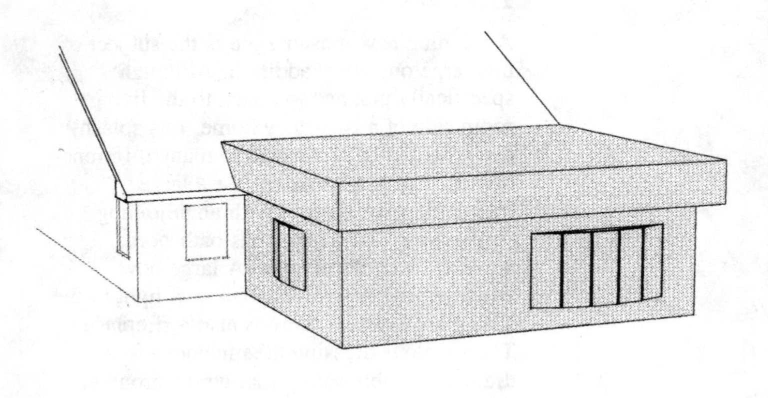

BRB14

A stunning and chic master bedroom suite is the subject of this 25′2″×18′4″ one-story addition. It can be added anywhere to an existing rear, side, or even front wall. Your main concern is closing of existing windows and where to place the door to the suite. The bedroom boasts an elegant tray ceiling and a beautiful bay window. The extraordinary bath includes a beautiful angled interior form, with his-and-her walk-in closets flanking a large soaking tub. There is a separate compartment for the water closet, plus a large stall shower and an eight-foot double vanity. The flat roof as shown is readily converted to a gable, hip, or shed as might be required or preferred. See Chapter 6 for available blockforms.

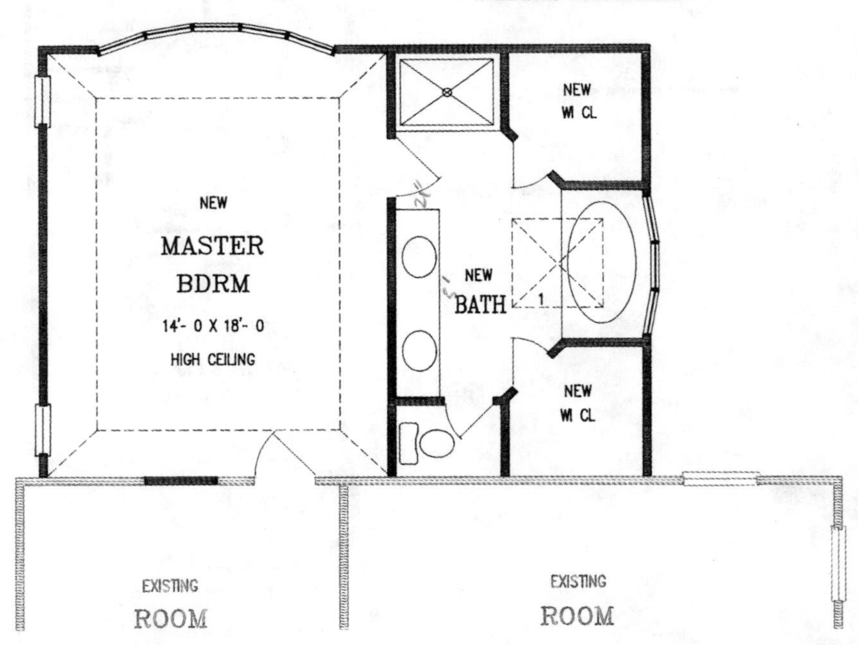

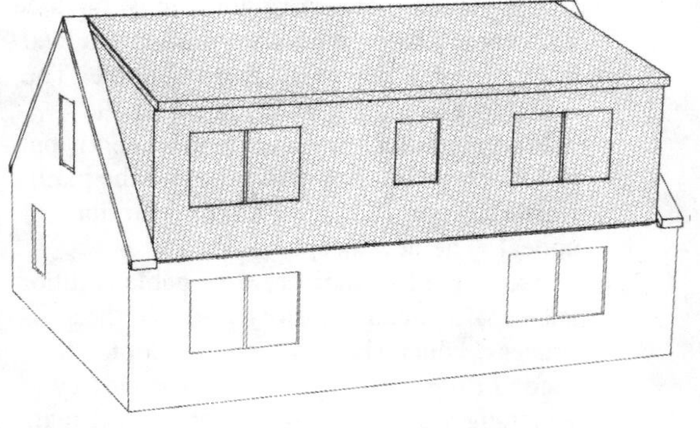

BRB12

This plan provides two new large bedroom[illegible] bath in a new dormered rear. The dormer sh[illegible] 29′4″ wide, and the existing home is 32′0″ wid[illegible] Both could be wider or smaller. This type of shed roof dormer is adaptable to almost any existing one- or one-and-one-half-story home; however, it is most suited to a home that already has a stand-up attic under a steeply pitched roof. Design issues you should be concerned with include keeping at least a small sliver of the old roof on each side to avoid the awkward, bulky-looking side form so often seen on such homes. Also, consider raising the ceiling alongside the stair on the first floor to achieve a more inviting entry to both floors.

ROOFLINES AVAILABLE:

ONLY AS SHOWN

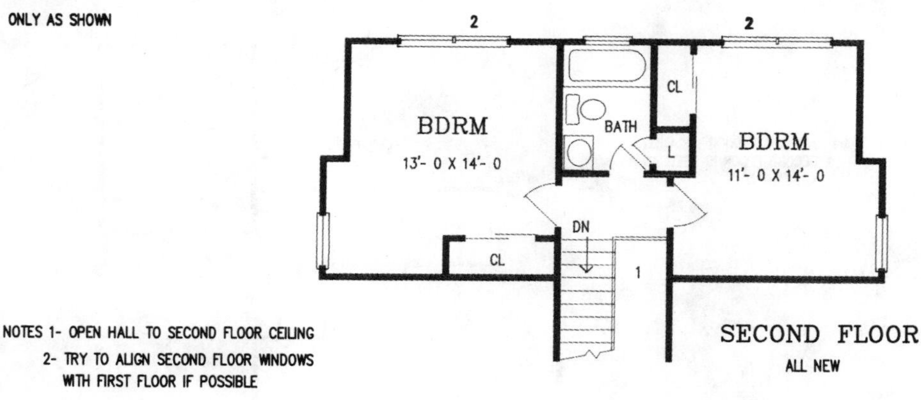

NOTES 1- OPEN HALL TO SECOND FLOOR CEILING
2- TRY TO ALIGN SECOND FLOOR WINDOWS WITH FIRST FLOOR IF POSSIBLE

SECOND FLOOR
ALL NEW

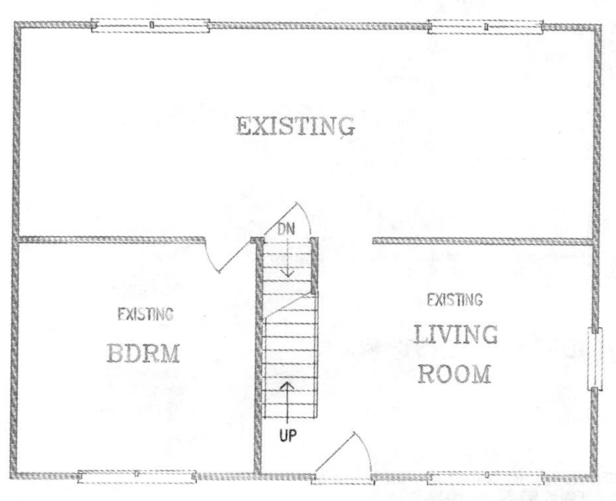

FIRST FLOOR

NOTE: CONSTRUCTION BLUEPRINTS CAN BE READILY MODIFIED TO ACCOMMODATE YOUR SPECIFIC HOUSE OR ROOM SIZES.

BRB23

Adding up on a one-story home must be done with care, particularly when the space added is small in relation to the existing home. The design presented here adds just a small bedroom and bath (or a mini-apartment), but it does so with a flair that provides the plain, flush one-story with a newfound exterior appeal. The new steeply pitched one-and-one-half-story reverse-gable addition adds a 4′0″ deep one-story porch in the process, which shelters the front door. The second floor could be readily expanded by widening the addition or dormering the rear. If used as a mini-apartment, the front area could support a small kitchenette.

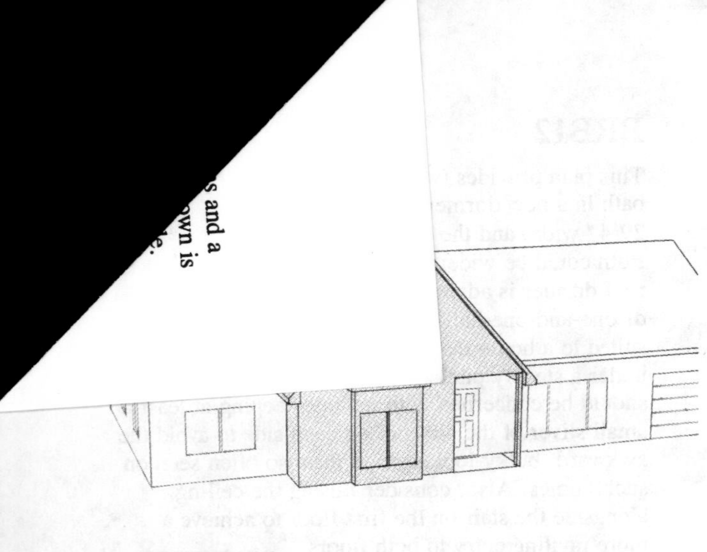

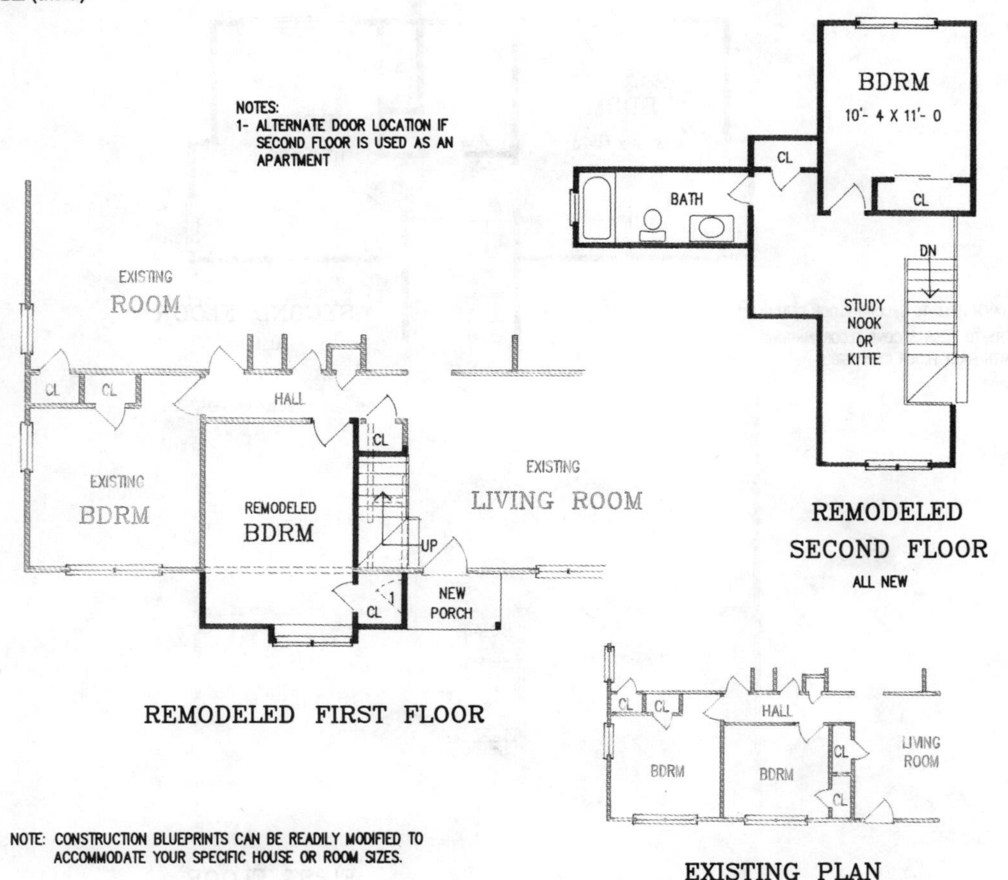

BRB19

One of the most popular homes for decades—the split level—is now one of the most popular remodeling subjects. The variety of types that were built offers a myriad of potential solutions, but common to virtually all split levels is the ability to place a large bedroom suite over the living room wing. However, this is also one of the most commonly flawed additions undertaken. Unless careful attention is given to existing rooflines, the addition could look awful. This plan shows you how to create a spectacular new master bedroom suite while enhancing your home's appearance. The suite includes a sensational bath and a lovely sitting area, all under a stylish vaulted ceiling.

ROOFLINES AVAILABLE:
GABLE (SHOWN), HIP

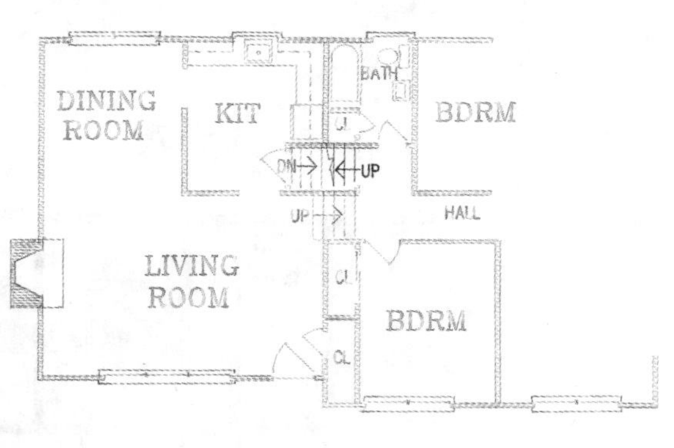

EXISTING PLAN

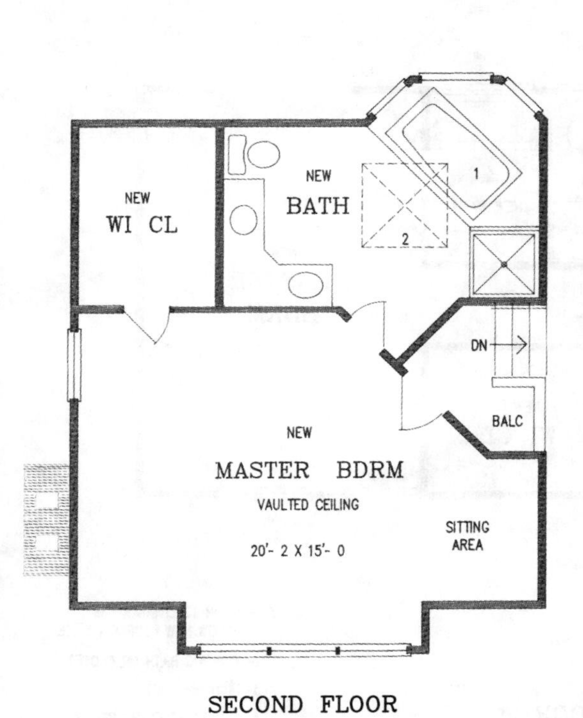

SECOND FLOOR
ALL NEW

NOTES 1- WHIRLPOOL TUB
2- SKYLITE

NOTE: CONSTRUCTION BLUEPRINTS CAN BE READILY MODIFIED TO ACCOMMODATE YOUR SPECIFIC HOUSE OR ROOM SIZES.

BRB18

This new bedroom suite makes an ideal master retreat. Tucked out of sight on its own level, with its own private bath and private balcony, it will be a joy to come home to. It is especially designed to be added above the living room wing of a typical split level, a place where the untrained designer is prone to make an aesthetic mistake. This addition actually enhances the lines of the home. The suite includes a generous bedroom, a compartmented bath with a whirlpool tub and double vanity, plus a large walk-in closet. This space is available in most split levels, and if it's not a bedroom you desire, it could be utilized as an apartment or a playroom.

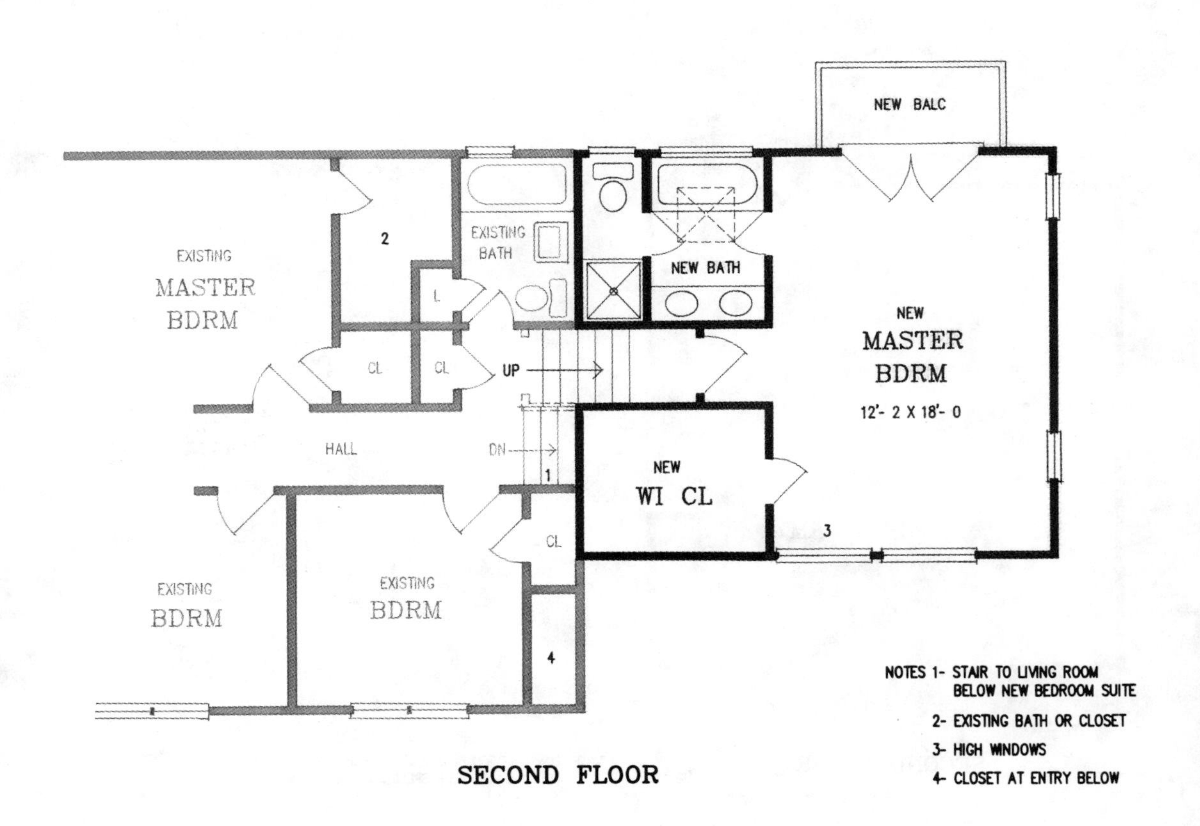

SECOND FLOOR

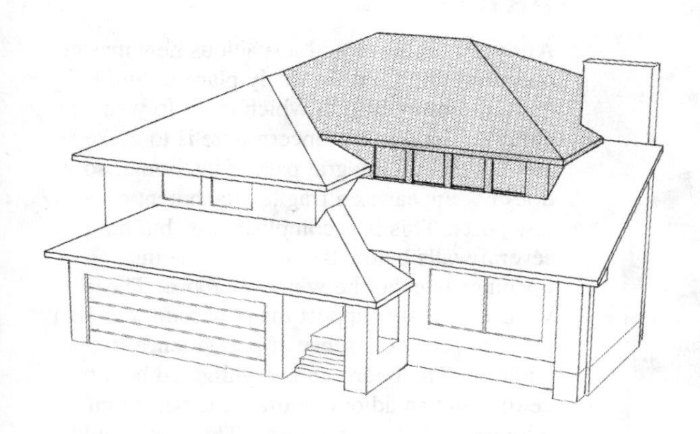

ROOFLINES AVAILABLE:
HIP (SHOWN)

BR007

The narrow-style, side-to-side, split-level or two-story home pictured usually has lots of high volume over the living room, dining room, and kitchen, but it is usually short a bedroom, and its minimal basement (if one exists at all) doesn't provide space for a recreation room. This architecturally well-composed addition provides the space for both. The trade-off is the loss of that high volume, but you could go up a few extra steps in the addition and create a nine- or even 10-foot-high ceiling below. A suggestion: use sound deadening board as an underlayment on the recreation room floor and an extra-thick carpet and pad.

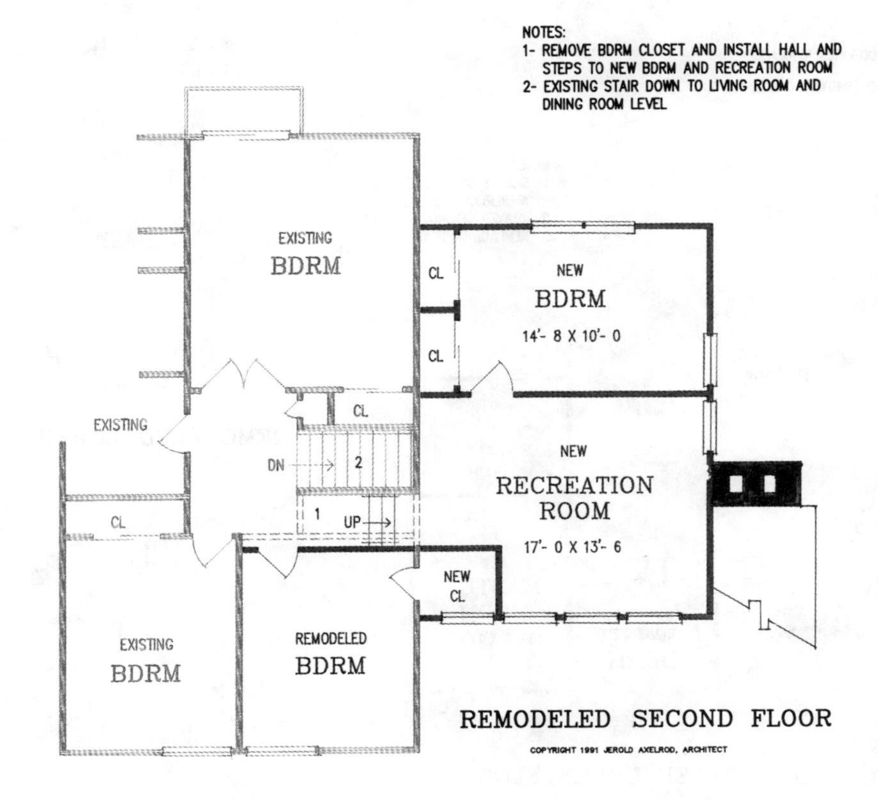

CONSTRUCTION BLUEPRINTS ARE AVAILABLE FOR THIS PLAN (AS WELL AS ALL OTHERS) AND INCLUDE A MATERIALS LIST AND INFORMATION ON HOW TO MAKE CHANGES TO PLANS

BRB25

Are you looking to add a spacious new master bedroom suite, but the likely place to build has a driveway below, which leads to your garage? The design concern here is to make the addition an integral part of the home, so it does not appear as a fragile attachment on a few posts. This is accomplished by building several walls below the bedroom, with openings to echo the windows above. The walls obviously support the new room but also serve to provide a sense of permanence to the addition. The bedroom is highlighted by a tray ceiling and an adjoining dramatic bathroom with an angled tub platform. This suite could be built, of course, without the basement.

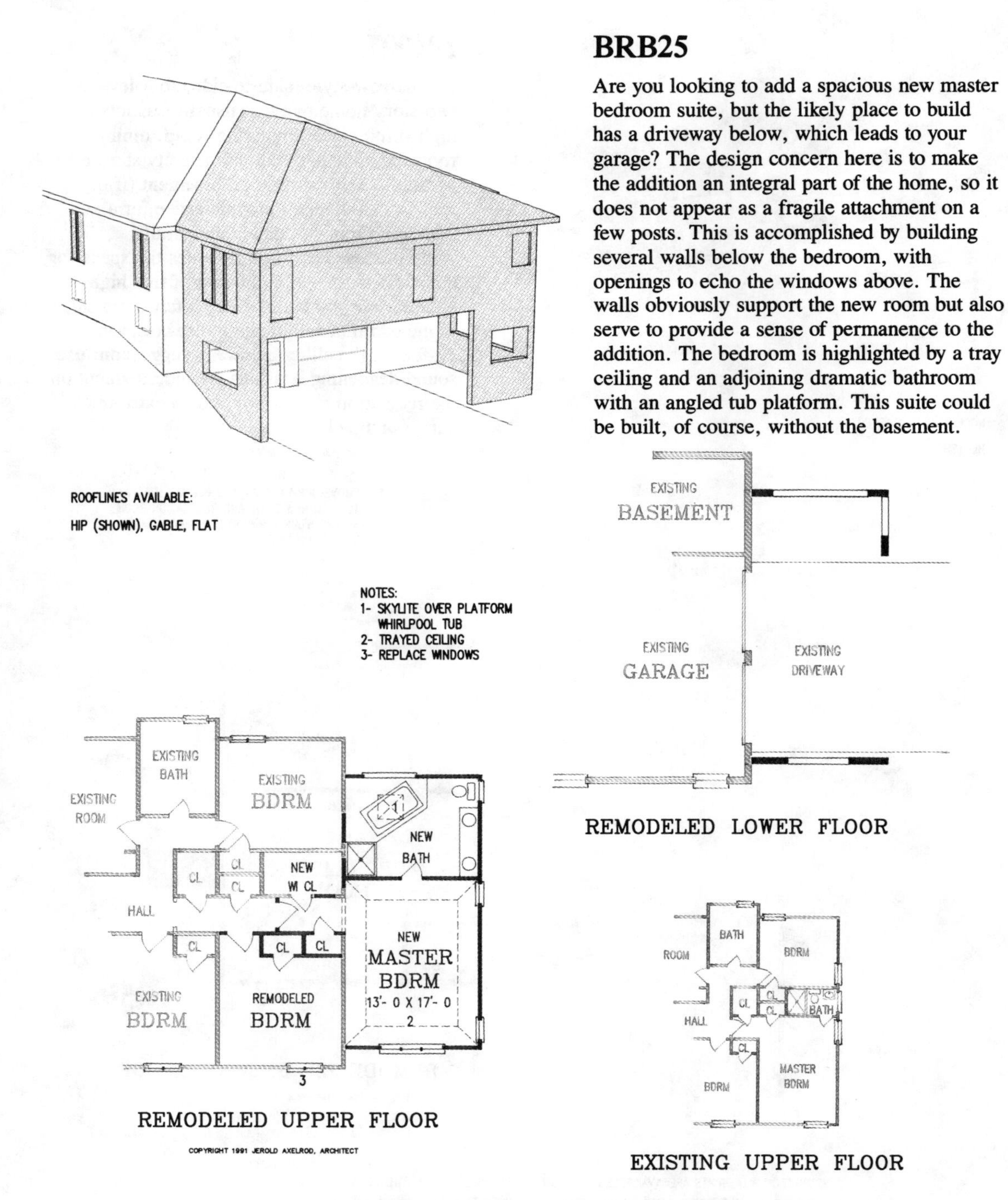

CONSTRUCTION BLUEPRINTS ARE AVAILABLE FOR THIS PLAN (AS WELL AS ALL OTHERS) AND INCLUDE A MATERIALS LIST

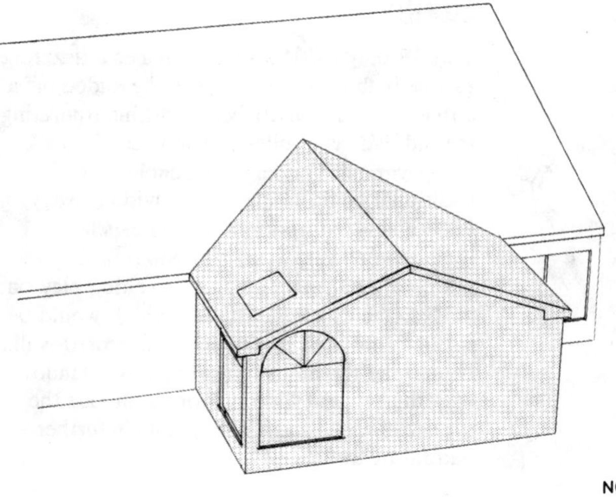

BRB07

A fabulous new master bath and room-size walk-in closet are the subjects of this 19′4″×11′8″ addition. A splendrous master suite is the ultimate result. The bedroom itself benefits from the removal of the existing bathroom and closet. The area from the old closet is ideal for built-ins, and the area from the old bath graciously enlarges the bedroom, creating an attractive sitting or dressing area. The master bath features a dramatic angled entry and a corner platform tub at a complementary angle. Modish corner windows above the tub and a vaulted ceiling add to the high fashion of this bathroom.

ROOFLINES AVAILABLE:

GABLE (SHOWN), HIP, SHED, FLAT

NOTES 1- REMOVE EXISTING MASTER BATH
2- REMOVE CLOSET-SPACE FOR BUILT-INS
3- WHIRLPOOL TUB ON PLATFORM

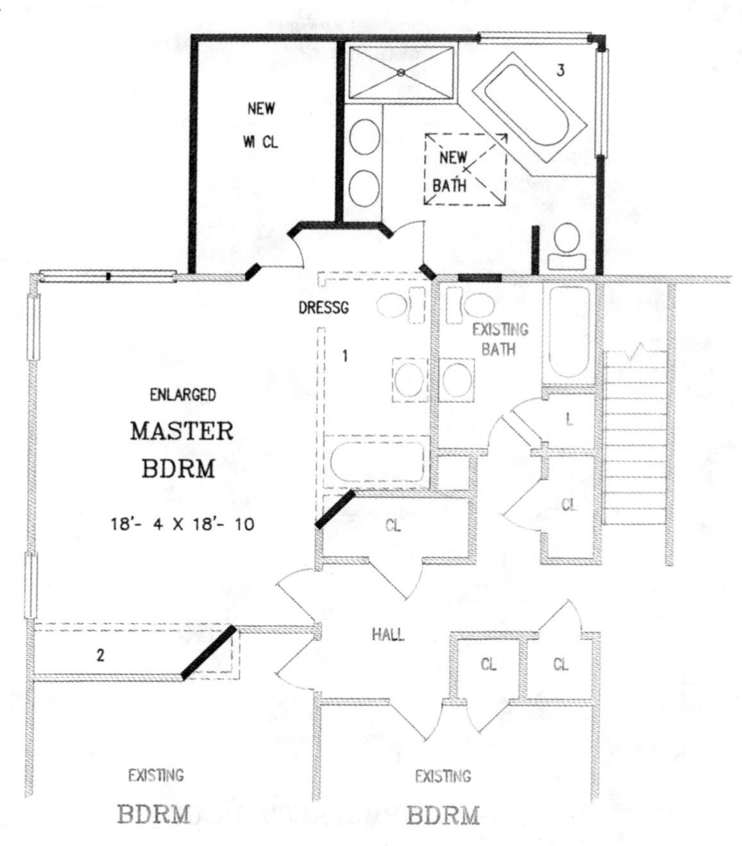

B0008

This 17′0″×14′6″ addition creates a dazzling garden bath that is designed to be added off a corner of your master bedroom, incorporating the old bath as a toilet and shower compartment. Garden walls enclose some additional outdoor space to provide privacy for the new bathroom, which boasts a spacious dressing area, a fabulous whirlpool tub, and a built-in sauna. An exciting play on angles creates spatial drama, which would be further enhanced by the use of mirrored walls behind the tub and vanity. The large window over the tub and sliding doors both face the private garden, and a skylight adds further natural light.

ROOFLINES AVAILABLE:
GABLE , FLAT

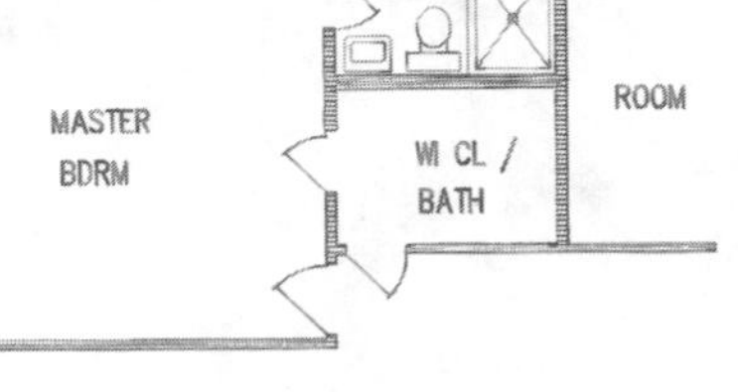

EXISTING PLAN

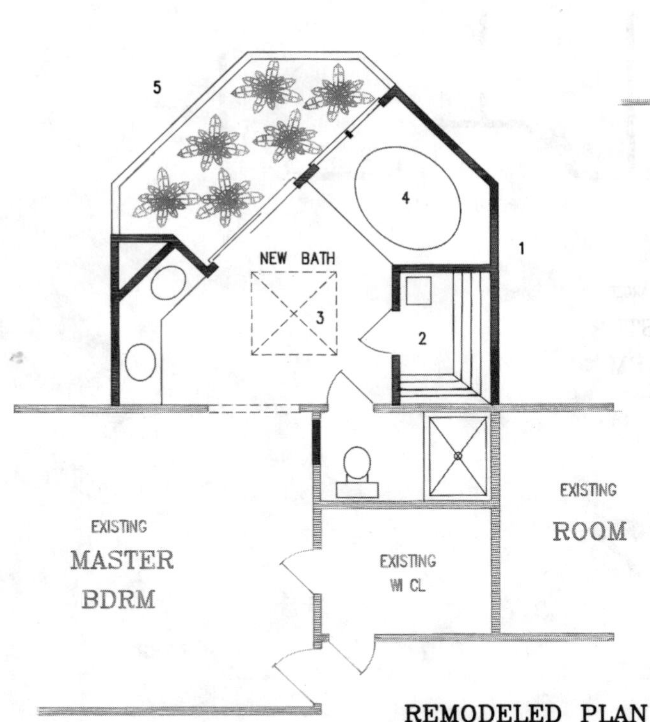

NOTES:
1- 14'- 6" REAR ADDITION
2- SAUNA
3- SKYLITE
4- WHIRLPOOL TUB
5- GARDEN WALLS FOR PRIVACY

REMODELED PLAN

LBRB1

Few major additions expand to the front due to lack of available front yard, but if you have the room, and your current home is a dated in-line ranch, you could create the stylish U-shaped design pictured. The 18-foot-deep addition places an inviting new great room at the front of the existing living room, and a smart new master suite in front of the existing bedrooms. It is an easy addition to stage and, when it is complete and the old front wall is removed, the new entertainment wing will become a spacious, open, flowing space that you can decorate and furnish to suit your lifestyle. The new master bedroom includes a private compartmented bath and a large walk-in closet. Both new rooms feature large bow windows and high trayed ceilings.

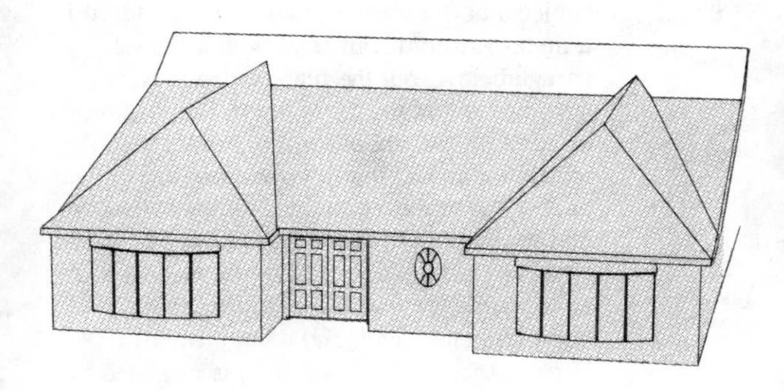

ROOFLINES AVAILABLE:
HIP (SHOWN), GABLE, FLAT

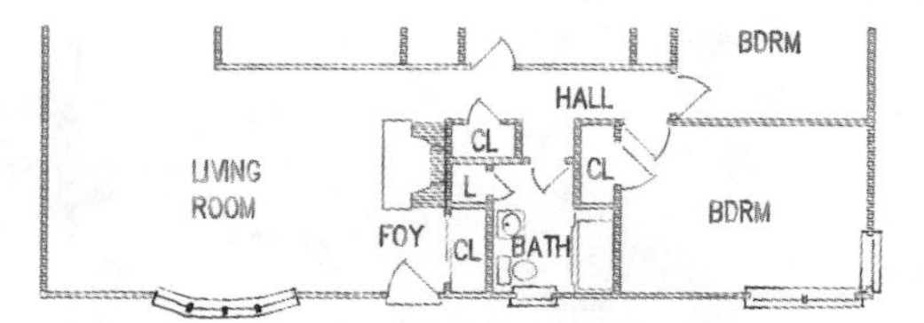

EXISTING PLAN

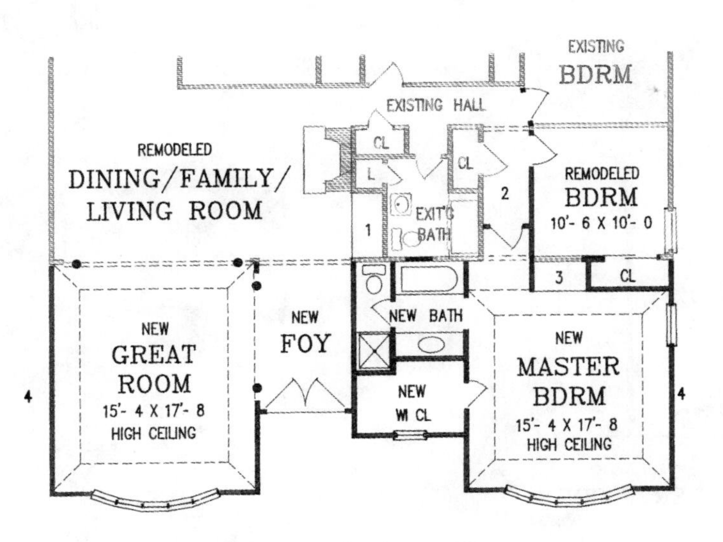

REMODELED PLAN

NOTES:
1- MEDIA UNIT LOCATION
2- EXTEND HALL
3- BUILT-INS
4- 18'- 0" ADDITION

BRG01

A fourth bedroom and a second garage are the subjects of this addition; the home pictured is a three-bedroom split level with a one-car garage below, but the plan is also perfectly suited to any multistory home. This is the type of addition that should marry perfectly to the existing home so that it appears as an original part of the design; matching siding, windows, and roofing are therefore essential. The fourth bedroom is shown as a new master bedroom suite, complete with its own private, compartmented bath and a large walk-in closet. Access to the bedroom is provided by simply extending the existing hall.

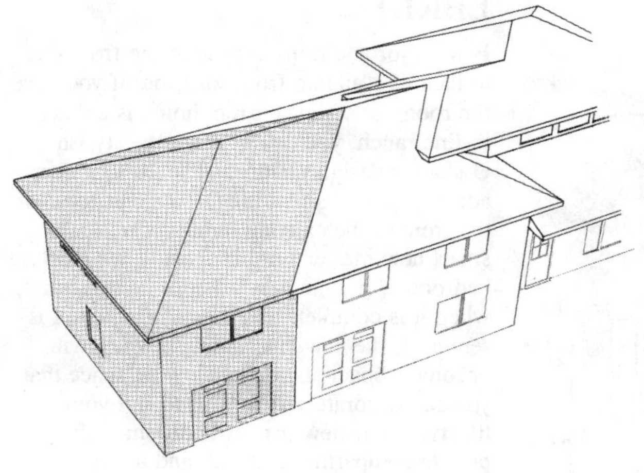

ROOFLINES AVAILABLE:
HIP (SHOWN), GABLE

NOTES:
1- REMOVE CLOSETS AND EXTEND HALL

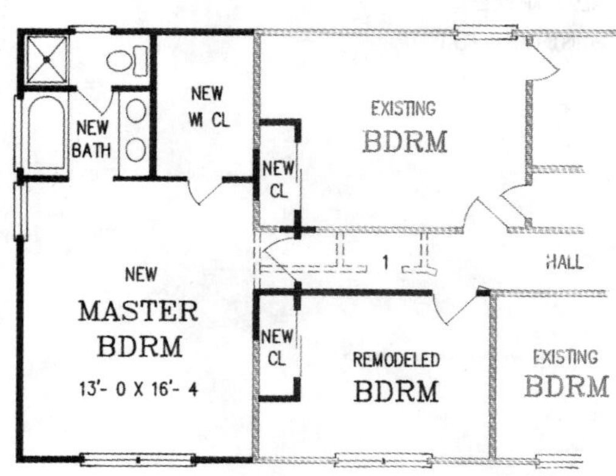

REMODELED SECOND FLOOR

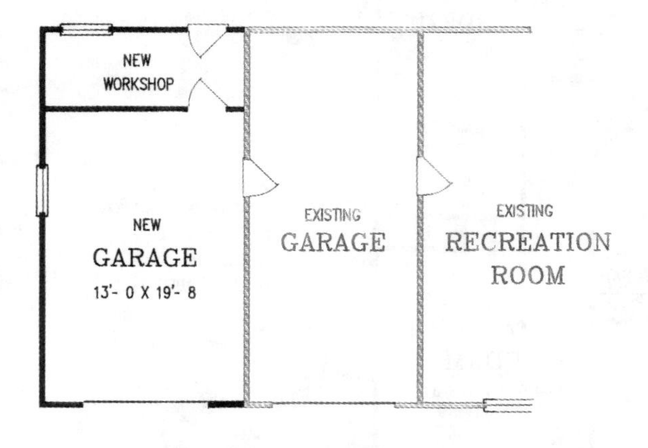

REMODELED LOWER FLOOR

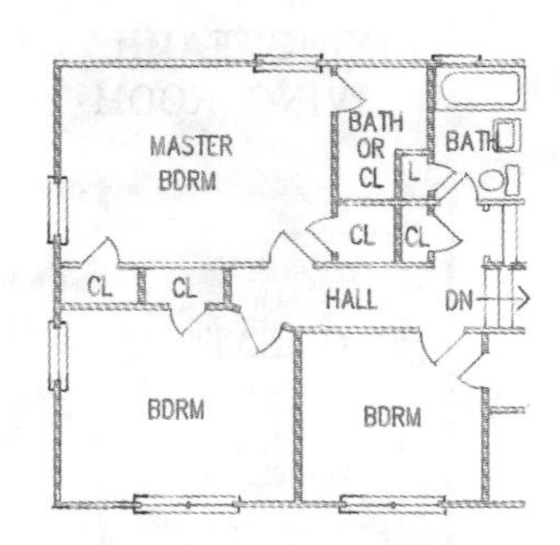

EXISTING SECOND FLOOR

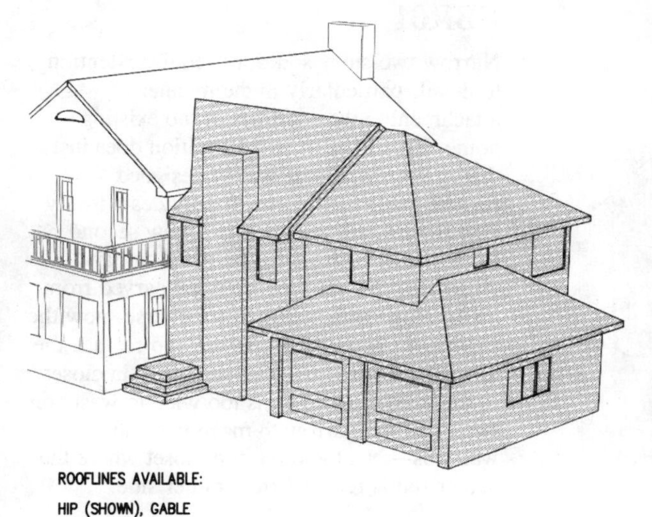

ROOFLINES AVAILABLE:
HIP (SHOWN), GABLE

FBRG1

Yours is a lovely, old two-story home, but it is missing a family room and garage. This clever architectural design shows you how to add both plus a fabulous new master suite above. Whether the addition is to the rear or side of your home, it will be an attraction. The family room is shown attaching to the rear of the stair hall and kitchen, but that might vary in your home. A fireplace with space for built-ins at either side is featured in the family room, as is a wall of windows and doors at the opposite side. The new master suite is an exquisite area; it includes a fireplace and bay window in the bedroom, plus a fabulous master bath, dressing area, and spacious walk-in closets.

NOTES:
1- WIDEN OPENING TO NEW FAMILY ROOM
2- CREATE TWO PASS-THRU OPENINGS
3- WHIRLPOOL TUB
4- SKYLITE
5- ADDITION IS 34'- 8" DEEP

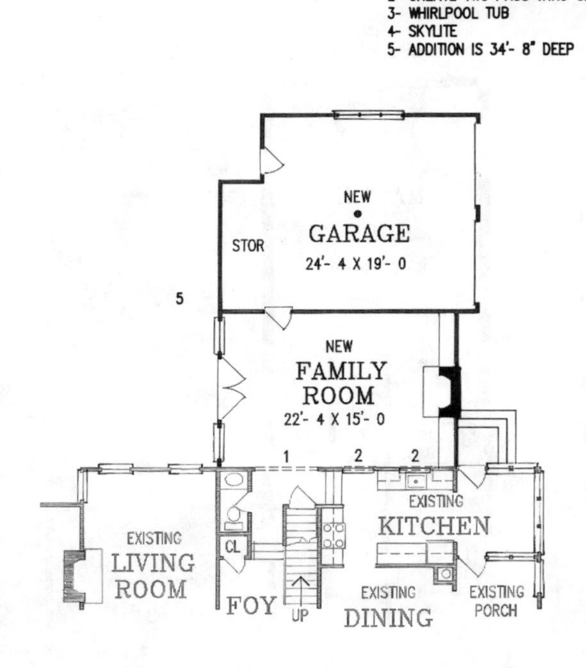

REMODELED FIRST FLOOR

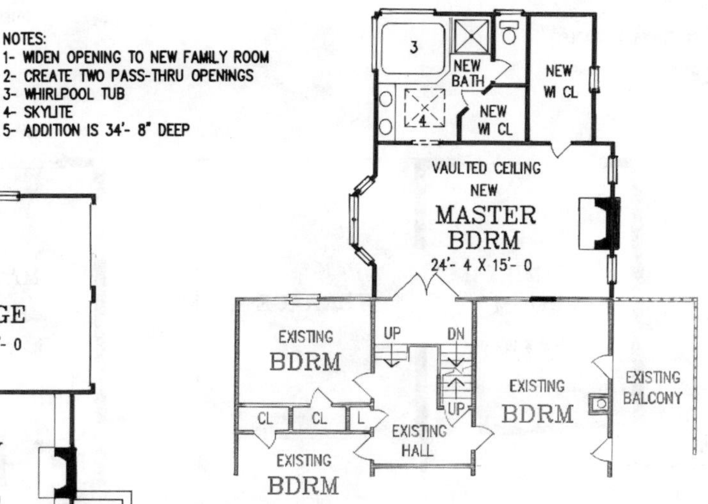

REMODELED SECOND FLOOR

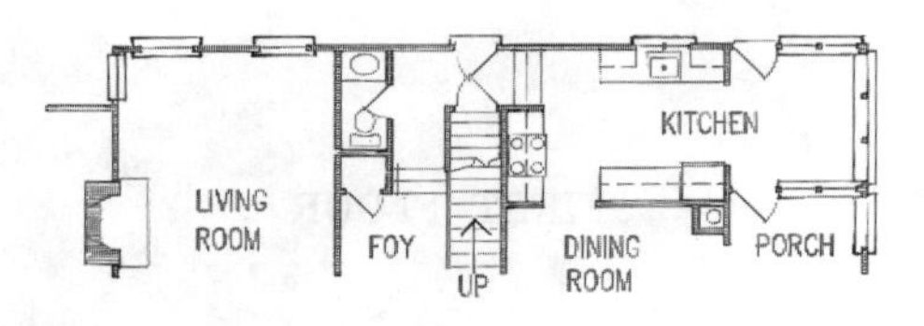

EXISTING FIRST FLOOR

NOTE: YOU MAY VIEW THESE FLOORS AS SEPARABLE: EITHER MAY BE BUILT AS A ONE STORY ADDITION TO BOTH A ONE OR TWO STORY HOME.

FBR01

Narrow two-story additions require attention to detail, particularly in the manner of attachment to the rooflines of the existing home. This 13′4″×16′8″ addition does just that in the shape of a crisply designed reverse-gable blockform. It provides a lovely first floor family room, plus a new second floor master bedroom. The bathroom and closets for the new bedroom are carved from an existing small bedroom (of course, now the old master bedroom is available to replace the lost small bedroom). The new walk-in closet in the master bedroom is too wide to waste on two rods, too narrow to make two full walk-ins—thus the three-rod closet where the center rod is reached from either side.

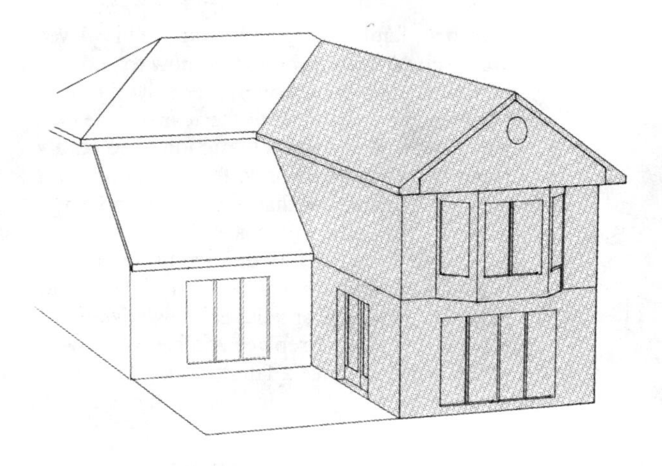

ROOFLINES AVAILABLE:
GABLE (SHOWN), HIP

NOTES 1- CLOSE EXISTING WINDOW
2- REMOVE SMALL BEDROOM LOCATED HERE
3- (3) RODS IN WALK-IN CLOSET

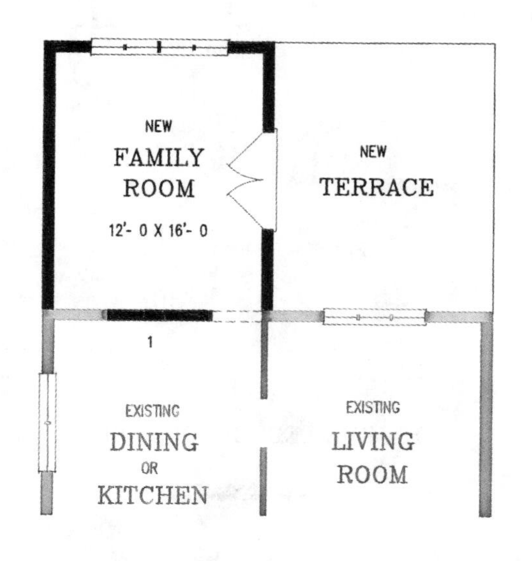

FIRST FLOOR

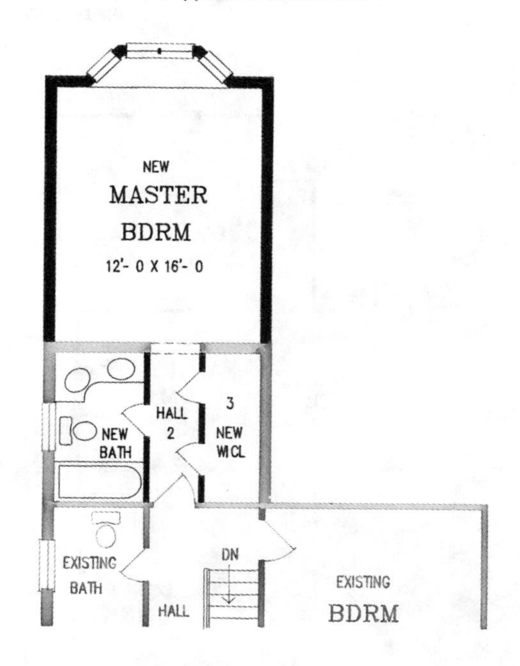

SECOND FLOOR

NOTE: YOU MAY VIEW THESE FLOORS AS SEPARABLE: EITHER MAY BE
BUILT AS A ONE STORY ADDITION TO BOTH A ONE OR TWO STORY HOME.

FBR03

A bright new sunroom and a classy new master suite are the subjects of this 21′6″ × 13′10″ two-story addition. Box bay windows on either side and triple French doors flanked by additional windows flood the sunroom with light, but do so with style. The master suite includes a lovely walk-in angle bay, a large walk-in closet, and a new-fashioned, compartmented bath with a platform tub and curved vanity. Attachment of such an addition is a major concern, especially as it impacts on the existing roof. Although not mandatory, this plan shows attachment to a wall at least four inches longer on either side. Such a four-inch offset allows the addition to be independent of the existing roof and sidewalls, making the matching of materials less of a concern.

ROOFLINES AVAILABLE:

HIP (SHOWN), GABLE, SHED, FLAT

NOTE: YOU MAY VIEW THESE FLOORS AS SEPARABLE: EITHER MAY BE BUILT AS A ONE STORY ADDITION TO BOTH A ONE OR TWO STORY HOME.

NOTES 1- IF UTILIZED AS FAMILY ROOM, REMOVE WINDOWS AND DOOR AND BUILD LOW WALL

2- REMOVE EXISTING SMALL REAR BEDROOM AND CONVERT TO MASTER BATH.

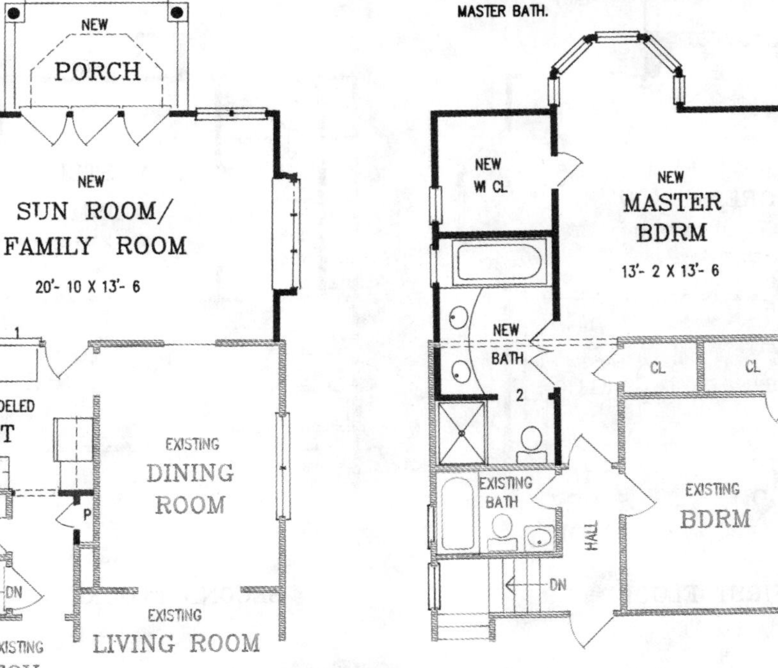

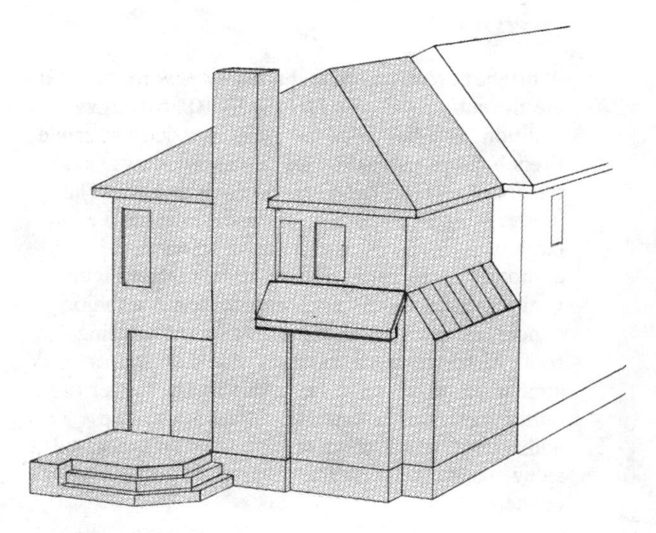

FBR04

A sensational great room and compact master suite are provided in this 22′0″×13′10″ two-story addition. As with all the two-story plans in the book, these floors can be pulled apart and viewed as separate one-story additions. The great room includes a fireplace and an adjacent alcove for a built-in media unit. A greenhouse roof is included at one end, and a floor-to-ceiling bay at the other end is ideally designed for table space—especially if it's adjacent to the kitchen, as indicated in the specific plan shown. The access to the new master suite from the top of the stairs (common in many older two-story homes) probably requires only the removal of a window at most.

ROOFLINES AVAILABLE:
GABLE (SHOWN), HIP, SHED, FLAT

1- GREENHOUSE ROOF OVER
2- SPACE FOR BUILT-IN MEDIA UNIT
3- EXISTING REAR WALL OF KITCHEN TO BE REMOVED AND LOW PARTITION INSTALLED.
4- CREATE OPENING BETWEEN LIVING AND GREAT ROOMS

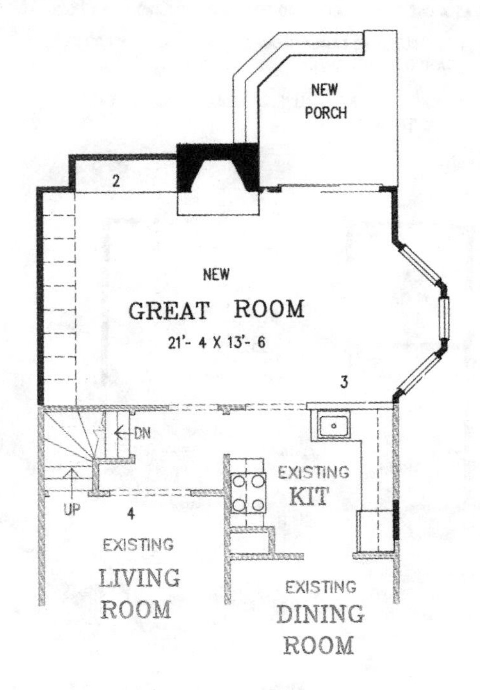

FIRST FLOOR

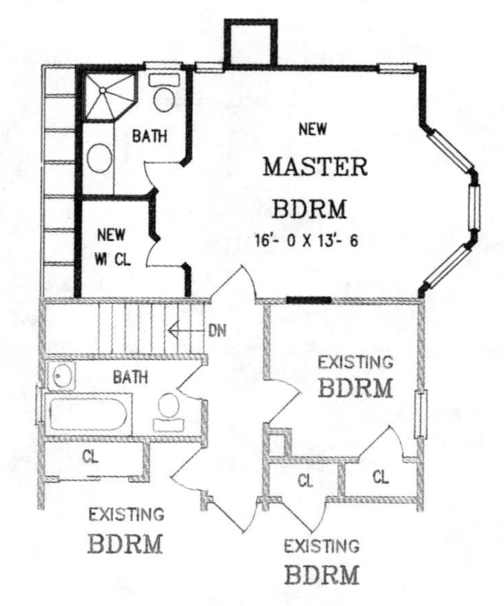

SECOND FLOOR

NOTE: YOU MAY VIEW THESE FLOORS AS SEPARABLE: EITHER MAY BE BUILT AS A ONE STORY ADDITION TO BOTH A ONE OR TWO STORY HOME.

KFBR1

An older one-story kitchen wing is removed in this design and is replaced by a 25′0″×12′0″ stylish two-story addition that creates a new kitchen and family room, plus a lovely new master suite. The new U-shaped kitchen is a working delight. A peninsula return is great for snacks and, while separating the kitchen, it does not disturb the visual connection to the family room. This plan, as virtually all of the two-story additions proposed in the book, could be viewed as two separate plans. If you love the new master suite, but don't need a new family room, the bedroom can be built alone, either as a one-story addition on a first floor or as a second floor addition over a porch (or whatever). Similarly, the new kitchen and family room could be constructed as a one-story addition without the second floor. Blueprints could be ordered to show any of these variations.

NOTES 1- REMOVE EXISTING ONE STORY KITCHEN WING
2- ADD LAV & CLOSET AND CREATE AN OFFICE ALCOVE
3- DINING ROOM BAY WINDOW ADDS NEEDED CIRCULATION SPACE

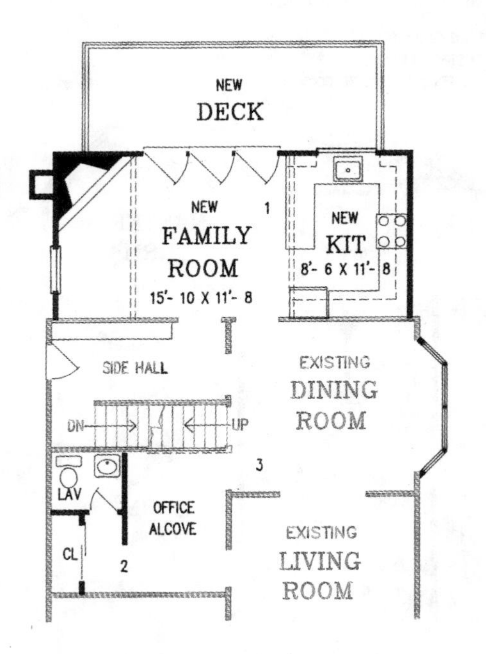

FIRST FLOOR

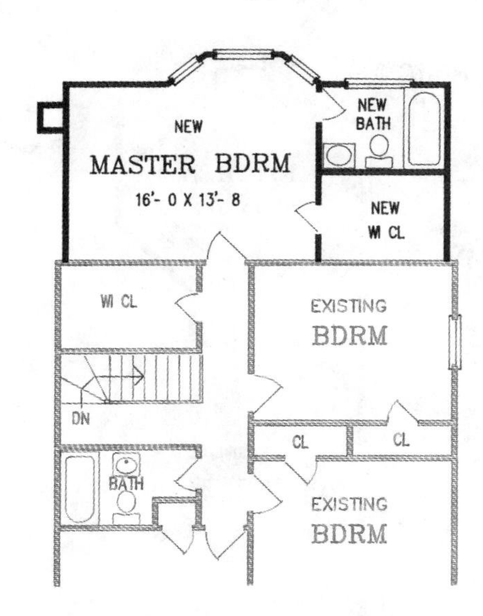

SECOND FLOOR

NOTE: YOU MAY VIEW THESE FLOORS AS SEPARABLE: EITHER MAY BE BUILT AS A ONE STORY ADDITION TO BOTH A ONE OR TWO STORY HOME.

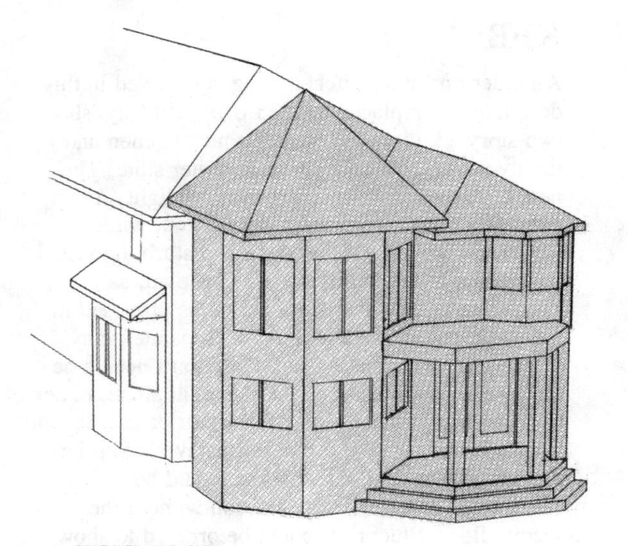

KFBR2

If angles turn you on, take a hard look here. This plan creates a stunning octagonally shaped new master bedroom above a fabulous octagonal kitchen. A complementary octagon-shaped platform tub is the focal point of the master bath and the overall effect is certainly distinctive. This two-story addition could be attached to many two-story homes but it is especially suited to a classic older home, where such forms already exist. In the plan shown, the addition is married to the rear of a narrow, hip-roofed, two-story. The old kitchen is removed and the space is combined with the breakfast area of the new addition to create a long family room. The addition also includes an attractive covered porch.

NOTE: YOU MAY VIEW THESE FLOORS AS SEPARABLE: EITHER MAY BE BUILT AS A ONE STORY ADDITION TO BOTH A ONE OR TWO STORY HOME.

NOTES 1- REMOVE EXISTING KITCHEN
2- OPEN EXISTING REAR WALL
3- NEW HALL FROM EXISTING BEDROOM

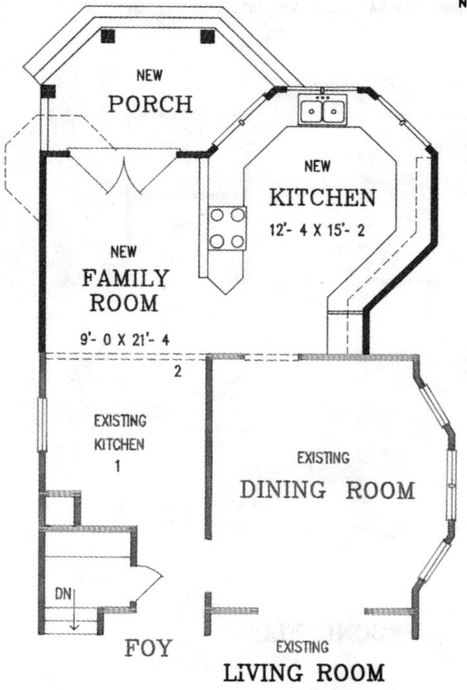

FIRST FLOOR

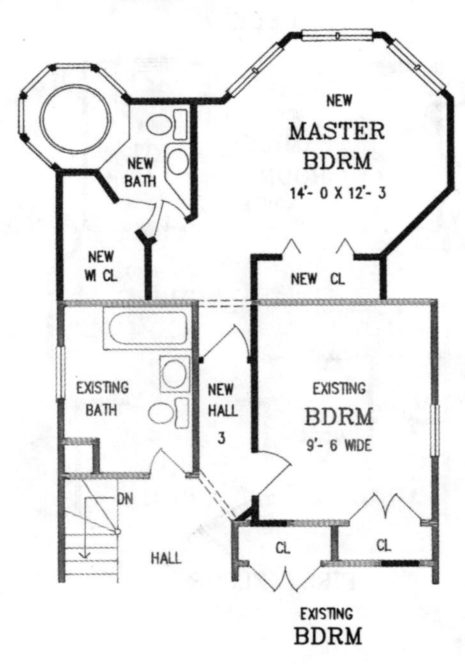

SECOND FLOOR

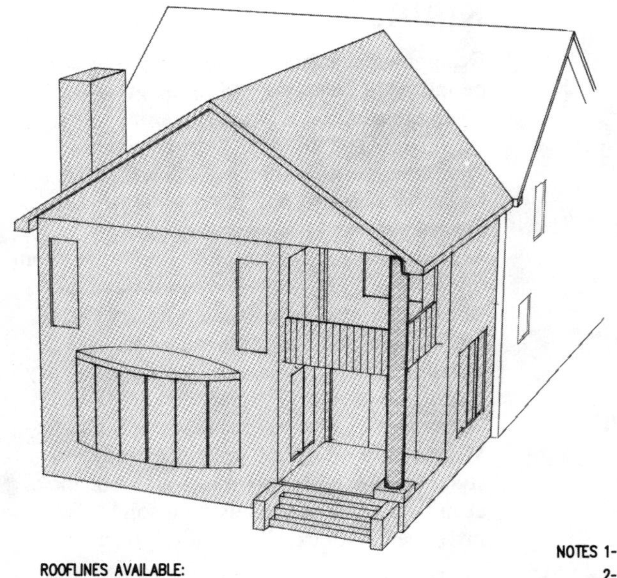

KFBR3

A luxurious new family room and breakfast room are crowned by a fabulous master suite in this 25′0″×18′4″ two-story addition. Fireplaces and covered decks are included on both floors, as are an abundance of windows and glass; French doors connect to both decks. The compartmented master bath with its whirlpool tub and separate shower plus two walk-in closets add to the luxurious quality of this addition. Although suited to attachment to virtually any two-story, the revisions shown to the specific 30-foot-wide two-story plan include a remodeled kitchen, a possible new powder room, and a smaller remodeled second floor hall bath. Access to the master suite is flexible depending upon your home.

ROOFLINES AVAILABLE:
GABLE (SHOWN), HIP, SHED, FLAT

NOTES 1- REMOVE EXISTING CLOSETS TO CREATE A CENTER HALL
2- POWDER ROOM CAN BE CARVED OUT OF KITCHEN
3- EXISTING BATH TO BE REDUCED IN SIZE AND REMODELED
4- OPEN WALL TO CREATE NEW PASS-THRU

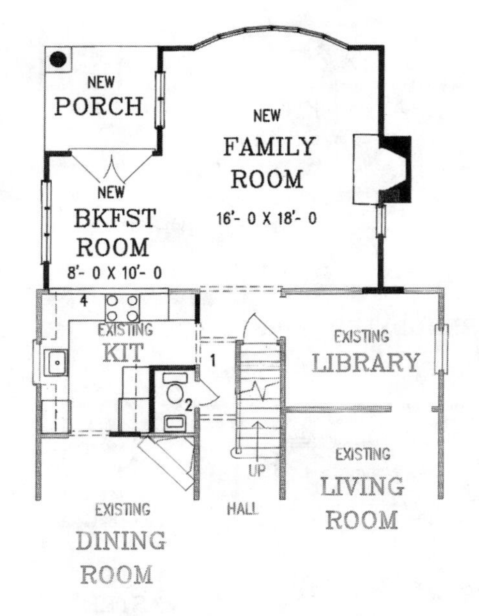

FIRST FLOOR

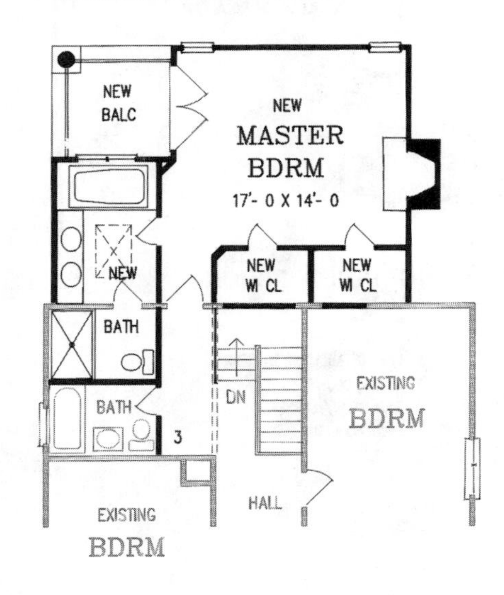

SECOND FLOOR

NOTE: YOU MAY VIEW THESE FLOORS AS SEPARABLE: EITHER MAY BE BUILT AS A ONE STORY ADDITION TO BOTH A ONE OR TWO STORY HOME.

KFBR5

Designing expansions to two-story homes can be infinitely complex, and particularly so, when the addition is primarily a one-story expansion. The designer must be concerned with how rooflines intersect the two-story, how they affect second floor windows, how the volume can be used advantageously on the second floor, and—even more important, from my point of view—how it all looks and feels. The plan on this page expands a very dated rear kitchen and breakfast room, adds a missing family room and a first floor laundry room, and turns an ordinary second floor bedroom into a marvelous new master suite. What's more, it does it creatively and with style. The new addition functions well, looks even better, and will likely be a solid investment with excellent return potential.

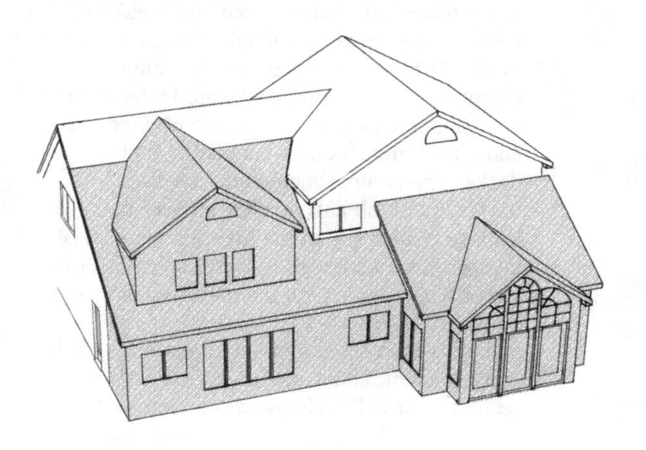

ROOFLINES AVAILABLE:
ONLY AS SHOWN

NOTE: CONSTRUCTION BLUEPRINTS CAN BE READILY MODIFIED TO ACCOMMODATE YOUR SPECIFIC HOUSE OR ROOM SIZES.

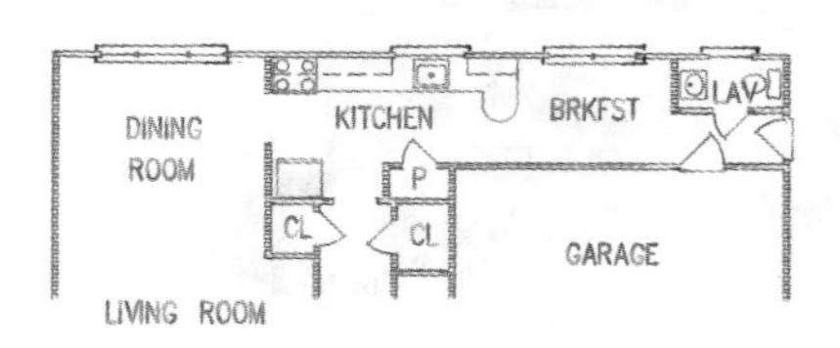

EXISTING FIRST FLOOR

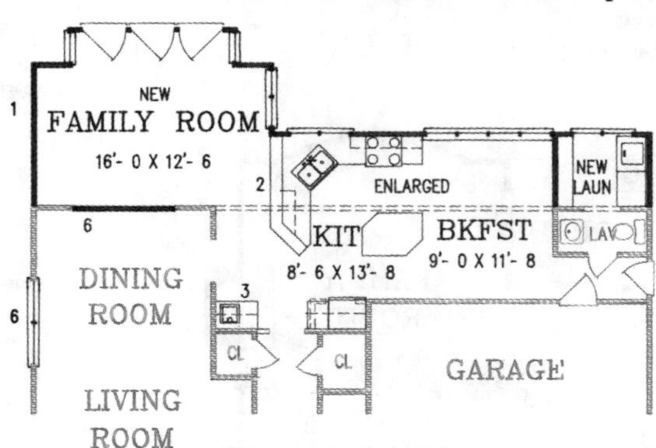

REMODELED FIRST FLOOR

NOTES:
1- 12'- 10" ADDITION TO REAR
2- SNACK COUNTER
3- BAR SINK / SERVER
4- WHIRLPOOL TUB
5- SKYLITE
6- RELOCATE WINDOWS TO SIDE WALL

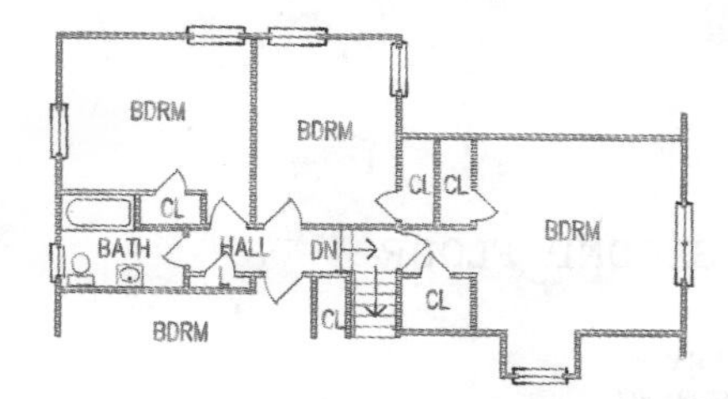

EXISTING SECOND FLOOR

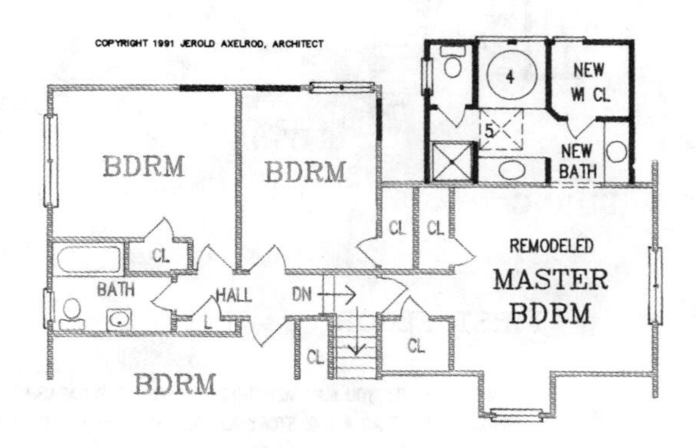

REMODELED SECOND FLOOR

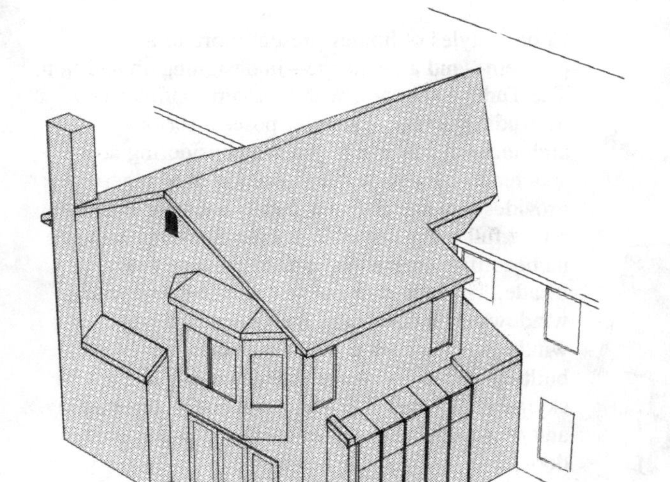

FBR06

This fabulous two-story addition provides you with a wonderful new family room on its first floor and a dramatic new master bedroom suite above. The plan is different from others shown, not only in its distinctive architectural styling, but also in the fact that it includes its own stair to the second floor. This stair enables ease of connection, since no second floor access is needed unless you desire it. And the plan is also very flexible as concerns the location and style of home it attaches to. Furthermore, the design of the new second floor has been carefully planned to reduce its impact on existing windows. The nuances of the design are lovely; study them all.

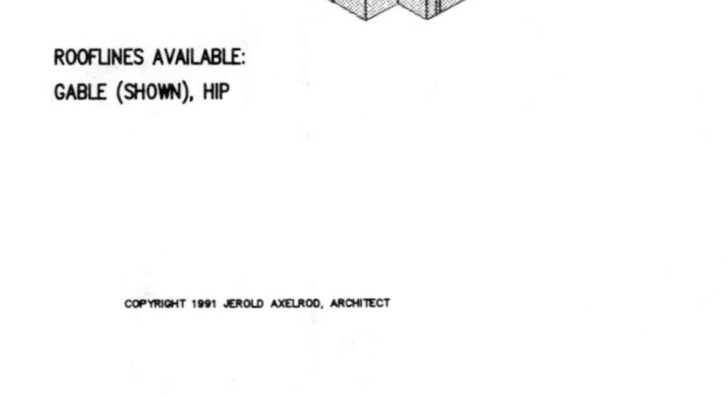

ROOFLINES AVAILABLE:
GABLE (SHOWN), HIP

EXISTING ROOMS

FIRST FLOOR ADDITION

NEW
MASTER
BDRM
13'- 0 X 16'- 0
NEW
WI CL
NEW
BATH
DN
BALC
NEW WI CL
4

EXISTING BEDROOMS

SECOND FLOOR ADDITION

NOTES:
1- SKYLITES
2- WET BAR
3- BUILT-IN MEDIA
4- CONNECTION TO SECOND FLOOR IS OPTIONAL
5- 18'- 8" ADDITION

FBR09

Certain styles of homes present more of a problem—and a challenge—in designing an addition. The Tudor two-story, with its steep rooflines and protruding corner vestibule, poses a serious architectural concern if you are considering an attachment alongside. The addition shown here provides a wonderful new family room or bedroom, plus a full bath. But what is even more appealing is its beautiful integration into the existing front facade. The new room is an exciting space with windows on three sides—including high front windows, under which you can install a wall of built-ins. There is also a vaulted ceiling and a skylight. The new roofline merges with the existing and even adds a three-foot overhang at the front door.

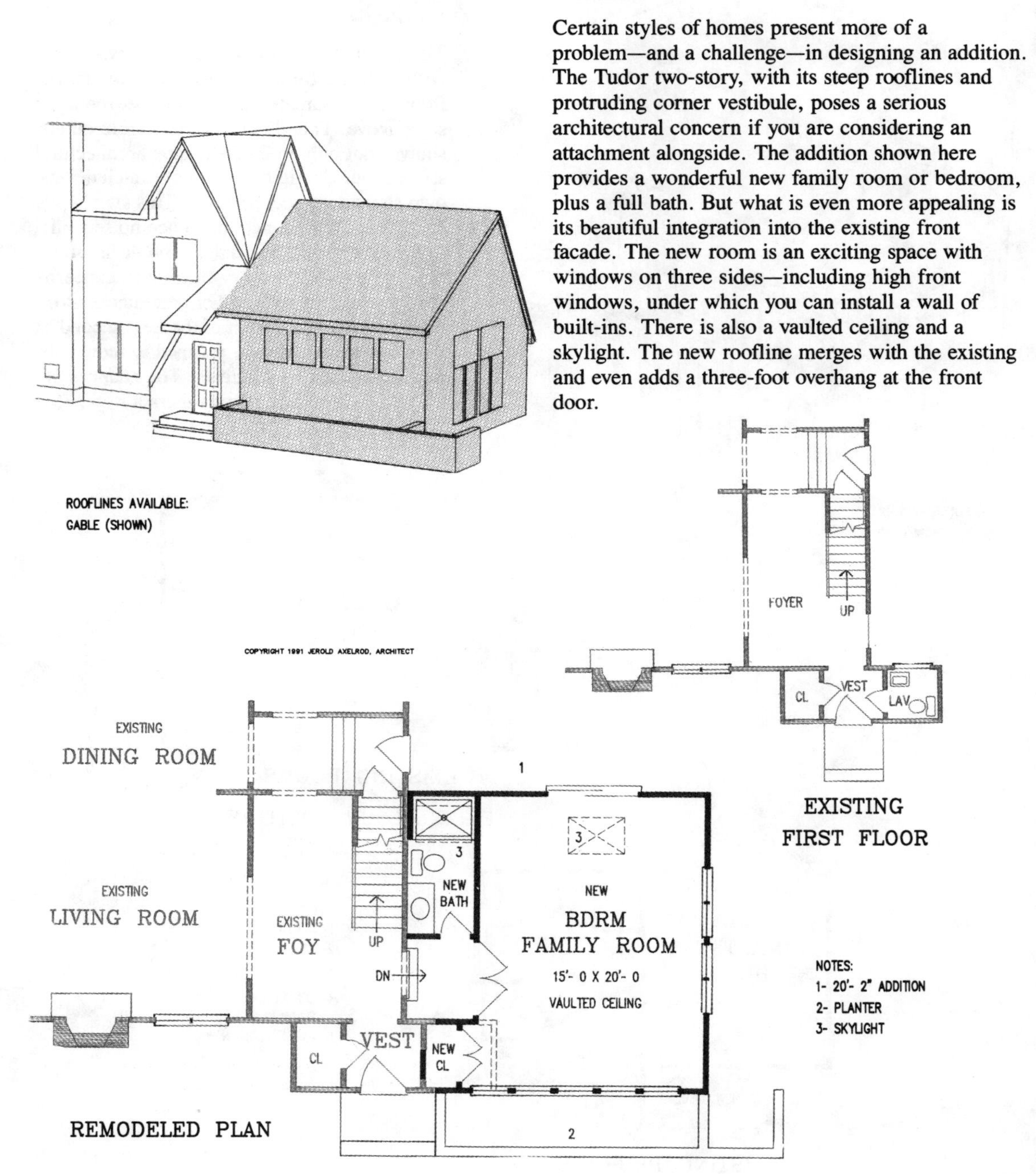

CONSTRUCTION BLUEPRINTS ARE AVAILABLE FOR THIS PLAN (AS WELL AS ALL OTHERS) AND INCLUDE A MATERIALS LIST AND INFORMATION ON HOW TO MAKE CHANGES TO PLANS

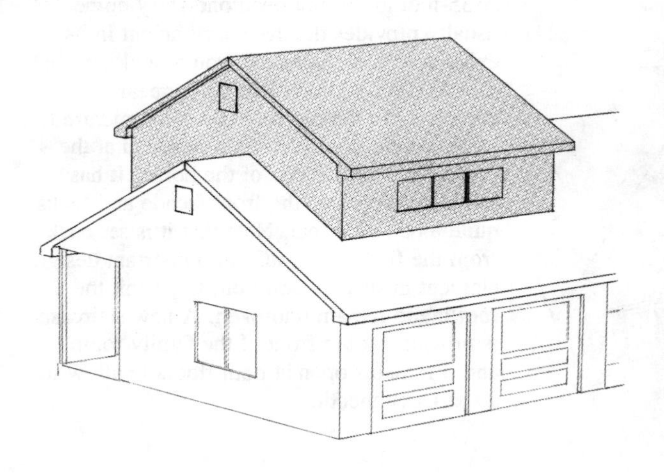

FBR07

This is one of several designs that makes use of the found space over a garage. Additional bedrooms or apartments are the subject of the other plans. This one provides for a large playroom, but it, too, could be turned into a bedroom. This particular plan is especially designed for the roofline of a one-story home. In adding over a garage in a one-story home, you have to be careful not to let the two-story element appear too bulky or too high. The solution is to set it back from all sides and to wrap it in the roof to give it the appearance of a dormer. The new stair to the playroom is shown entering from a family room or breakfast room, but the connection could also be from other rooms, depending on your home.

ROOFLINES AVAILABLE:
GABLE (SHOWN)

SECOND FLOOR

ALL NEW

EXISTING PORCH

EXISTING LAUN

LAV

EXISTING FAMILY/ BRKFST ROOM

1

UP

2

EXISTING KITCHEN

EXISTING GARAGE

FIRST FLOOR

NOTES:
1- NEW STAIR TO SECOND FLOOR
2- COLUMN MAY BE REQUIRED
3- LOCATION FOR OPTIONAL BATH
4- LOCATION FOR OPTIONAL WALK-IN CLOSET

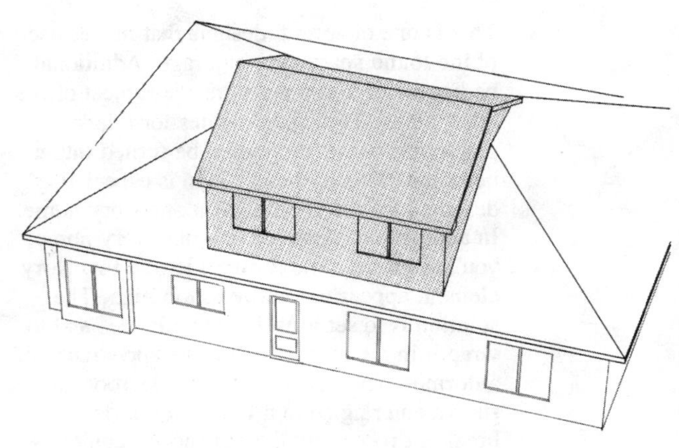

F0010

A 35-foot to 40-foot-deep one-story home usually provides the necessary height in its existing roofline to enable you to add a partial second floor addition that will appear attractive. In the hip-roofed version pictured, a large second floor playroom is added at the middle part of the rear of the home. It has minimal impact on the front facade and looks quite nice in the rear. Note that it is set back from the first floor wall—an important design element in such an addition, to permit the roofline to remain unbroken. A new staircase is provided at the front of the family room, and it remains open at both floors to allow for a spatial connection.

ROOFLINES AVAILABLE:
GABLE & HIP (SHOWN)

NOTE: CONSTRUCTION BLUEPRINTS CAN BE READILY MODIFIED TO ACCOMMODATE YOUR SPECIFIC HOUSE OR ROOM SIZES.

SECOND FLOOR
ALL NEW

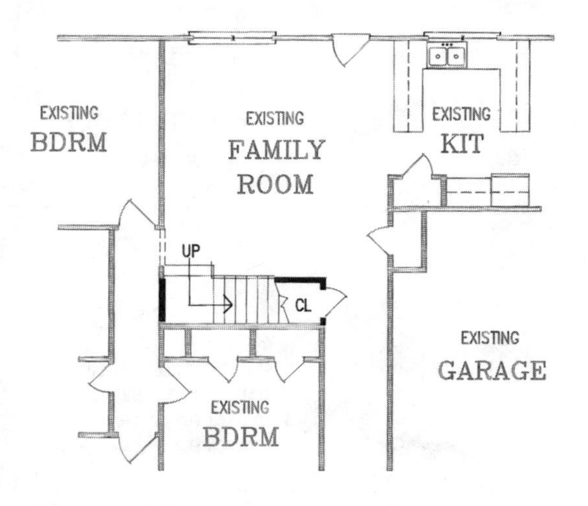

REMODELED PLAN

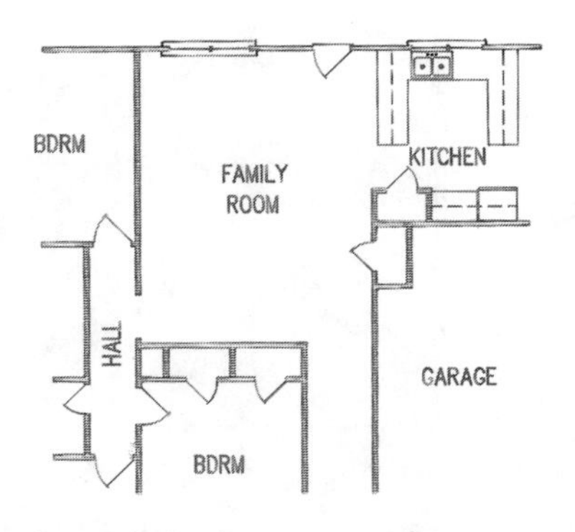

EXISTING PLAN

F0004

Need a family room, but the only place to attach it is to the rear of your kitchen? This plan shows you how to do just that. Attachment to the kitchen means a potential loss of light to the kitchen, therefore the need to visually open it to the family room. A wide counter, which serves both as a pass-through and as a snack counter, is an excellent idea. It's also a good place for the sink, as this part of the kitchen will benefit most from borrowed light from the new family room. The family room shown includes a vaulted ceiling, skylights, and a corner fireplace. If you prefer a centered fireplace, see plan F0001 on page 109. Alternate roof lines are readily available; refer to the blockforms in chapter 6.

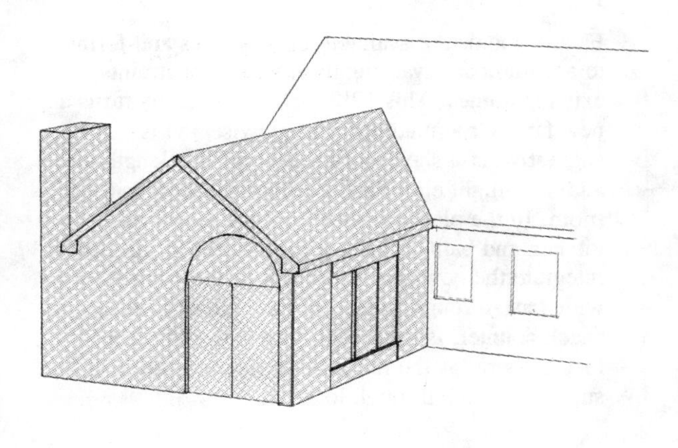

ROOFLINES AVAILABLE:
GABLE (SHOWN), HIP, SHED, FLAT

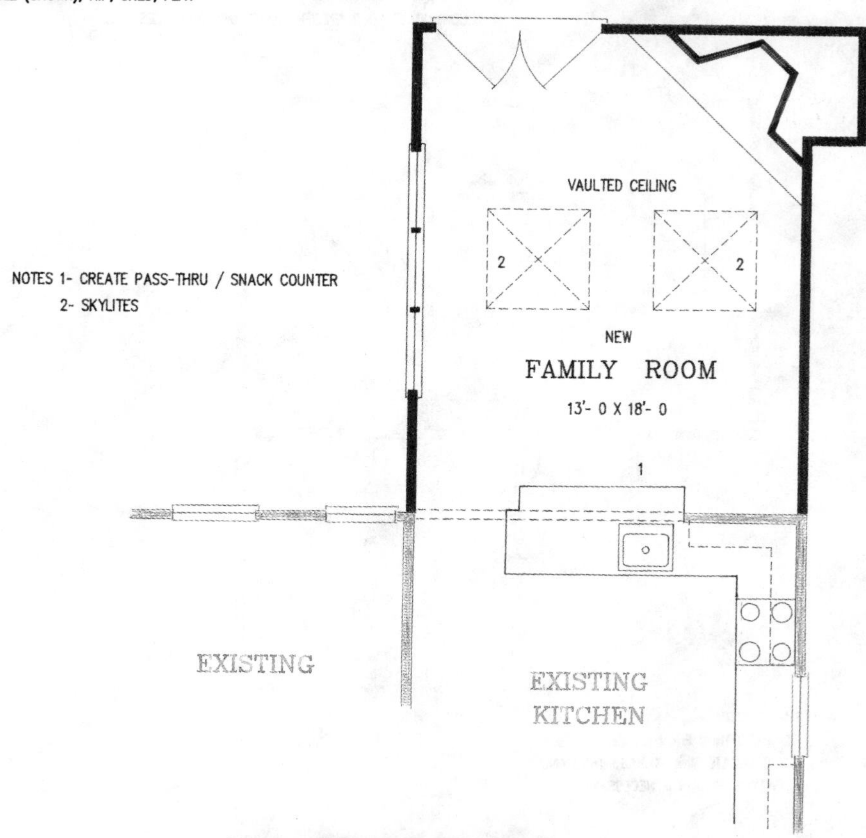

NOTES 1- CREATE PASS-THRU / SNACK COUNTER
2- SKYLITES

NOTE: CONSTRUCTION BLUEPRINTS CAN BE READILY MODIFIED TO ACCOMMODATE YOUR SPECIFIC HOUSE OR ROOM SIZES.

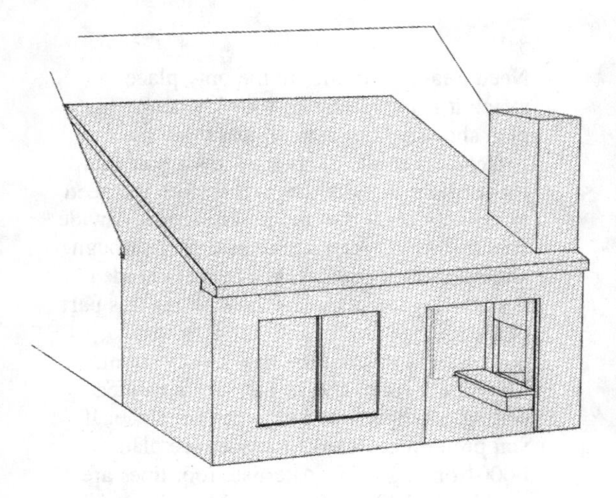

LINES AVAILABLE:
(SHOWN), GABLE, HIP, FLAT

F0003

Family rooms are available in all shapes and forms to accommodate varying desires, lot constraints, or existing homes. This 17′8″×12′4″ addition turns a new family room addition lengthwise and is suggestive of a shed roof as shown. This lengthwise addition might eliminate windows in more than one room. In the plan, the attachment is to the rear of a kitchen and bath. Since the design goal is to visually integrate the new family room with the kitchen, a wide pass-through, which can double as a serving or snack counter, is suggested. The sink should be located here, as the range or refrigerator are not suited to a pass-through location.

NOTE: CONSTRUCTION BLUEPRINTS CAN BE READILY MODIFIED TO ACCOMMODATE YOUR SPECIFIC HOUSE OR ROOM SIZES.

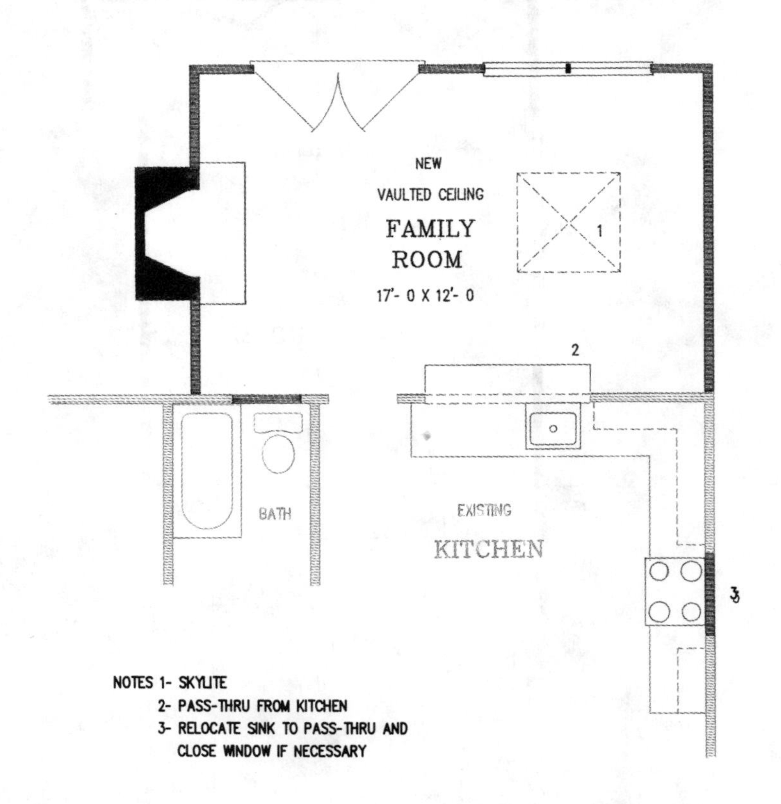

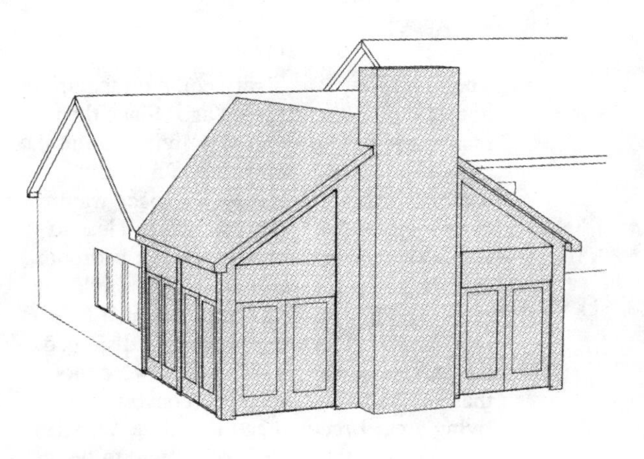

ROOFLINES AVAILABLE:

GABLE (SHOWN), HIP, SHED, FLAT

F0006

A fabulous new great room, 21′6″×14′4″, is the subject of this room addition. It is shown as attached to the rear kitchen and adjoining dining room of a modest split level, but it could attach to almost any style or type of home. Whether it's a one-story, a two-story, or a split-level home, the new room is a stunning space. It boasts six sets of sliding glass doors that wrap around all three sides, a centered fireplace, and a dramatic vaulted ceiling. When added as shown, it is suggested that the existing rear wall of the dining room be removed. The new space will now visually merge with the existing rooms, enhancing their capability to function as one continuous space.

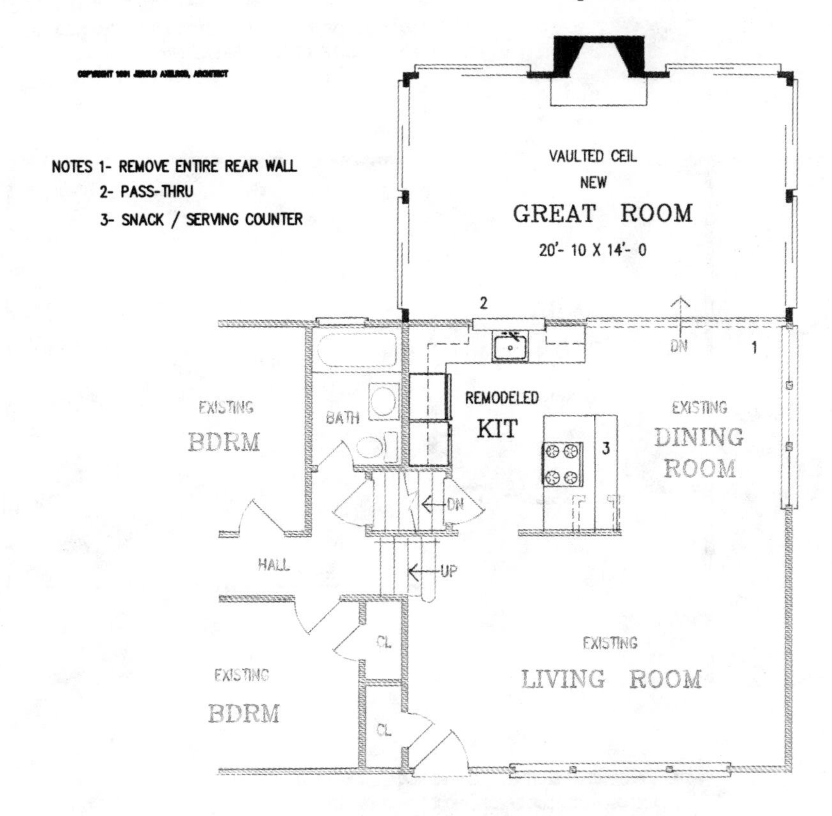

NOTE: CONSTRUCTION BLUEPRINTS CAN BE READILY MODIFIED TO ACCOMMODATE YOUR SPECIFIC HOUSE OR ROOM SIZES.

F0002

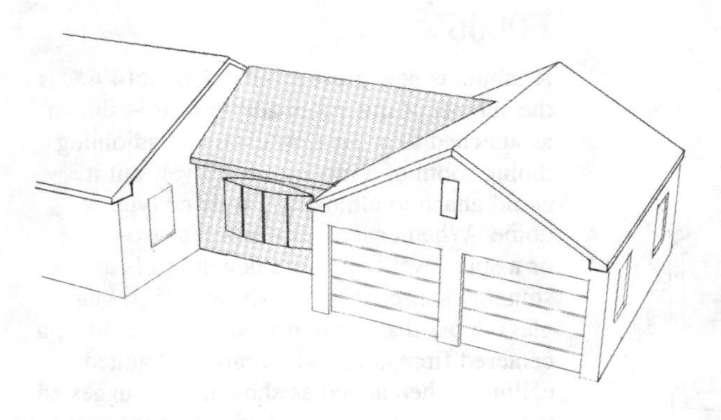

The breezeway was a very common feature of homes built 30 to 40 years ago. Since then, many have been converted to living space, the most common use being that of a family room. If yours still exists, it is space waiting to be claimed for better use. Other ideas to consider include a sunporch, guest bedroom, or even a new master suite, which would require adding space as well (see plan PBR03 on page 80). The breezeway often disjoined the garage by using a different type of roof than on the home; when you convert it to living area, I recommend installing a new roof, such as the gable roof shown, to tie the elements together.

NOTES 1- REMOVE EXISTING BREEZEWAY ROOF
2- SIZE OF NEW FAFILY ROOM DETERMINED BY SIZE OF OLD BREEZEWAY

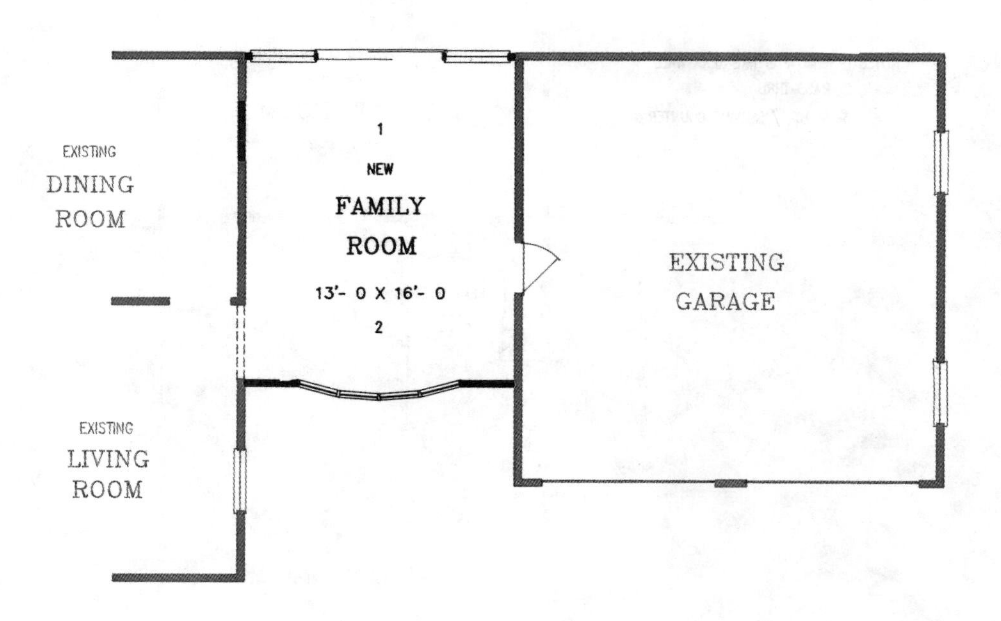

NOTE: CONSTRUCTION BLUEPRINTS CAN BE READILY MODIFIED TO ACCOMMODATE YOUR SPECIFIC HOUSE OR ROOM SIZES.

F0001

A common existing condition: a little-used patio slab behind a garage that adjoins an L-shaped living and dining room. A common need: a family room. The solution: remove the patio (or porch, if one exists) and construct an inviting, reverse-gable family-room addition. The roofline may need a *cricket* (see construction details in chapter 11 and glossary), but that's OK. The centered fireplace, flanked by tall, circular-capped windows, all under a vaulted ceiling, will create a very modish yet tasteful interior space. Suggestion: Open the connection to the existing house as wide as possible to visually connect the new family room. Further, if you are concerned about the furnishability of the dining room, consider an expansion, such as shown in plans KD001 and FDG01 on pages 128 and 139.

ROOFLINES AVAILABLE:
GABLE (SHOWN), HIP, SHED, FLAT

NOTE: CONSTRUCTION BLUEPRINTS CAN BE READILY MODIFIED TO ACCOMMODATE YOUR SPECIFIC HOUSE OR ROOM SIZES.

NEW
FAMILY
ROOM
13'- 6 X 18'- 0
VAULTED CEILING
1

EXISTING
DINING
AND
LIVING ROOM

EXISTING
GARAGE

NOTES 1- REMOVE PORCH OR PATIO IF ONE EXISTS

F0008

There are times when conditions dictate an atypical solution. For the striking new family room pictured here, it might have been a limiting rear yard (swimming pool or hill at one side) or a desirable orientation. However, beyond these there must also be a willingness to be different and unique. This modern family room works exceedingly well with the kitchen—and the entire house—but it clearly does not try to mimic the forms of the existing home. The room features a high ceiling, a fireplace, and a wall for built-ins. The kitchen is completely remodeled and opened to the new family room with a high snack counter separating the two.

ROOFLINES AVAILABLE:
FLAT (SHOWN)

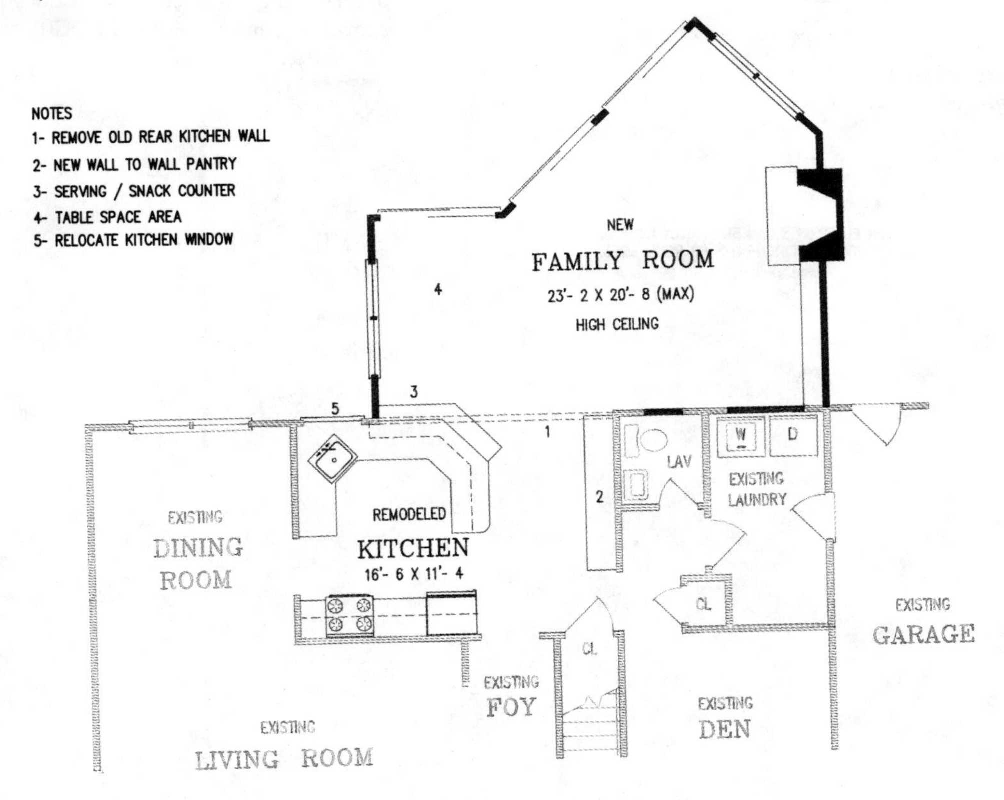

FIRST FLOOR

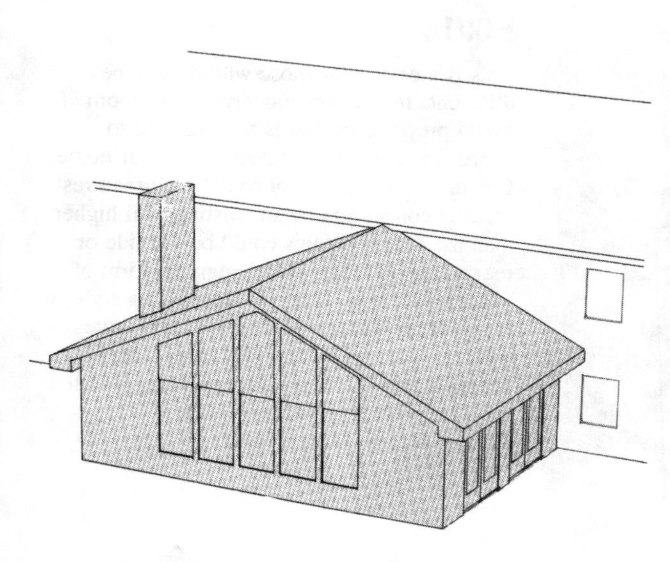

ROOFLINES AVAILABLE:
GABLE (SHOWN)

F0012

If your demands include a large contemporary family room, you should study this design. Enclosed within its walls is a 30-×-20-foot space dominated by a high vaulted ceiling, four large skylights, a fireplace wall flanked by built-ins, and a stunning panorama of glass. Two sets of sliding doors at one side and a dramatic window wall on the gable end bathe the room in natural light. This is one of those designs that is not specific to any given existing house, but which you can readily attach to almost any one- or two-story home. The plans can be readily adjusted in size as required and can be adapted to the requirements of any specific attachment.

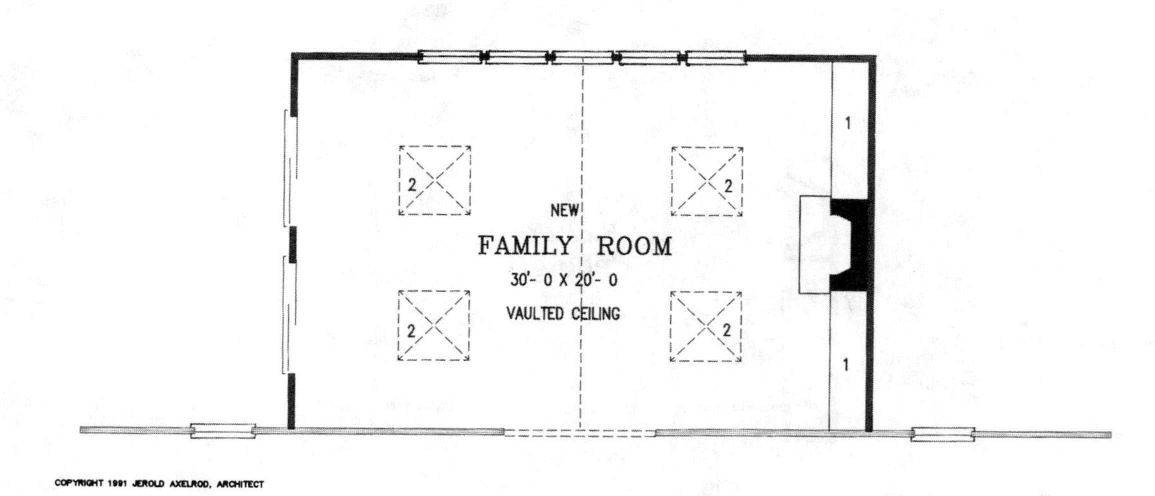

EXISTING HOUSE

NOTES:
1- BUILT-INS, MEDIA UNIT
2- SKYLITES

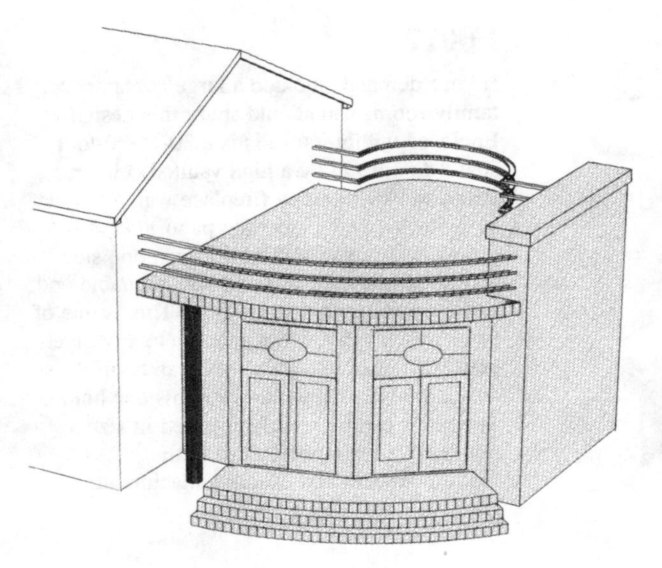

F0014

This is a design for those who dare to be different. It is a very modern family room of grand proportions that is not intended to mirror the existing architecture of your home. It is designed with a flat roof which requires that the connection be an existing wall higher than the new roof; this could be the side or rear of any two-story. Although the form of this addition may depart from existing style, it is important that materials blend and unite with the existing home. The room features a beautiful semicircular seating alcove, a wall for a fireplace and built-ins, plus an 11-foot-high ceiling throughout.

ROOFLINES AVAILABLE:
FLAT (SHOWN)

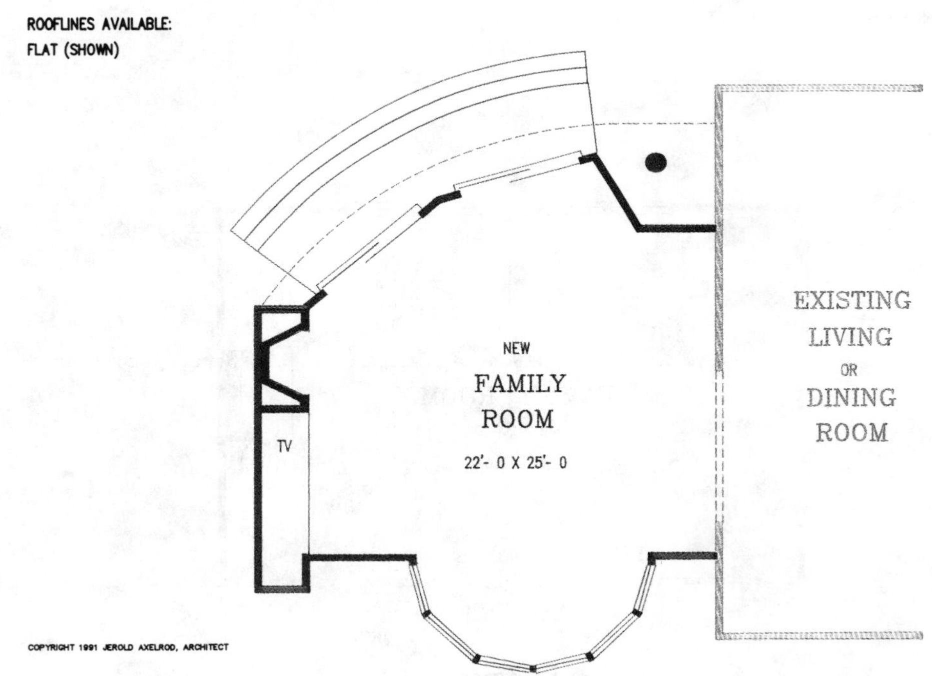

CONSTRUCTION BLUEPRINTS ARE AVAILABLE FOR THIS PLAN (AS WELL AS ALL OTHERS) AND INCLUDE A MATERIALS LIST & INFORMATION ON CUSTOMIZING OR CHANGING THE PLANS

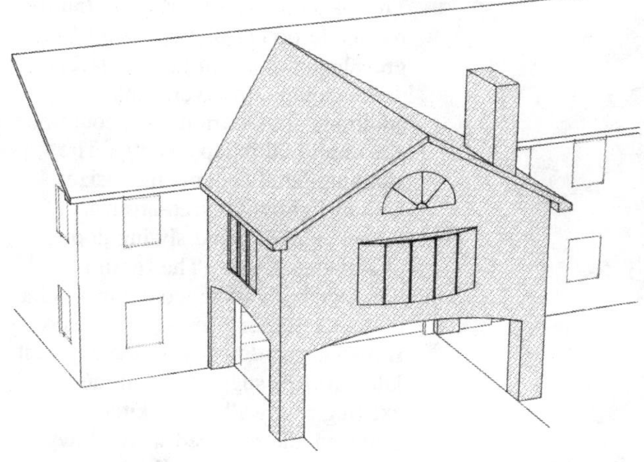

F0011

Is the place where you are considering adding a new family room situated above an existing lower-level driveway? If so, look here. Whether it is the side, rear, or even the front of your home, this beautiful design shows you how to add a nice-sized room, complete with all the desired features of a family room, without disturbing the required access to your garage. The new room possesses both a lovely exterior and interior character, including a bow window, a vaulted ceiling, and a built-in fireplace with recessed locations for media equipment on each side. The fireplace, of course, would be a pre-fab variety. Skylights add natural light.

ROOFLINES AVAILABLE:
GABLE (SHOWN), HIP, SHE T

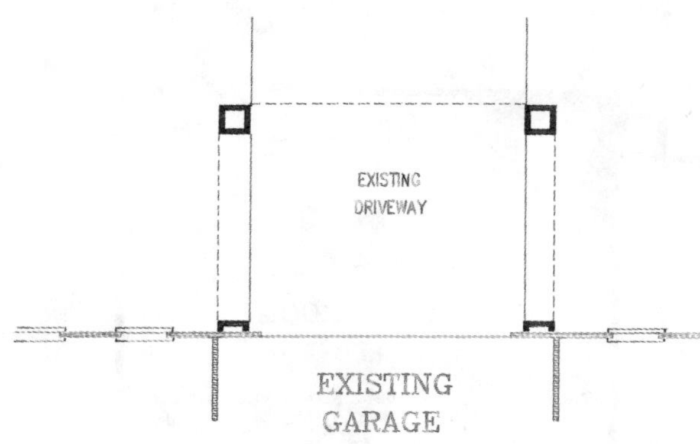

LOWER FLOOR

NOTES:
1- SPACE FOR BUILT-IN MEDIA
2- SKYLITES
3- WIDTH OF OPENING AND LOCATION DEPENDS ON EXISTING HOUSE

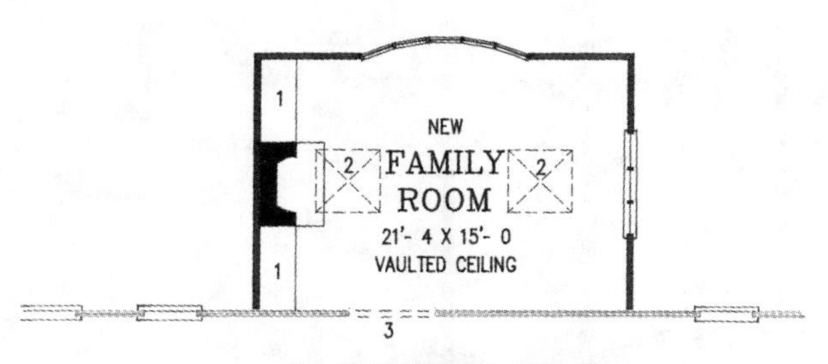

EXISTING ROOMS

UPPER FLOOR

NOTE: CONSTRUCTION BLUEPRINTS CAN BE READILY MODIFIED TO ACCOMMODATE YOUR SPECIFIC HOUSE OR ROOM SIZES.

FD001

This handsome 21′0″×13′4″ family room addition also shows a 5′6″ greenhouse addition to an adjacent dining room. However, both additions are separable and could be accomplished independently. The stunning family room emphasizes natural light with wrap-around corner windows and sliding doors, plus two skylights. The room is designed with a vaulted ceiling and a dramatic corner fireplace, and is shown a step down from the adjacent kitchen. It is suggested that the existing rear wall of the kitchen be removed and replaced with a low wall. The dining room is shown enlarged by the four-bay greenhouse, which makes for a very chic addition.

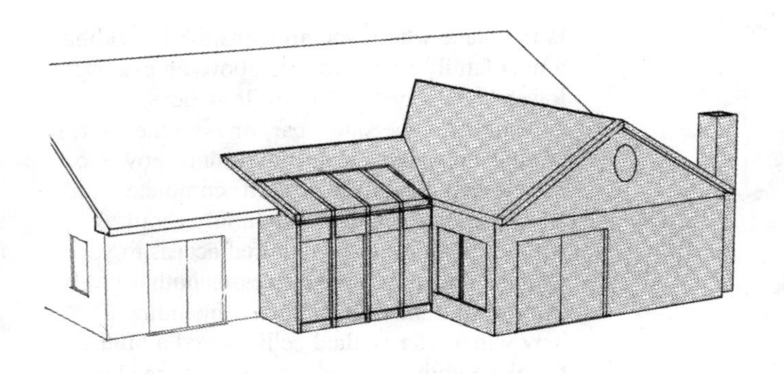

ROOFLINES AVAILABLE:
GABLE (SHOWN), HIP, SHED, FLAT

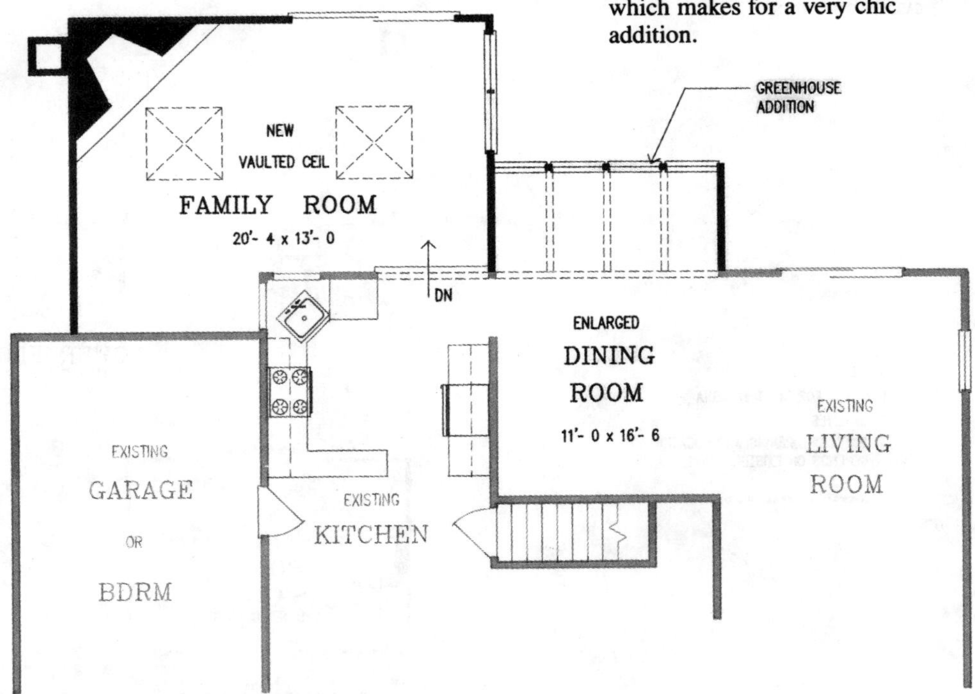

CONSTRUCTION BLUEPRINTS ARE AVAILABLE FOR THIS PLAN (AS WELL AS ALL OTHERS) AND INCLUDE THE DESIGN OF: STRUCTURAL HEADERS REQUIRED TO REMOVE REAR WALLS.

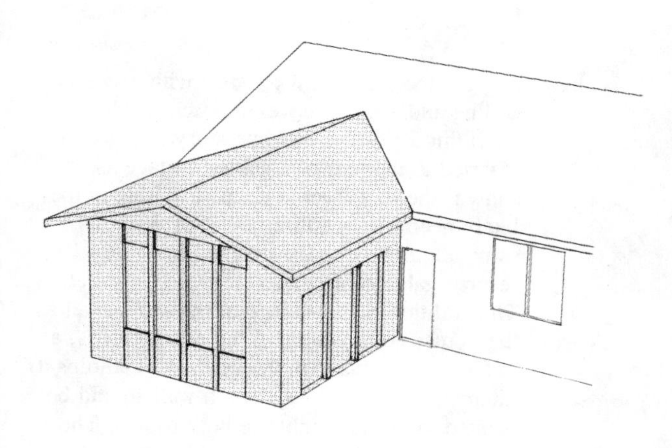

KF003

This fashionable 14′8″×14′4″ family room addition can stand by itself as a smart addition, or it can be combined with an adjacent remodeled kitchen, as shown. The family room boasts a 10-foot-high ceiling and lots of glass, including high transomed windows and French doors to the patio area. The family room ties to an existing kitchen via a split stairway that leads to the outside, as well as to the basement—a common condition in many homes. This allows the high ceiling, while maintaining the same roof line. The remodeled kitchen, open to the family room with its angled serving/snack top and the balconied breakfast area, makes the entire space a gorgeous renovation.

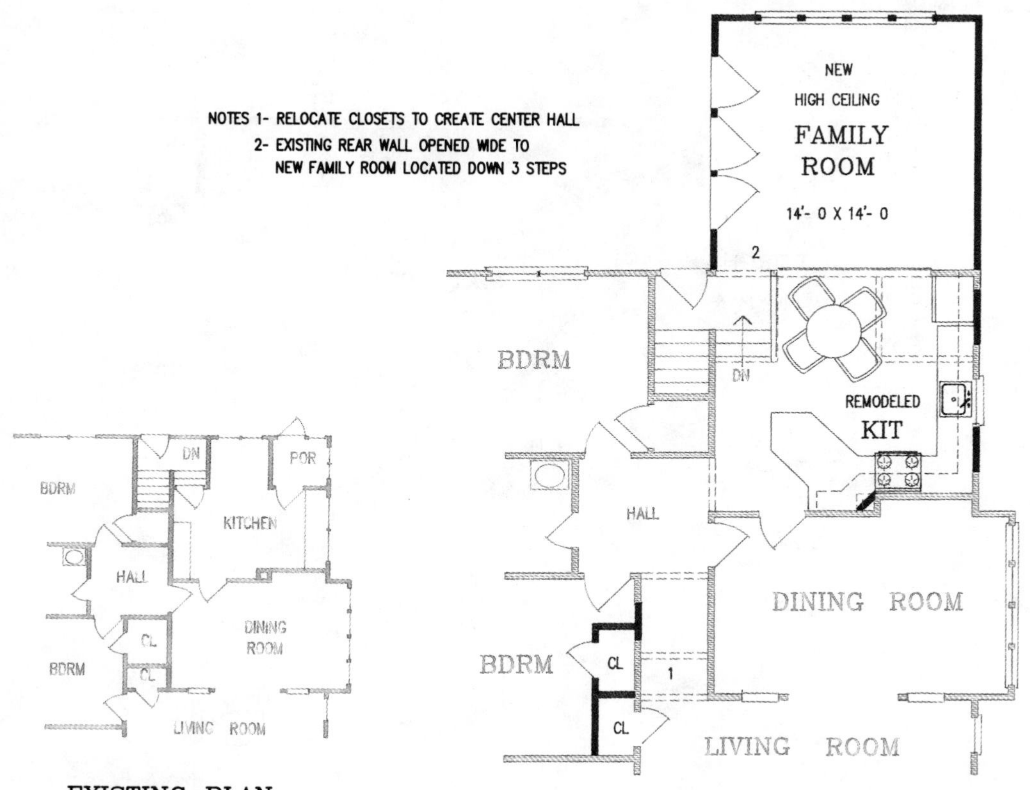

EXISTING PLAN

REMODELED PLAN

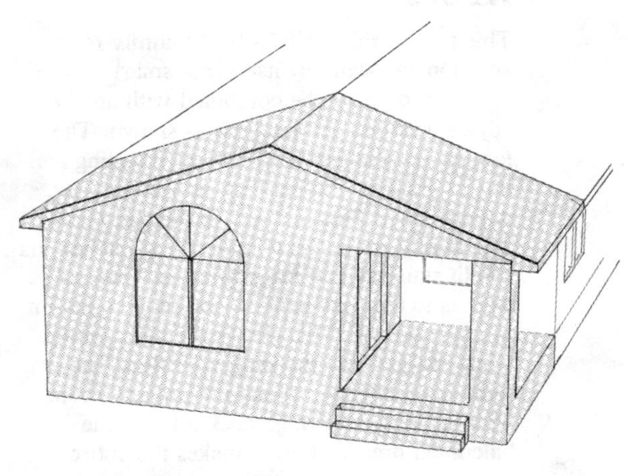

KF002

A great room (or family room) with a vaulted ceiling and a lovely covered porch results from this 25′6″×12′4″ one-story addition. If married to the rear of a gabled cottage as shown, the roofs would align, resulting in the likely need for an entire new roof. But this same addition is easily attached to a reverse-gable roof or two-story wall, as well. The addition is maintained at the same level as the home, not down the three or four steps, as in some other plans in the book. If it adjoins a kitchen, as shown, the kitchen wall should be opened to visually enhance both rooms. The side bay creates a delightful table space, if that's in your program.

NOTES 1- TABLE SPACE
2- GREAT ROOM COULD INCLUDE PORCH AREA
3- SKYLITE

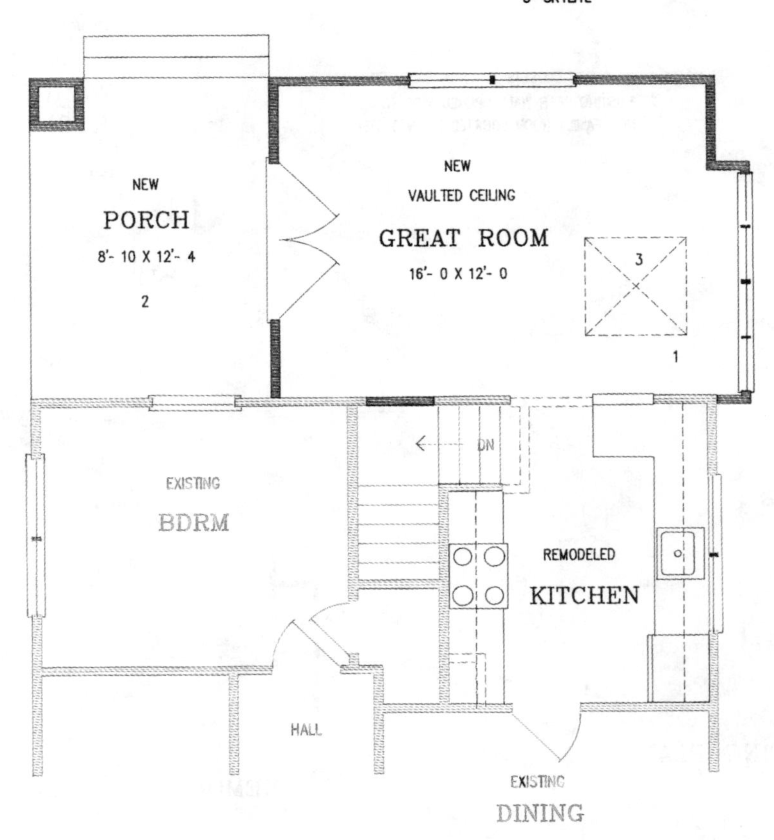

KF008

Twenty years ago, this was a top-selling family room and kitchen, and to many, especially if you are coming from a small home, that still may be so. But, if you have lived with it for 20 years, you might be dreaming of a fashionable new family living center that integrates the kitchen and family room. This plan accomplishes that with style. An 8′4″×10′0″ addition to the side of the family room provides a bright, sunny space for a table. Walls are removed, and the entire area is visually connected. The new, expanded kitchen is opened to the family room where a dramatic angled serving and snack counter is located. For an even more expansive plan see plan KF009 on page 118.

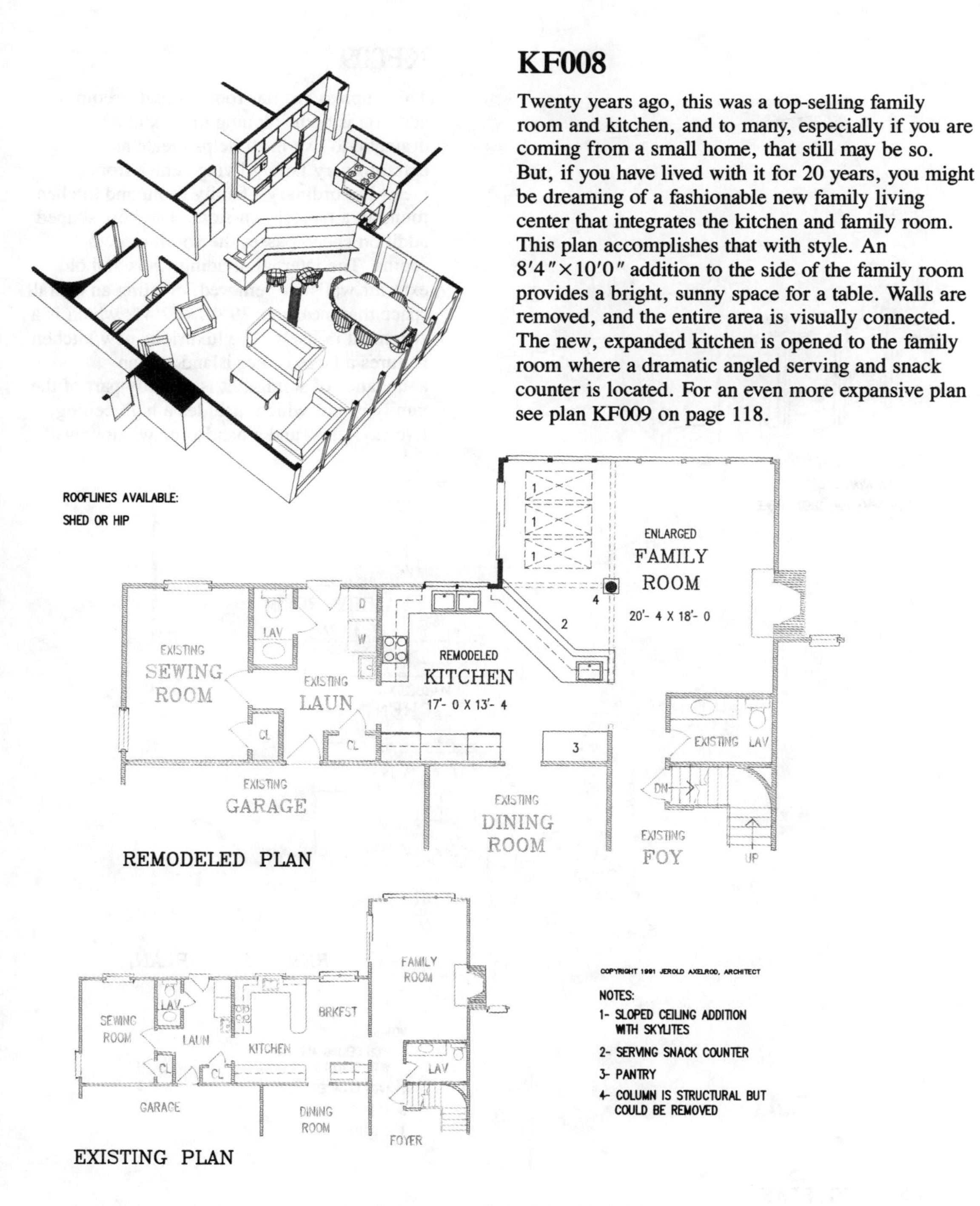

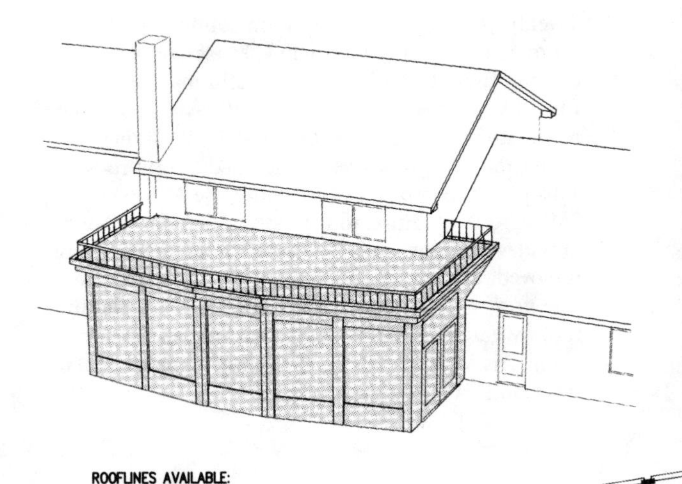

KF009

This impressive, flat-roofed family-room addition is both stunning to look at and dramatic to live in. It helps create an extraordinary family living center from a nice—but ordinary—family room and kitchen found in a typical two-story. The bow-shaped addition encompasses the width of both rooms. The internal dividing walls and old exterior walls are removed, creating an overall space that measures 29′4″×27′4″, which is a statement in itself. The luxurious new kitchen features a large center island and an abundance of storage. It is visually part of the family room, which includes a high ceiling, five skylights, and a handsome window wall.

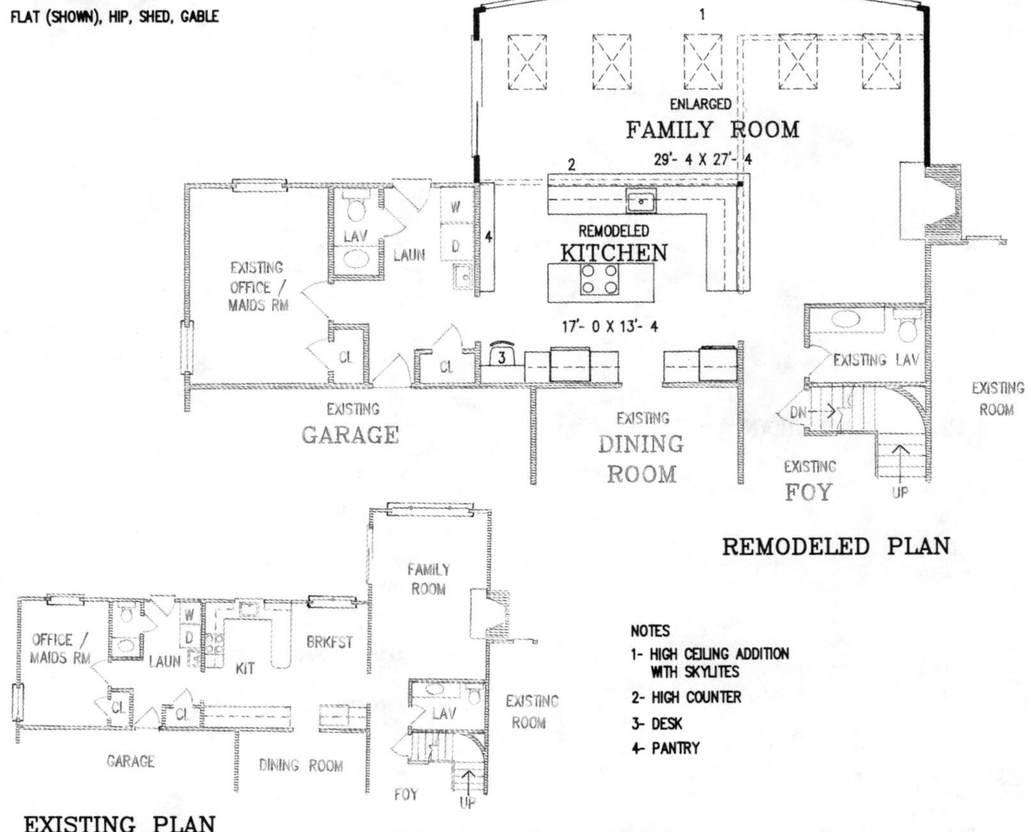

CONSTRUCTION BLUEPRINTS ARE AVAILABLE FOR THIS PLAN (AS WELL AS ALL OTHERS) AND INCLUDE STRUCTURAL DESIGN FOR REMOVING THE EXISTING REAR WALL

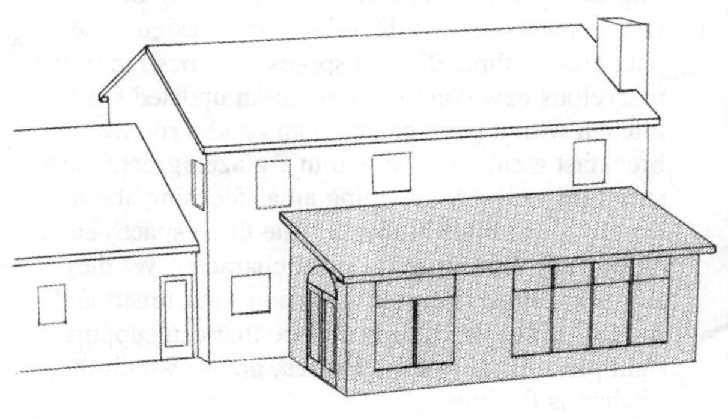

KF012

You know that the family room, kitchen, and breakfast room of your center-hall two-story are OK, but you now want something much larger and more in fashion. This architectural design accomplishes that. It provides an addition to the family room that virtually doubles it in size and also adds enough space for a bright new breakfast room—which, in turn, frees up enough area to create a fabulous new kitchen. The kitchen now boasts corner windows over a corner sink, a large center island, and a pantry wall combined with stylish angular shapes. The area of the addition visually flows to the newly opened interior space, creating a wonderful spacious feeling throughout.

ROOFLINES AVAILABLE:
ONLY AS SHOWN

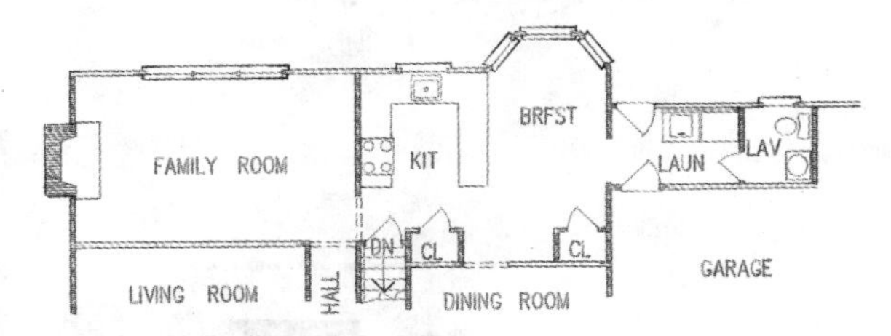

EXISTING FIRST FLOOR

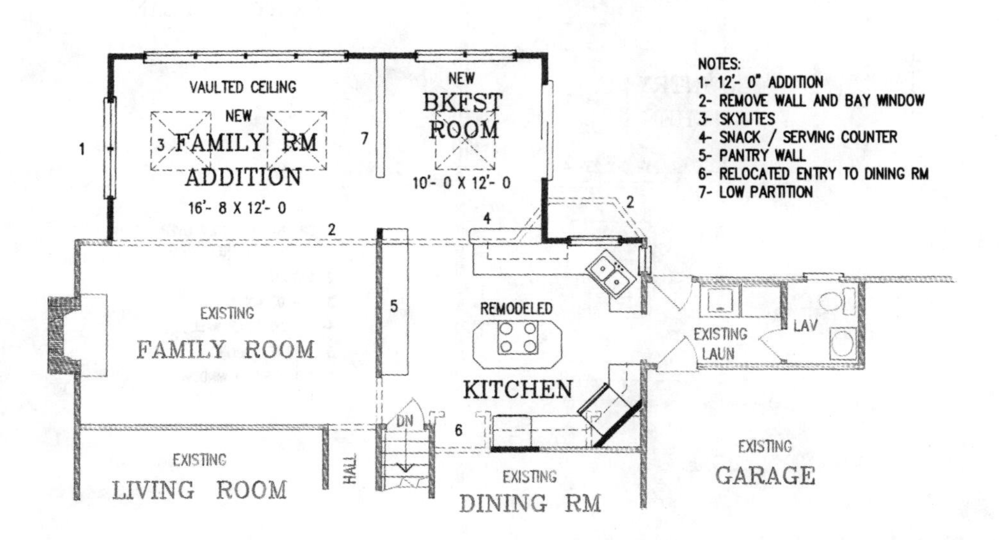

NOTES:
1- 12'- 0" ADDITION
2- REMOVE WALL AND BAY WINDOW
3- SKYLITES
4- SNACK / SERVING COUNTER
5- PANTRY WALL
6- RELOCATED ENTRY TO DINING RM
7- LOW PARTITION

REMODELED FIRST FLOOR

CONSTRUCTION BLUEPRINTS ARE AVAILABLE FOR THIS PLAN (AS WELL AS ALL OTHERS) AND INCLUDE A MATERIALS LIST AND INFORMATION ON HOW TO MAKE CHANGES TO PLANS

KF011

Expanding a dated kitchen into something other than just a larger one was the goal of this design. The plan creates three distinct spaces that mesh into one marvelous new country kitchen: an updated working kitchen with a large center island and a roomy breakfast area, situated within a glazed greenhouse structure, and a cozy sitting area, focusing about a fireplace and built-in media. The three spaces each enjoy their distinct and special character, yet they merge together to create a spacious and functional family living center. It is a place that can support multiple simultaneous activities, and in which the kitchen is the hub.

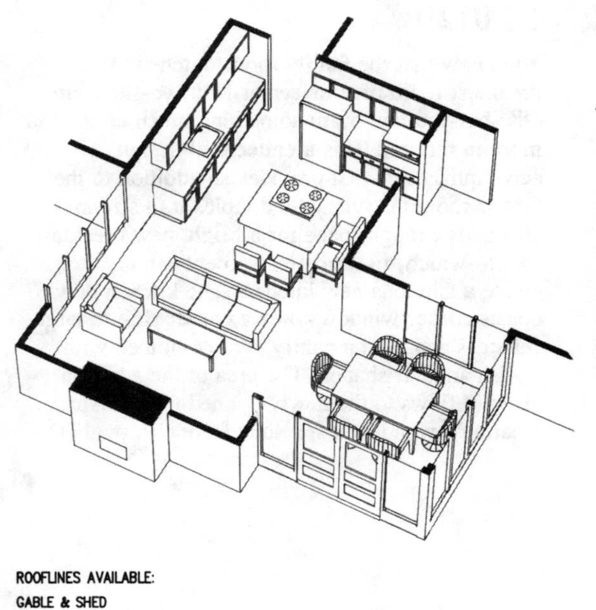

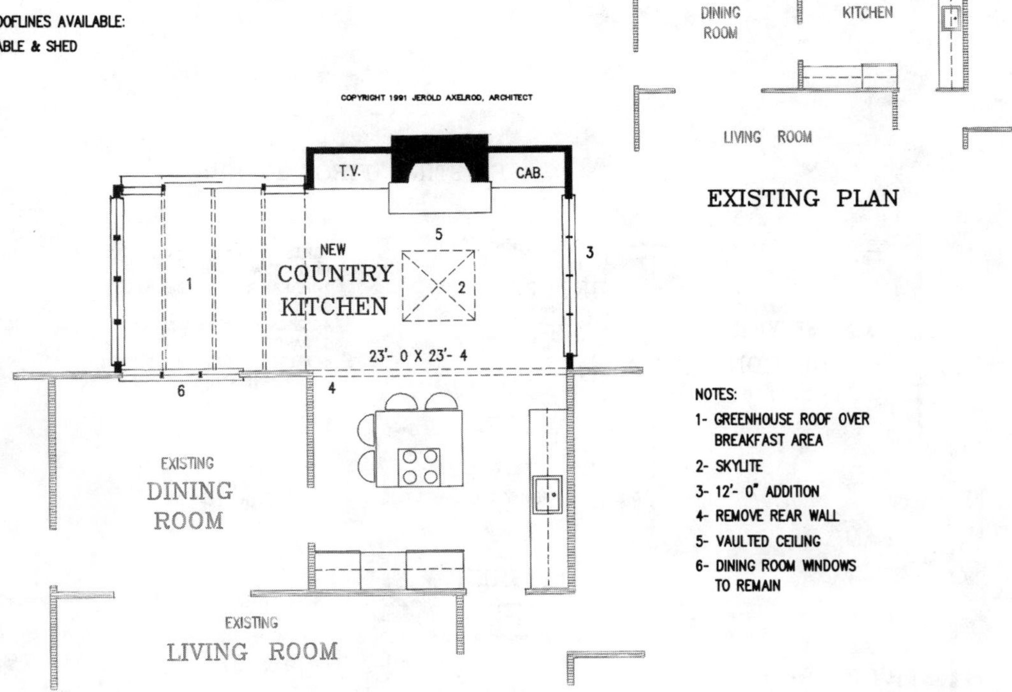

CONSTRUCTION BLUEPRINTS ARE AVAILABLE FOR THIS PLAN (AS WELL AS ALL OTHERS) AND INCLUDE FIREPLACE AND GREENHOUSE MANUFACTURER'S INFORMATION

KF010

A stylish new family room, a new laundry room, and a remodeled kitchen are the subjects of this design. The location is to the rear of the kitchen and dining room of a conventional L-shaped living/dining room. This is the type of plan found very frequently in split-level and one-story homes, and even many two-story homes. The new family room is designed to focus its seating about a built-in media center that wraps around a corner. The dining room window is relocated to the side to gain wall space for the family room. Finally, the kitchen is opened up and visually connected to the new family room, and the new laundry also serves as a mud room, providing access to the rear yard.

ROOFLINES AVAILABLE:
SHED & FLAT

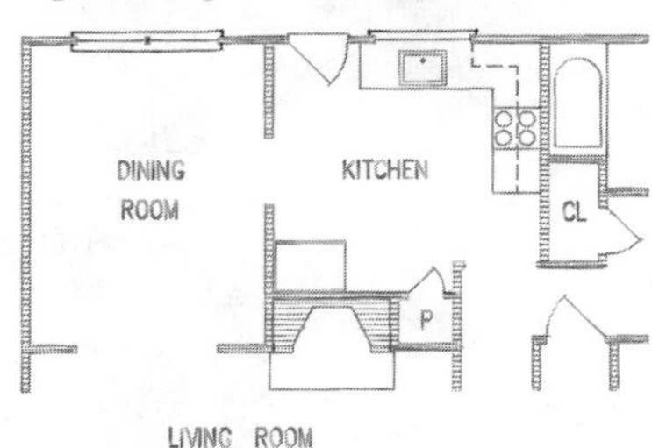

EXISTING PLAN

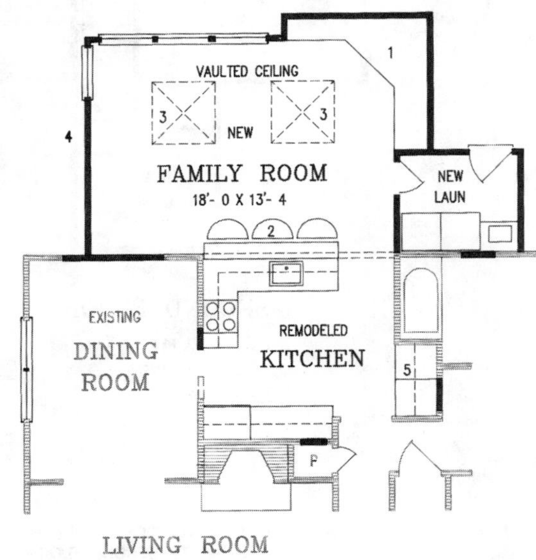

REMODELED PLAN

NOTES:
1- BUILT-IN MEDIA CENTER
2- SNACK COUNTER
3- SKYLITES
4- 14'- 8" ADDITION
5- WALL OVEN AND CABINET IN NEW RECESS

ROOFLINES AVAILABLE:
ONLY AS SHOWN

KBR01

Problem: You would love a new master suite, but yours is a small, one-story home with no room to expand out. And, if you could modernize your kitchen at the same time, that would be a bonus. This plan shows you how to do both. The solution, working within the existing gable rooflines, adds a fabulous new master suite upstairs, and remodels the existing kitchen to create a stylish new room with a small breakfast nook to boot. Access to the new partial second floor is over the existing basement stair. The new master suite includes a skylit bath, a walk-in closet, and a lovely balcony.

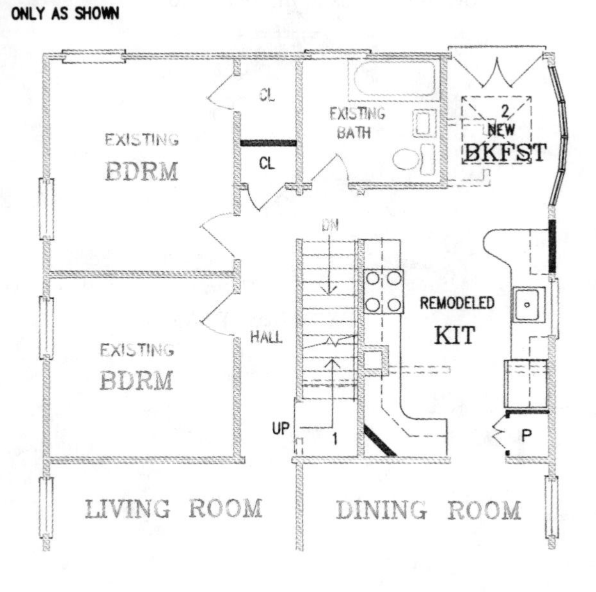

FIRST FLOOR
REMODELED

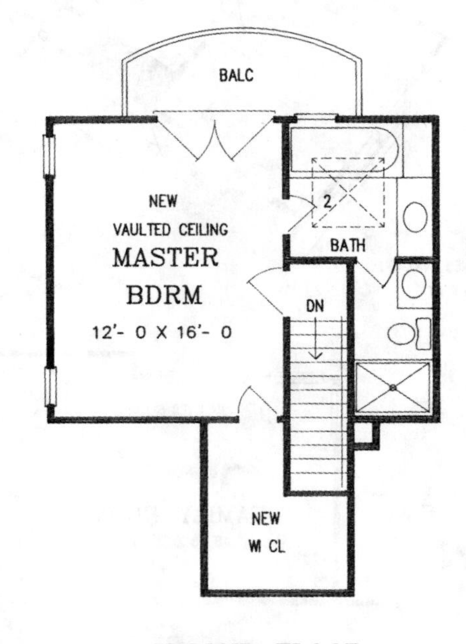

SECOND FLOOR
ALL NEW

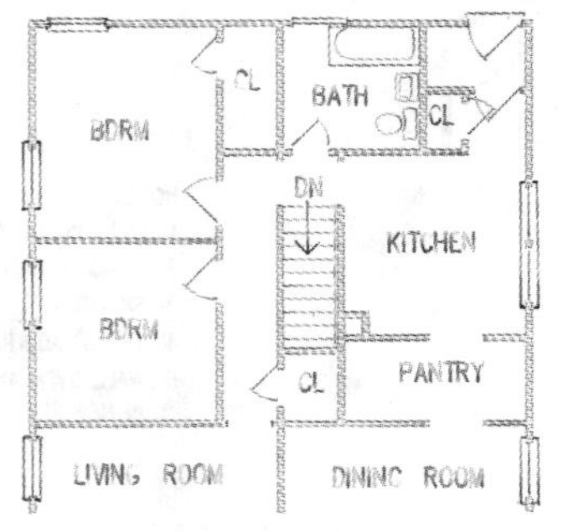

EXISTING PLAN

NOTES 1- REMOVE CLOSET AND SET STAIR TO SECOND FLOOR

2- SKYLITES

KEB01

A large new kitchen and breakfast room are the prime reasons to remodel this two-story home. An expanded master bath is another interest. But how will you enlarge your bath if the master bedroom is not close to the kitchen so you can take advantage of space over that contemplated expansion? Well, create a new first floor space behind the existing laundry and adjacent to the new breakfast room, so that you can expand the master bath. This area can serve as a sewing room, office, or even a modern exercise room. The new kitchen is a sensational, dramatic space, with skylights in the new extension and a fabulous center island. The expanded master bath is equally delightful.

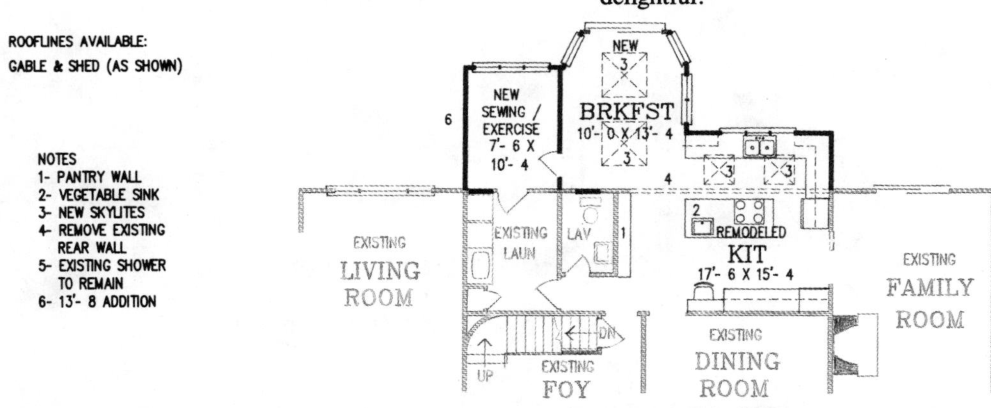

REMODELED FIRST FLOOR

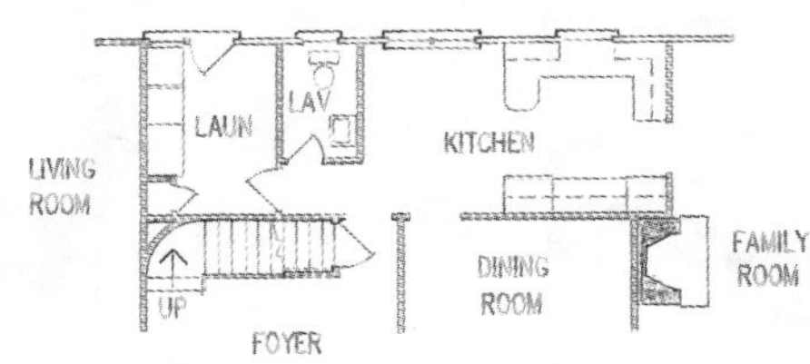

EXISTING FIRST FLOOR

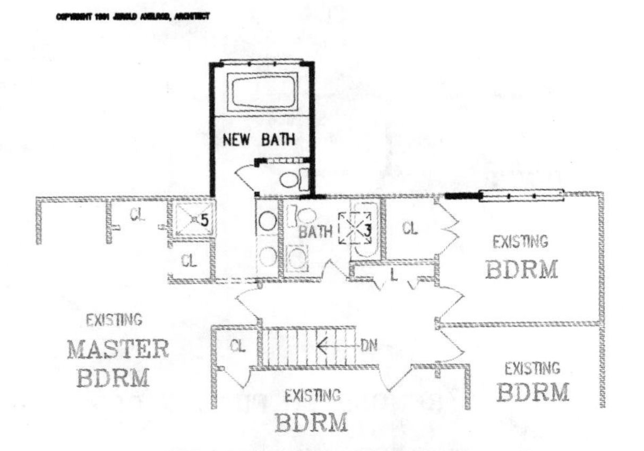

REMODELED SECOND FLOOR

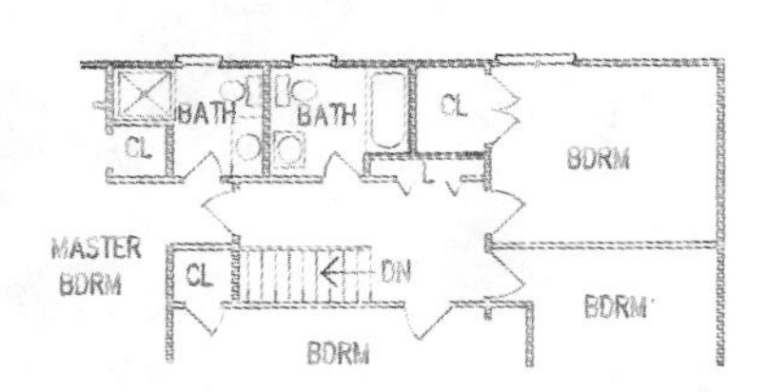

EXISTING SECOND FLOOR

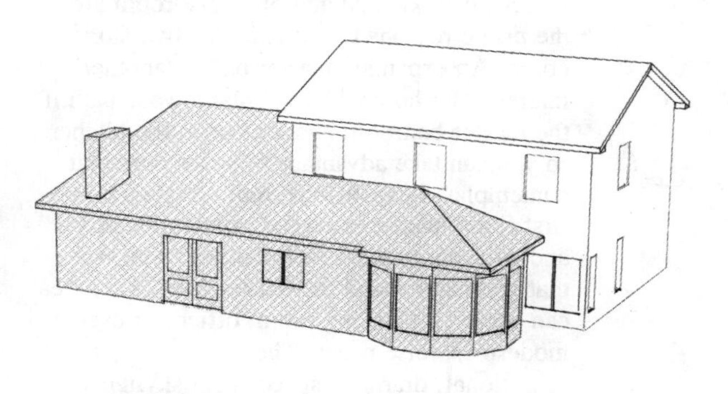

KFD03

A beautiful rotunda-shaped breakfast room with wrap-around windows, a bright new kitchen, and a spacious family room are the subjects of this design. It is planned as a one-story addition to the rear of a side-hall two-story home. Low walls separate the new family room and kitchen, which augments a visual integration across the entire new addition. Skylights, a vaulted ceiling, and a fireplace with space for built-ins at either side are further highlights of the family room. Some secondary benefits or bonuses include an enlargement of the dining room into the space formerly occupied by the kitchen, new large dining room windows, a new lavatory, and two extra closets.

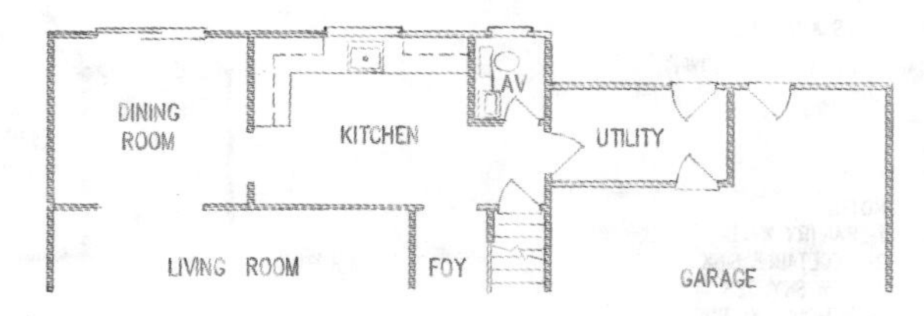

EXISTING FIRST FLOOR

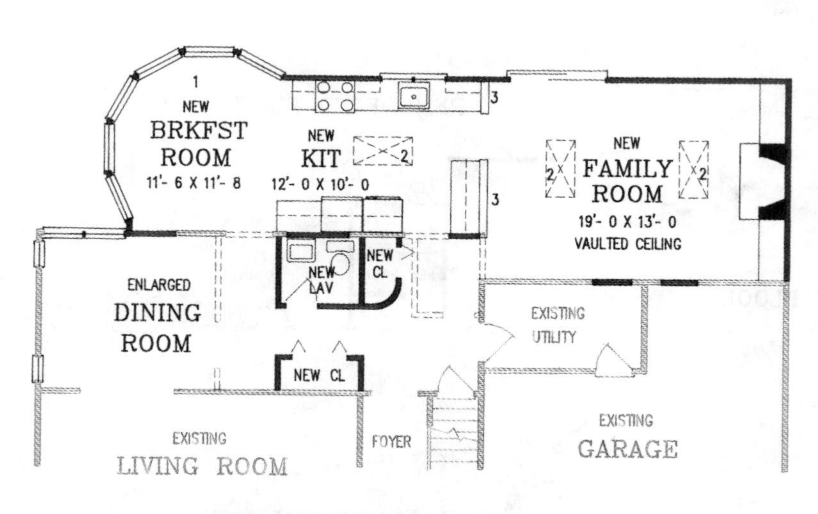

REMODELED FIRST FLOOR

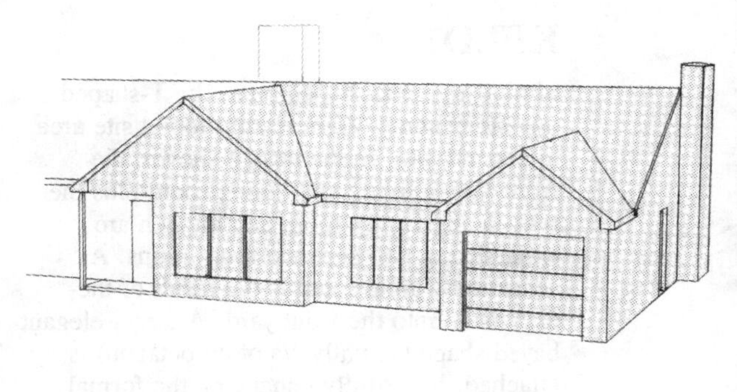

KFLD1

This 23′8″ side addition adds a fabulous new family room, a dining room, and a garage to a one- or one-and-one-half-story home. It is one of the more specific plans, designed to attach to the side of a kitchen and living room, with the plan showing how to remodel these rooms and the bath as well. If your existing basement stair is infringing into the kitchen area, consider moving it as indicated. The resulting plan gives you a covered porch, a defined entrance foyer, a nontraffic living room, and a great kitchen with a center island—plus the new family room, dining room, and garage. The addition also extends approximately nine feet forward of the front line of the home, but the construction drawings could be modified to move the new addition back if necessary.

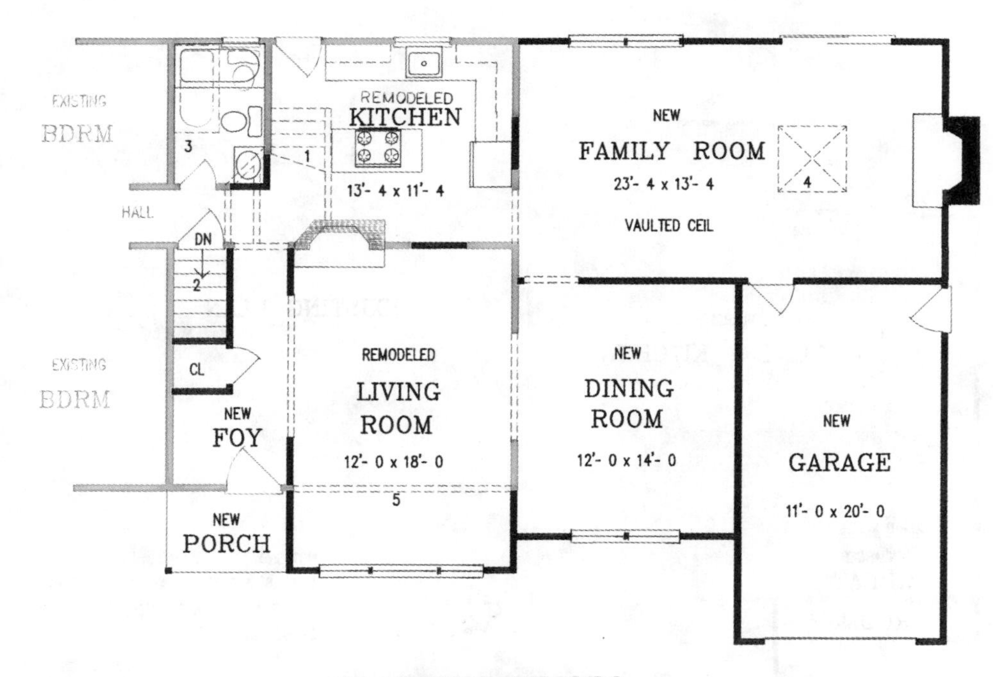

NOTES 1-REMOVE EXISTING BASEMENT STAIR
2-RELOCATED STAIR
3-REMODELED BATHROOM
4-SKYLITE
5-EXISTING FRONT WALL OF LIVING ROOM TO BE REMOVED

KFLD2

Similar to the L-shaped plan, the T-shaped one-story home offers readily found site area to expand into. In the design shown, the front-facing living and dining room plus the rear-facing family room and kitchen are expanded in two separate renovations. A sensational great room is the result of the expansion into the front yard. A large, elegant bayed shape (actually 5/8 of an octagon) is attached, beautifully enhancing the formal rooms. The addition to the rear yard is only a modest 6′8″, but it squares off the family room and kitchen, enabling it to now function as one large country kitchen. The kitchen itself benefits from a stylish new remodeling.

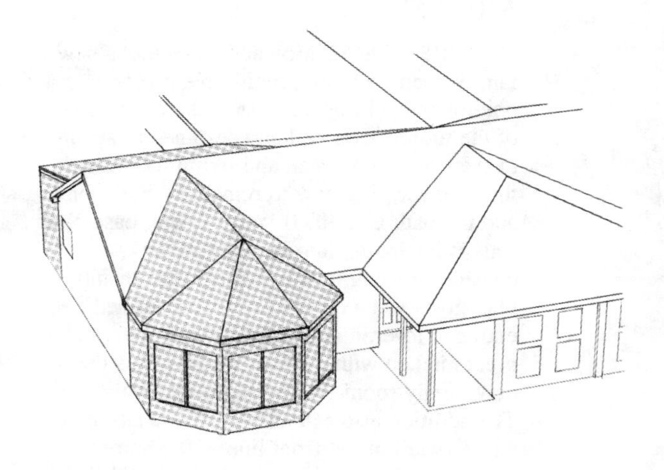

ROOFLINES AVAILABLE:

GABLE & HIP AS SHOWN

ENLARGED
COUNTRY KITCHEN
33'- 0 X 16'- 4
2
3
P
CL
FOY
21'- 0 X 22'- 4
ENLARGED
GREAT
ROOM
GARAGE
PORCH

REMODELED PLAN

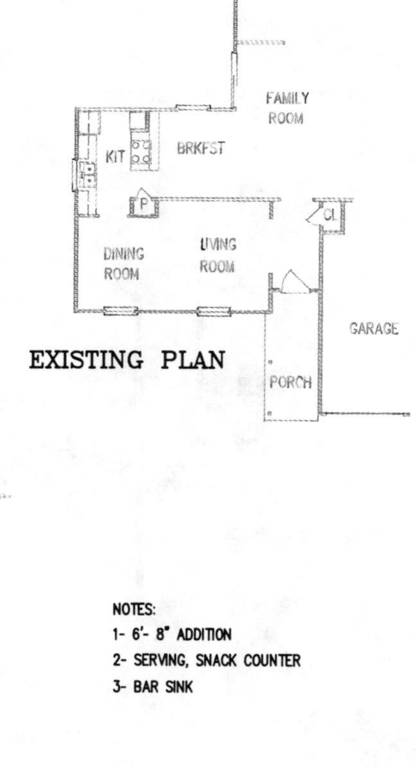

EXISTING PLAN

NOTES:
1- 6'- 8" ADDITION
2- SERVING, SNACK COUNTER
3- BAR SINK

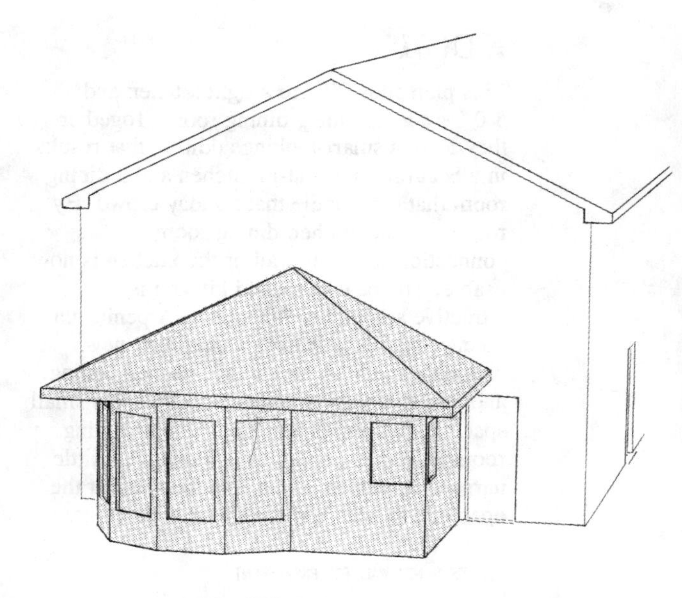

ROOFLINES AVAILABLE:

HIP (SHOWN), GABLE, SHED, FLAT

K0004

Existing side-facing kitchens pose expansion problems; going out to the rear usually means rearranging the entire house, which, depending on the quality of the home—and your budget—is a serious consideration. Going out the side is an excellent idea as long as you have the room. This picturesque, hip-roofed, 10′6″ wide by 17′6″ deep addition assumes you have the room. It results in a great kitchen and an equally beautiful and spacious rotunda-style breakfast room. Although the older part of the kitchen is still broken by the constraints of existing doorways, that part serves mainly for pantry, storage, and a desk. The new working part of the kitchen is a cheerful, bright, delightful space.

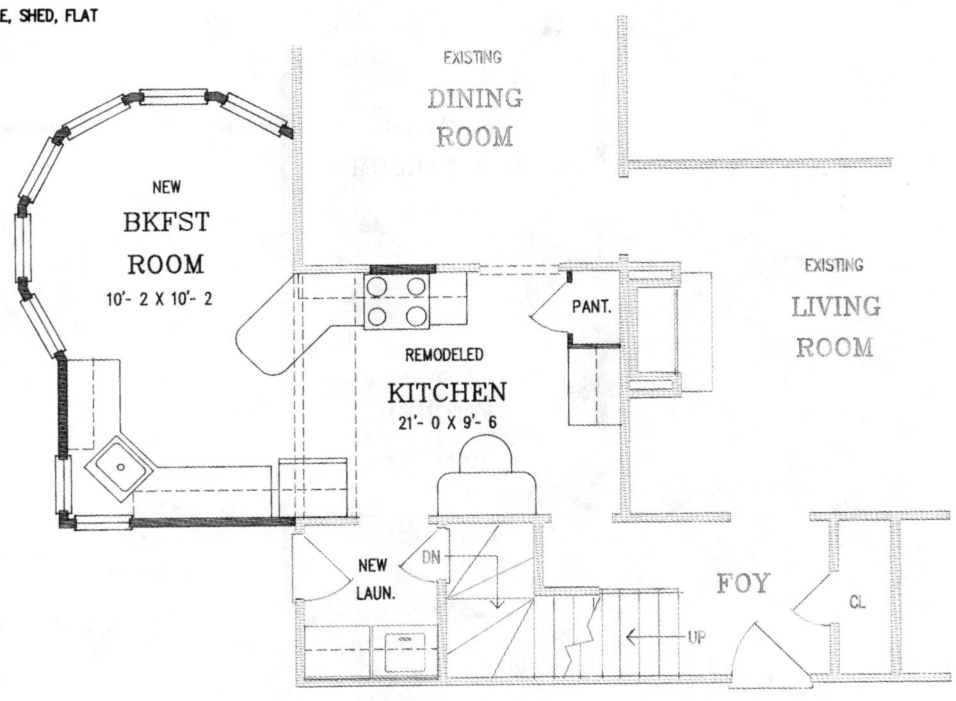

KD001

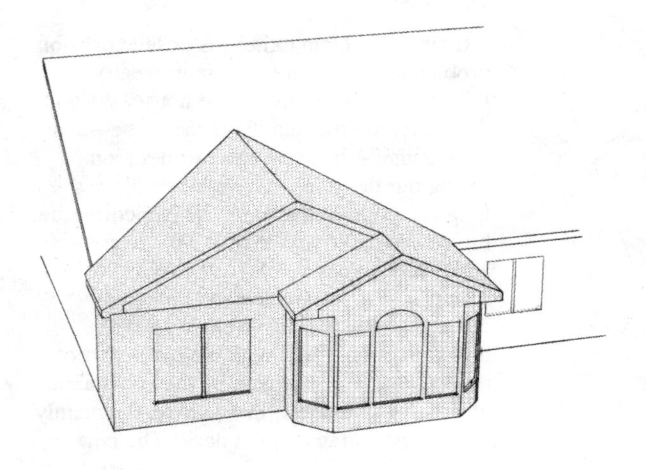

This plan adds 9′0″ to a tight kitchen and 5′0″ to the adjoining dining room. Together they form a smart-looking addition that results in a beautiful new eat-in kitchen and a dining room that can handle that holiday crowd. By relocating the kitchen/dining room connection, the rear wall of the kitchen is now usable. The new U-shaped kitchen is attractive and highly functional. A peninsula counter separates it from a charming new breakfast room. A bayed end, vaulted ceiling, and contemporary window wall give this small space an inviting character. The new dining room is long enough, but still may be a little narrow; therefore, add a five-foot wall at the open end to accommodate your breakfront.

ROOFLINES AVAILABLE:
GABLE (SHOWN), HIP, SHED, FLAT

NOTES 1- NEW WALL FOR BREAKFRONT
2- RELOCATE ENTRY TO DINING ROOM

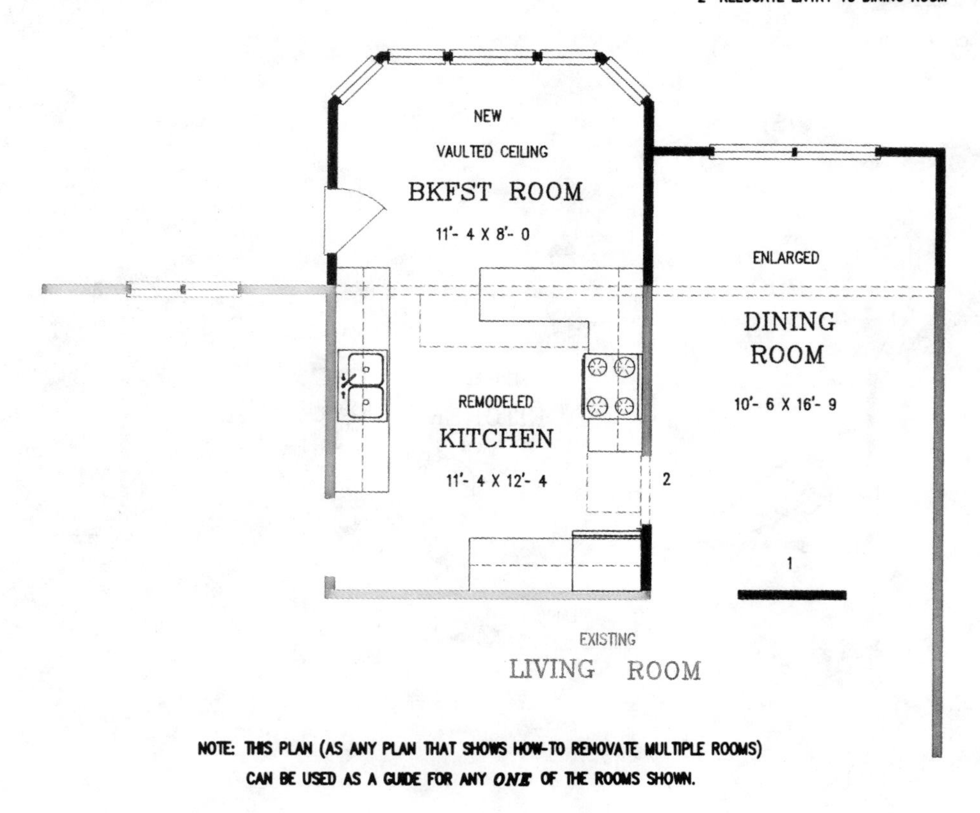

NOTE: THIS PLAN (AS ANY PLAN THAT SHOWS HOW-TO RENOVATE MULTIPLE ROOMS) CAN BE USED AS A GUIDE FOR ANY *ONE* OF THE ROOMS SHOWN.

KD002

Problem: an ample kitchen with no table space and a dining room that would be OK if it didn't have to double as a circulation route to the den and garage. Solution: a 10′0″ addition out the rear that solves both problems. The result also significantly helps improve the kitchen itself. The new U-shaped layout is perfection in function and attraction; the angled corner sink is located close to new windows, and the wide peninsula counter allows you to communicate with the folks at the breakfast table. Sliding doors, a skylight, and a vaulted ceiling enhance the new breakfast room. The new dining room is formal and separate from the new gallery that has been created to satisfy the circulation function.

ROOFLINES AVAILABLE:
GABLE (SHOWN), HIP, SHED, FLAT

NOTE: THIS PLAN (AS ANY PLAN THAT SHOWS HOW-TO RENOVATE MULTIPLE ROOMS) CAN BE USED AS A GUIDE FOR ANY *ONE* OF THE ROOMS SHOWN.

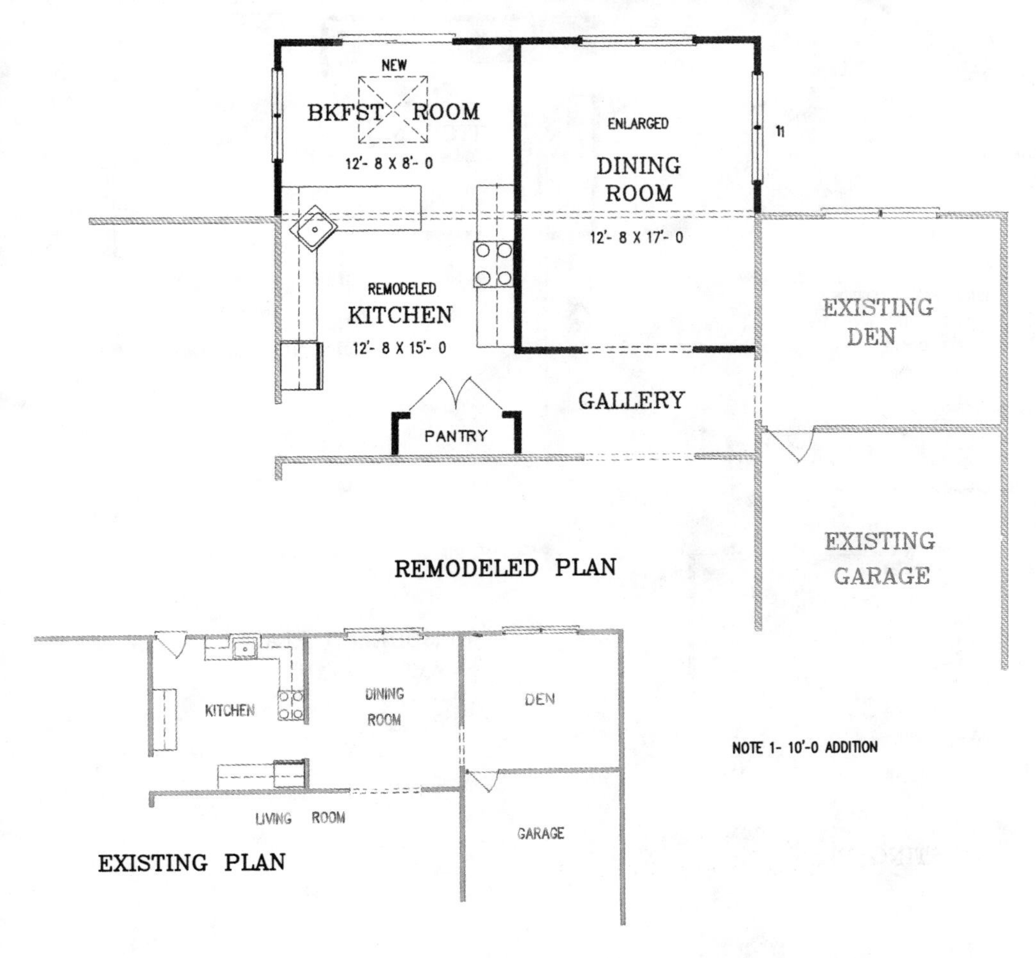

KD006

Given: a conventional L-shaped living/dining room with an adjoining old kitchen. Goals: a bright, new, cheerful kitchen and a larger dining room. This plan is for the person who would like some table space in the kitchen, but whose real preference is an expansive formal dining room. The interesting solution developed here expands the dining room into the former kitchen space, and, in doing so, creates an impressive, open-planned living/dining room. The new kitchen is created in a 12′8″×10′4″ rear addition. Cabinets are in the form of an "L," with a corner sink under corner windows and sliding doors lead to a new terrace behind the dining room. This is an easy addition to stage because the kitchen is completely new.

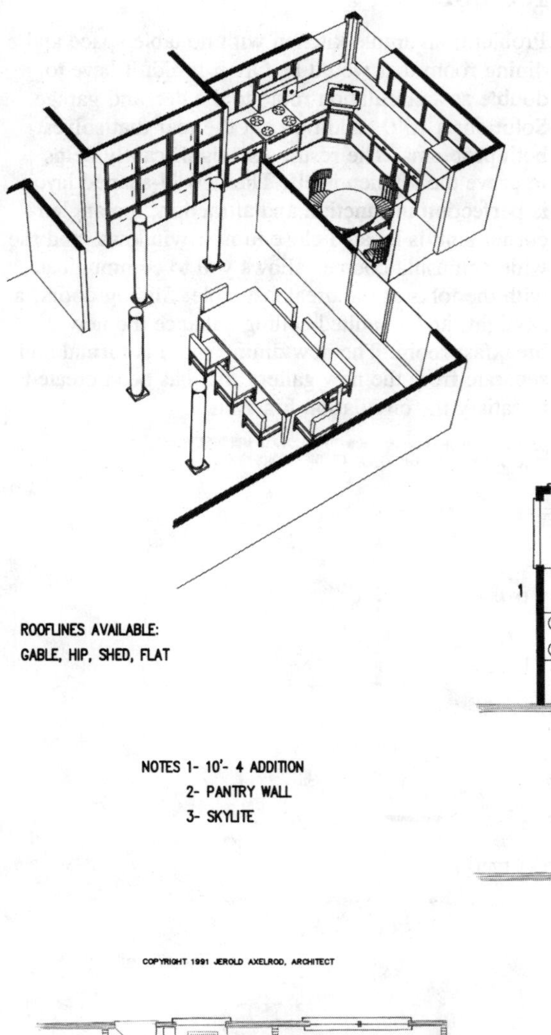

NEW
KITCHEN
3
12'- 0 X 10'- 0
NEW
TERRACE
1
2
ENLARGED
DINING ROOM
15'- 0 X 11'- 8
EXISTING
LIVING ROOM

REMODELED PLAN

KITCHEN
DINING
ROOM
LIVING ROOM

EXISTING PLAN

KF004

A really tight kitchen that is difficult to work with is enlarged and remodeled in this plan, and a delightful sunroom is tucked into the corner. Doorways to the hall, dining room, basement stair, and a pantry make the rear wall of this kitchen almost useless. The new kitchen is created in an L-shape and a serving/snack counter is designed along the circulation corridor, thereby making use of this area. Cozy table space is provided under the new large window. A beautiful new sun room or porch is proposed for the corner behind the dining room. The old windows in the dining room should remain and new windows should be added from the kitchen, facing into the porch and bringing borrowed light to both rooms.

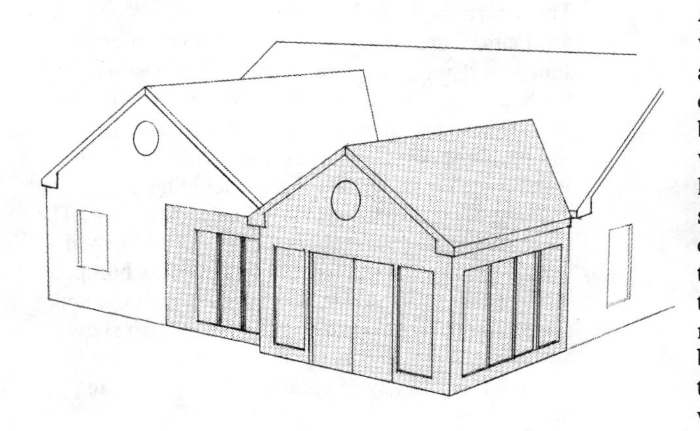

ROOFLINES AVAILABLE:
GABLE(SHOWN), HIP, SHED, FLAT

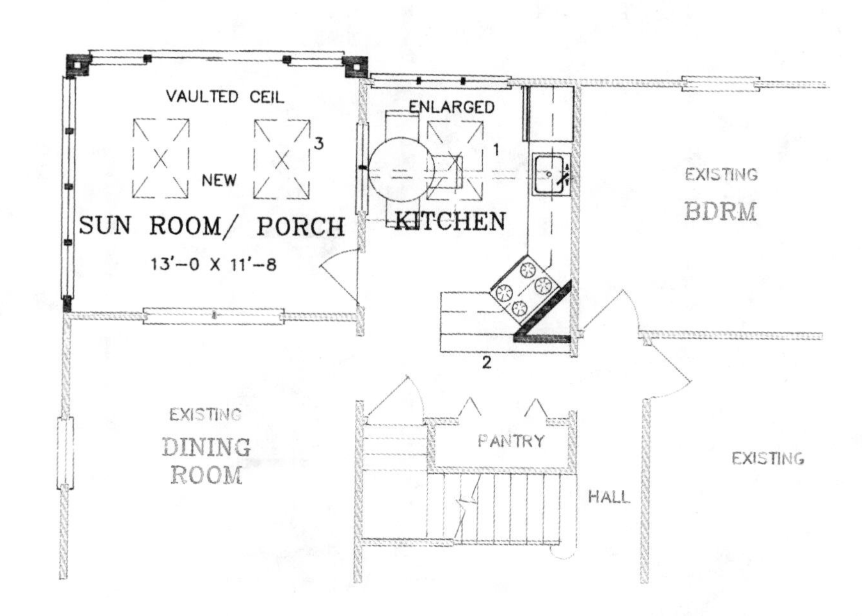

NOTES 1–REMOVE EXISTG PORCH
2–SERVING/SNACK COUNTER
3–SKYLIGHT (TYP)

KD004

This charming little 11′10″×8′6″ addition provides for a sunny breakfast room, or a sun porch or sunroom if you prefer, off an adjacent existing kitchen. If you have a small corner recess adjacent to your kitchen, the plan would neatly attach as shown. Otherwise, it would attach flush to the rear. Windows abound on two sides, including a decorative curved-top window on the long side. The room also includes a vaulted ceiling. The kitchen itself is shown remodeled; the rear wall is bumped out two feet to allow for a countertop with the sink under a new large window. The layout also shows an interesting angled range at an inside corner. Consider this in any kitchen you are considering remodeling.

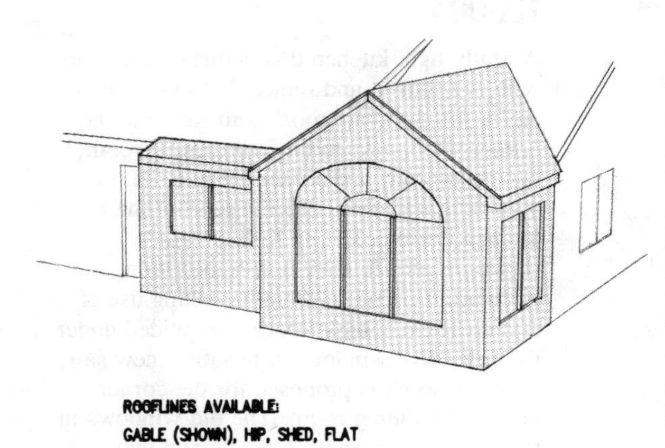

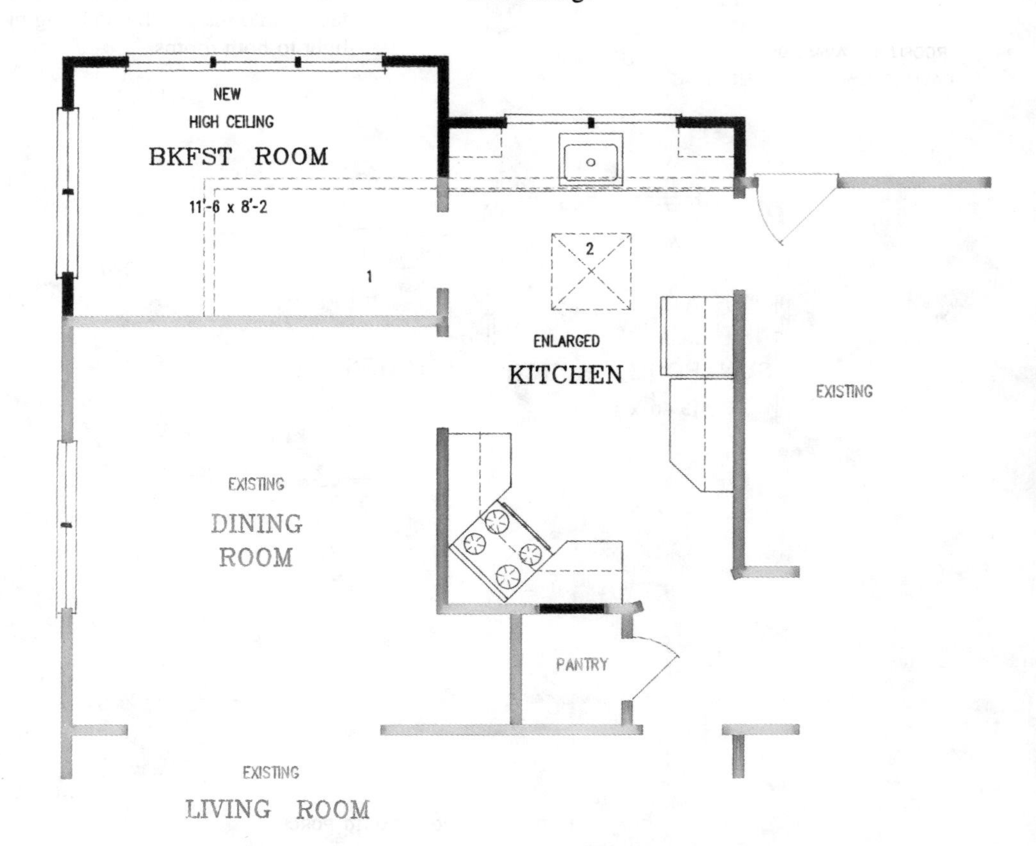

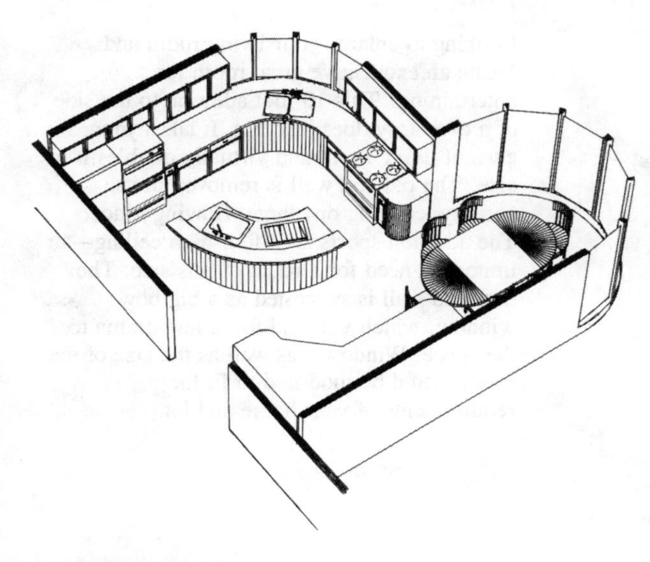

K0011

This might be dubbed your dream kitchen—and rightfully so. It is a very expansive (almost 22 ft. × 22 ft.) addition that creates a dazzling new kitchen and breakfast room. The design theme is curves. The kitchen rounds a corner in a quarter circle, includes a curved island and snack counter, and culminates in a semicircular wrap-around window wall in the breakfast area. An area for built-ins, including space for a pantry and television, is included in the design. This is one of the plans that is less specific as to where you attach it to your home; that is for you to resolve, including the question of which rooflines to plan. Plan F0014 on page 112 is a family room addition of similar architectural design.

ROOFLINES AVAILABLE:
GABLE, HIP, SHED, FLAT

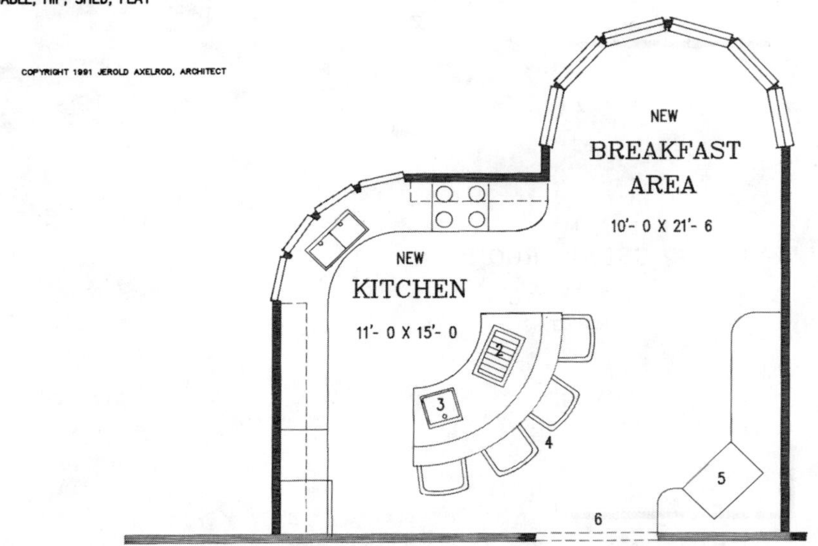

EXISTING ROOM

NOTES:
1- 21'- 11" DEEP ADDITION
2- BARBECUE
3- VEGETABLE SINK
4- SNACK COUNTER
5- PANTRY AND BUILT-IN TV UNIT
6- OPENING TO HOUSE VARIES

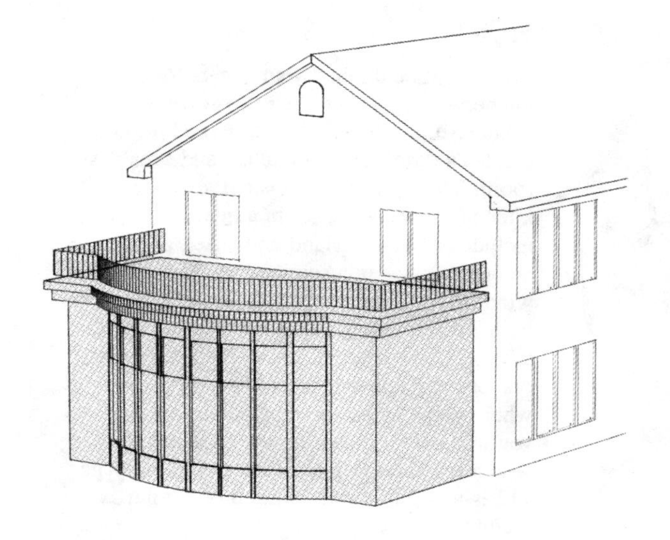

L0001

Looking to enlarge your living room and create an exquisitive great room for entertaining? This 10-foot addition to the side of a two-story does just that. It takes your current living room and virtually doubles it in size. The old end wall is removed in its entirety, creating one large, flowing space. The addition sports a 12-foot-high ceiling—an important need for a room of this size. The new end wall is suggested as a big bow window, which will add focus and drama to the space. Windows, as well as the size of the room, could be modified to fit the requirements of your home and lot.

ROOFLINES AVAILABLE:
FLAT (AS SHOWN)

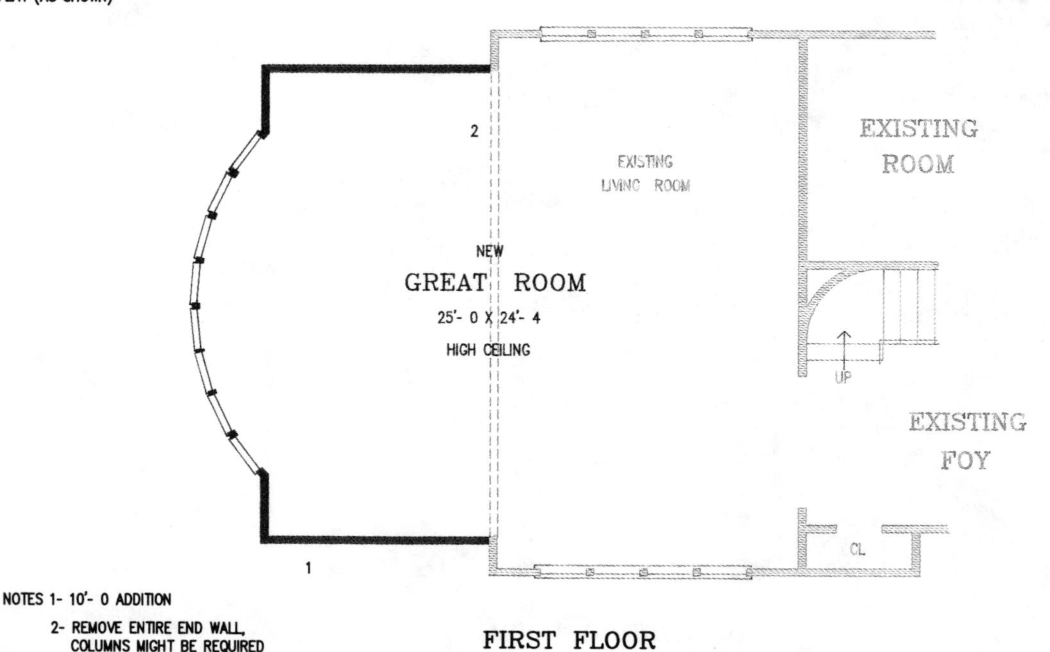

NOTES 1- 10'- 0 ADDITION
2- REMOVE ENTIRE END WALL, COLUMNS MIGHT BE REQUIRED

FIRST FLOOR

CONSTRUCTION BLUEPRINTS ARE AVAILABLE FOR THIS PLAN (AS WELL AS ALL OTHERS) AND INCLUDE STRUCTURAL DESIGN FOR REMOVING END WALL AND A MATERIAL LIST

L0002

The existing L-shaped plan as pictured usually allows expansion into the front yard. What a wonderful way to expand the entertainment wing and give your home a new striking appearance, too! A beautiful reverse-gable addition adds a generous amount of footage to the existing living room. When the old wall is removed, the area of the new space is such that it could now function as a multi-use great room. The main entrance to the home is also significantly enhanced in this design with the addition of another reverse-gabled space. Both elements include half-round windows and vaulted ceilings, and together they form a highly attractive new front facade of architectural merit.

ROOFLINES AVAILABLE:

ONLY AS SHOWN

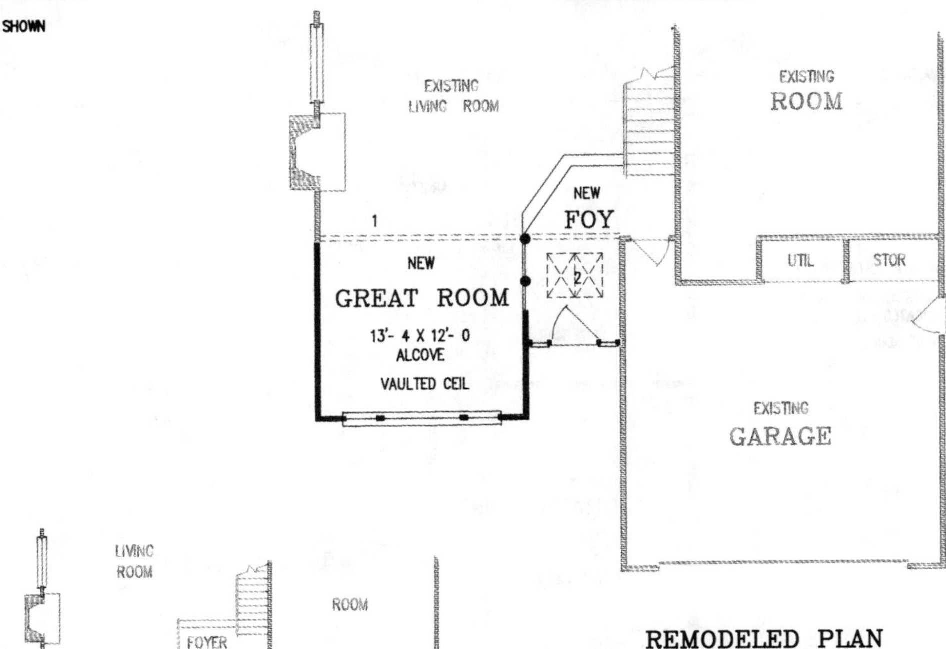

LIVING ROOM

FOYER

ROOM

UTIL STOR

GARAGE

EXISTING PLAN

REMODELED PLAN

NOTES: 1- REMOVE FRONT WALL
2- SKYLITES IN VAULTED CEILING

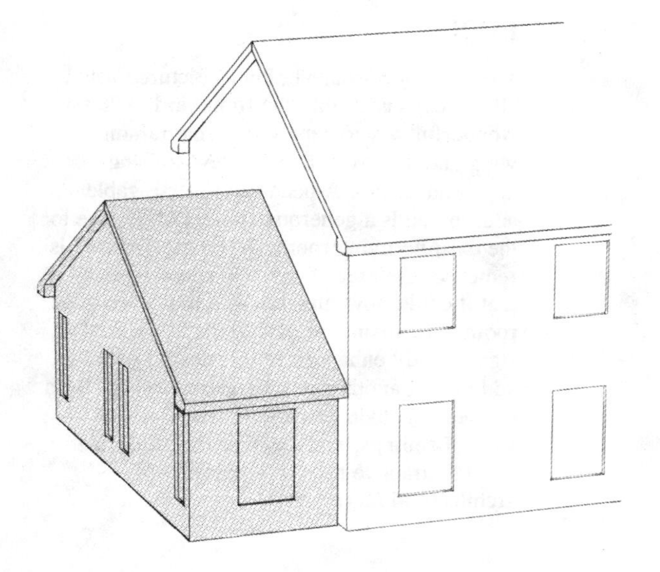

LD001

Given: the side of a two-story containing an L-shaped living and dining room. Program requirements: expand the dining room and add an office or library. This 8′0″×24′4″ addition helps achieve both. The new office borrows two feet from the seldom-used living room, creating a 10-foot office/library. If desired, an outside door could be installed to this office, and the double doors to the living room could be closed. The newly expanded dining room is an impressive size, certainly capable of handling the holiday dinner. A word of caution: the eight-foot addition is OK if the two-story section is no more than 27−28 feet in length. If it is longer, the eight-foot section is too small.

ROOFLINES AVAILABLE:

GABLE (SHOWN), HIP, SHED, FLAT

NOTES 1- REMOVE ENTIRE END WALL
2- OUTSIDE DOOR COULD BE LOCATED HERE
3- 8'-0" ADDITION

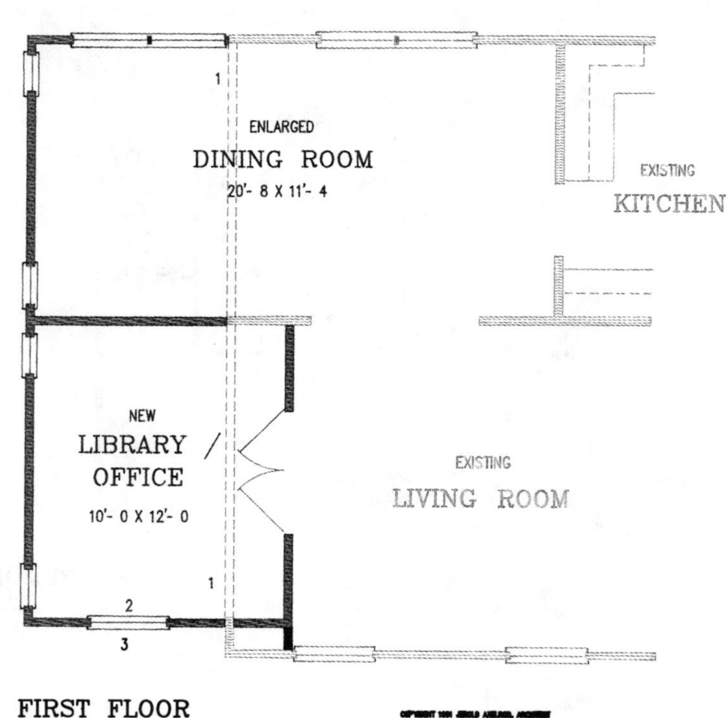

FIRST FLOOR

CONSTRUCTION BLUEPRINTS ARE AVAILABLE FOR THIS PLAN (AS WELL AS ALL OTHERS) AND INCLUDE A DETAILED LIST OF BUILDING MATERIALS

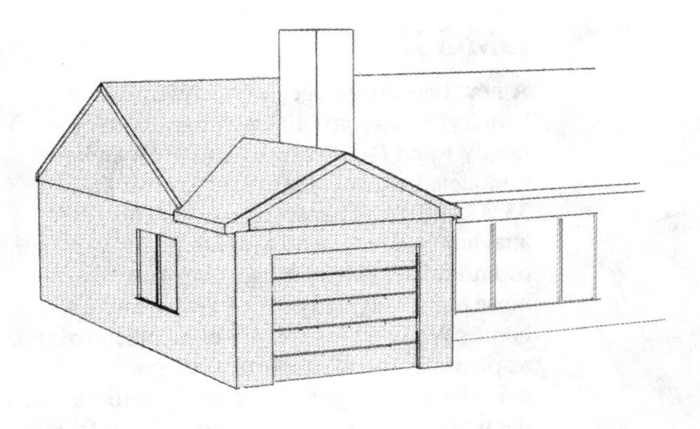

DMG01

An attached garage, laundry/mud room, and expanded dining room are the subjects of this trim 12′4″×32′4″ addition. If your lot permits, the garage could be further expanded to accommodate two cars. The approximately six-foot addition to the dining room is a welcome by-product, since all too often an L-shaped dining room is too tight for the few occasions when it is really needed. The extra space also makes it a little more comfortable for traffic flow to the mud room and garage. The laundry/mud room provides space for a closet plus counter space for either a sink or for folding and ironing. If just a dining room addition is all you are looking for see plan D0001 on page 224.

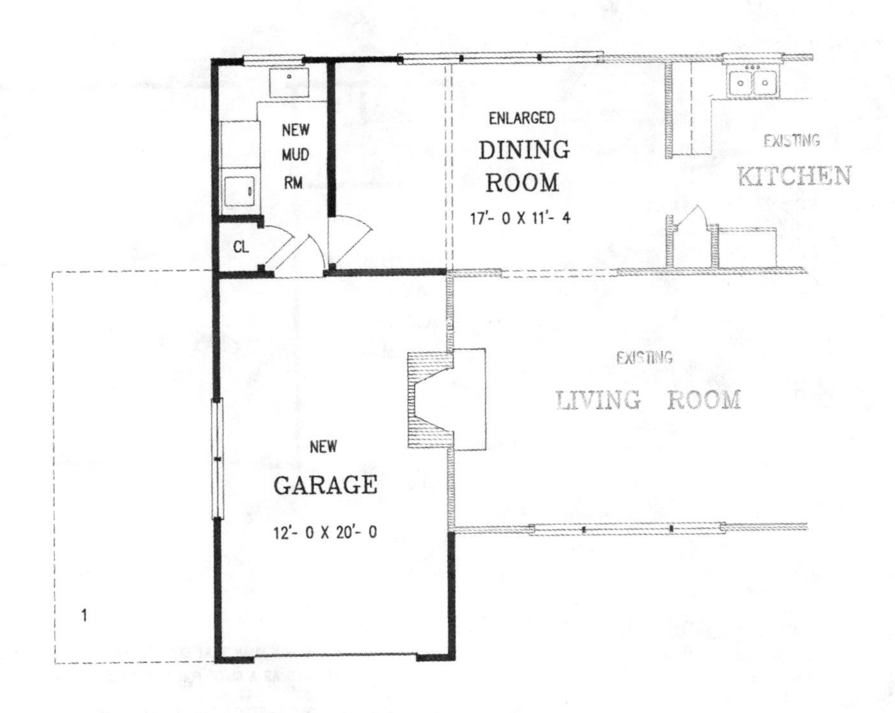

NOTE 1- ALTERNATE 2 CAR GARAGE

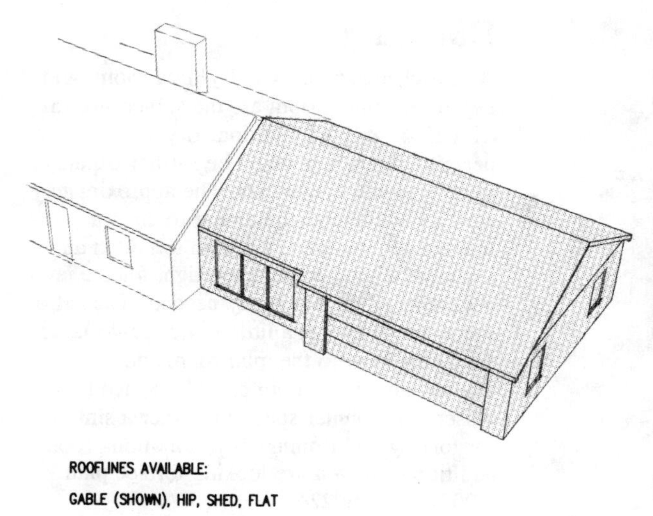

DMG02

A new two-car garage, a convenient laundry/mud room, a new dining room or family room (your choice as to use), and a remodeled kitchen are the subjects of this 33′8″×22′8″ one-story addition. The attachment shown is to the living room/kitchen side of a one-story home; however, it is equally suited to any multistory home. While you might prefer as much space as possible, the all-important design consideration suggests varying the setbacks for the dining (or family) room and garage from the original line of the house. Openings in the old end wall allow visual integration between the living room and the new dining or family room and direct circulation from the kitchen to the new laundry room and garage.

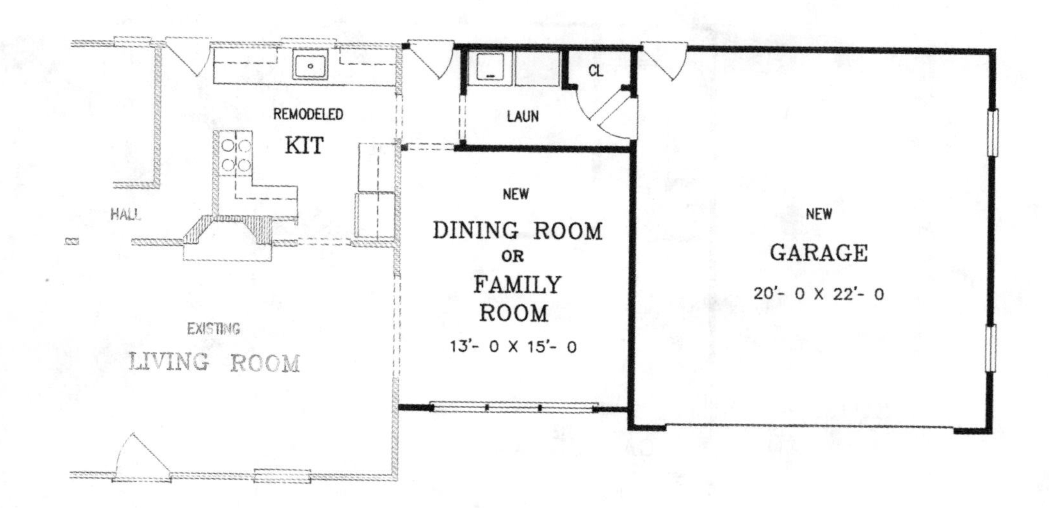

NOTE: THIS PLAN (AS ANY PLAN THAT SHOWS HOW-TO RENOVATE MULTIPLE ROOMS) CAN BE USED AS A GUIDE FOR ANY *ONE* OF THE ROOMS SHOWN.

NOTE: CONSTRUCTION BLUEPRINTS CAN BE READILY MODIFIED TO ACCOMMODATE YOUR SPECIFIC HOUSE OR ROOM SIZES.

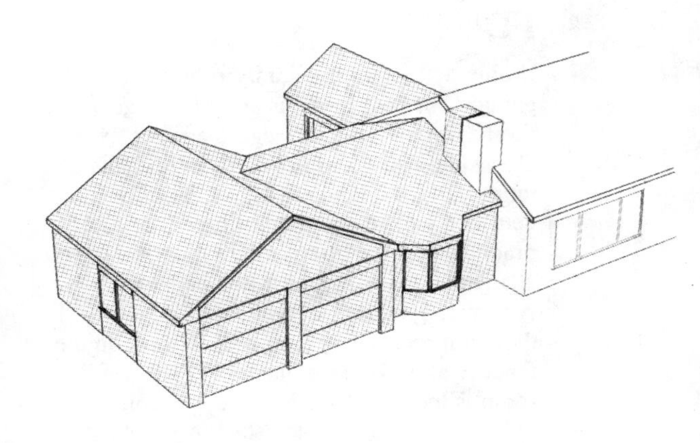

FDG01

A large front-to-rear family room, attached two-car garage, and enlarged dining room are attractively combined in this expansive addition to the side of an L-shaped living and dining room. The addition is very flexible and is suited for any style or type of home, including a one- or two-story or a split-level. This addition shows the importance of programming. If all you intended to add was the family room and garage, you might not have considered that in order to provide an uncramped connection between the kitchen and family room, the dining room would suffer badly. Thus the rear extension for the dining room. The plan also shows a new fireplace attaching to the rear of the existing living room fireplace.

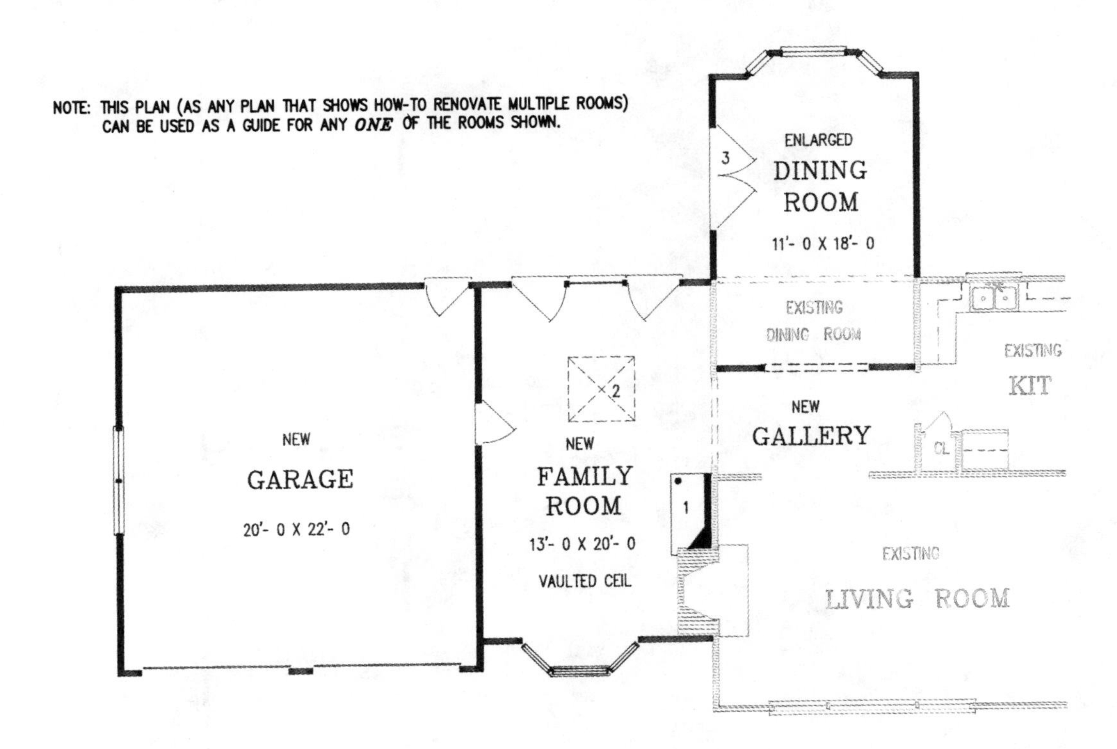

FDG02

Problem: You want a family room and a garage, but your home is an old cottage with a lovely front porch; although the space exists alongside, can you do this gracefully? It's a tough call. The better solution might be a detached garage and a rear family room, but practical considerations might dictate that you attach them. If so, follow this plan carefully. A reverse gable is the best solution, even though it requires special roof flashing in the form of a *cricket* (see glossary). The family room is located at grade since it is connected at the landing of the side stair. A dramatic new balconied breakfast room is created, which visually connects the new family room to the remodeled kitchen.

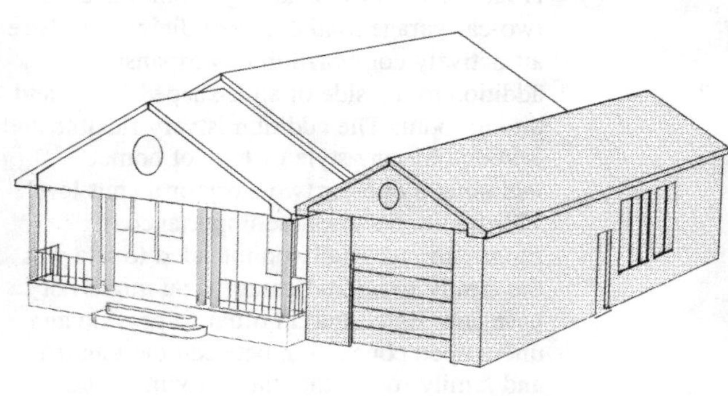

ROOFLINES AVAILABLE:
GABLE (SHOWN), SHED, FLAT

NOTE 1- FAMILY ROOM IS LEVEL WITH STAIR PLATFORM AT OLD SIDE DOOR

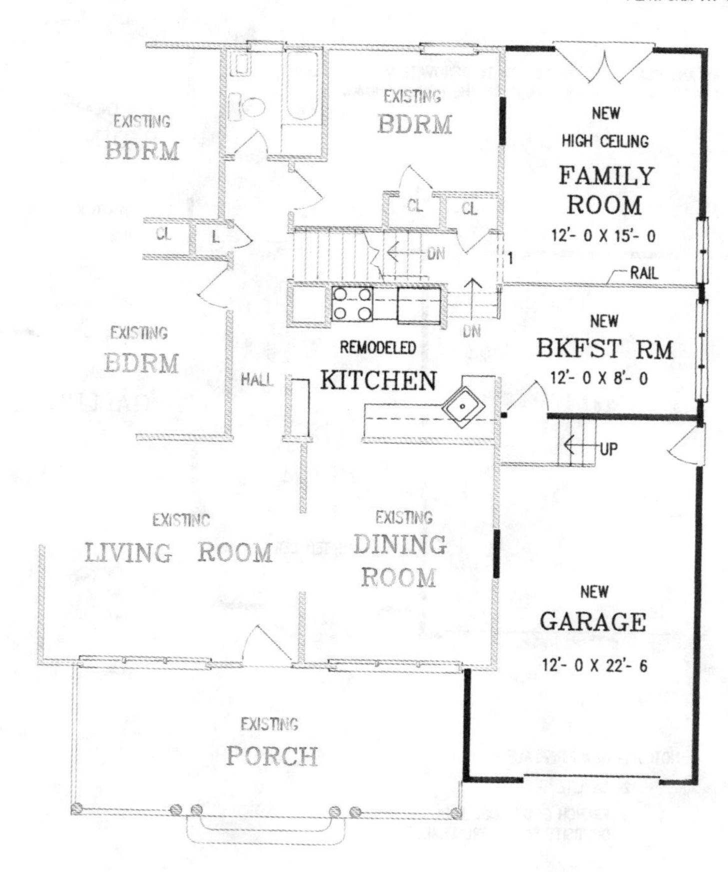

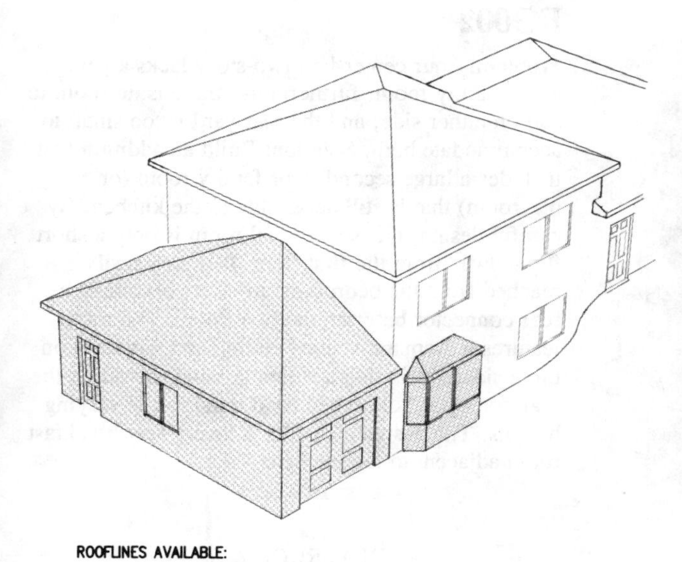

ROOFLINES AVAILABLE:

HIP (SHOWN), GABLE, SHED, FLAT

FG001

One of the more obvious ways to enlarge a family room in a split-level is to break through into the adjoining garage. However, if your lot permits, a new garage (one- or two-car) can also be added alongside. Such an addition usually enhances the lines of the typical split-level by providing an aesthetic balance to the one-story living room wing at the opposite side. This plan shows you how to add the garage and expand the family room. The entire wall can easily be removed between the family room and the current garage, as this is usually a nonbearing wall. New doors to the rear yard are placed in the rear wall, and a large bay window is designed in place of the old garage door.

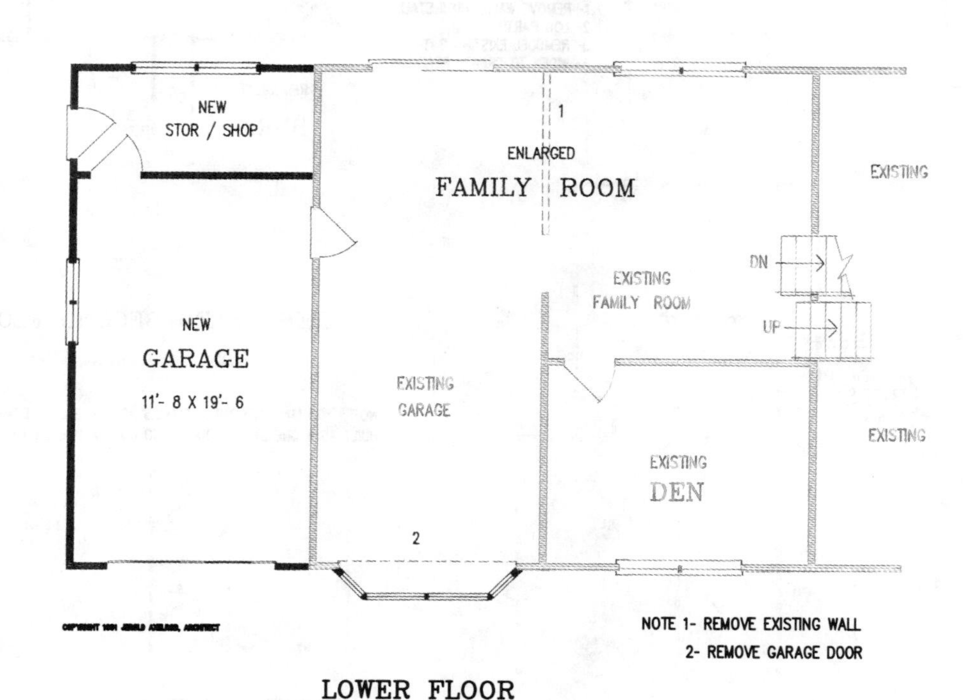

LOWER FLOOR

FG002

Problem: Your center-hall two-story lacks a garage and a family room; furthermore, there is no room to add on either side, and the rear yard is too small to accommodate both. Solution: Build an addition that includes a large second floor family room (or playroom) that is still accessible to the kitchen. By careful design, this sensational room is only a short flight away from the first floor. It is also easily reached from the bedrooms, in effect becoming a rear connector between the two floors. The room features a dramatic vaulted ceiling and windows on three sides, including a stunning window wall at the rear comprised of arched head windows of varying heights. The design also adds a lovely new breakfast room adjacent to the kitchen.

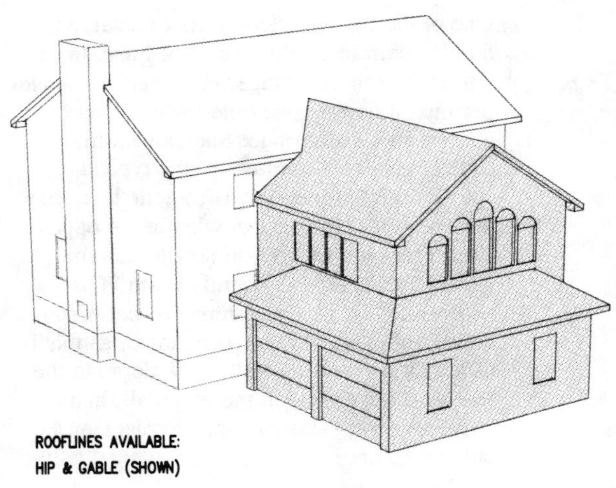

ROOFLINES AVAILABLE:
HIP & GABLE (SHOWN)

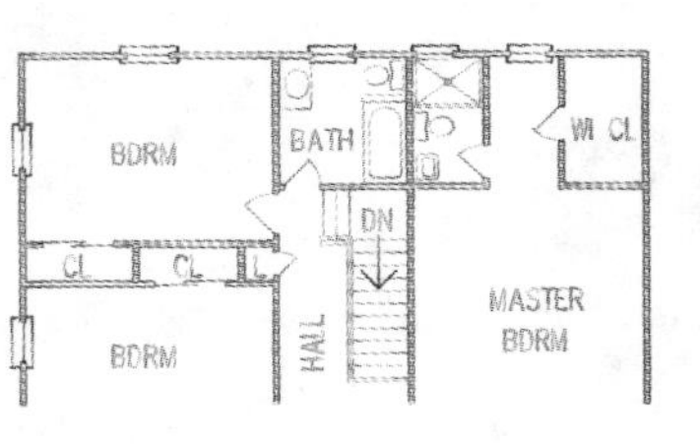

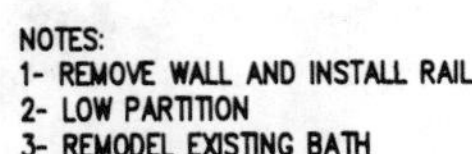

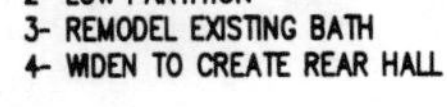

NEW
PLAYROOM/
STUDIO/APT
17'- 0 X 18'- 0
VAULTED CEILING
DN. TO KITCHEN
REMODELED
BDRM
DN
4
3
WI CL
DRSNG
DN
CL
CL
L
EXISTING
MASTER
BDRM
EXISTING
BDRM
HALL

REMODELED SECOND FLOOR

NOTE: YOU MAY VIEW THESE FLOORS AS SEPARABLE: EITHER MAY BE BUILT AS A ONE STORY ADDITION TO BOTH A ONE OR TWO STORY HOME.

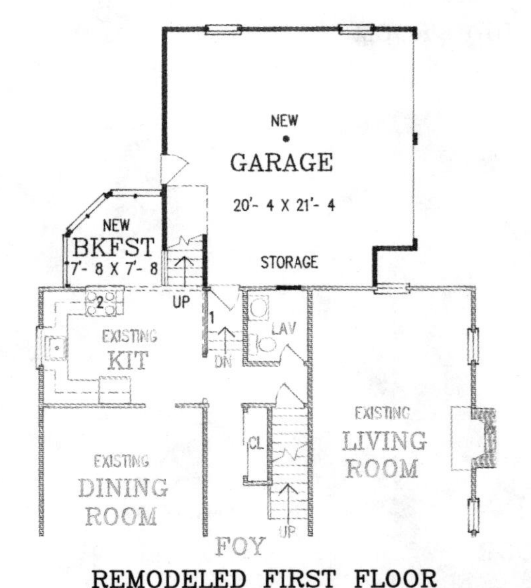

REMODELED FIRST FLOOR

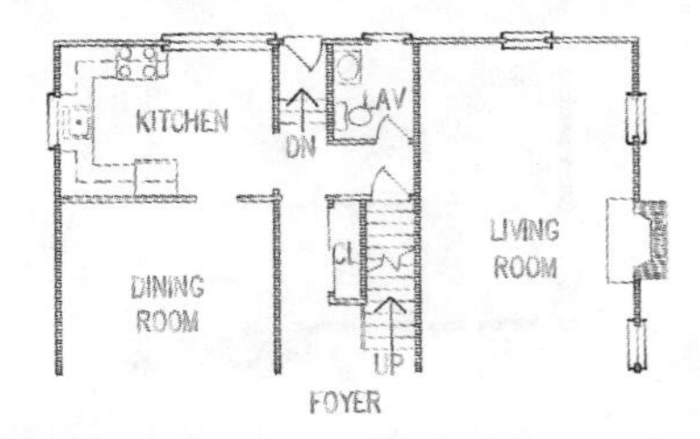

EXISTING FIRST FLOOR

CONSTRUCTION BLUEPRINTS ARE AVAILABLE FOR THIS PLAN (AS WELL AS ALL OTHERS) AND INCLUDE A MATERIALS LIST

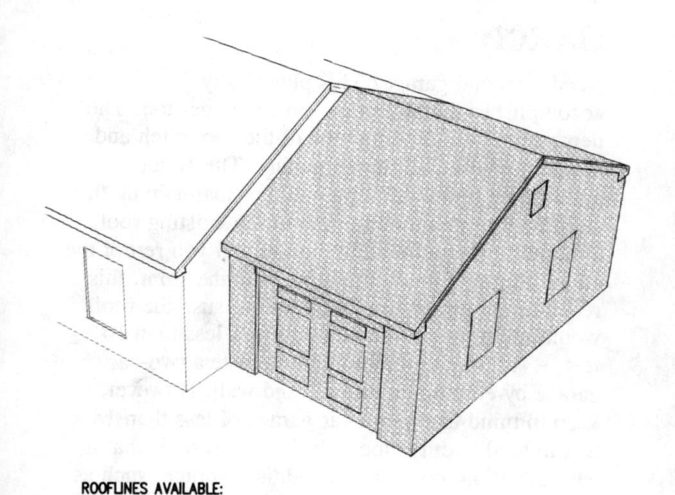
ROOFLINES AVAILABLE:
GABLE (SHOWN), HIP, SHED, FLAT

GAR02

A simple one- or two-car attached garage is the only subject of this plan. The garage shown can be attached to the side of any house type (one-story, two-story, split-level, etc.). A mirror image puts it on the opposite side. As simple as it seems, it is easy to botch this one up. The setback from the front wall is an important design consideration. The elements of a garage door do not mirror any other part of the front elevation and, as such, deserve a separate definition. Also, construction benefits of the setback include the elimination of the need to perfectly match roofing and siding, and a greater flexibility in establishing the level of the garage floor. On a one-story design, as shown, if the rear is set flush to the existing home, the roofs do have to perfectly match unless you make the rear wall lower.

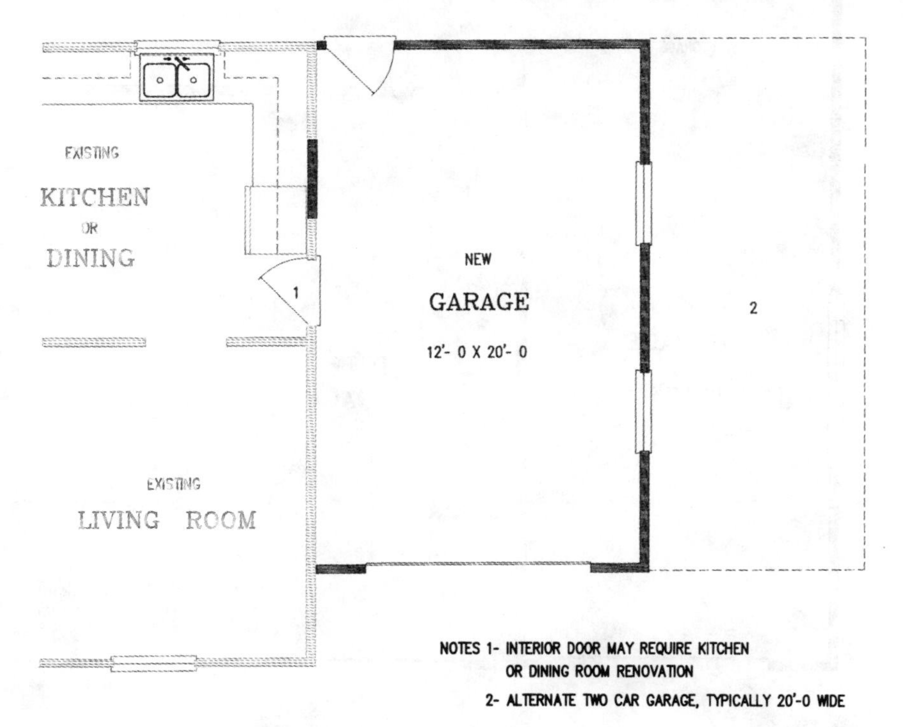

NOTES 1- INTERIOR DOOR MAY REQUIRE KITCHEN OR DINING ROOM RENOVATION
2- ALTERNATE TWO CAR GARAGE, TYPICALLY 20'-0 WIDE

NOTE: CONSTRUCTION BLUEPRINTS CAN BE READILY MODIFIED TO ACCOMMODATE YOUR SPECIFIC HOUSE OR ROOM SIZES.

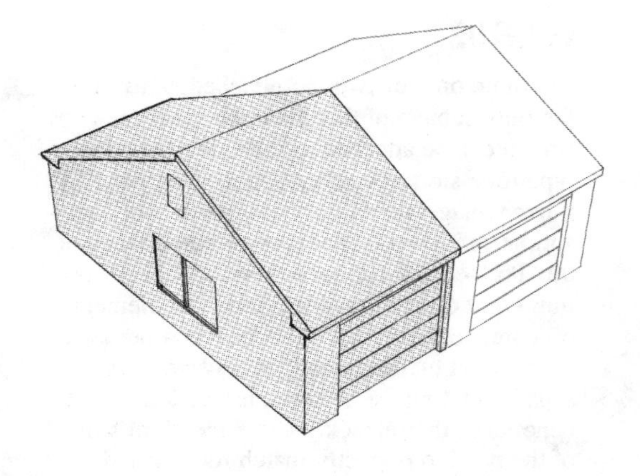

ROOFLINES AVAILABLE:
GABLE (SHOWN), HIP, FLAT

GAR01

Need a second garage? This plan neatly accomplishes that and adds some storage, too. The depth of storage is a function of the roof pitch and the level of the floor in the garage. The better appearance of this addition calls for maintaining the same roof pitch and ridge line of the existing roof, although this would likely require that you reroof the entire house. If you make a break at the front, this requirement would be eliminated because the roofs would not align. If you can only add less than 10 feet, it still might be possible to create a two-car garage by eliminating the old end wall. However, keep in mind that a two-car garage of less than 18 feet in total width is too tight for comfort. If that is your situation, consider just adding storage, such as shown in plan STR01 on page 231.

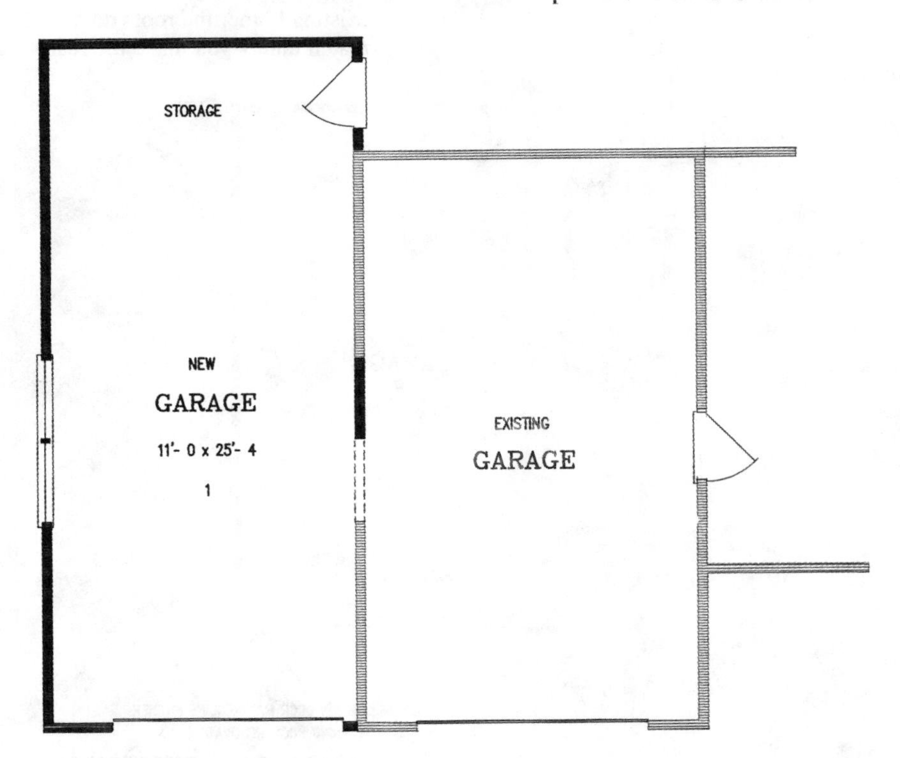

NOTE 1- GARAGE WIDTH AND DEPTH WILL VARY AND IS A FUNCTION OF AVAILABLE YARD SPACE AND THE PITCH OF THE ROOF

GAR05

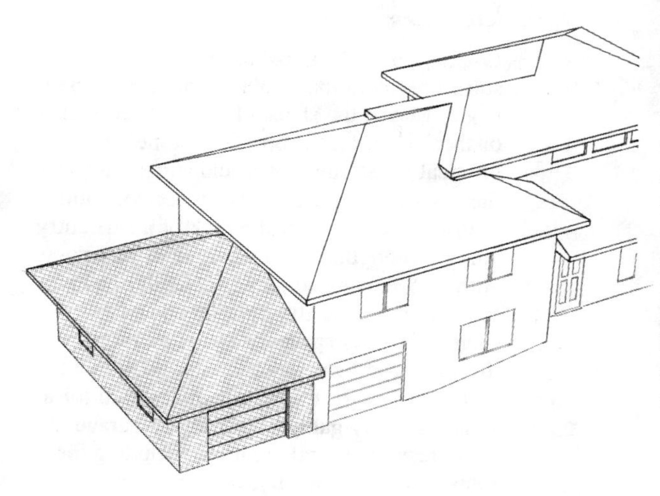

This plan accomplishes one thing: It adds a 13-foot-wide one-car garage to a split-level or two-story home. It retains the existing garage, thereby creating a two-car garage (as contrasted to plan FG001 on page 141, which added a somewhat narrower garage and also converted the old garage to enlarge the family room). As with the other plan, this addition does significantly enhance the lines of the home. Make sure to align the garage door head with the existing opening, and re-side the entire front to match. There is no need to remove the old garage wall.

ROOFLINES AVAILABLE:
HIP (SHOWN), GABLE

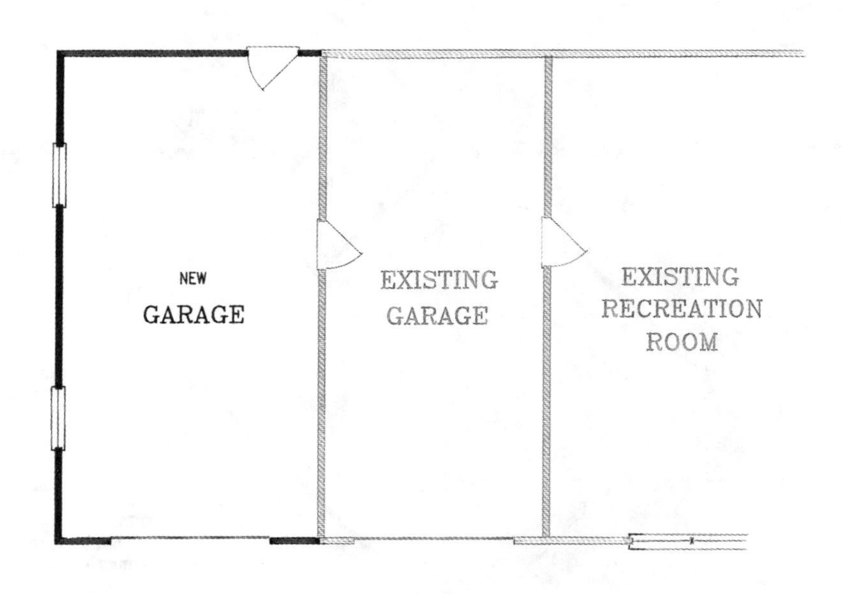

LOWER FLOOR

CONSTRUCTION BLUEPRINTS ARE AVAILABLE FOR THIS PLAN (AS WELL AS ALL OTHERS) AND INCLUDE A MATERIALS LIST & INFORMATION ON CUSTOMIZING THE PLANS

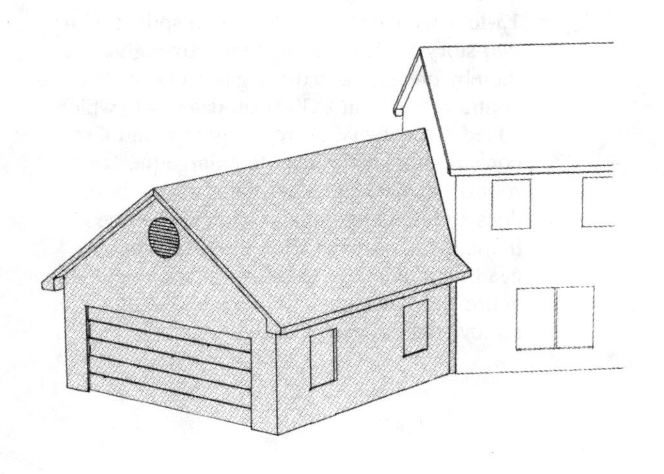

GAR04

Looking to attach a two car garage? Give some consideration to other than a 90 degree angle. Maybe the shape of your lot requires it, or there is a grade drop-off or some other physical constraint that would suggest turning the garage. One of the other more common reasons is a desire for the look of a side entry garage where there is a shortage of space. Although the angled garage actually requires more yard for the structure (about 10 feet more), the total space for structure plus driveway can be achieved in about 40 to 42 feet, which is 10 feet less than required for a true side-entry garage. The angled garage plan also presents several bonuses, including the found space for laundry, lavatory, pantry, storage, etc., as shown, as well as a different aesthetic appeal.

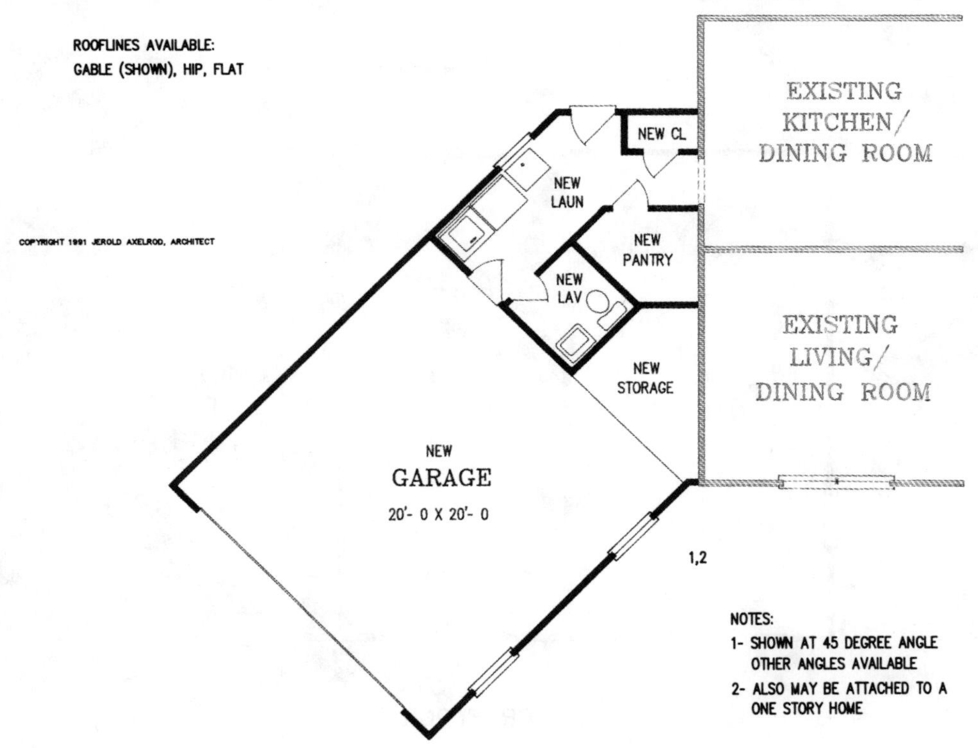

CONSTRUCTION BLUEPRINTS ARE AVAILABLE FOR THIS PLAN (AS WELL AS ALL OTHERS) AND INCLUDE LAYOUT DIMENSIONS FOR VARIOUS ANGLES AND A MATERIALS LIST

GAR03

A plain, characterless, L-ranch is turned into a sparkling Spanish-styled court home in this plan. The open site area of the L is space waiting to be better utilized. In this design, a new garage is built flush with the front wall, and a large new room is built alongside. This room is ideal for use as an office, although it is also perfect as a studio, playroom, or spacious bedroom suite. The former garage is space that can be converted to many uses, such as an apartment, a new master suite, or a fabulous new family room. A stylish wall is constructed at the front to enclose the new courtyard and helps create the southwestern facade.

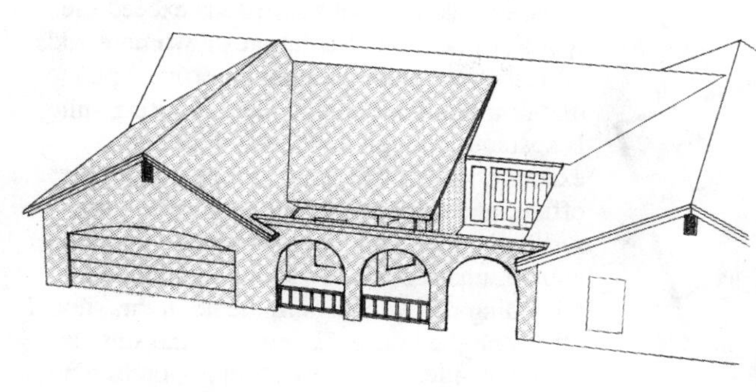

ROOFLINES AVAILABLE:
GABLE (SHOWN)

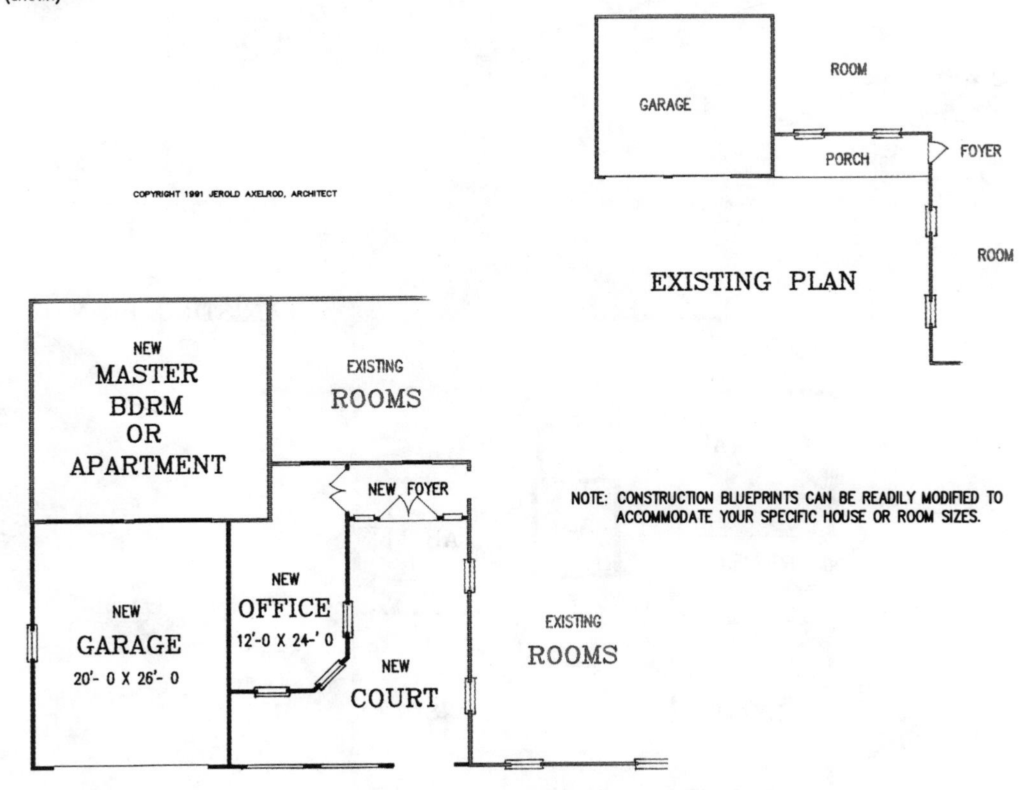

APT06

Your goal is to create a professional suite from a two-car garage, but your needs exceed the space available. The solution shown here adds 10 feet to the side, creating an office equal to that of a three-car garage. The resulting suite is spacious enough to accommodate two examination (or treatment) rooms, a private office, and a laboratory, plus reception and waiting space. The garage doors are closed, of course, and the windows in the garage, plus the siding and windows in the addition, should all match the house. Entrance to the suite is from the side, via a new covered porch. This entrance would likely be convenient to additional parking, which is usually required by most zoning ordinances.

NOTES:
1- 10'- 0" ADDITION TO SIDE
2- WINDOW

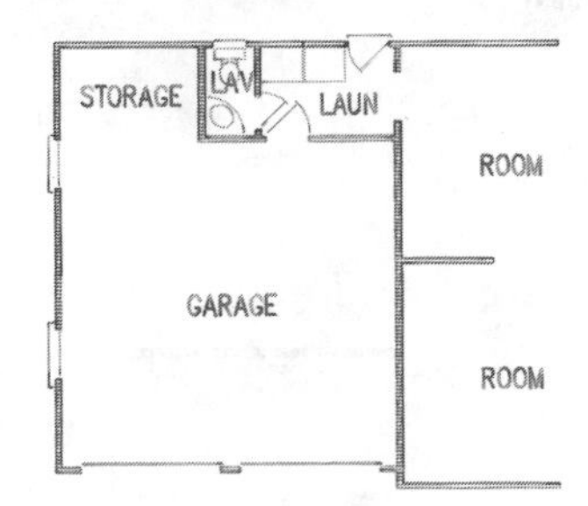

EXISTING PLAN

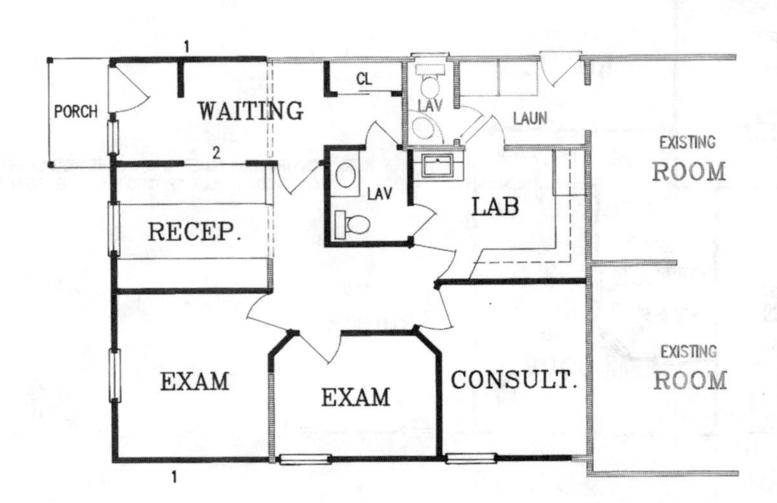

REMODELED PLAN

HAPT2

Finding or creating space to house elderly parents is an increasingly requested need expressed by many homeowners. Where to put them is often an extremely difficult decision. Since stairs pose a problem and adequate first floor space usually does not exist, an addition is a frequent consideration, especially if it's a full apartment you are seeking. This 20′4″×28′0″ addition is proposed at the rear of a deep garage. It provides direct access to the house through the laundry room, yet functions very independently with its own side entrance and access to the rear patio.

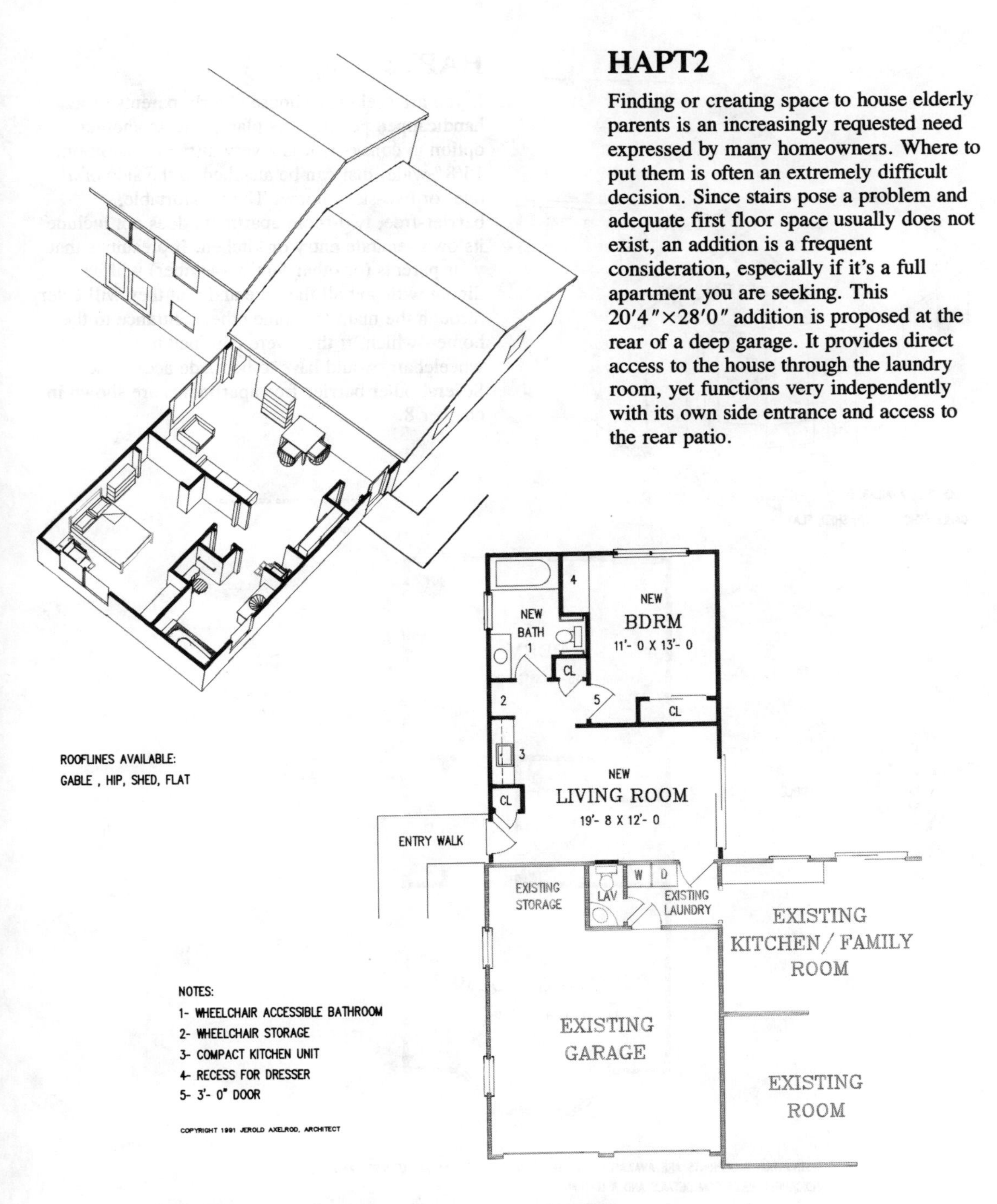

CONSTRUCTION BLUEPRINTS ARE AVAILABLE FOR THIS PLAN (AS WELL AS ALL OTHERS) AND INCLUDE HANDICAPPED BATHROOM DETAILS AND A MATERIALS LIST

HAPT3

If you are seeking to house elderly parents or a handicapped person, this plan presents another option to consider. It is a very attractive addition, 15′8″ wide, that can be attached to the side of a one- or two-story home. The comfortable, barrier-free, two-room apartment does not include its own separate entry or kitchen. It presumes that your parents (or other family member) will be dining with you all the time and that they will enter through the main (or some other) entrance to the home—which, if they were confined to a wheelchair, would have to be made accessible. Several other barrier-free apartments are shown in chapter 8.

ROOFLINES AVAILABLE:
GABLE (SHOWN), HIP, SHED, FLAT

NEW
BDRM
15'- 4 X 12'- 0

EXISTING
ROOM

NOTES:
1- 4'- 0" HALL
2- 3'- 0" DOORS
3- WHEEL-CHAIR ACCESSIBLE BATHROOM

CL

NEW
BATH
3

2

1

NEW
CL

EXISTING
DINING / LIVING /
OR FAMILY ROOM

NEW
LIVING ROOM
15'- 4 X 13'- 0

CONSTRUCTION BLUEPRINTS ARE AVAILABLE FOR THIS PLAN (AS WELL AS ALL OTHERS) AND INCLUDE HANDICAPPED BATHROOM DETAILS AND A MATERIALS LIST

APT01

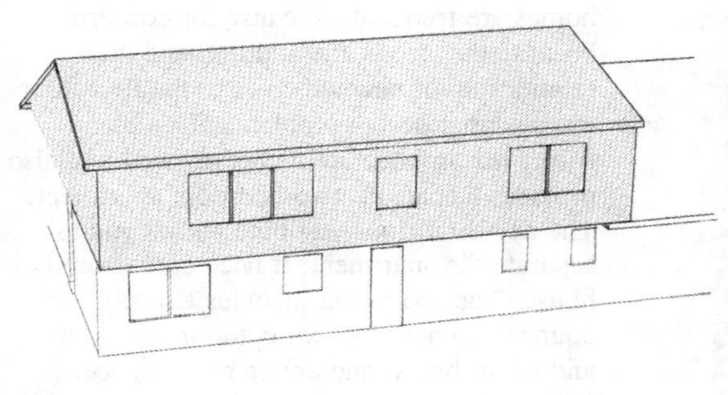

Problem: You're looking to create a small apartment, but you don't have the lot area to go out. Look up; it might be possible to create an apartment in a partial second floor. See where your basement stair is located. Could you put a stair to the second floor above it? This design shows how to do it. The existing plan shown is the rear of a one-story home that is well-suited to this addition. The apartment is 40 feet across and cantilevers two feet beyond the first floor wall to gain additional area. The cantilever has the added advantage of permitting greater design freedom in the sizing and placement of windows at the second floor, since alignment with first floor windows is not a prerequisite.

ROOFLINES AVAILABLE:
ONLY AS SHOWN

NOTES 1- REMOVE POSSIBLE CLOSET OVER EXISTING BASEMENT STAIR
2- COMPACT KITCHEN UNIT

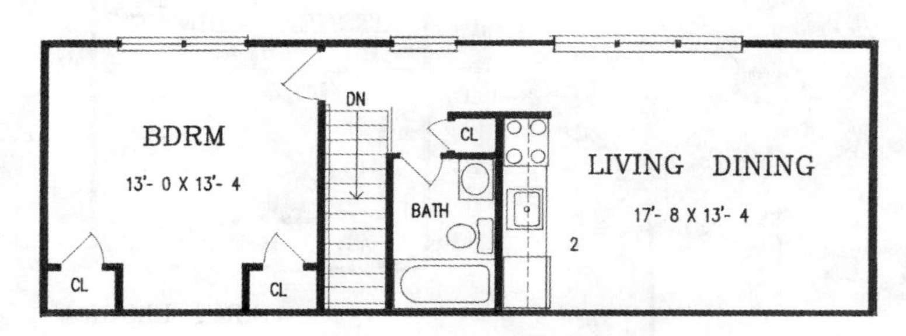

SECOND FLOOR
ALL NEW

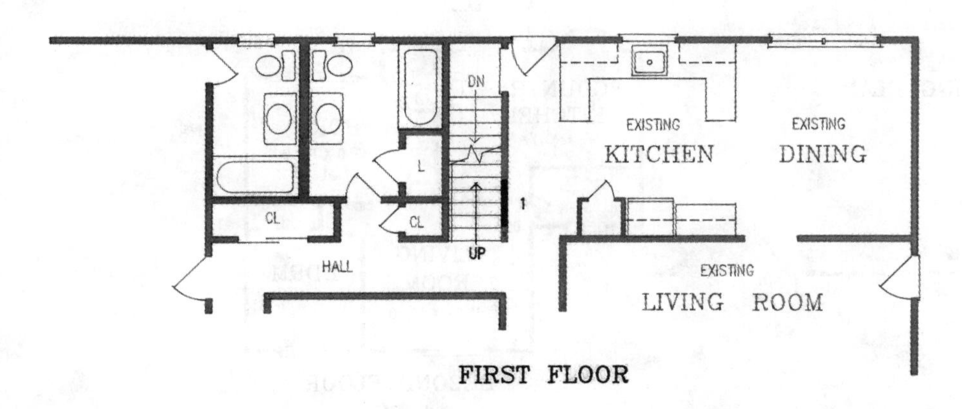

FIRST FLOOR

CONSTRUCTION BLUEPRINTS ARE AVAILABLE FOR THIS PLAN (AS WELL AS ALL OTHERS) AND INCLUDE THE DESIGN OF:
DETAILS ON HOW TO MINIMIZE DISTURBANCE TO THE EXISTING FIRST FLOOR

APT07

Second floor additions on long one-story homes are frequently a cause for concern because the size of the addition and the relationship of new windows to the first floor are potential design problems. This plan solves the architectural concerns well and also provides a spacious two-bedroom apartment. The new first floor vestibule allows you to separate the apartment, if necessary, from the house. The apartment includes a lovely country kitchen, its own outdoor deck, one and a half baths, and ample closet space. It utilizes the space over an existing basement stair to create the stair up to the apartment. If your stair is not in this location, you would have to take three feet from the bedroom or closets located here.

ROOFLINES AVAILABLE:
GABLE (SHOWN), HIP

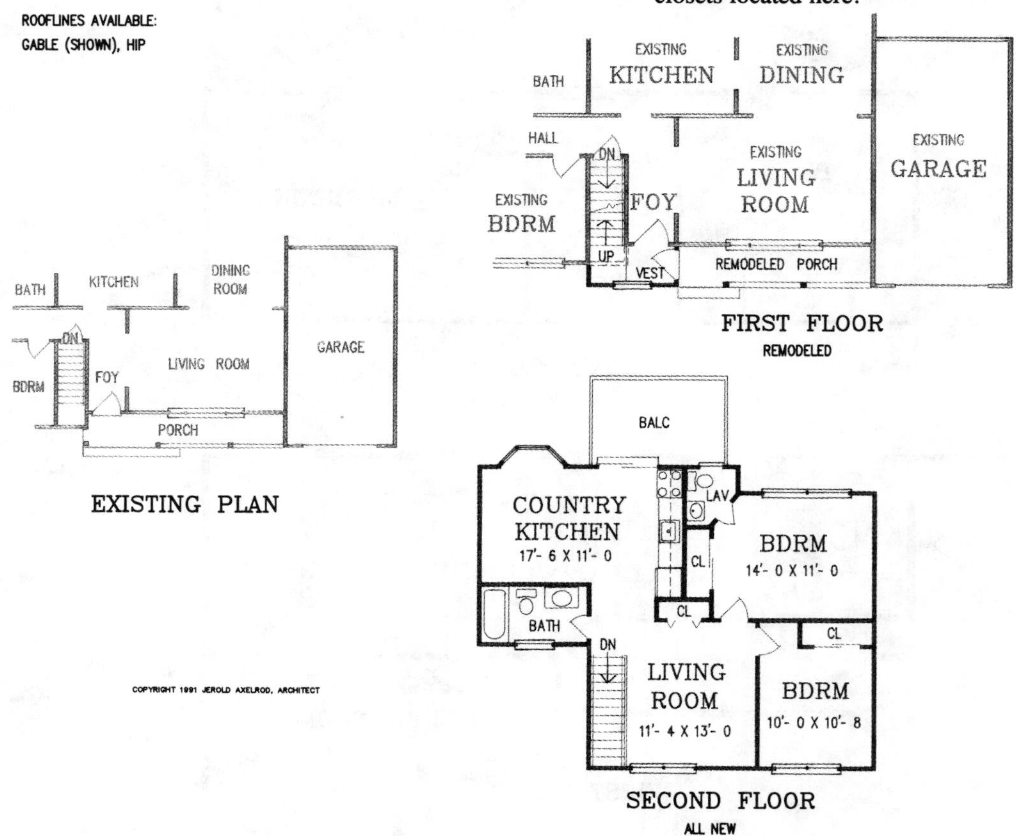

CONSTRUCTION BLUEPRINTS ARE AVAILABLE FOR THIS PLAN (AS WELL AS ALL OTHERS) AND INCLUDE A MATERIALS LIST AND INFORMATION ON HOW TO MAKE CHANGES TO PLANS

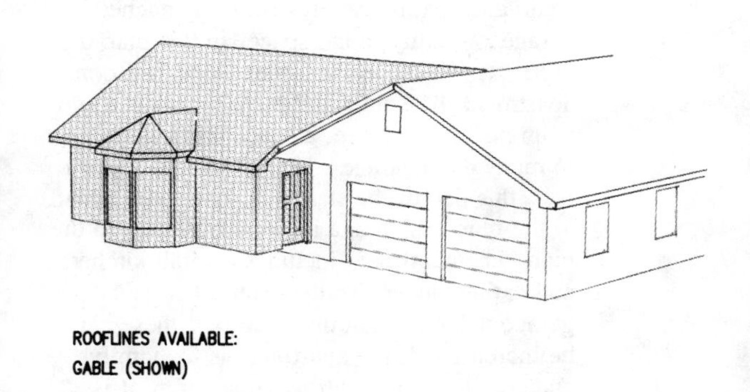

ROOFLINES AVAILABLE:
GABLE (SHOWN)

APT04

The yard behind a garage is an ideal location for a new ground-floor apartment because this area is frequently underutilized, there are no rooms affected at this location, and its proximity to the driveway provides reasonably private access. The comfortable one-bedroom apartment shown includes a large living room, a U-shaped kitchen, and a lovely bayed dining area. An entrance foyer with two closets, a dual entrance bath, and a large walk-in closet are other features of note. The blank wall facing the rear yard behind the house should maintain privacy for the owners. Compare this feature and other aspects of the design with plan HAPT2 on page 149, which is designed for elderly parents, and you will note distinct differences.

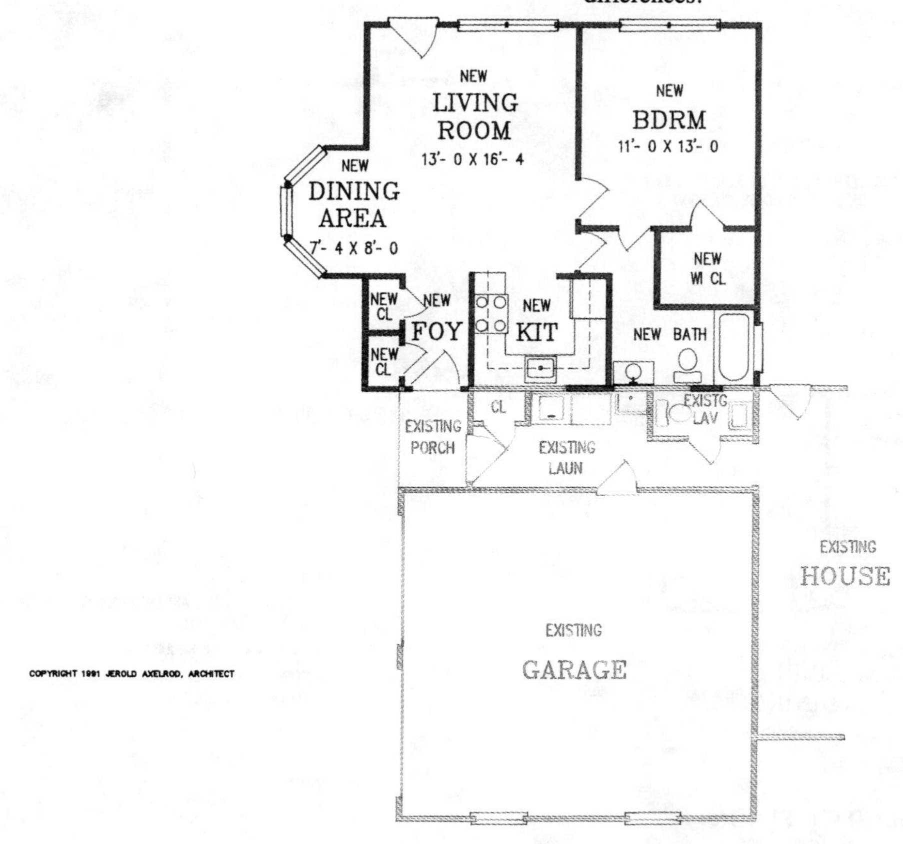

FIRST FLOOR

APT02

As I have demonstrated before—and will show again and again—the area over an attached garage is readily found space. In this plan the 20′0″×26′0″ space becomes a one-bedroom apartment. The stair to the apartment is taken from a storage alcove, a space quite common in many deep garages. If your garage doesn't have this depth, the stair will have to be added to the rear, which is a minor adjustment to the plan. The apartment includes a small kitchen and eating alcove. To the extent that your garage is larger than that pictured, these can be increased. If the apartment is for family and you desire an internal connection, it is possible to connect to an existing second floor bedroom or extend the second floor hall.

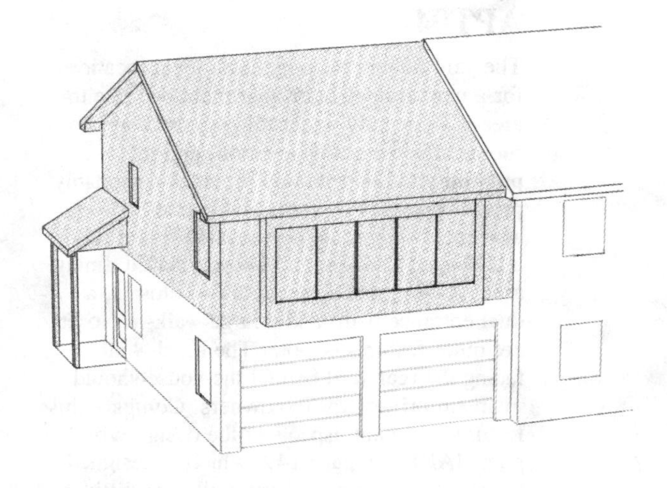

ROOFLINES AVAILABLE:
GABLE (SHOWN), HIP, FLAT

NOTE: CONSTRUCTION BLUEPRINTS CAN BE READILY MODIFIED TO ACCOMMODATE YOUR SPECIFIC HOUSE OR ROOM SIZES.

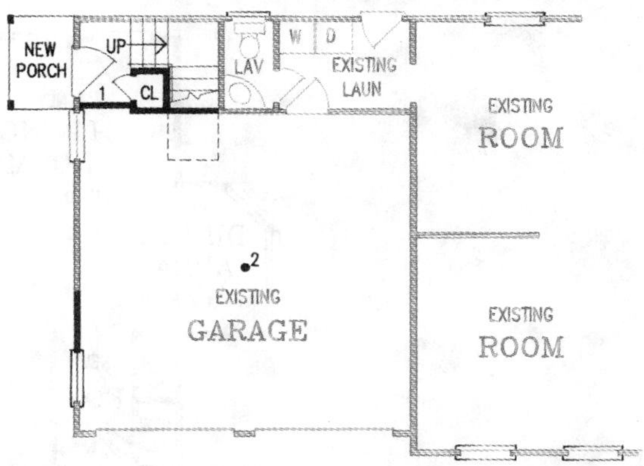

REMODELED FIRST FLOOR

SECOND FLOOR
ALL NEW

NOTES:
1- FORMER STORAGE AREA CONVERTED TO NEW STAIR HALL
2- COLUMN MAY BE REQUIRED
3- CONNECTION TO EXISTING SECOND FLOOR IS POSSIBLE

APT03

The area above this 34′6″ deep garage wing is found space for a spacious, elegant apartment. The only ground floor area needed is a 3′6″ addition for the stair to the apartment. An interesting bonus is the use of the space below this stair to provide a second stair to the basement—and it has direct entrance to the outside. The apartment is housed in a distinctive-looking roofline, featuring intersecting reverse-gable forms. Half-round windows add exterior interest as well as interior charm. A spacious living/dining room and an equally large bedroom with plenty of closets are the noteworthy aspects of this delightful apartment.

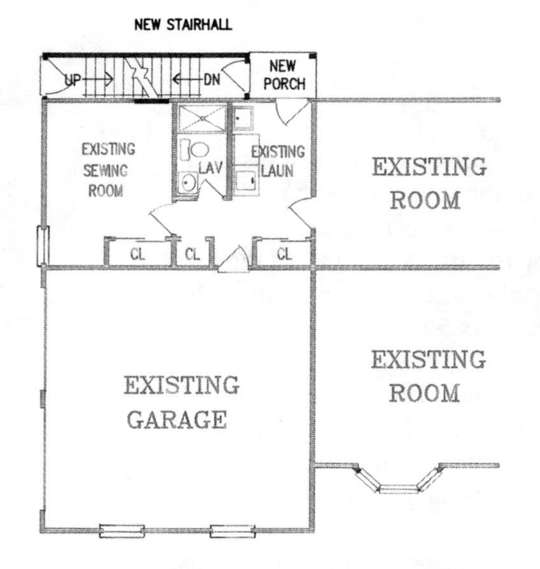

FIRST FLOOR

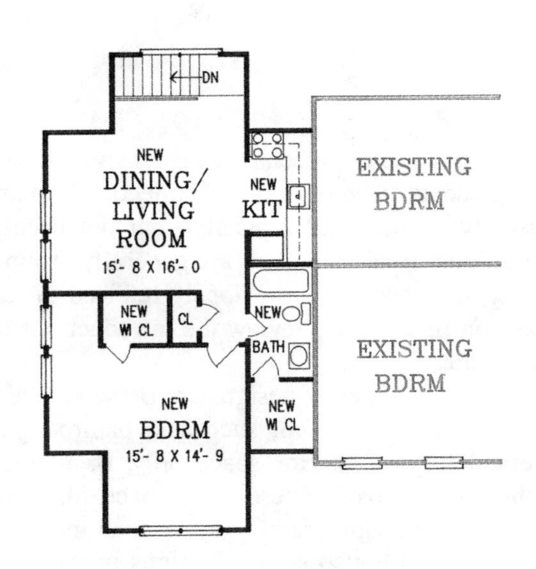

SECOND FLOOR

CONSTRUCTION BLUEPRINTS ARE AVAILABLE FOR THIS PLAN (AS WELL AS ALL OTHERS) AND INCLUDE A MATERIALS LIST

8 CHAPTER

Bumps, bays, extensions & interior remodeling

The last chapter presented an extensive assortment of room additions. This chapter presents an equally varied collection of plans, but the scope of these projects is usually smaller and typically—but not necessarily—less costly than room-size additions. Again the plans are generally grouped by room type, and all of the plan viewing conventions previously referred to still apply, so if you are reading this section first, please review the introductions to chapters 5 and 7 before trying to view these plans.

Several types of designs are presented here. The most prominent is a design for one specific room, such as a bathroom or a kitchen. The plan could be a remodeling of interior space only, or it might include a small addition, or a "bump" or "bay." These small space additions can have an effect on the interior space that is significantly greater than their area.

There are also several designs in this section that remodel multiple rooms. However, these are not new room additions, but extensions to existing rooms. They are typically under eight feet. If you are seeking larger additions, see chapters 7 and 10. This chapter is a collection of smaller projects.

Some of the designs pictured here are also not necessarily specific to a given room. For example, there are plans for front porches, vestibules, and even some plans that show you how to add natural light to the interior of your home—even some that can light up a basement. A larger number of these plans are pictured with interior perspective views because they do not add much, if any, new exterior space.

It is also possible that some of these designs will be incorporated within, or together with larger room additions presented in the prior chapter. These designs could also become integral components to a new facelift to your home, as pictured in the next section, or even a part of a whole-house renovation, as shown in chapter 10.

Space additions of this nature do have one common inherent difficulty: they are more difficult to live with during construction than a new-room addition, so extra planning is necessary to help you cope with the project. A discussion on staging is included in chapter 11.

P0007

Today's new home is likely to have a front porch, as that has been in vogue and is likely to remain popular for some time. The porch brings nostalgia, charm, and a theme of relaxation, while it also has the useful function of providing shelter to the entry. And it is still a place to sit and watch the world. But you don't have to move to find a porch. It is easy to add one to your current home. You will have to check the zoning ordinances, especially if your home is currently at the minimum setback. Many ordinances allow a one-story open porch to encroach into a required front yard. This porch will adapt to either a one- or two-story home.

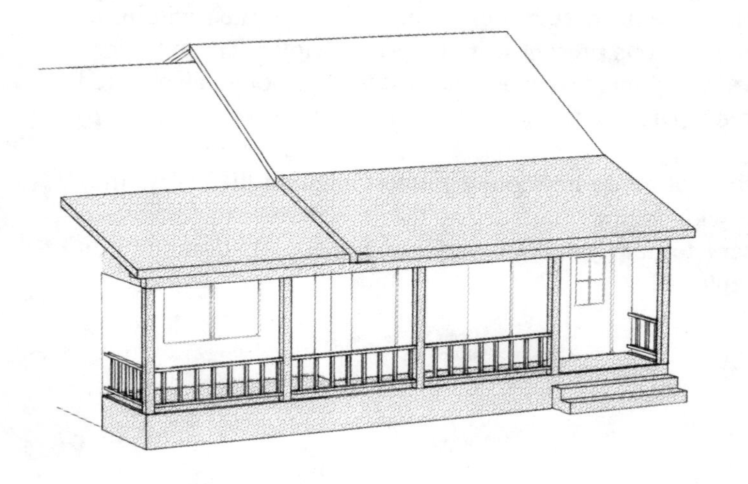

ROOFLINES AVAILABLE:
SHED (SHOWN), HIP, FLAT

CONSTRUCTION BLUEPRINTS ARE AVAILABLE FOR THIS PLAN (AS WELL AS ALL OTHERS) AND INCLUDE DETAILS ON HOW TO CONSTRUCT THE PORCH AND A LIST OF MATERIALS.

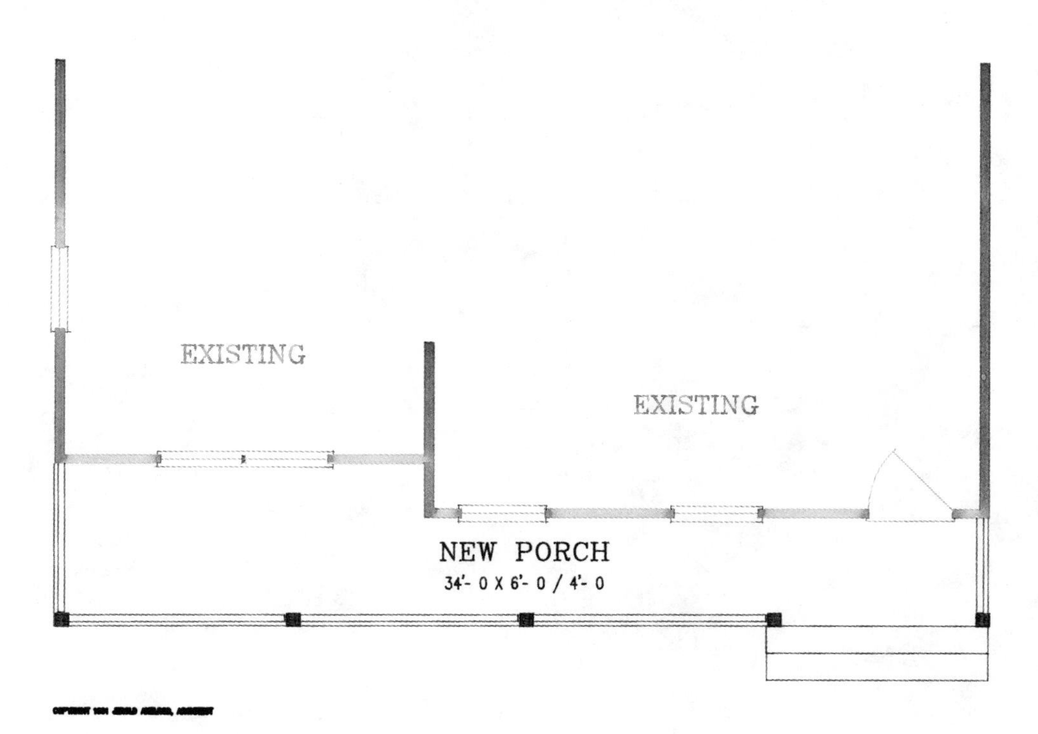

P0013

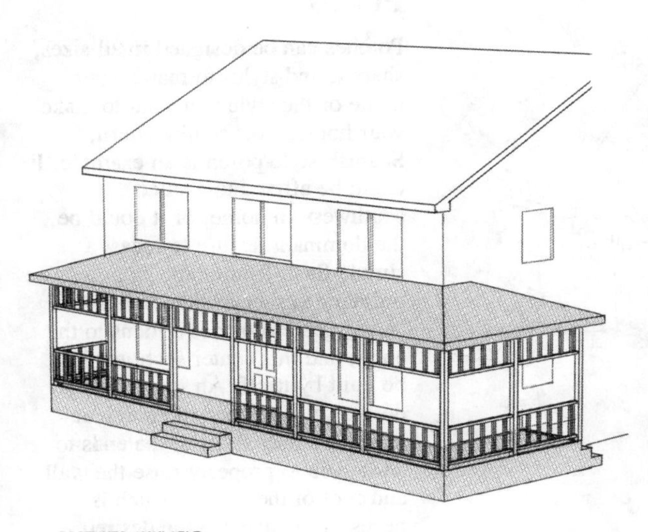

Simple, flush-front facades dominated the styling of many two-story homes built 15 to 30 years ago, and in some regions they are still popular. There is, however, a definite stylistic change occurring. It is called *country* because it relies on the hominess and warmth associated with rural life. The country front porch is "in" and—because the style is not new and flashy but old and laid back—it is likely to be with us for a long time. Therefore, if you're dreaming of a wrap-around, country-style porch, go ahead and do it. I don't believe you will have to worry that it will be out of fashion when you decide to sell. In fact, in some regions, you might be out of fashion with your old flush-front.

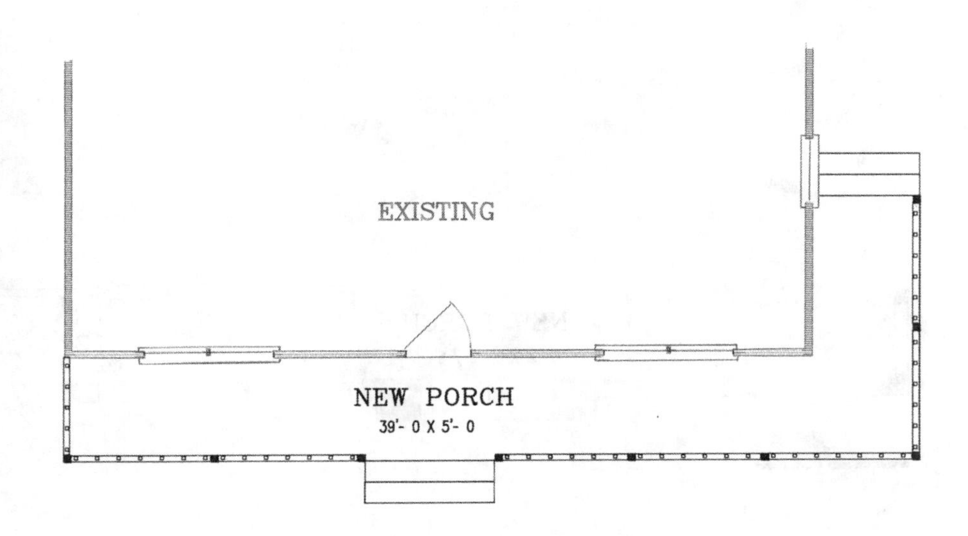

CONSTRUCTION BLUEPRINTS ARE AVAILABLE FOR THIS PLAN (AS WELL AS ALL OTHERS) AND INCLUDE PORCH FRAMING DETAILS AND A MATERIALS LIST.

P0010

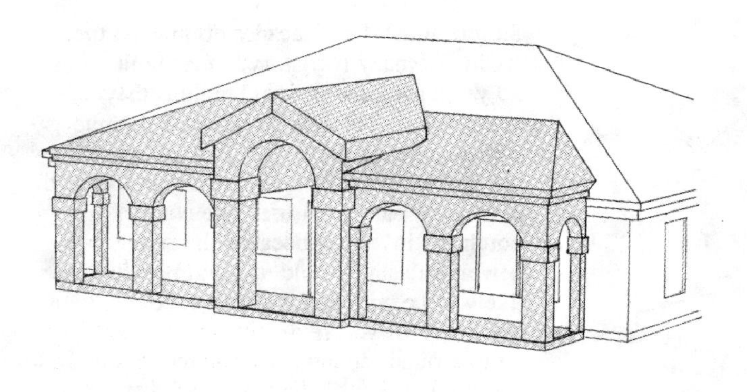

ROOFLINES AVAILABLE:
HIP & GABLE (SHOWN), SHED, FLAT

Porches can be designed in all sizes, shapes, and styles to match your home or the style you want to make your home. This southwestern, Spanish-style porch is an example. It could be affixed to a stucco southwestern home, or it could be the dominant feature to recast a simple flush-front home from ordinary to spectacular. A raised center gable provides a focus to the entry, and this center section could be built by itself. An important design tip is to keep the porch at least three feet short of the ends to allow you to properly raise the wall and roof of the porch, which is necessary to achieve the desired effect.

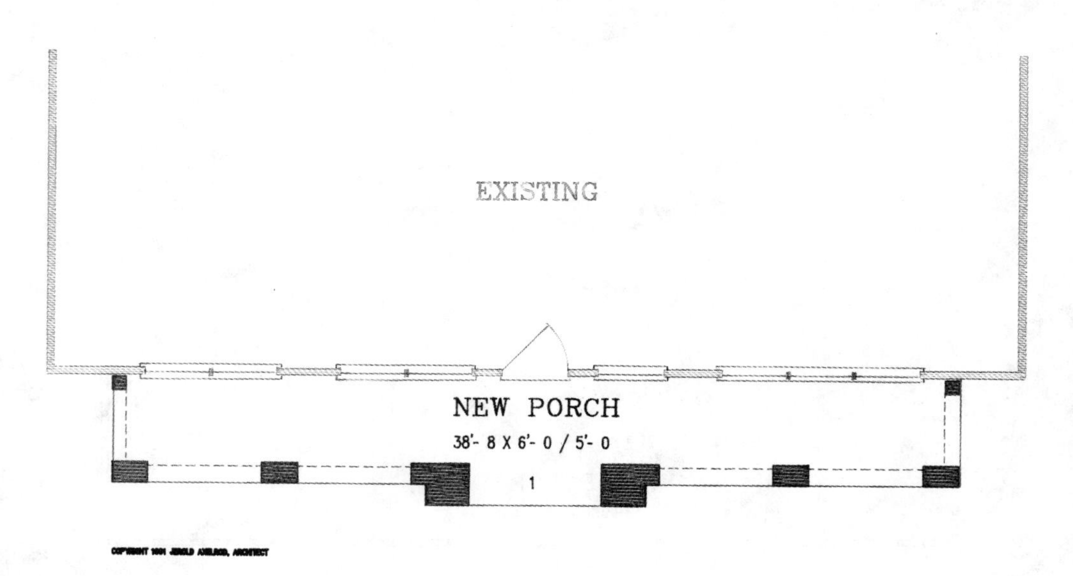

NOTE 1- CENTER SECTION COULD BE BUILT ALONE

NOTE: CONSTRUCTION BLUEPRINTS CAN BE READILY MODIFIED TO ACCOMMODATE YOUR SPECIFIC HOUSE OR ROOM SIZES.

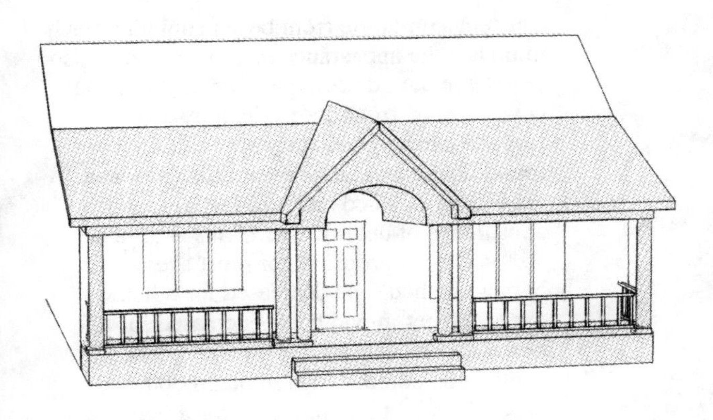

ROOFLINES AVAILABLE:
GABLE AND SHED (SHOWN)

P0008

This beautiful, continuous front porch creates a more formal, stately entrance than the rambling shed roof in P0007 or the country look in P0013. Heavy, round columns and the reverse-gable center section help delineate this understated formality. As with any of these porches, the main impetus is a desire to change or update the exterior of your home—thus the variety of styles to help create the particular mood or feeling you are seeking. The porch may become the focus for change, and you should review the new facades shown in chapter 9 for further ideas. Unless otherwise indicated, most of the porches shown in this chapter are adaptable to both one- or two-story homes.

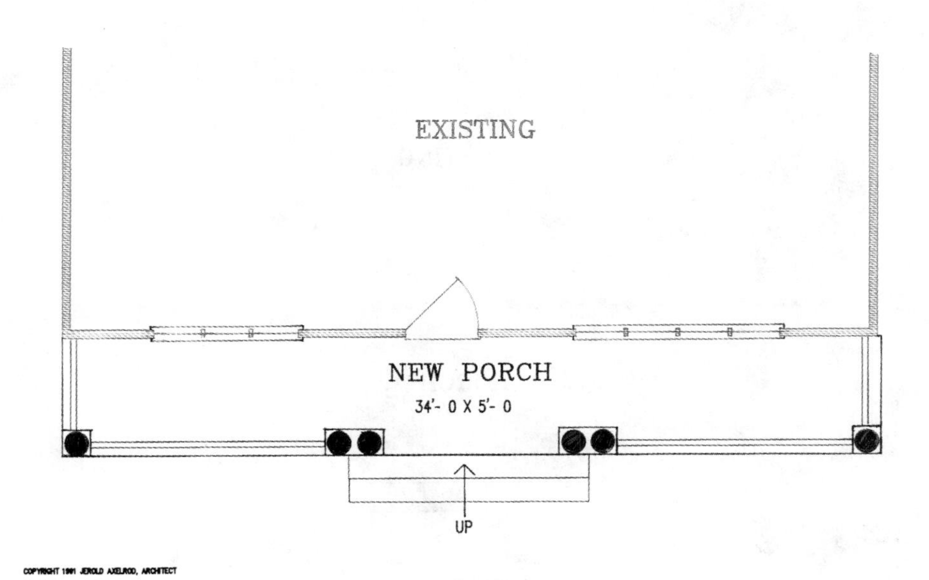

NOTE: CONSTRUCTION BLUEPRINTS CAN BE READILY MODIFIED TO ACCOMMODATE YOUR SPECIFIC HOUSE OR ROOM SIZES.

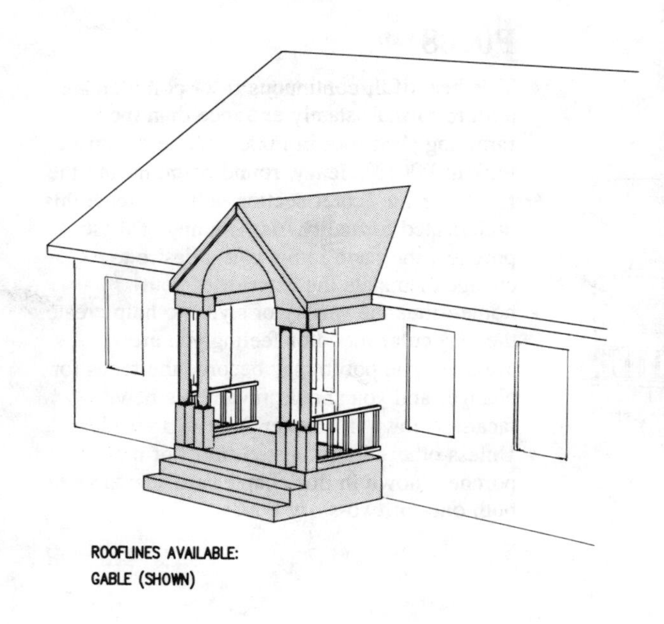

P0009

This engaging little front porch could do much to update the appearance of your home. It also will serve the practical purpose of providing cover to your front door. The attractive reverse-gable arched entryway, clustered round columns, and wooden rails have been exquisitely detailed and will foster an initial appearance of quality. The trim 9-×-5-foot size will be appropriate for most average homes. It should be decreased for a home under 30 feet in width and increased for a home greater than 45 feet in width. The continuous porch shown in plan P0007 on page 158 might be preferable for a home less than 30 feet wide.

EXISTING

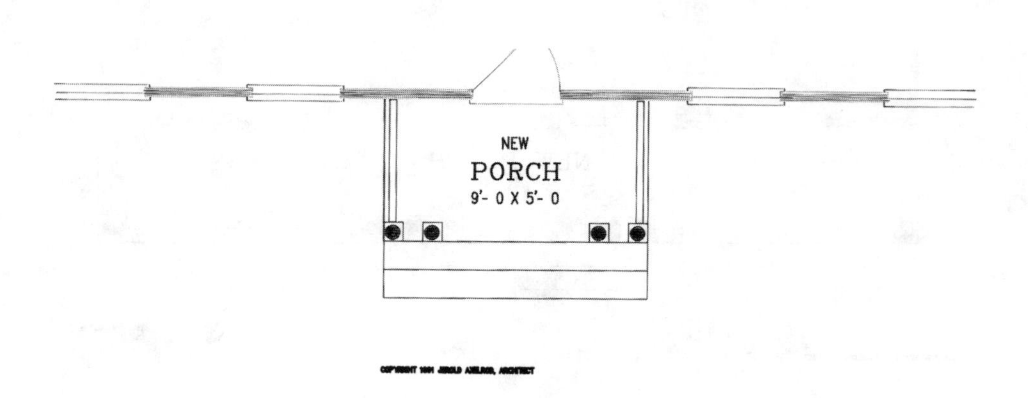

CONSTRUCTION BLUEPRINTS ARE AVAILABLE FOR THIS PLAN (AS WELL AS ALL OTHERS) AND INCLUDE PORCH DETAILS AND A LIST OF MATERIALS

P0011

Tired of that cold blast of arctic air in winter when someone opens the front door? Well, then, you should consider installing a vestibule. A vestibule serves as an air lock or *decompression chamber*, keeping the cold air outside your living spaces. As with a front porch, it is necessary to check your zoning ordinance to see if you can build one as-of-right, or if you might need a variance. The lovely looking 7′8″-×-4′10″ reverse-gable vestibule pictured here is adaptable to most one- and two-story homes. On a two-story, it is important to verify the height of the windows over the front door to see if the vestibule fits without having to make window changes.

ROOFLINES AVAILABLE:
GABLE (SHOWN)

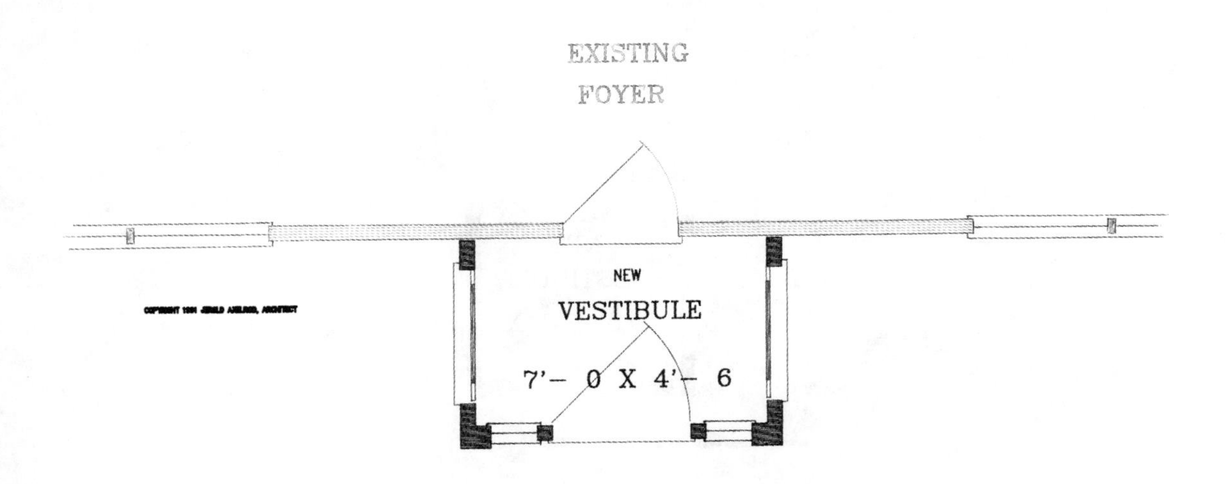

CONSTRUCTION BLUEPRINTS ARE AVAILABLE FOR THIS PLAN (AS WELL AS ALL OTHERS) AND INCLUDE A MATERIALS LIST AND WINDOW AND DOOR MANUFACTURER INFORMATION.

P0012

This 6′4″-×-4′10″ flat-roof vestibule is the most modest of the vestibules shown in the book. It is comprised of three doors, one in the front and two fixed ones on the sides. The glazed doors provide an abundance of light while attractively enclosing this space. The flat roof is capped by a low decorative railing. On a two-story home with deep second floor windows, this would be the more practical solution. This vestibule, as with any of the vestibules or porches, also performs a secondary function—that of dressing up the facade of the home it is attached to. It becomes a cost-effective method to effect a modest facade change.

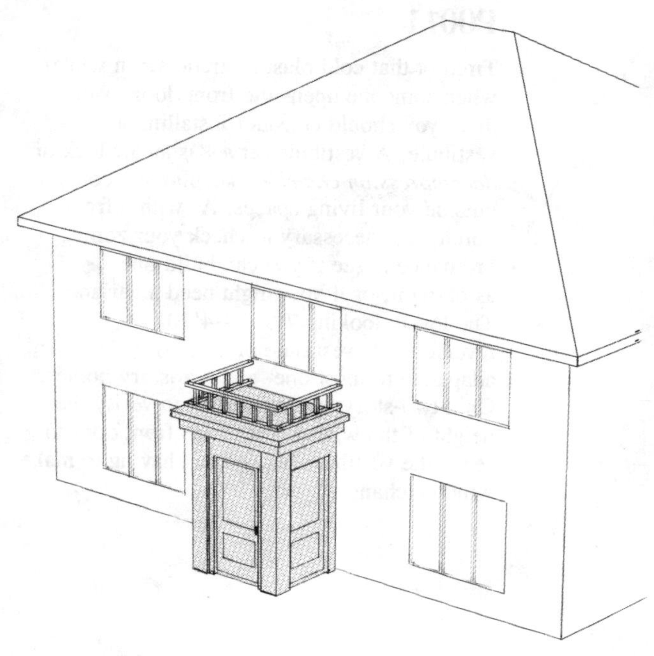

ROOFLINES AVAILABLE:
FLAT (SHOWN)

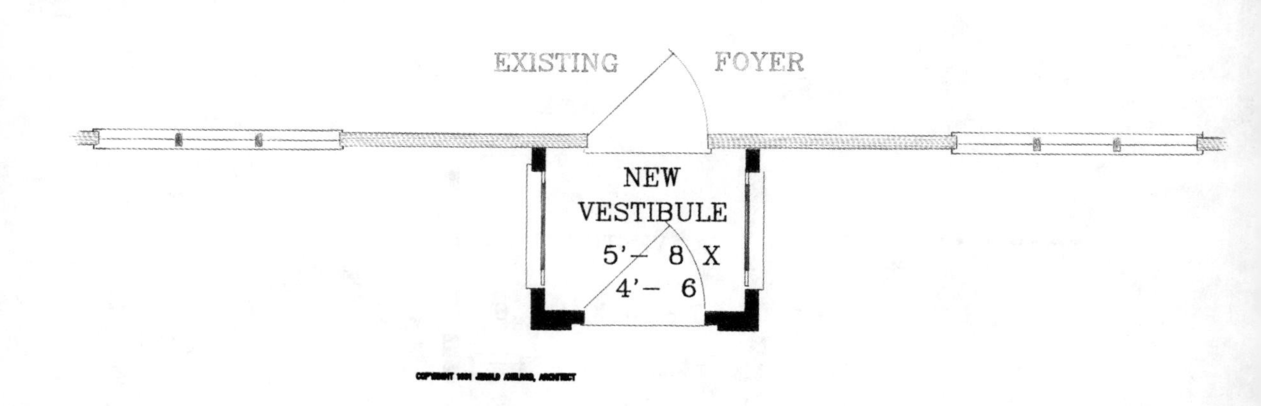

NOTE: CONSTRUCTION BLUEPRINTS CAN BE READILY MODIFIED TO ACCOMMODATE YOUR SPECIFIC HOUSE OR ROOM SIZES.

P0015

This tasteful glass-enclosed vestibule is designed specifically for a center-hall two-story home. It is exceptionally well-suited to a hip-roofed home, with matching bay windows and a recess at the front door, as shown. It will, however, be appropriate to any flush-front two-story home. The enclosure utilizes four glazed doors, with the side ones being fixed in place. The space is covered with a hip roof designed to complement the bays (if they exist). A truly stunning option might be a glazed roof in the shape shown. If, however, yours is a very sunny location that gets hot in summer, that would not be advisable.

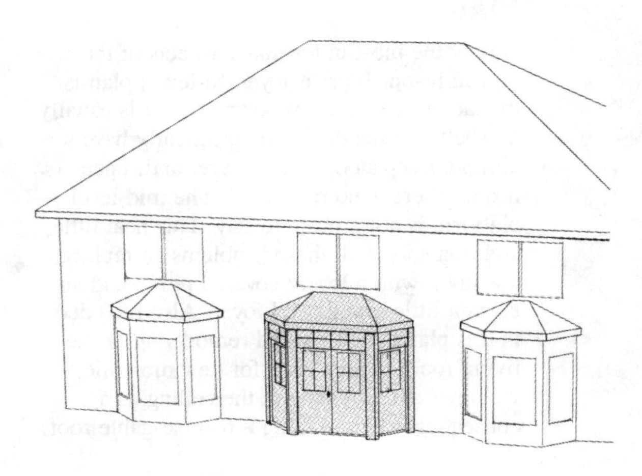

ROOFLINES AVAILABLE:

HIP (SHOWN)

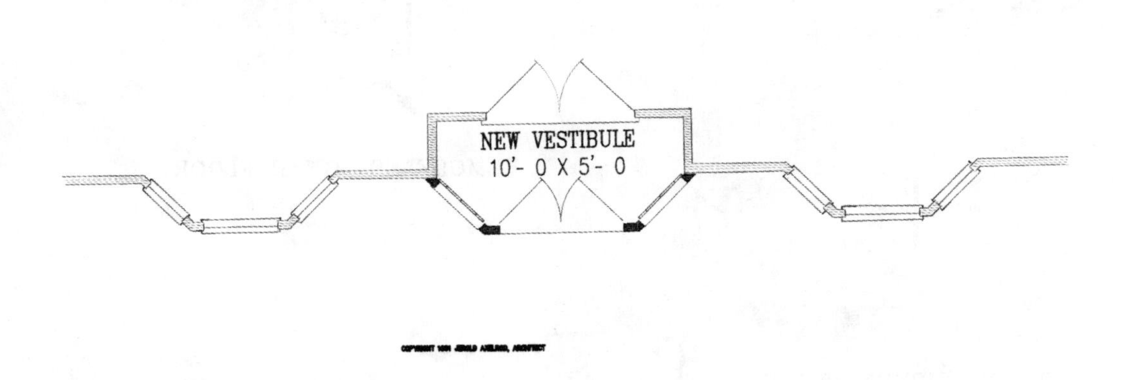

CONSTRUCTION BLUEPRINTS ARE AVAILABLE FOR THIS PLAN (AS WELL AS ALL OTHERS) AND INCLUDE A LIST OF MATERIALS AND WINDOW AND DOOR MANUFACTURER INFORMATION.

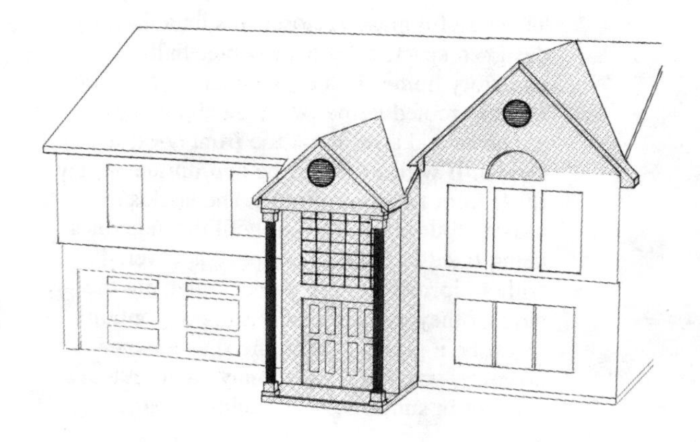

P0019

One of the most unfortunate aspects of the typical hi-ranch (split-foyer, bi-level) plan is the lack of an attractive entry. There is usually no shelter at the door, you frequently have to climb a steep stoop to get there, and, once inside, there is no true foyer. The mid-level platform is not a place to stay. This neat little addition solves all these problems; it replaces the stoop with a lovely covered porch and an elegant little grand-level foyer. Also included in this plan is a suggested remodeling of the living room, which calls for new dramatic windows and a raising of the ceiling as a consequence of installing a reverse-gable roof.

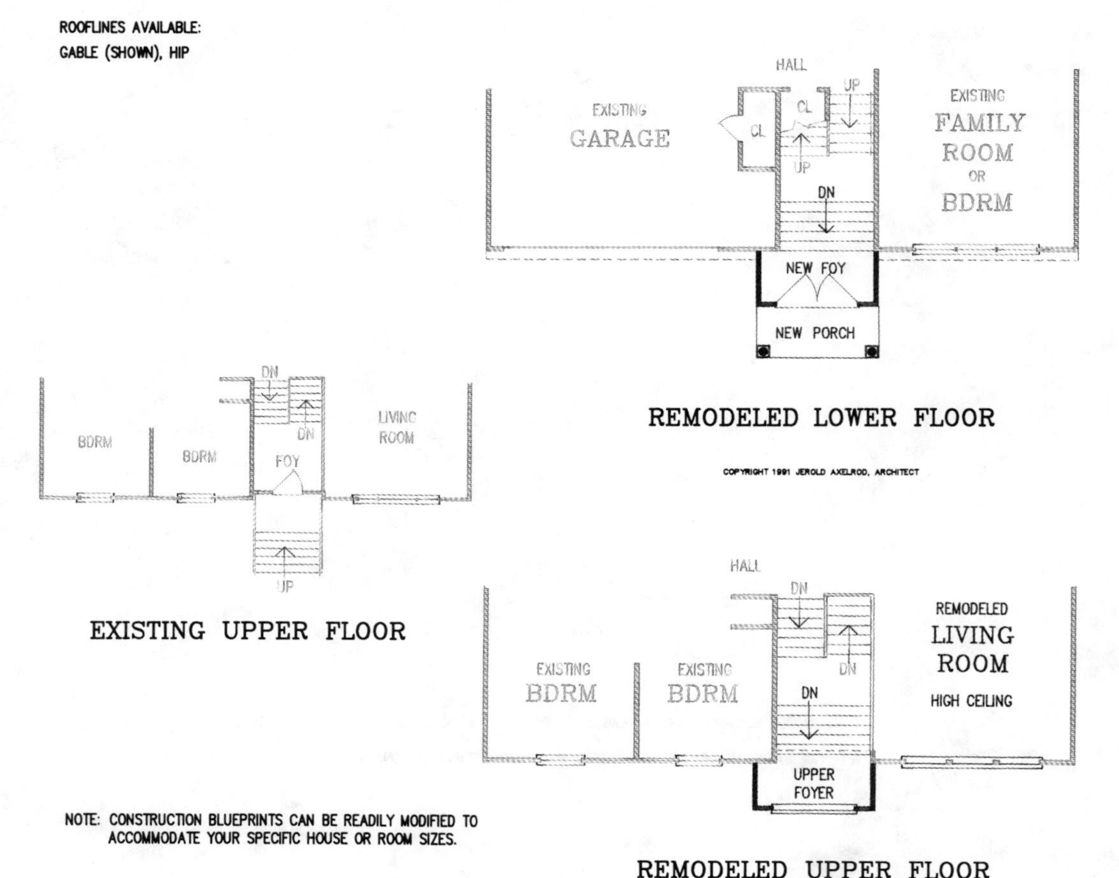

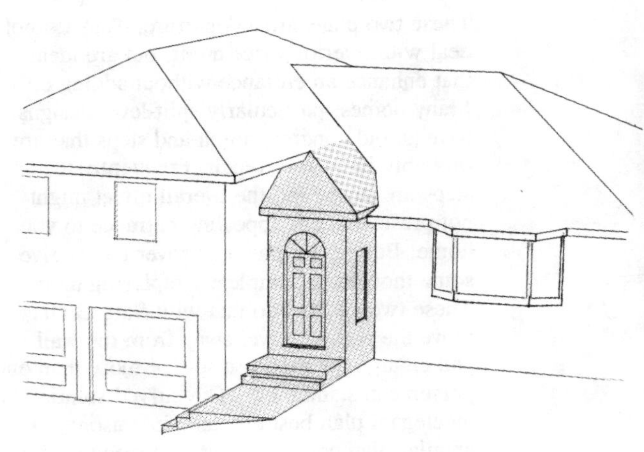

P0016

There are some additions that look so natural it is a wonder they were not part of the original home. This new little vestibule is an example. Designed specifically for the L-shaped split-level, it creates an attractive, separately defined entrance foyer, potentially freeing your living room from serving this double duty. It is placed down two steps, which will reduce the height of the outside stoop. This lower floor also allows the height for a half-round window over the new door, while creating a more inviting entrance. A window at the side further enhances the space. For some new ideas on entrance steps see plans SML02/03 on the next page.

ROOFLINES AVAILABLE:
HIP (SHOWN), GABLE, SHED

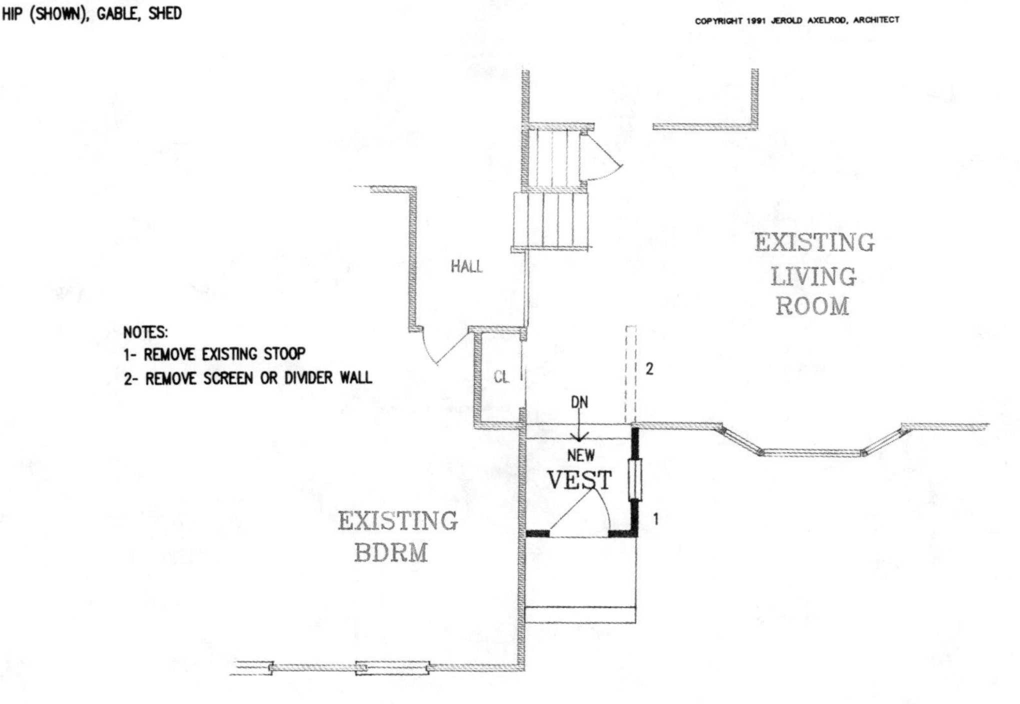

CONSTRUCTION BLUEPRINTS ARE AVAILABLE FOR THIS PLAN (AS WELL AS ALL OTHERS) AND INCLUDE A MATERIALS LIST

SML02/SML03

These two plans are a departure. They do not deal with interior space at all, but are ideas that enhance an entrance without adding on. Many homes, particularly split-level designs, have an old concrete porch and steps that are probably in need of repair. Frequently the steps are steep, and the overall effect might not present a very appealing entrance to your home. Before you patch or cover them, give some thought to completely replacing them. These two designs do that with flair. Both move the path of travel away from the wall and enlarge the platform so that more than one person can stand there. The curved solution is an elegant plan best executed in masonry. The angular plan could be built as a wood deck or in masonry.

EXISTING PLAN

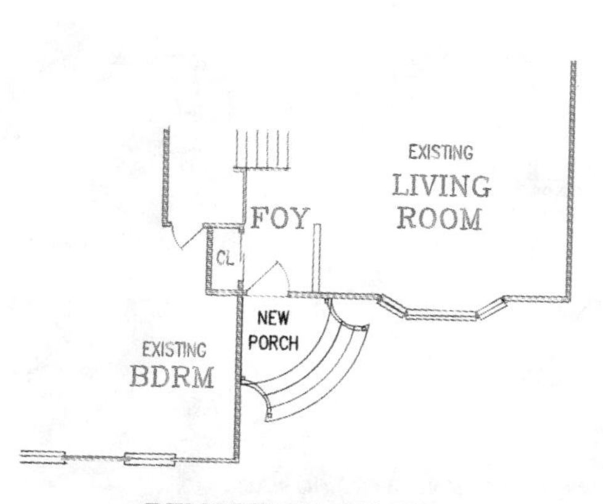

REMODELED PLAN
SML02

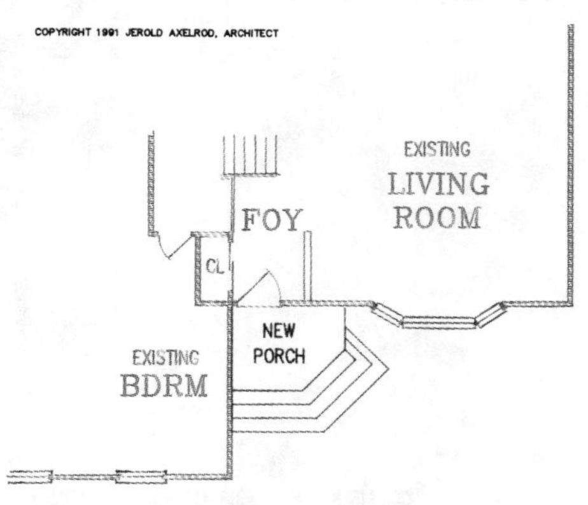

REMODELED PLAN
SML03

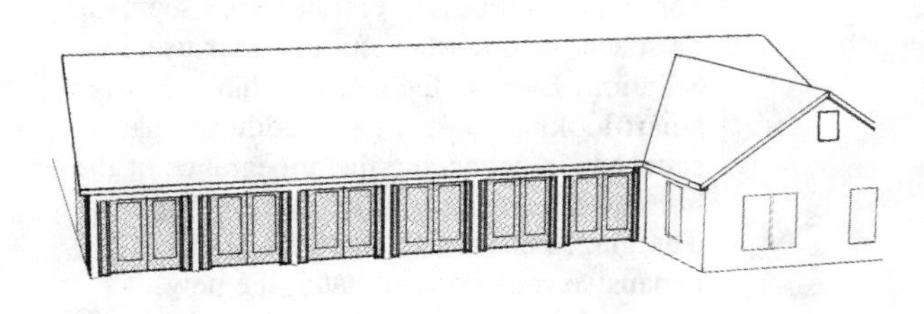

P0021

Is the front of your home adorned by a long, covered porch that you have often thought might be better utilized? This design shows you how to incorporate the porch into your home. It creates a continuous loggia, parts of which can be added to the actual floor area of the adjoining room. In front of other rooms, such as a kitchen, cost considerations might dictate leaving the existing wall. Even in that situation, though, the interior room benefits from natural light that the new loggia has provided in the form of continuous skylights and glazed doors. The nuances of your home would determine the best utilization of the newfound space.

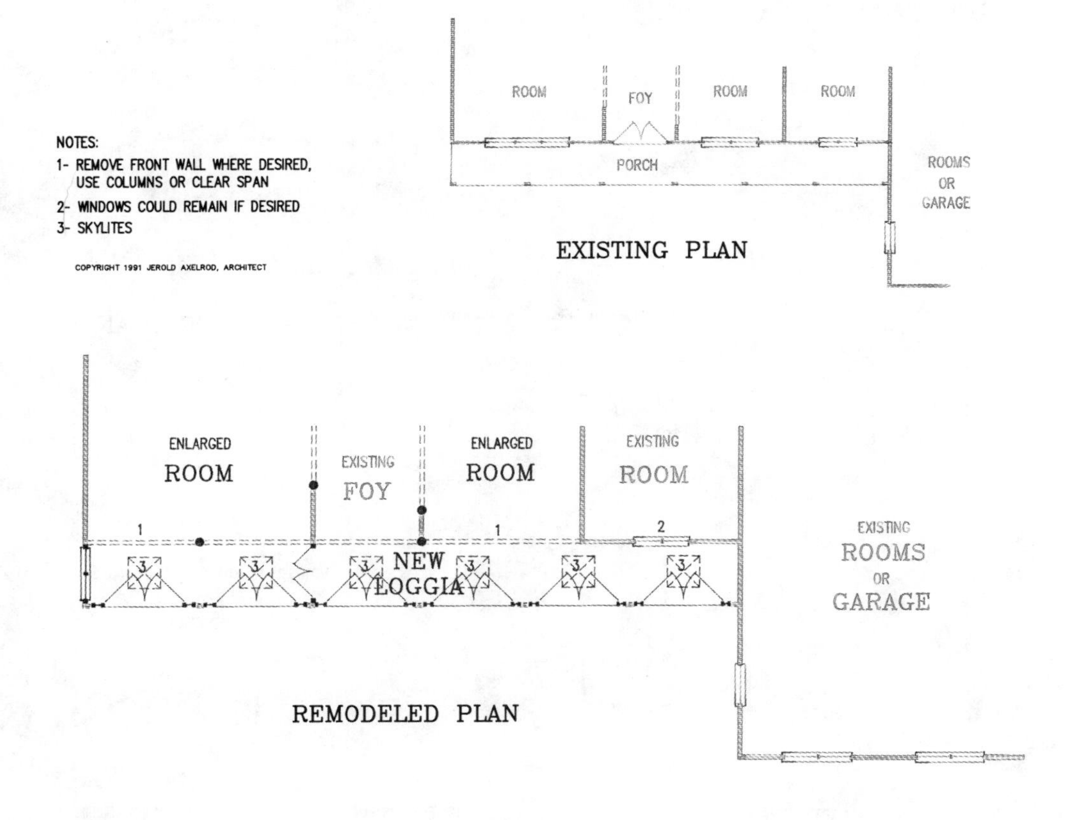

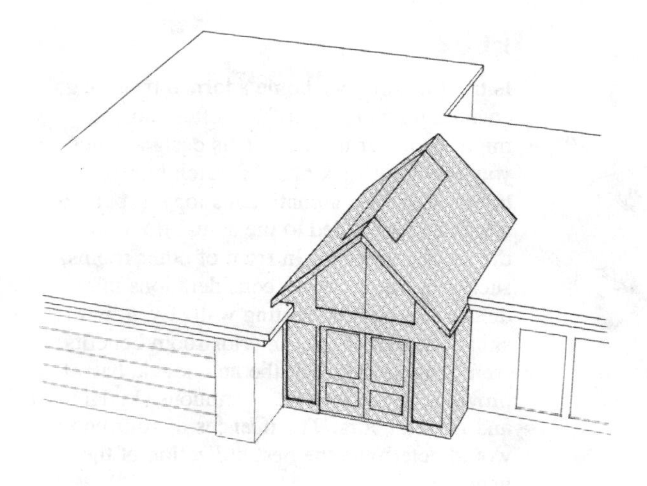

ROOFLINES AVAILABLE:
GABLE (SHOWN), HIP, FLAT

PF001

Problem: an uninspired, flat-roofed one-story home with a recessed entrance that seems to be space waiting to be put to better use. Solution: Enclose the court with a smart-looking reverse-gable addition that succeeds in enhancing the appearance of the home while benefiting the rooms it adjoins. The glazed ends of this gable and an expansive roof skylight bathe the new enclosed foyer with natural light, which also serves to visually enhance the adjoining living and dining rooms. The dramatic new front to rear interior vista will undoubtedly generate many positive comments.

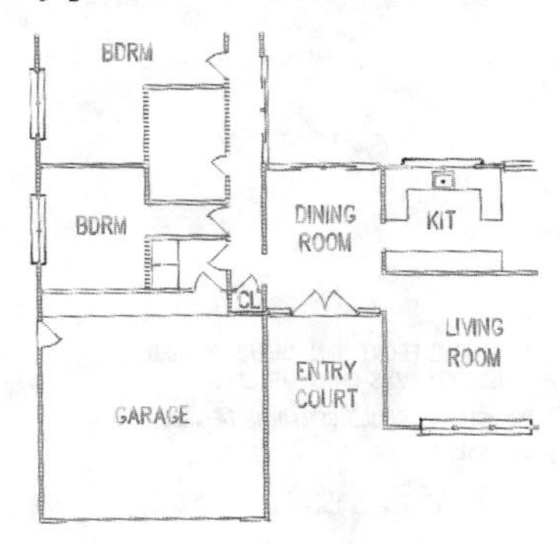

EXISTING PLAN

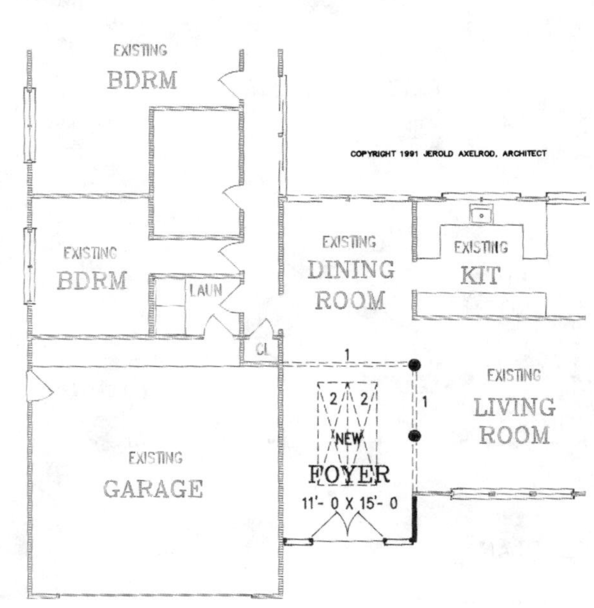

REMODELED PLAN

NOTE: CONSTRUCTION BLUEPRINTS CAN BE READILY MODIFIED TO ACCOMMODATE YOUR SPECIFIC HOUSE OR ROOM SIZES.

ROOFLINES AVAILABLE:
GABLE (SHOWN), HIP, SHED

P0014

You keep driving by that brand-new two-story, and envying its dramatic main entrance. Well, if your home is a simple, flush-front, center-hall two-story, envy no more, because this modest change will create the new look you are seeking. A 13′0″-×-3′6″ deep, reverse-gable bump is created in the plan shown. Practical benefits include a recessed, sheltered entry and a 3′6″ increase to the second floor room above. An alternative might be to remove the room over the foyer and create a dramatic two-story foyer. This addition requires minimal change to the existing home, yet could accomplish much for others to envy.

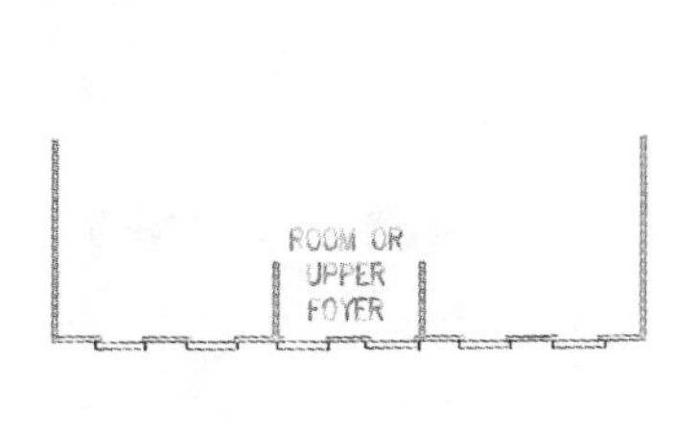

EXISTING SECOND FLOOR

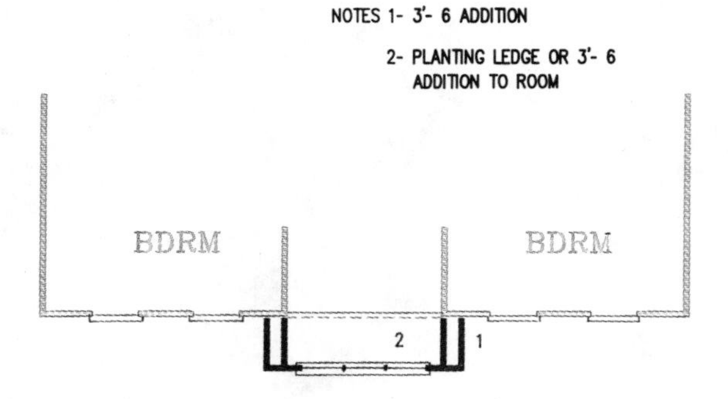

REMODELED SECOND FLOOR

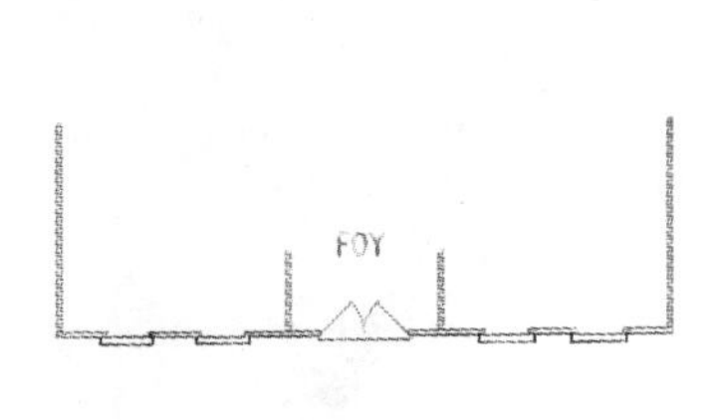

EXISTING FIRST FLOOR

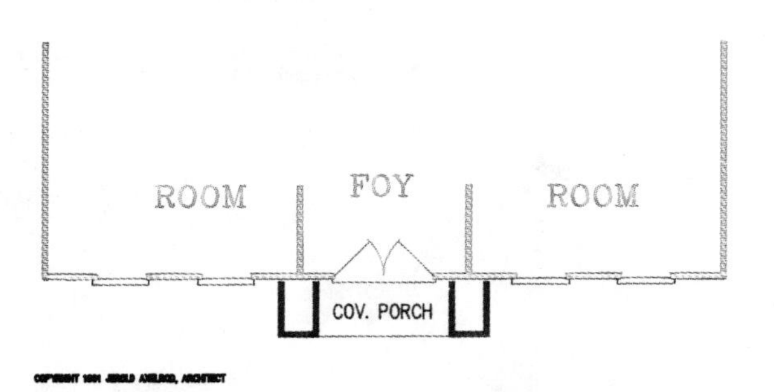

REMODELED FIRST FLOOR

PF002

ROOFLINES AVAILABLE
GABLE (SHOWN)

A dramatic new entrance is the main focus of this design. Two small three-foot-deep closets are bumped out at each side of the first floor foyer and extend two stories in height, where they terminate in a rounded arch under a reverse gable roof. A fabulous window occupies the entire space over the new double-door entry. The floor over the foyer has been removed, and this window floods the new two-story-high foyer with natural light. The overall effect is stunning, and it provides a 30-year-old home with a new presence. There is a room removed at the second floor, of course, which is the trade-off. Maybe a new first floor master suite is what you wanted anyway. There are a number of those to review in chapter 7.

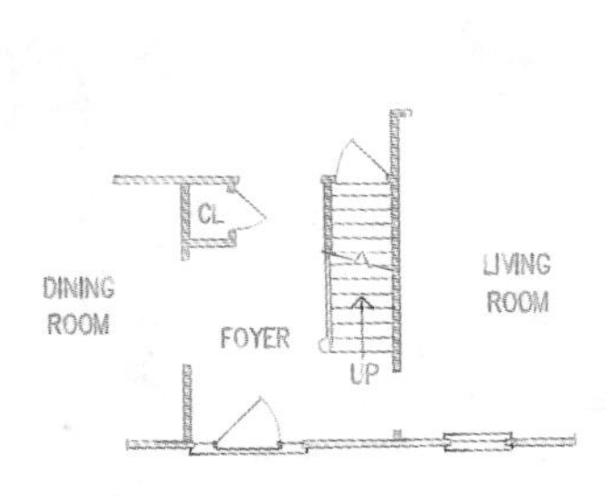

EXISTING FIRST FLOOR

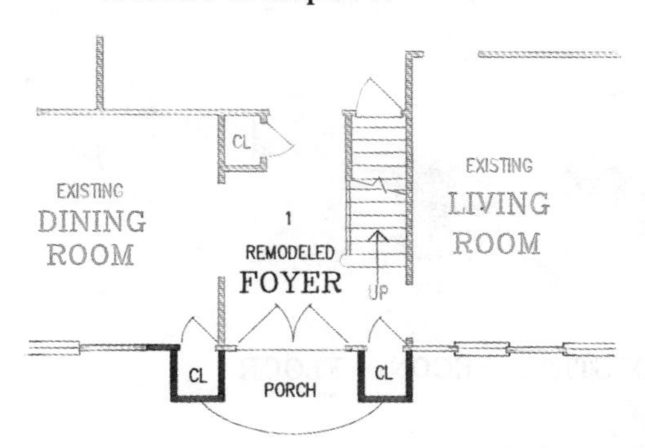

REMODELED FIRST FLOOR

NOTES
1- OPEN CEILING ABOVE
2- RAIL
3- ALTERNATE: INCREASE ADJOINING BEDROOM

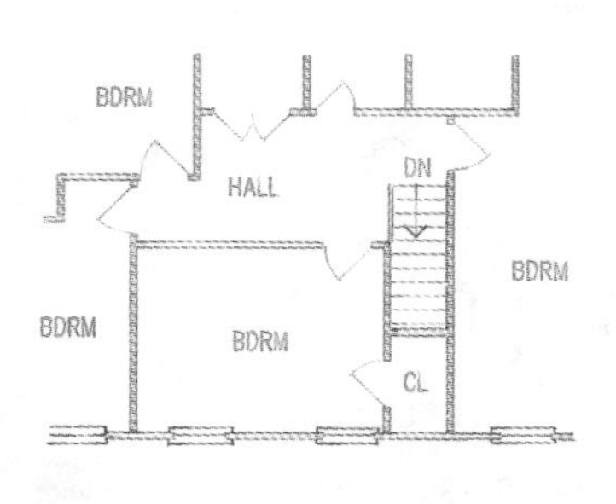

EXISTING SECOND FLOOR

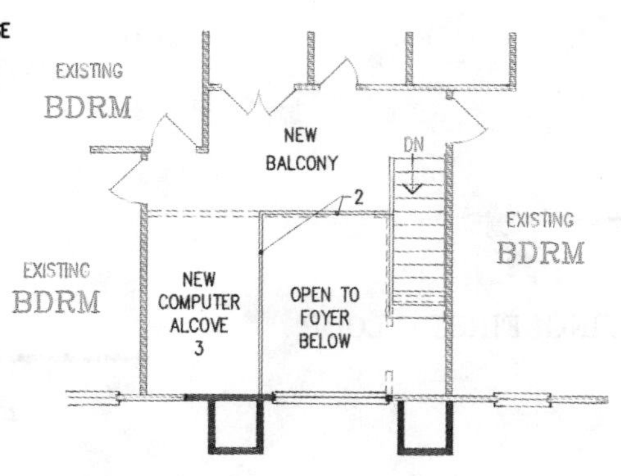

REMODELED SECOND FLOOR

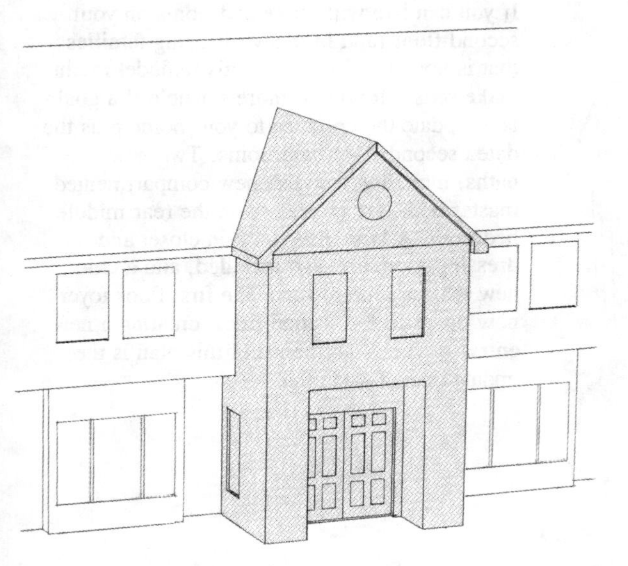

PFB01

Enlarging and modernizing second floor front bathrooms is a thorny design problem. Plan PFB02 on page 174 accomplishes it by removing a bedroom and relocating the baths to the rear. This plan achieves similar results, without the loss of the bedroom, by adding a 7-foot-deep, two-story extension to the front. It obviously requires the ability to build in front, which might require a variance. The results, however, are worth it—a sparkling new master bath and dressing room. A remodeled front facade plus an enhanced entrance foyer, with an appealing visual opening to the second floor, are extra benefits. A word of caution during construction: If you want a functioning second floor bath, you will have to stage them one at a time, doing the hall bath first.

ROOFLINES AVAILABLE:
GABLE (SHOWN), HIP, SHED, FLAT

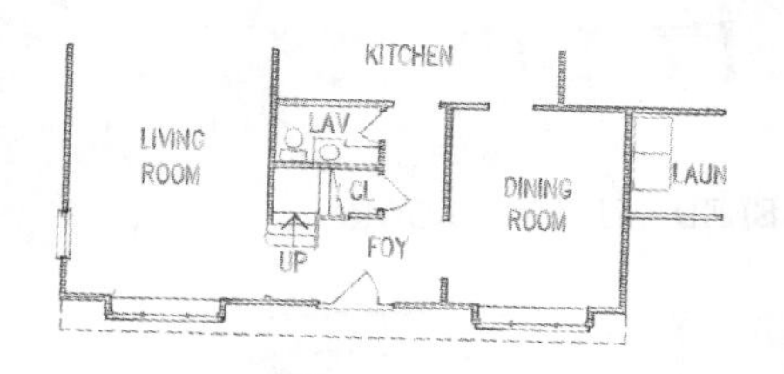

EXISTING FIRST FLOOR

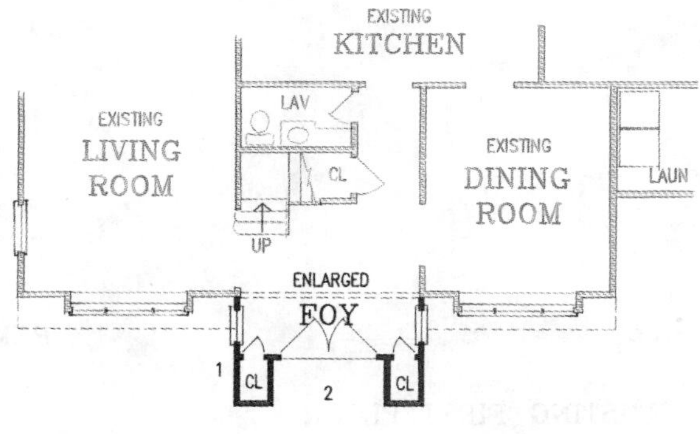

REMODELED FIRST FLOOR

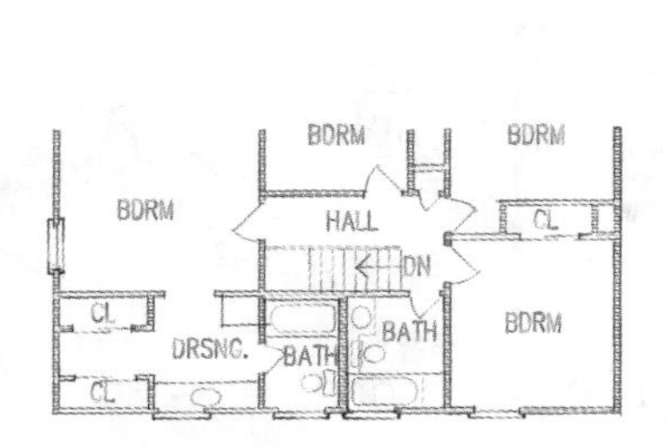

EXISTING SECOND FLOOR

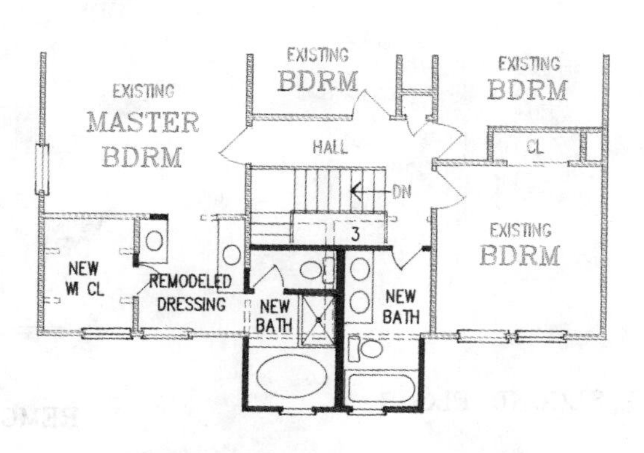

REMODELED SECOND FLOOR

PFB02

If you can live with three bedrooms on your second floor (and in many maturing families that is now possible), then this remodel might make sense. It will be more sensible if a goal is to update the entrance to your home plus the dated second floor bathrooms. Two new baths, including a stylish new compartmented master bath, are created from the rear middle bedroom. A large new walk-in closet and dressing room are also included, and a chic new master suite results. The first floor foyer now opens to the second floor, creating a new entry. A significant bonus of this plan is the updated new facade that results.

ROOFLINES AVAILABLE:
HIP (SHOWN), GABLE

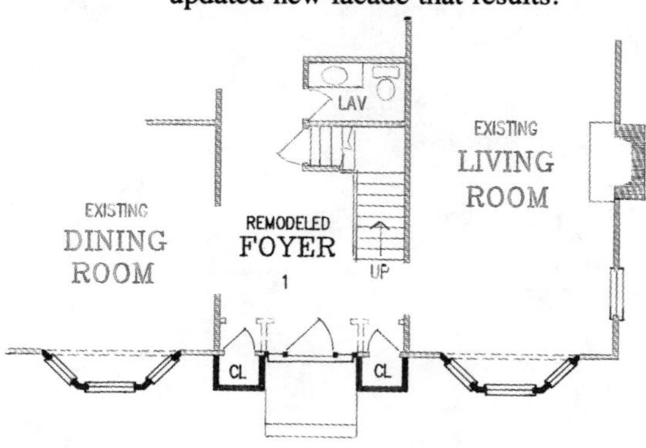

REMODELED FIRST FLOOR

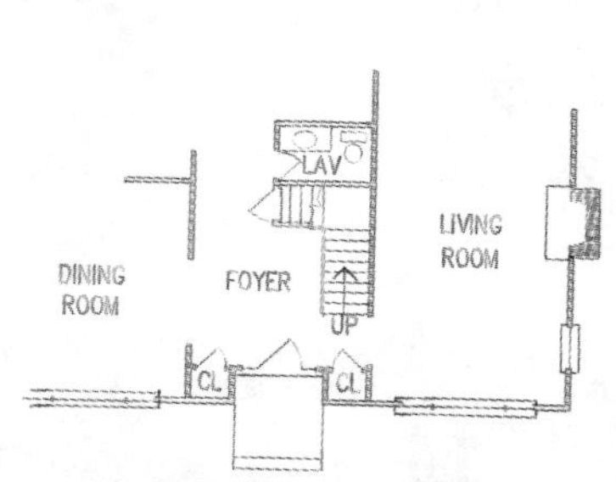

EXISTING FIRST FLOOR

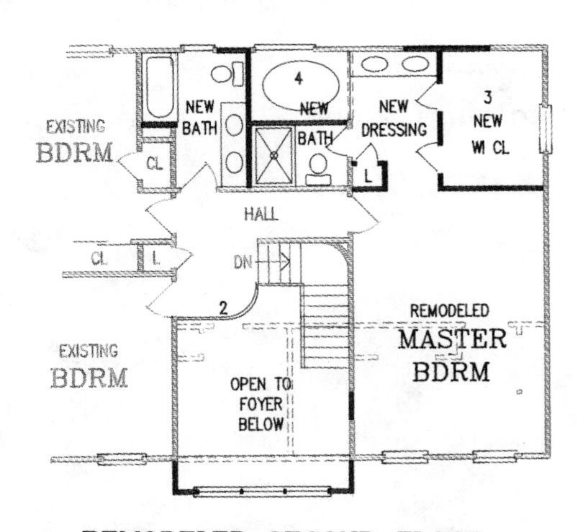

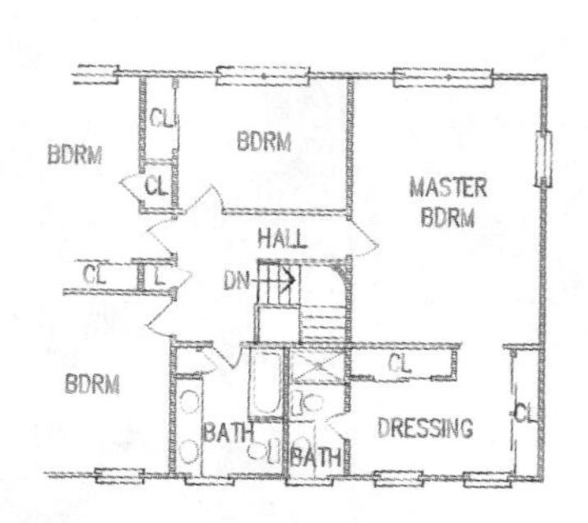

EXISTING SECOND FLOOR

REMODELED SECOND FLOOR

CONSTRUCTION BLUEPRINTS ARE AVAILABLE FOR THIS PLAN (AS WELL AS ALL OTHERS) AND INCLUDE A MATERIALS LIST

B0001

The following three plans deal with another common problem: How do you create two baths (or 1 1/2) where there's presently only one hall bath? The plans shown are extremely popular layouts derived from typical one-story, split-level, or bi-level (hi-ranch) homes. If this is your need, one of these should be fairly close to the layout of your home. The plan pictured on this page shows how to gain an extra half bath, without any exterior additions, from an existing bath only 11 feet deep. As you can see, the extra half bath is gained for the master bedroom, with very little change. With a small exterior bump of two feet, you could get two full baths. (See the plans that follow.)

NOTE: CONSTRUCTION BLUEPRINTS CAN BE READILY MODIFIED TO ACCOMMODATE YOUR SPECIFIC HOUSE OR ROOM SIZES.

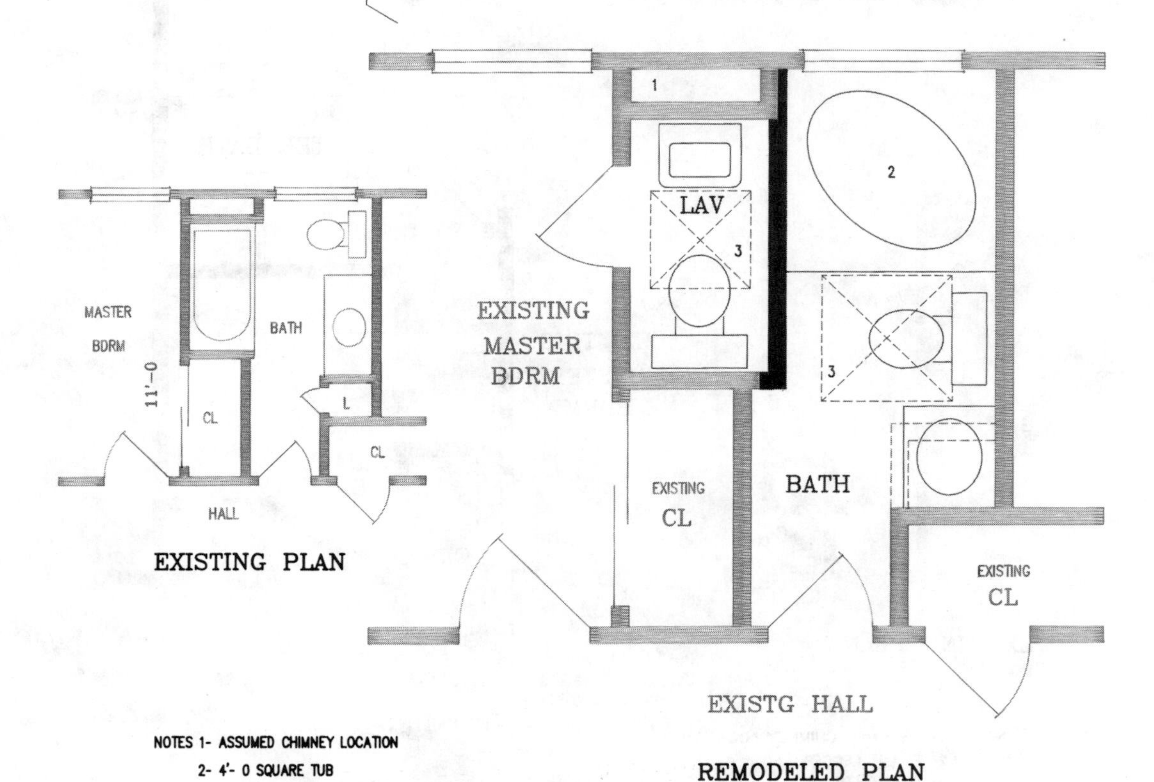

B0002

This plan creates two full tub baths from the one 13-foot-deep hall bath. It requires a small exterior bay to be popped out and does result in the loss of some closet area for the master bedroom. However, there is a respectable gain of wall space in the bedroom, which, with the use of built-ins or additional chests, could more than compensate the small loss of closet. Common to many homes of this variety is a chimney or duct chase located in this area. If one does not exist in your home, you could gain back some of the closet lost in this layout, or reduce the size of the bay. The master bath created in this design includes its own tub bath with a raised plant ledge alongside.

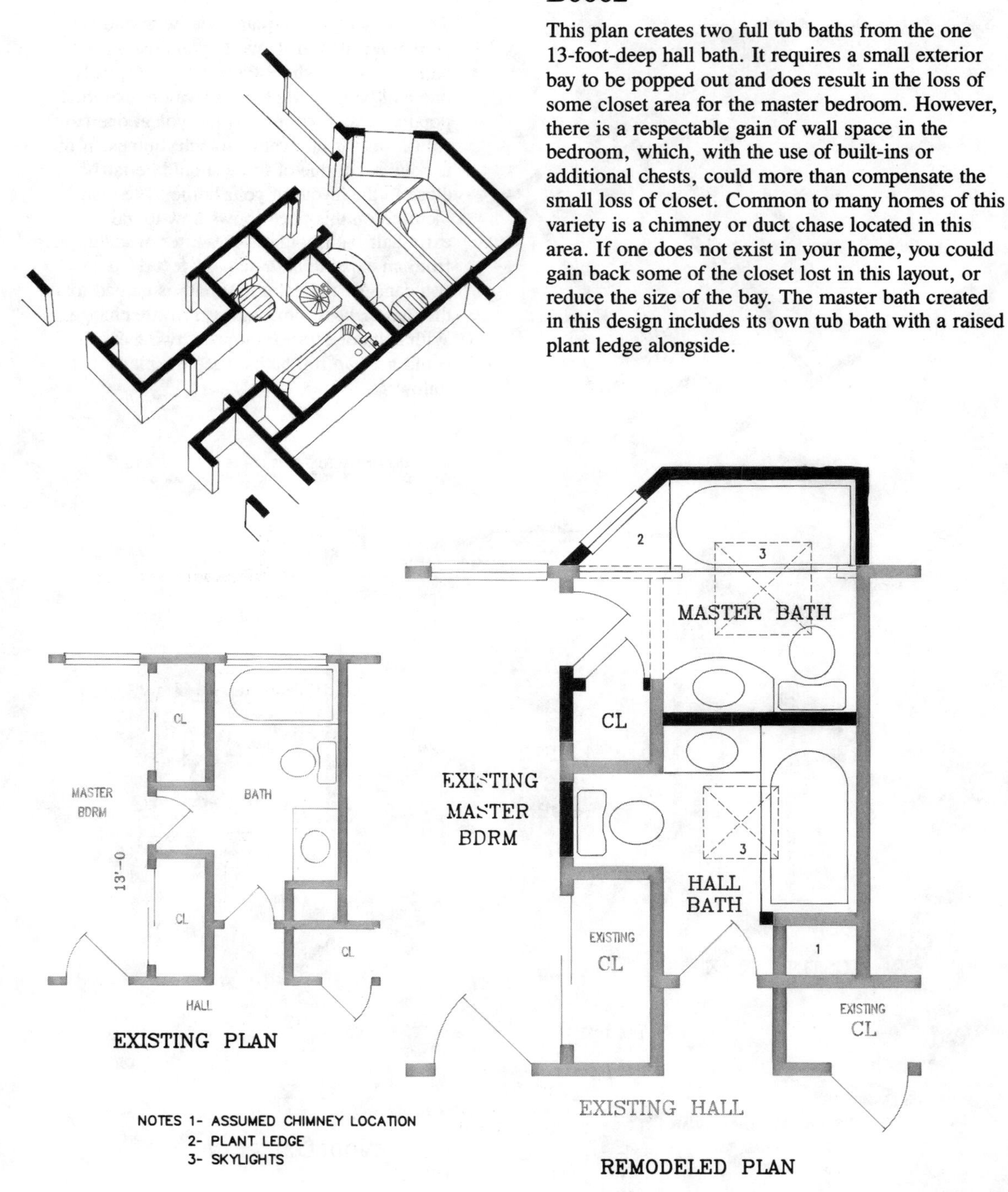

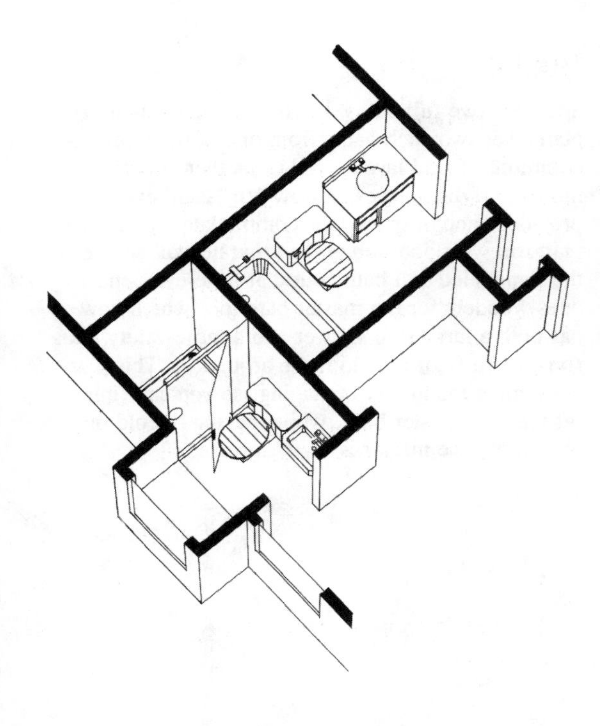

B0003

The third version creates two full baths from a 12-foot hall bath. The new master bath is a shower bath with a small box bay bumped out of the wall that adds visual enhancement to the space. In all three plans, the end result is two (or one and a half) new modern baths with 30″-×-30″ skylights in each, if possible. One of the major concerns of all of these plans is the staging problem. Because of the tight areas, none of these permit the luxury of finishing one bathroom while leaving an old one alone for daily use. All require gutting and remodeling the entire space at one time, which calls for a well-coordinated construction schedule to reduce the length of time you are inconvenienced.

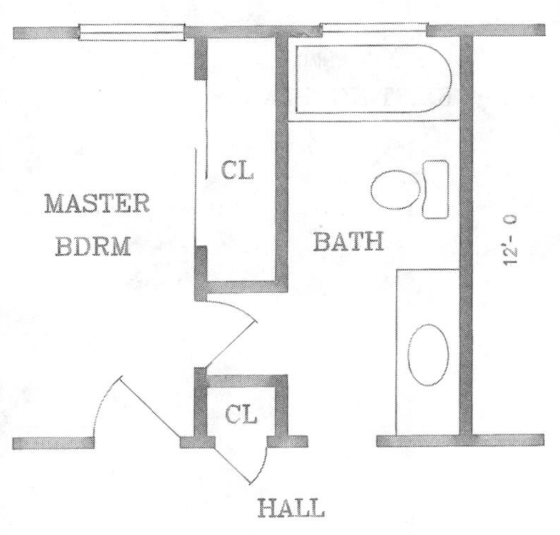

EXISTING PLAN

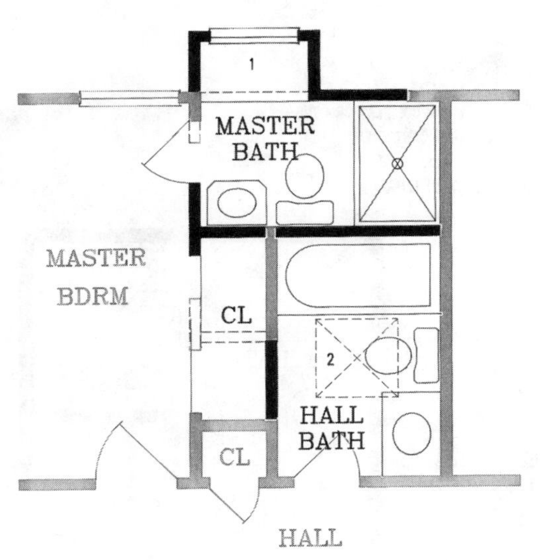

REMODELED PLAN

NOTES 1- WINDOW SEAT
2- SKYLIGHT

NOTE: CONSTRUCTION BLUEPRINTS CAN BE READILY MODIFIED TO ACCOMMODATE YOUR SPECIFIC HOUSE OR ROOM SIZES.

B0004

You have two full baths, but they are out-dated. This plan takes two such baths from one of the more commonly found layouts and turns them into modern, stylish places. A new 1′6″ cantilevered bay provides enough space to accommodate a luxurious platform whirlpool tub and a separate stall shower in the remodeled hall bath. But that little extra space does wonders for the master bath too, which now has both a large stall shower and a large vanity, plus some extra floor area to move around in. This plan does enjoy the luxury of staging, so you can finish off the new master bath, while leaving the old one alone until the master is ready for use.

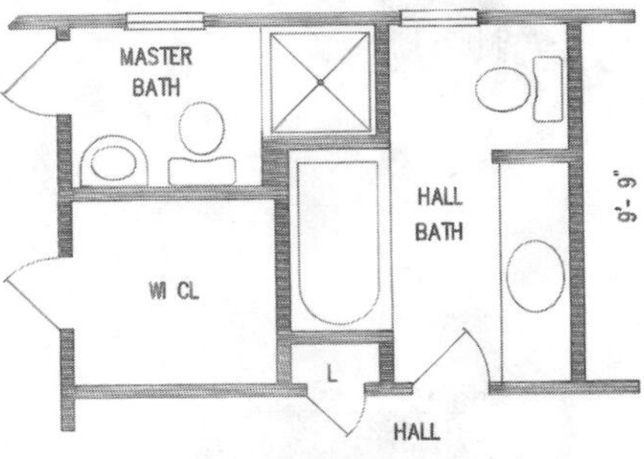

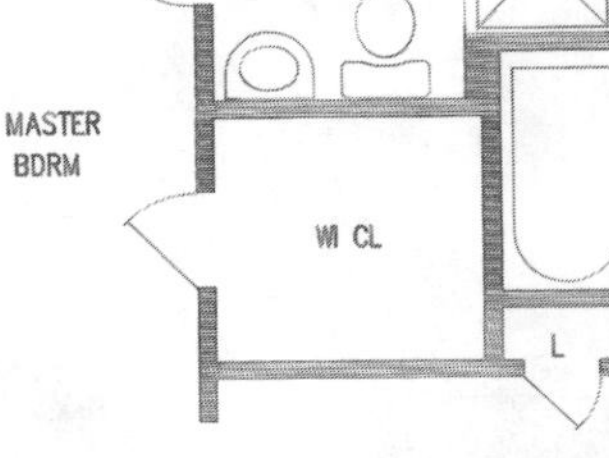

EXISTING PLAN

NOTES 1- SKYLITES
2- WHIRLPOOL PLATFORM TUB
3- 1'- 6 CANTILEVER
4- SHOWER SEAT
5- PLANT LEDGE

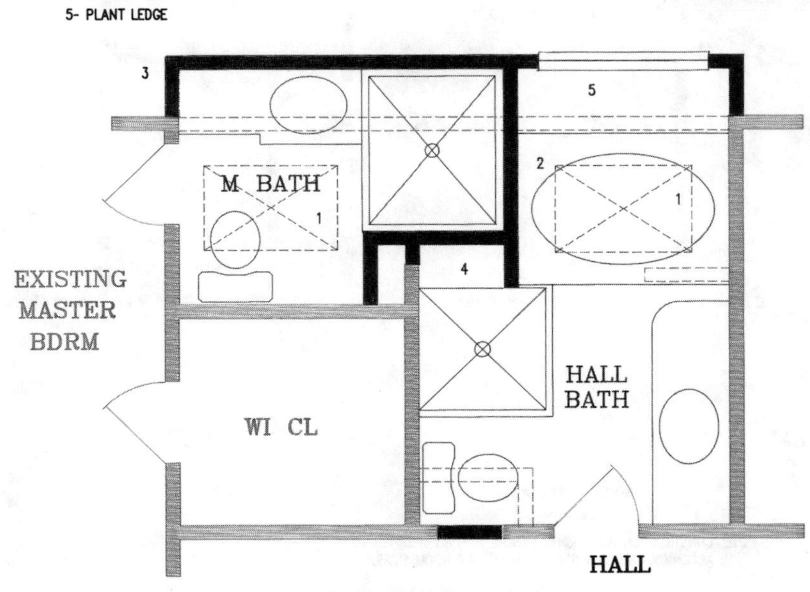

REMODELED PLAN

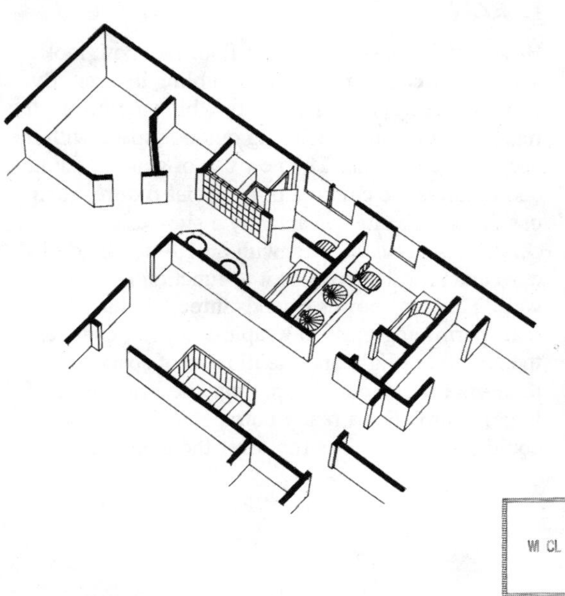

PLAN# B0006

The art of finding space requires that you consider your attic. If your home is a one-story design with a fairly steep roofline, the attic over your children's bedrooms could become a find indeed. The design illustrated on this page shows you how to take two small bedrooms and virtually double them in size by creating a loft area accessible by a ladder. This is not for everyone, but if you are not inhibited or restricted by code, the solution pictured could be a lot of fun. The lofts extend out over the hall below and could extend even further over the rooms on the opposite side. The loft could provide study, play, or sleeping space. Whatever uses you ascribe to the loft leave the bedroom to satisfy the others. The view shows alternate layouts for each bedroom.

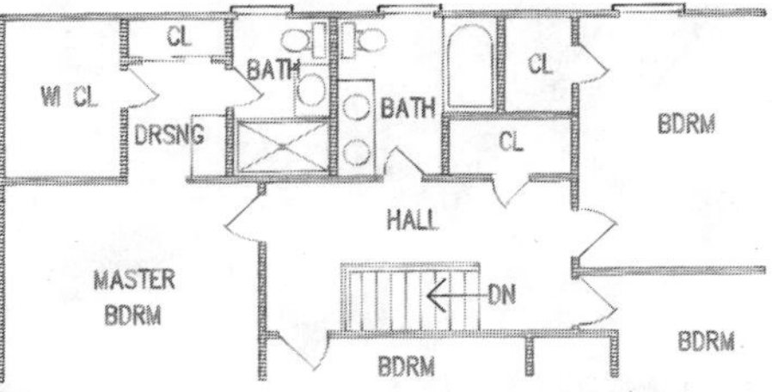

NOTES 1- REDUCE THIS BEDROOM BY 4'- 0"
2- STALL SHOWER WITH SEAT AND GLASS BLOCK WALL

EXISTING PLAN

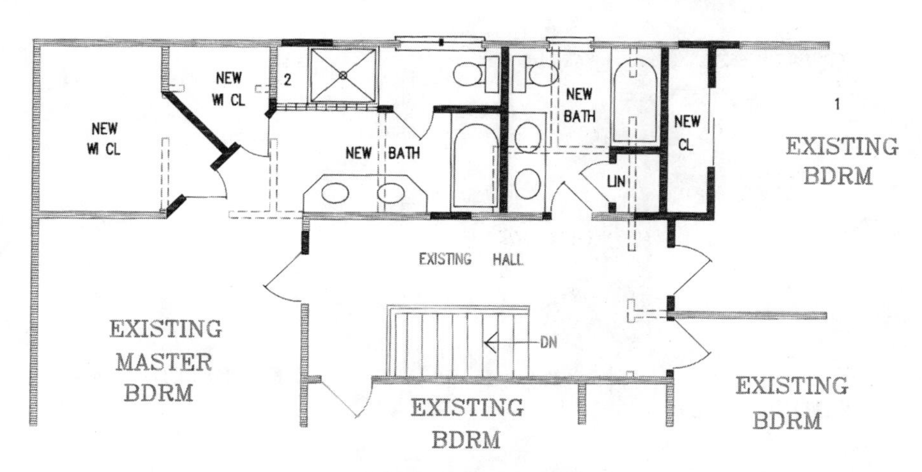

REMODELED PLAN

CONSTRUCTION BLUEPRINTS ARE AVAILABLE FOR THIS PLAN (AS WELL AS ALL OTHERS) AND INCLUDE A MATERIAL LIST AND DETAILS ON STAGING THE PROJECT

B0007

How do you make a tiny first floor hall bath look and feel twice as big without doubling its size? By creating the illusion of space by "blowing open" the rear with glass and enclosing outdoor space with the use of garden walls. The new use of glass at the tub end requires the creation of a special fixture; such use of the "fixture" is actually a site-built ceramic-tile stall shower, with an 18-inch-high tiled curb and seat that enables it to function as a tub as well. This tub/shower extends three feet into the rear, permitting glass to wrap the corner, enhancing the amount of light and the illusion of space. The three-foot bump-out also permits the installation of a larger vanity. For a really outrageous look, you could consider a glass roof over the addition.

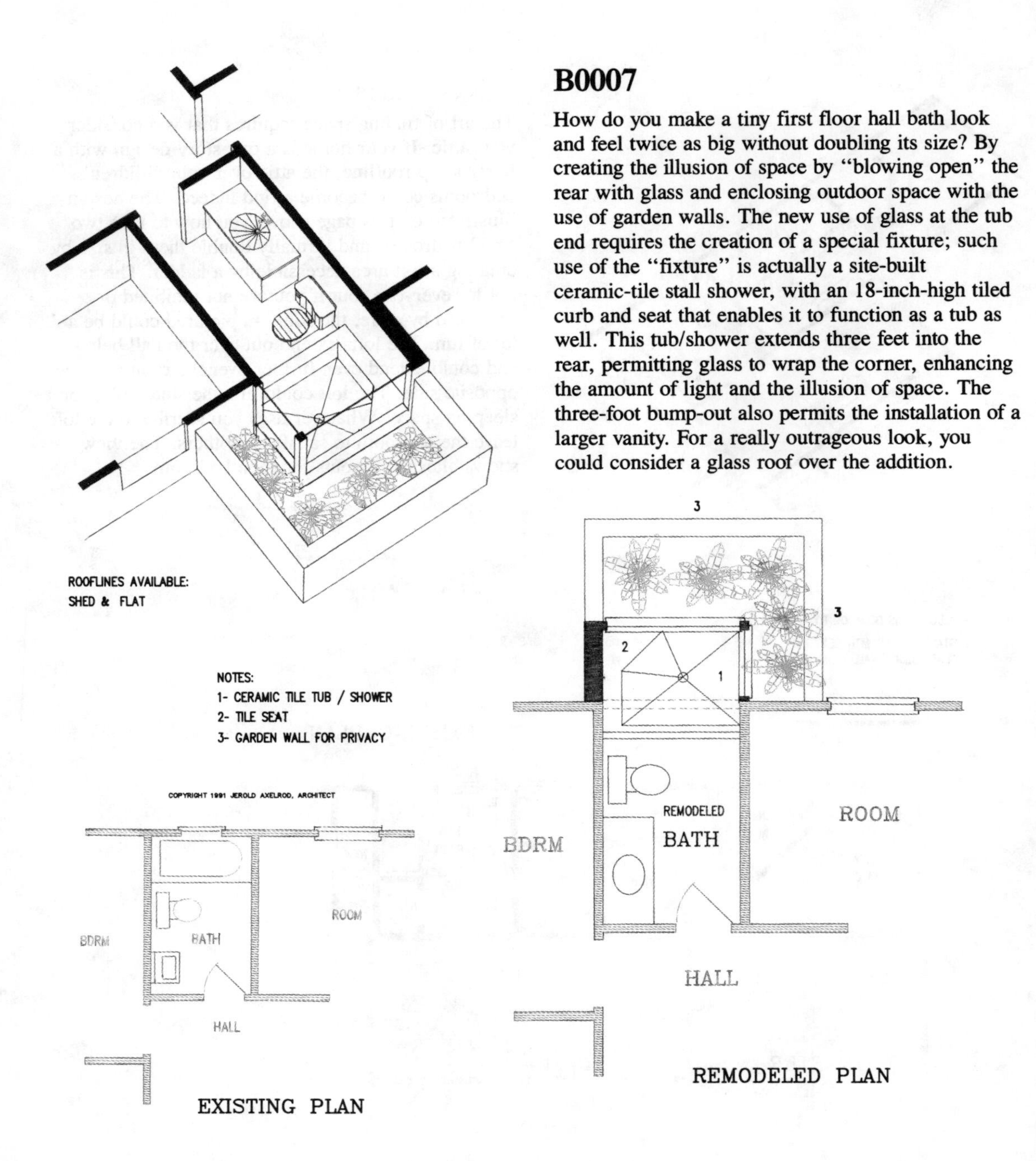

CONSTRUCTION BLUEPRINTS ARE AVAILABLE FOR THIS PLAN (AS WELL AS ALL OTHERS) AND INCLUDE FRAMING INSTRUCTIONS FOR CUSTOM TUB / SHOWER & FOR THE ROOF EXTENSION

B0009

A time-worn modest master bath is turned into an engagingly attractive new space in this modest, 4′4″ deep garden bath addition. The new room now sports a separate stall shower plus a whirlpool tub and an expanded vanity. A fixed light at the side of the vanity plus French doors bring an abundance of natural light to the room and visually expand the space to the rear. Garden walls are proposed to provide the requisite privacy, eliminating the need to heavily shade the glass. You could also consider adding a skylight to further enhance the space.

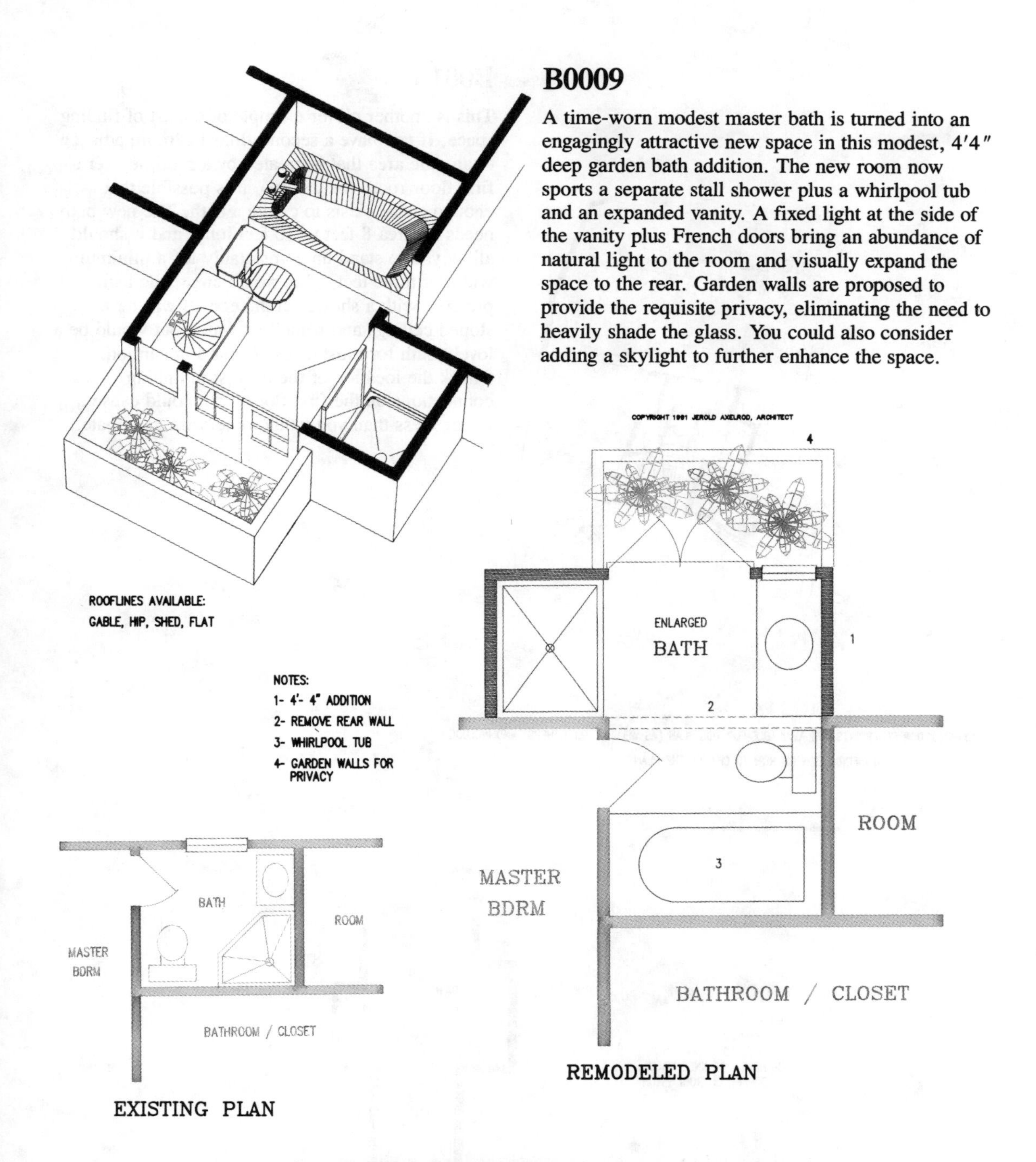

CONSTRUCTION BLUEPRINTS ARE AVAILABLE FOR THIS PLAN (AS WELL AS ALL OTHERS) AND INCLUDE PLUMBING FIXTURE MANUFACTURER'S INFORMATION AND A MATERIALS LIST

B0010

This is another clever example of the art of finding space. If you have a second floor bedroom adjacent to an attic area that is created by a roofline over a first floor room (or garage), it is possible that enough space exists to create a bath. The new bath needs an area 8 feet to 10 feet long, and it should allow you to stand up comfortably for a minimum width of three feet. This would create the bath pictured with a shower at the end. It will have sloped ceilings and some low walls, but would be a lovely bath for most of us. A word of caution: Check the location of the nearest plumbing connections on the first floor, or it could cause a larger mess than you would otherwise anticipate.

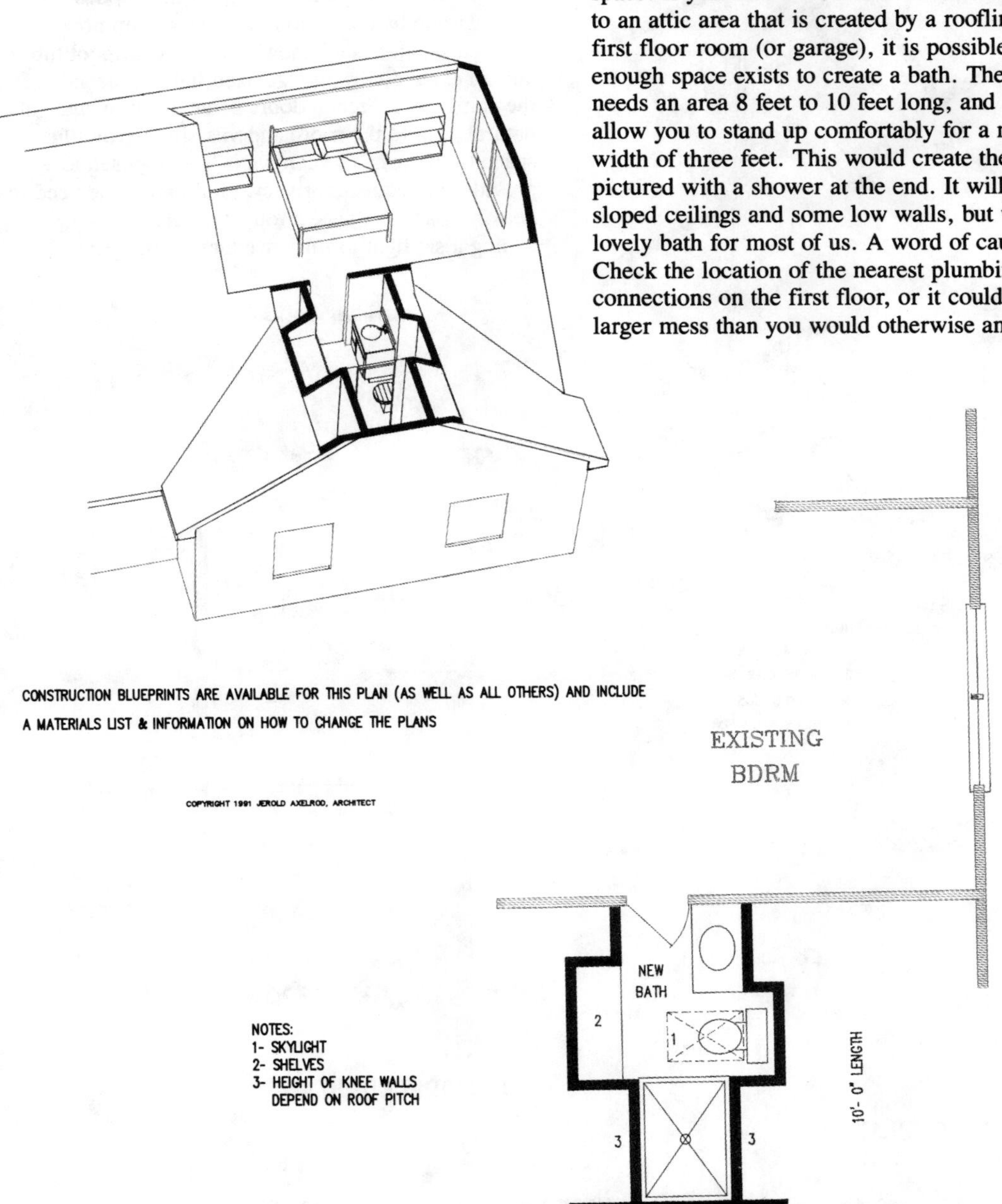

EXISTING SECOND FLOOR

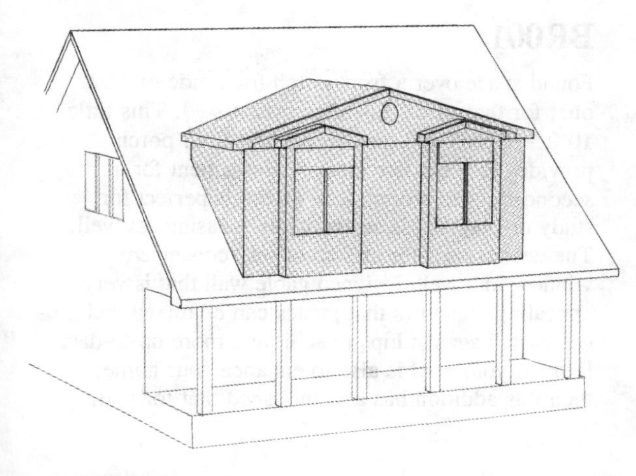

BR002

This attractive-looking second floor "bump" is certain to please if you need a little extra elbow room for two small second floor bedrooms. The addition is located over an existing first floor porch common to many older two-story homes, and replaces a tiny dormer with a 19′8″-×-8′0″ dormered addition. Each bedroom gains an alcove approximately 7′6″×7′6″ and a nice-sized closet. The alcoves are ideally furnishable as play or study spaces to complement the sleeping areas. The reverse-gable rooflines should blend well with the existing roof.

ROOFLINES AVAILABLE:
GABLE (SHOWN), HIP, SHED, FLAT

NOTE 1- REMOVE EXISTG FRONT WALL AND SMALL FRONT DORMER

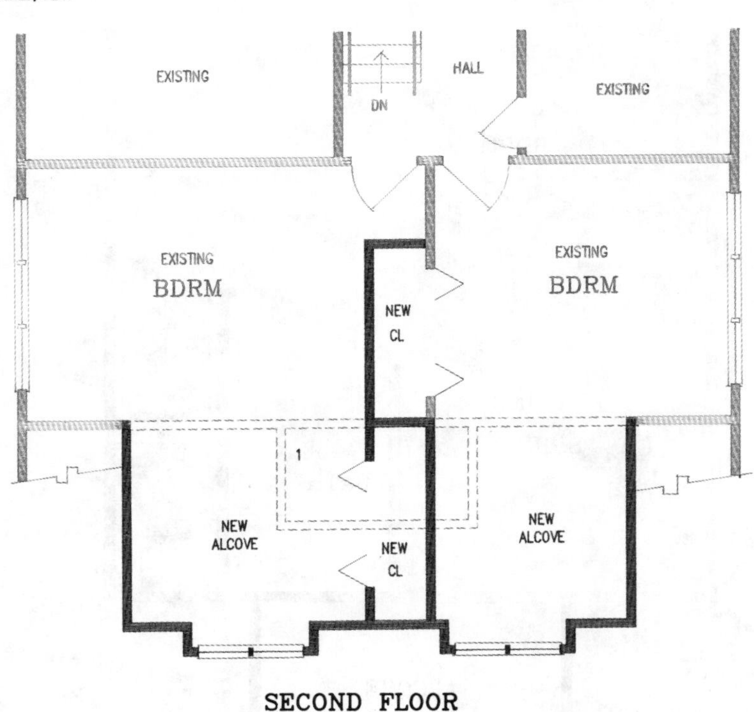

CONSTRUCTION BLUEPRINTS ARE AVAILABLE FOR THIS PLAN (AS WELL AS ALL OTHERS) AND INCLUDE THE DESIGN OF:
THE REVERSE GABLE ROOFLINES

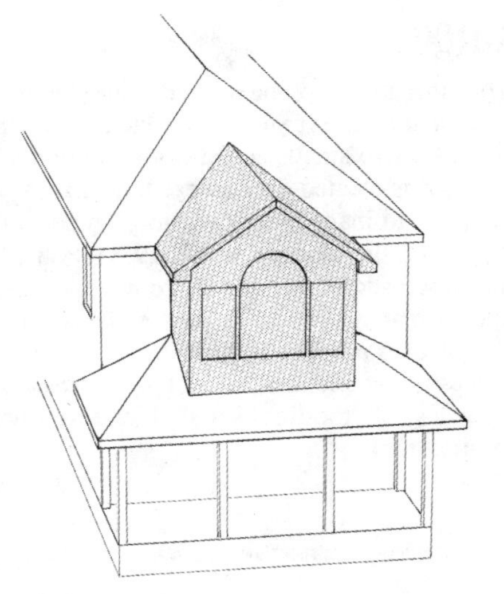

ROOFLINES AVAILABLE:
GABLE (SHOWN), HIP, SHED, FLAT

BR001

Found space over a front porch (or a side or rear one, for that matter) is often overlooked. This little 10′0″-×-5′6″ addition over a first floor porch provides an attractive alcove enlargement for a second floor bedroom. The alcove is perfect for study or play and is aesthetically pleasing, as well. The design calls for an attractively conceived window in a well-designed gable wall that is very appealing. It shows that gables can be mixed and contrasted against hips to achieve a more up-to-date look. If your goal is also to enhance your home, then this addition has accomplished that for you, too.

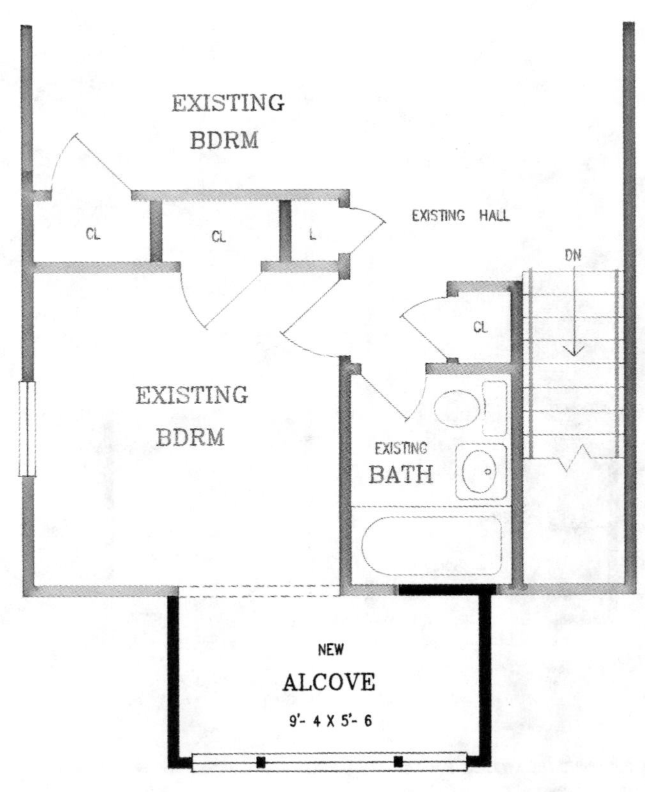

SECOND FLOOR

BR003

As you may have noticed by now, one of my goals is to show alternatives and to get you thinking that way. There are a mutlitude of solutions to each remodeling need, and each solution is likely to be different as concerns aesthetics, function, and cost. This plan, plus the next two, present three different solutions to modifying a one-and-one-half-story roofline (cottage, cape, etc.) that has two tiny shed roof dormers. The goal is to provide a new look, more light, and possibly more space to the bedrooms. This first plan links the two dormers together. There is a significant gain in living space to the two bedrooms and an aesthetic improvement to the home. An elliptical window provides light to the stair hall.

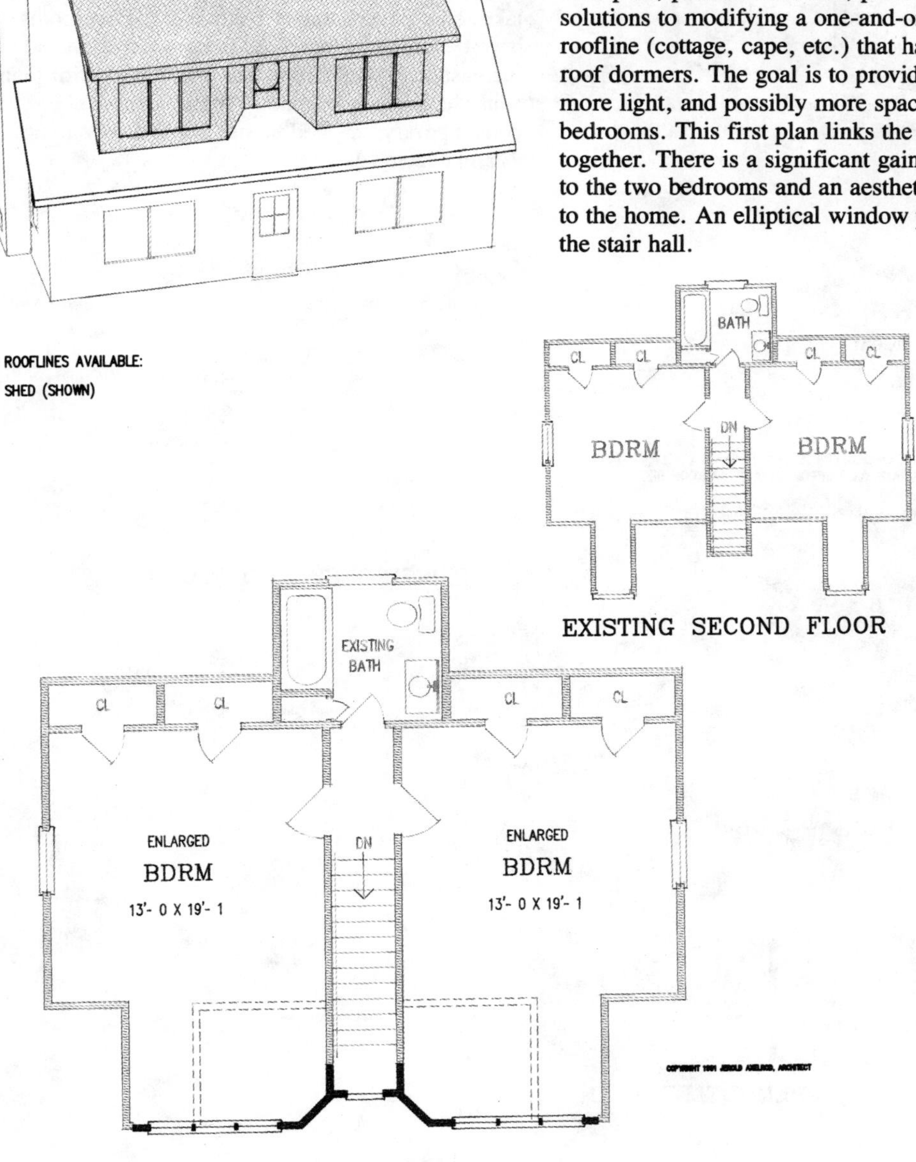

EXISTING SECOND FLOOR

REMODELED PLAN

BR004

This second alternative roofline change replaces the existing narrow shed dormers with two new extra-wide reverse-gable dormers. Each dormer features a triple window with a stylish half-round light over; the overall effect is one of updating, where, by simply changing these dormers, the home takes on a new character. Although the bedrooms have benefited by increased light and size, the actual increase in area is a little less than in the prior plan. With the addition of a new porch you have a contemporary, updated home. See FAC34 on page 272 in chapter 9.

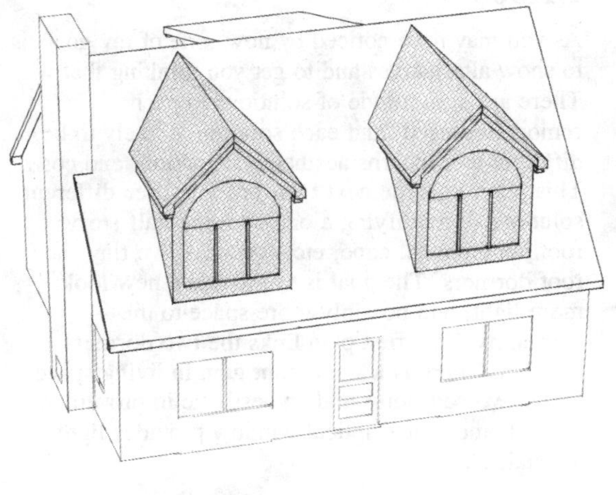

ROOFLINES AVAILABLE:
GABLE (SHOWN), HIP, SHED, FLAT

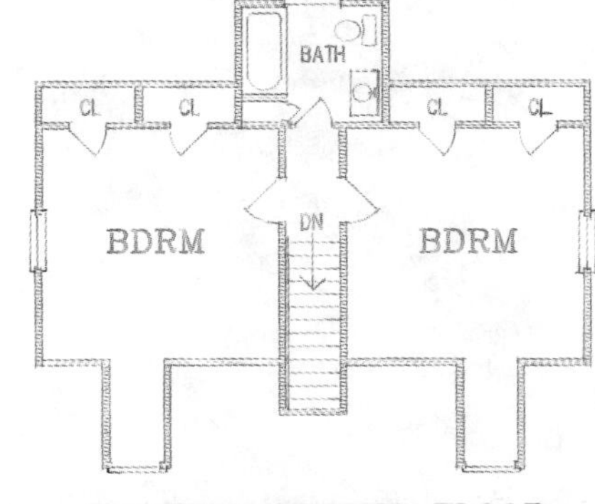

EXISTING SECOND FLOOR

NOTE: CONSTRUCTION BLUEPRINTS CAN BE READILY MODIFIED TO ACCOMMODATE YOUR SPECIFIC HOUSE OR ROOM SIZES.

EXISTING BATH

CL CL L CL CL

DN

ENLARGED BDRM

ENLARGED BDRM

NEW ALCOVE 6'- 10 X 6'- 0

NEW ALCOVE 6'- 10 X 6'- 0

REMODELED PLAN

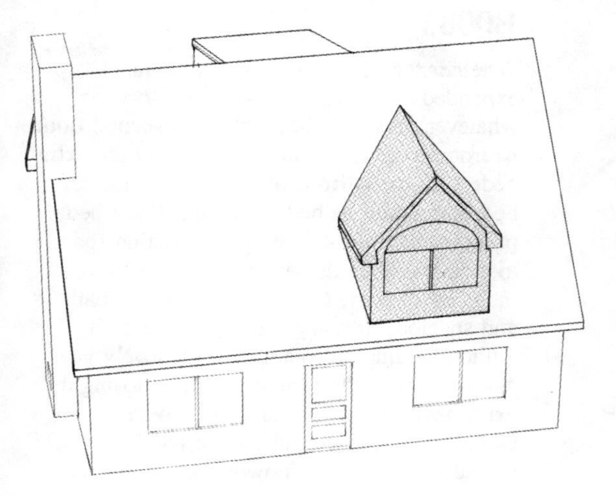

BR005

The third plan effects a significant transformation of several areas. It actually does not provide any gain in living area—there is, in fact, a slight loss—but it results in an important change to the exterior form and the interior spaces, as well. Both dormers are removed and the bedroom on just one side is increased by the addition of a large reverse gable, which has been designed as a gable wall rather than a dormer. This is not mirrored on the opposite side, where the bedroom has been replaced with a dramatic open loft. Part of the ceiling of the living room below is removed and an expansive vaulted ceiling results.

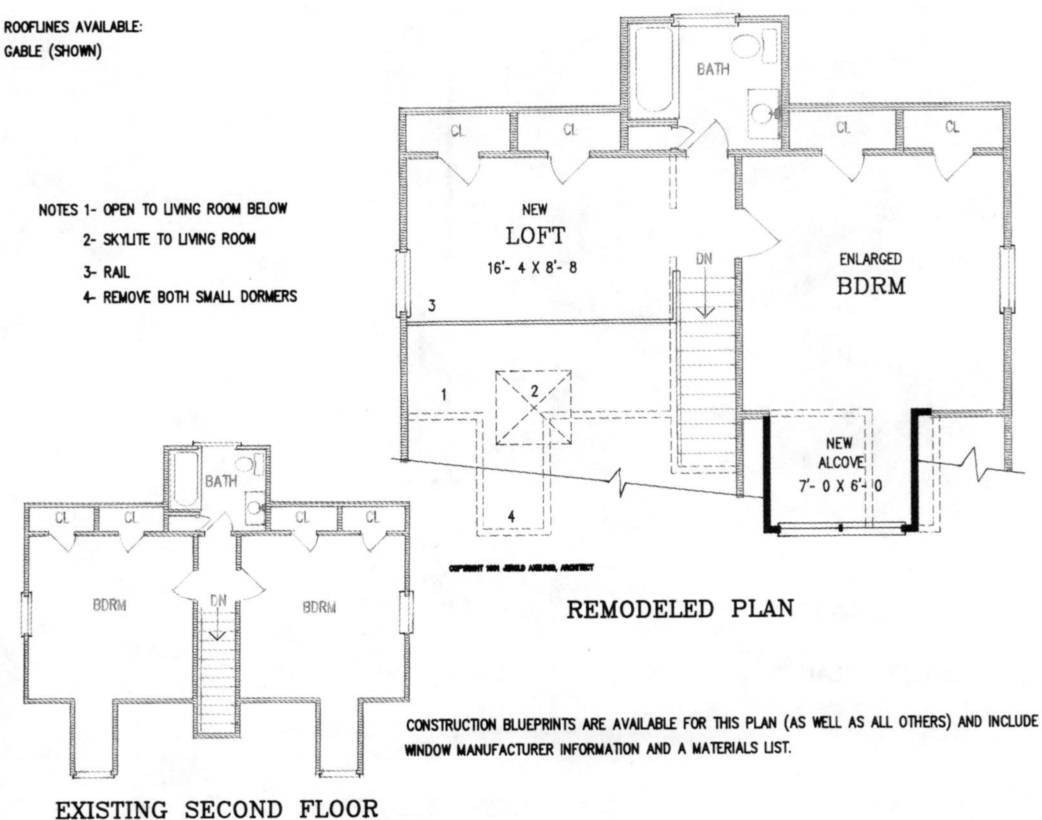

EXISTING SECOND FLOOR

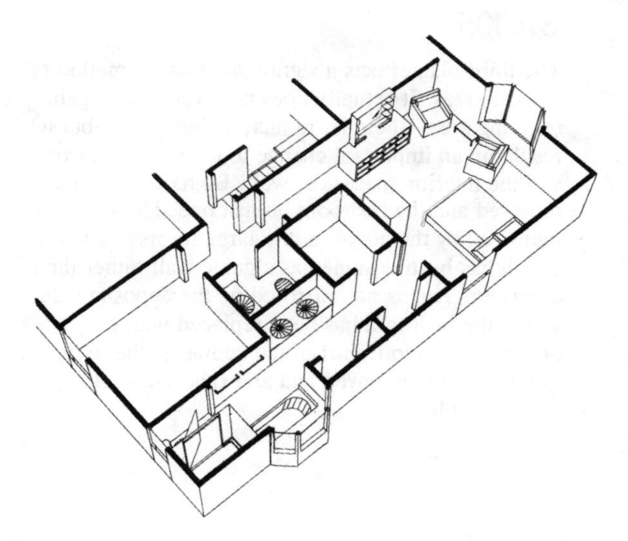

B0005

Whether they call your home farm-ranch, expanded cape, one-and-one-half-story, or whatever, it is your belief that the second floor bedrooms are ample for the kids, but the extra bedroom across from your first floor master bedroom could be better utilized. What better purpose than redistributing the existing space to create a fabulous new master suite that includes a smart, new, compartmented bath and spacious dressing area? And we do it without adding on (other than the lovely new bay window at the tub) and without losing the extra bedroom. The bedroom becomes a useful home office, and the old hall bath is scaled down to a tidy powder room. A great plan, if it fits your needs.

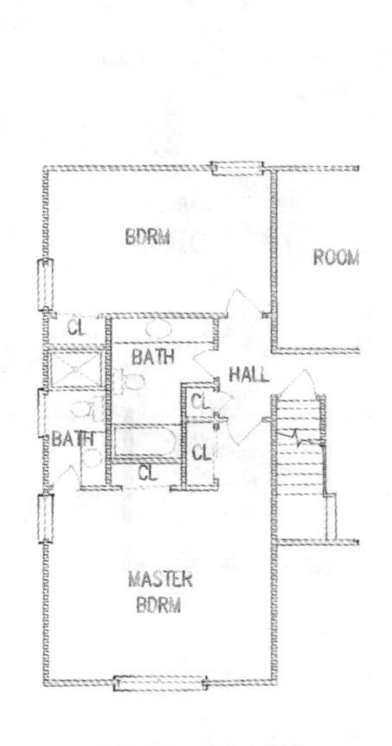

EXISTING PLAN

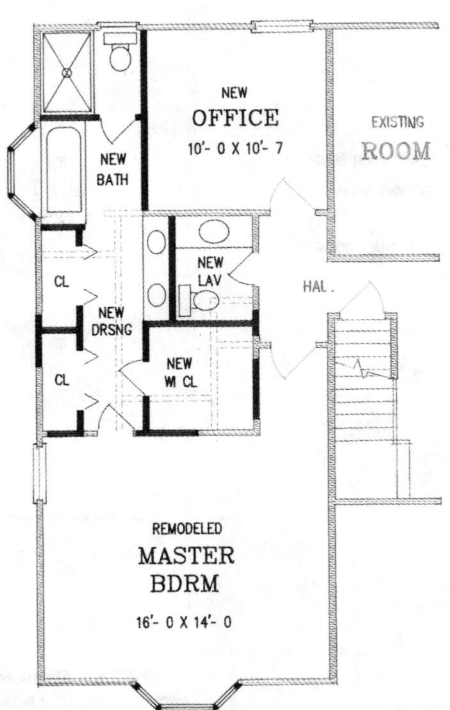

REMODELED PLAN

BRB22

That extra first floor bedroom next to yours is seldom used anymore, and its space is very appealing. That is particularly so, since you really would love to have a new private bath and dressing area, with all the modern accoutrements. This plan shows you how to do it. Since the old hall bath is no longer needed, it is converted to a powder room for guests. The space left over is added to help create a fabulous new master bath. A separate stall shower is located in one section of the two-compartment bath. A new angle bay window encloses the rear of a luxurious whirlpool tub, which shares space with a dressing-top vanity in the new dressing room. The old closet is replaced by a spacious new walk-in closet, and its former space is now ideal for built-ins.

NOTES:
1- WHIRLPOOL TUB IN STEPPED PLATFORM
2- BUILT-IN

NEW LAV
NEW BATH
NEW DRSNG.
NEW WI CL
CL
HALL
REMODELED MASTER BDRM
FOY
CL

BATH
CL
BDRM
CL
HALL
MASTER BDRM
CL
FOY
CL

REMODELED PLAN

EXISTING PLAN

CONSTRUCTION BLUEPRINTS ARE AVAILABLE FOR THIS PLAN (AS WELL AS ALL OTHERS) AND INCLUDE A MATERIAL LIST

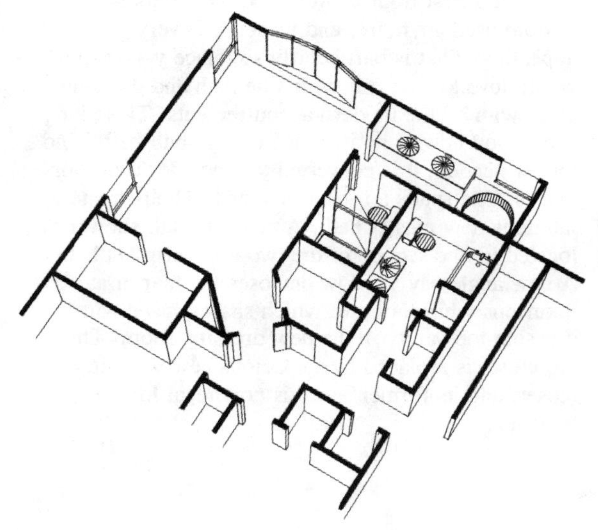

BRB13

This minimal 5′4″ deep one-story addition is expressly designed to attach to the rear of a typical master bedroom and two standard baths. Though modest in area, it provides all the room necessary to create an up-to-date, splashy new master suite. The hall bath remains, but the old master bath is remodeled to create a two-compartment room—one for a stall shower and toilet, and the new addition part for a large round whirlpool tub and a long dual-basin vanity. The old bath and closet areas are also remodeled to create two new walk-in closets, and the entrance to the suite is changed to a dramatic new angle. Construction blueprints provide the addition as a simple shed roof, but you could have it changed to a gable or hip as may be necessary to suit your home.

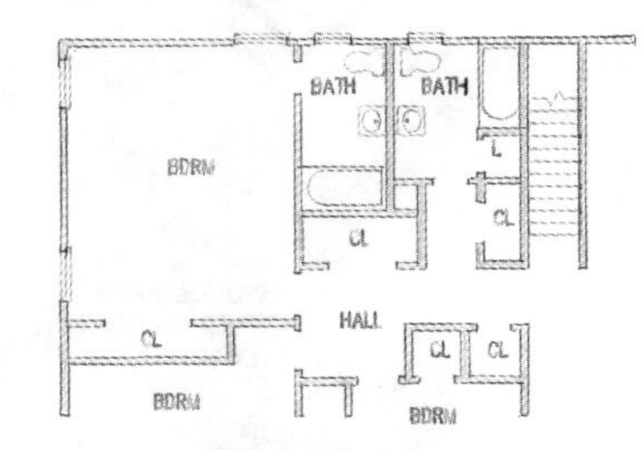

EXISTING PLAN

NOTES 1- REMOVE EXISTING TUB AND SINK
2- ADD ADDITIONAL SINK IF SPACE EXISTS
3- WHIRLPOOL TUB
4- SKYLIGHTS

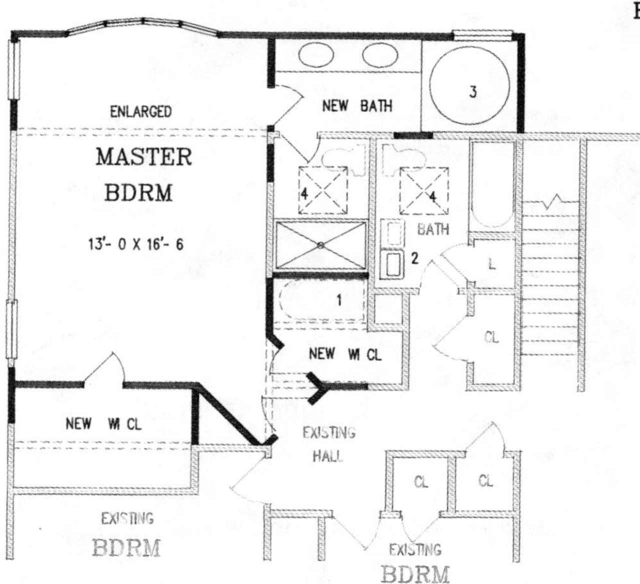

REMODELED PLAN

BRB21

The fabulous new master bath shown here is a marvelous example of how the art of finding space, as discussed in chapter 3, can produce a remodeling project that far exceeds expectations. Your house is a wide two-story with an attached garage. The master bedroom is situated on the garage side of the second floor, and its dated bathroom and dressing area is cramped and ripe for remodeling. By utilizing the found space over the rear corner of the garage for an expansive new walk-in closet, the former closet and dressing area become available to create the smart new tub and vanity section of the new compartmented bath. The old shower area and water closet remain, although they'll likely be replaced.

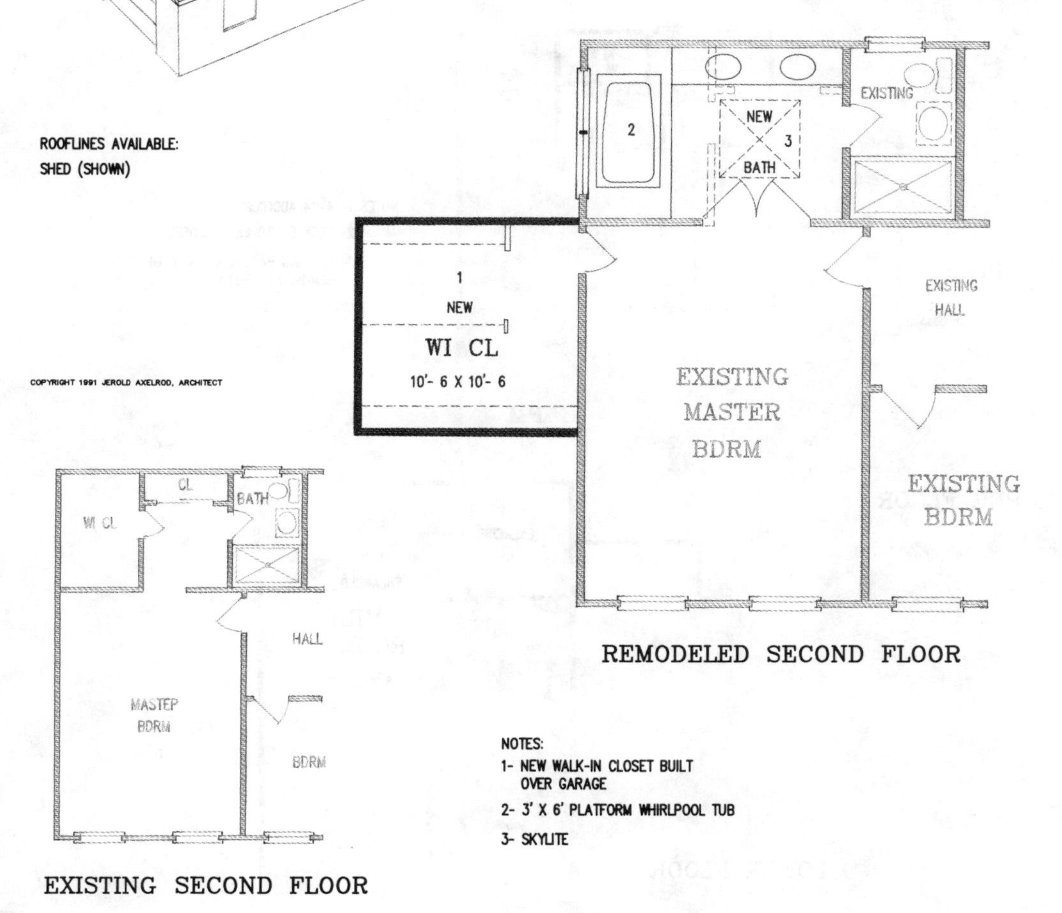

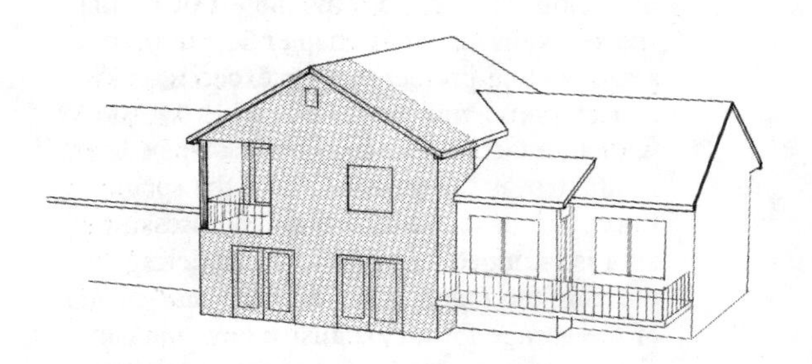

FBR05

In my pursuit to provide you with studied alternatives, this book presents three types of solutions to the split-level where three bedrooms share one bath. The least costly—those plans that create an extra bath from the existing space—were explored earlier in this chapter. Elsewhere in the prior chapter are various plans that add an entire new master bedroom suite, either up or out. This plan is a middle road. It adds a compartmented new bath, dressing alcove, and walk-in closet to the existing master bedroom and does so with a little style. A covered private deck is part of the plan, as is an expanded lower level family room. The addition measures 22′4″×8′4″.

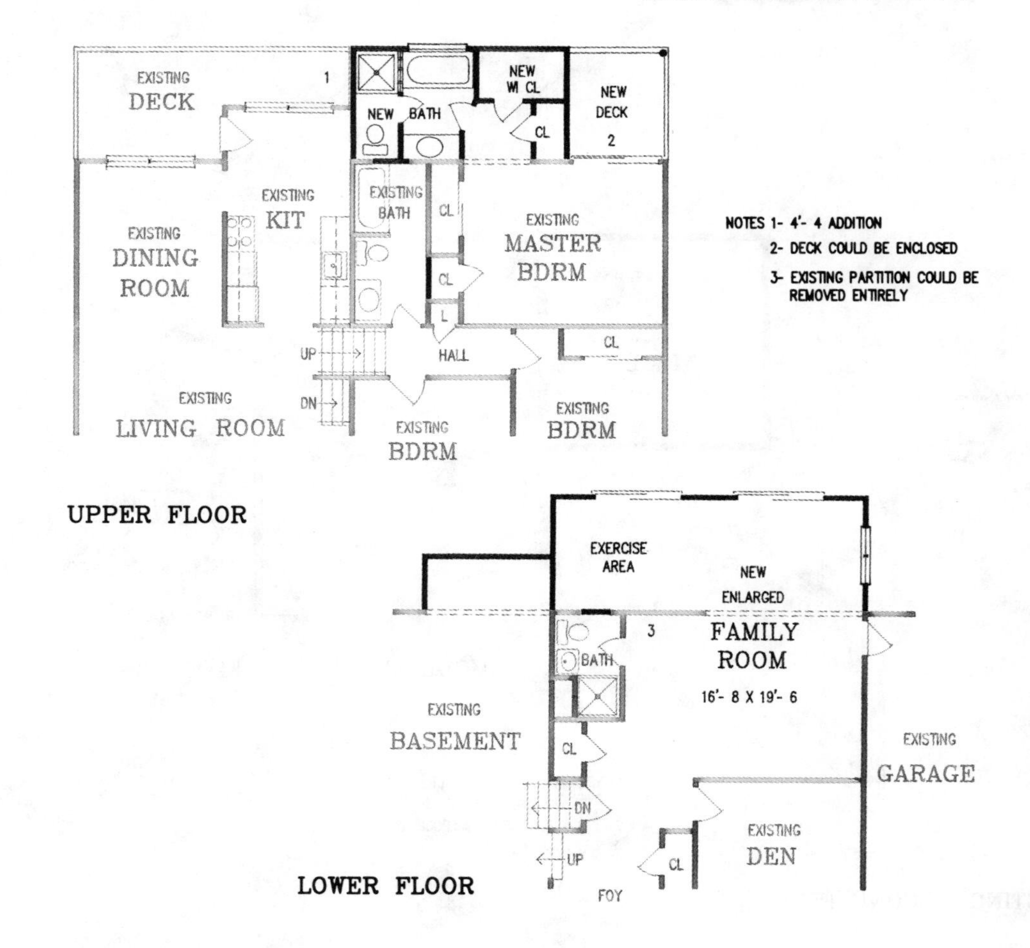

KB001

What is the best way to create a master bath and expand the kitchen in a split-foyer (hi-ranch, bi-level) plan? The best way probably involves pushing out the lower level, as shown in plan KBR02 on page 195, but if budgeting is a real concern, the plan pictured here accomplishes the task as economically as possible. The 5′4″ deep addition provides a new master bath with its own tub (a stall shower could be substituted if desired), and adds enough space to create a lovely island kitchen. This remodeled kitchen provides two places for informal meals. The first is the snack counter at one side of the island; the other is a spacious, skylit breakfast area under a new triple window.

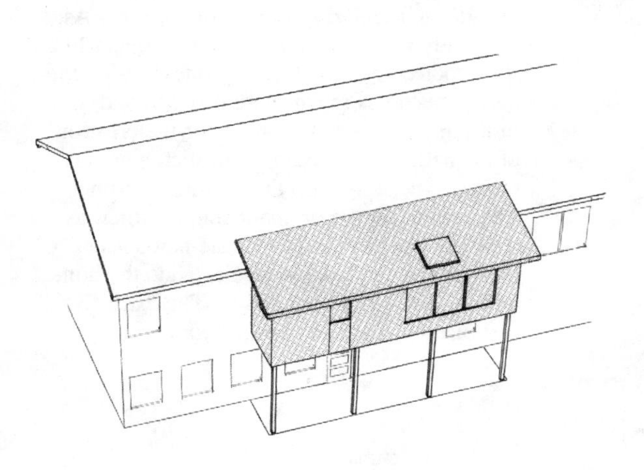

ROOFLINES AVAILABLE:
SHED (SHOWN), GABLE, HIP

NOTES:

1- 5'- 4 ADDITION
2- REMOVE EXISTING REAR WALL
3- NEW KITCHEN ISLAND
4- ALTERNATE: STALL SHOWER
5- SKYLITE

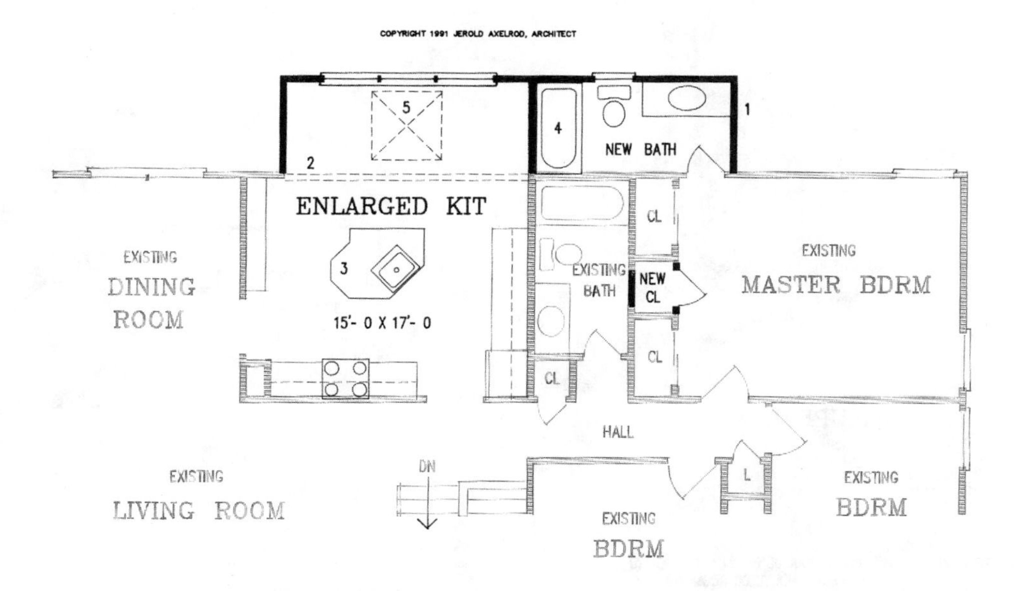

UPPER FLOOR

KB002

A desire for both an expanded kitchen and a second bath certainly ranks high on many homeowner's wish lists. The solutions are infinite, particularly in one-story homes. As you study multilevel plans—and particularly a shed roofed split-level, as the one shown—the design becomes much more difficult and limiting. The design solution for the creation of a small, 5-foot "bump" on such a home requires meshing with the existing roofline. The results can be dramatic and exciting, as shown. Special windows in the new master bath and new breakfast area provide the home with a refreshing appearance. The extra five feet also enhance the family room.

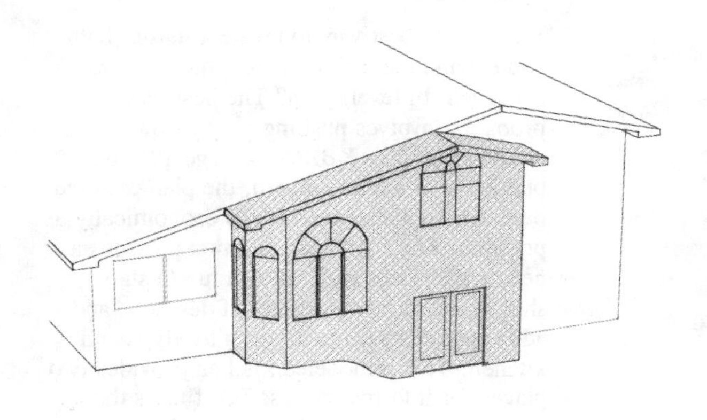

ROOFLINES AVAILABLE:
ONLY AS SHOWN

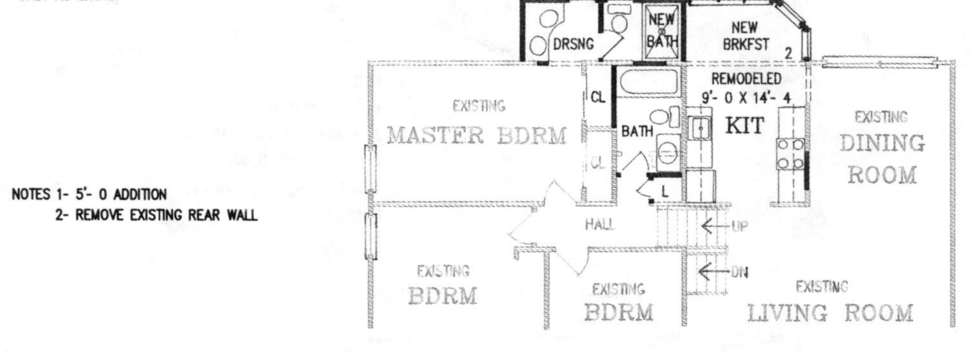

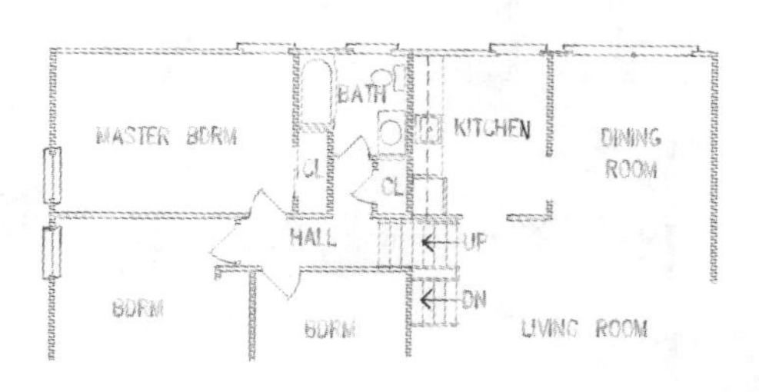

EXISTING UPPER FLOOR

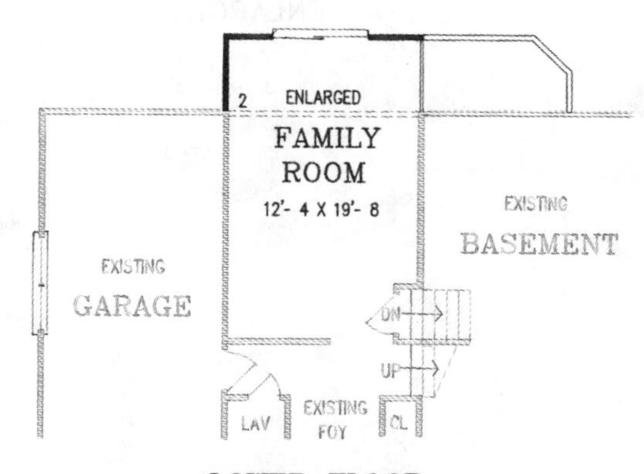

LOWER FLOOR
REMODELED

NOTE: CONSTRUCTION BLUEPRINTS CAN BE READILY MODIFIED TO ACCOMMODATE YOUR SPECIFIC HOUSE OR ROOM SIZES.

KBR02

ROOFLINES AVAILABLE:
GABLE (SHOWN), HIP, SHED, FLAT

Adding to a high ranch (also known as a raised ranch, bi-level, or split foyer, depending on where you live) adds the question of what to do with the lower floor. Many times we design to cantilever out or build on posts when doing a renovation to such a home. This plan, however, adds eight feet to both floors and accomplishes benefits for both. The prime beneficiaries are a new, private master bath and an expanded kitchen with a lovely outside breakfast corner. The dining room has been bumped out a few feet, and a new large deck is proposed. The lower level is enhanced with a new workshop off the garage and an extension to the family room, which makes for a great exercise or play area.

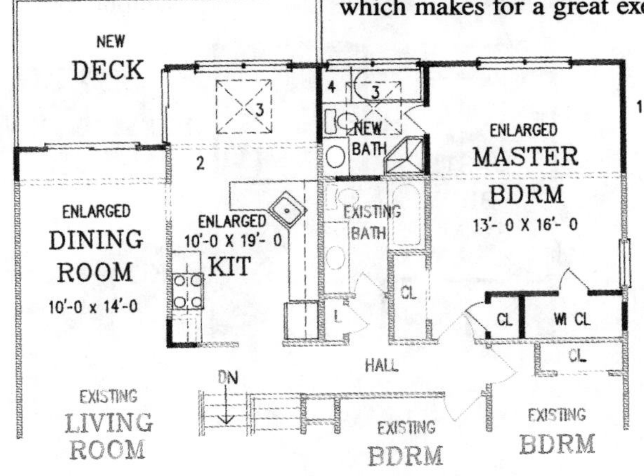

UPPER FLOOR

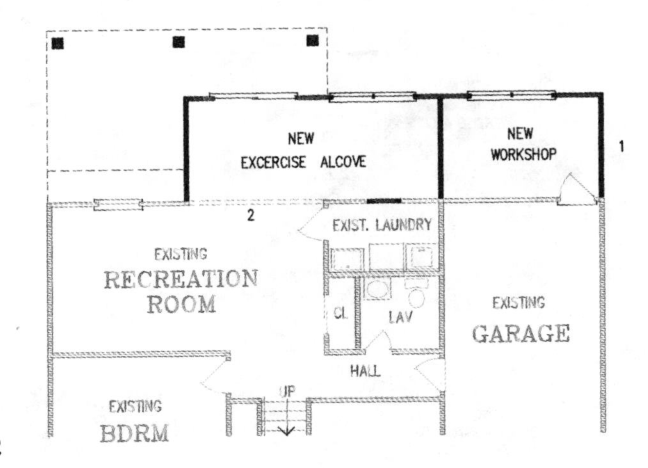

NOTES 1- 8'- 0 ADDITION
2- REMOVE EXISTING REAR WALL
3- SKYLITES
4- PLANT LEDGE

LOWER FLOOR

ROOFLINES AVAILABLE:
SHED (SHOWN)

KFBR4

Adding to a split-level appears easy, but many homeowners have found—after the fact—that the aesthetics of their addition eluded them. The split-level demands unyielding attention to roof forms, and the shed-roofed version shown here demands even more. But the results can be pleasurable and even dramatic. Eight feet have been added across the entire rear of this home, and it is seamlessly married to the existing roofline. A wonderful new master suite with a luxurious private bath is one of the main improvements. A new breakfast room enables the addition of more cabinets in the remodeled kitchen. "Extras" include the covered porch, enlarged family room, and sunroom or exercise alcove.

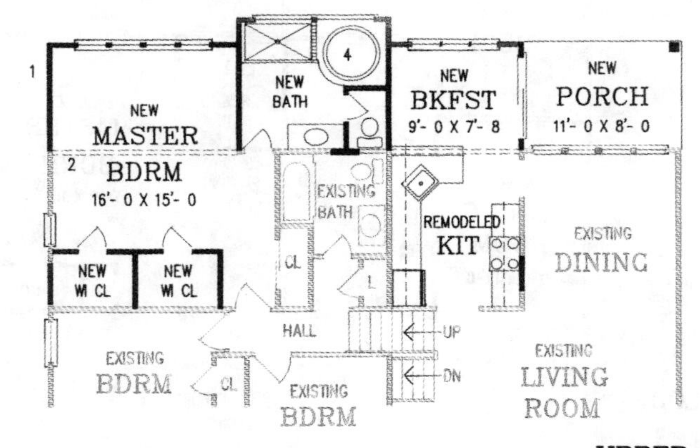

UPPER FLOOR

NOTES
1- 8'- 0 ADDITION
2- REMOVE EXTERIOR EXISTING WALL
3- ROOM OR ALCOVE COULD BE ENCLOSED
4- WHIRLPOOL TUB
5- RAIL AT BKFST RM

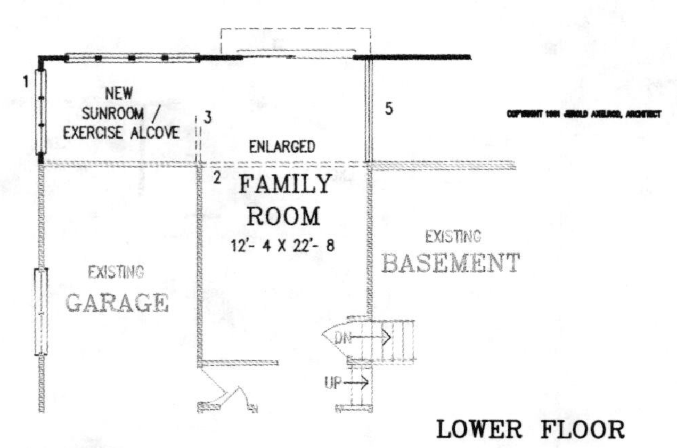

LOWER FLOOR

NOTE: CONSTRUCTION BLUEPRINTS CAN BE READILY MODIFIED TO ACCOMMODATE YOUR SPECIFIC HOUSE OR ROOM SIZES.

KBR03

The rear of this 30-year-old one-story looks like it has had its share of additions, but they do not flow together. This latest 5′10″ addition across the rear ties the kitchen to the family room and adds a new rear entry, laundry, master bath, and dressing area to create a thoroughly up-dated plan. The connection from the enlarged and newly remodeled kitchen to the remodeled family room is through a new rear foyer located behind the existing dining room. Bonuses in the plan include the visual enlargement of the dining room, the enlarged master bedroom, and the wet bar in the family room, which replaces a misplaced lavatory that may no longer be needed, since there are now two full baths.

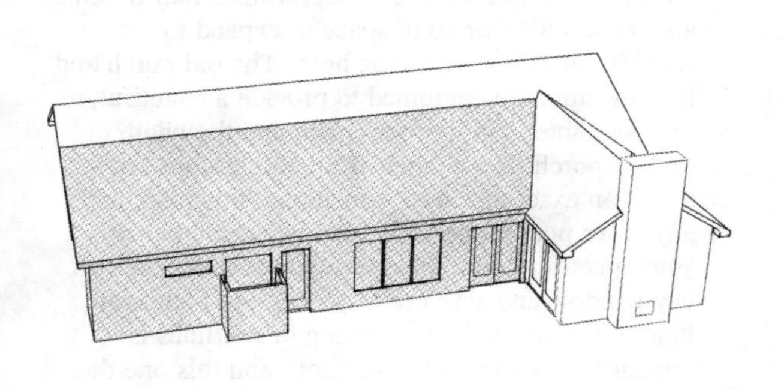

ROOFLINES AVAILABLE:
SHED (SHOWN)

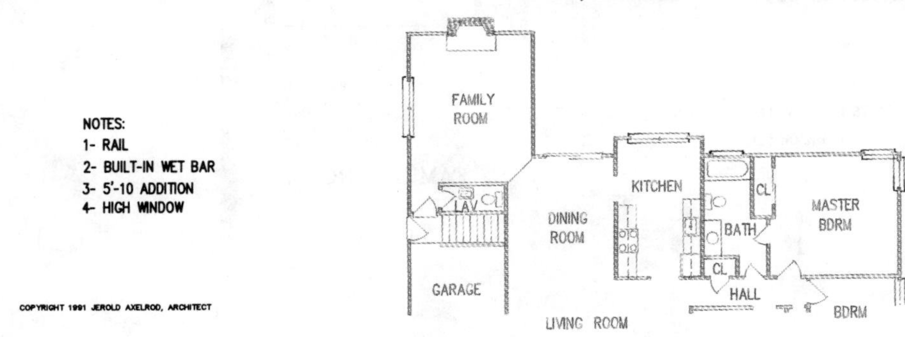

EXISTING PLAN

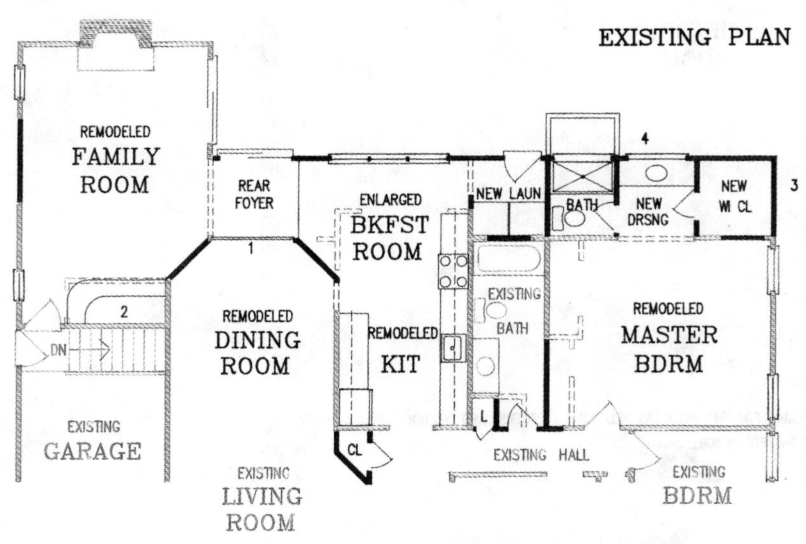

REMODELED PLAN

F0005

Do you have a charming older home without a family room, but it has an underutilized rear porch and there's 8′6″ or so of space to expand to the rear? Then take a hard look here. The old porch and the new area are combined to provide a beautiful, reverse-gabled family room matching the width of the old porch. Remember, if the dimensions shown aren't an exact match to your home, this plan, as any other plan in the book, can be adapted to fit your specific dimensions or requirements. The rear bay window and side French doors add bountiful light to the room. The marrying of rooflines is crucial to any successful addition, and this one does it admirably.

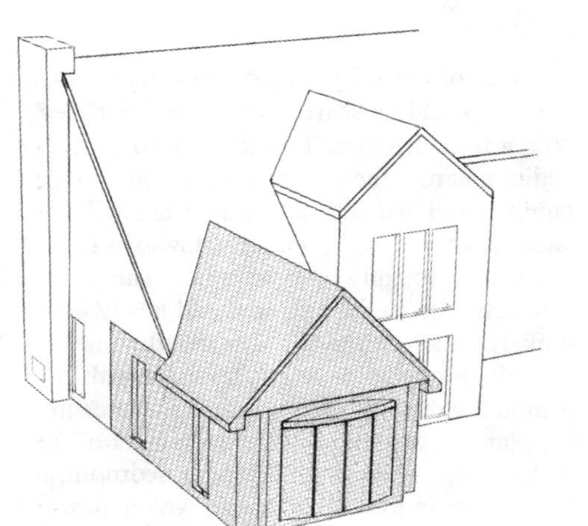

ROOFLINES AVAILABLE:
GABLE (SHOWN), HIP, SHED, FLAT

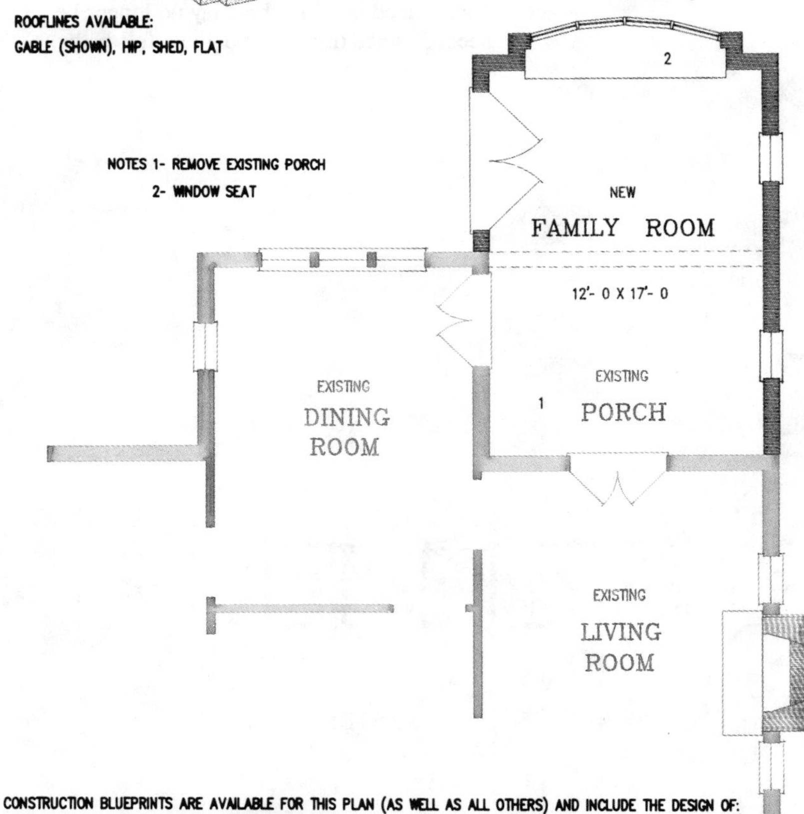

CONSTRUCTION BLUEPRINTS ARE AVAILABLE FOR THIS PLAN (AS WELL AS ALL OTHERS) AND INCLUDE THE DESIGN OF:
THE HEADER NECESSARY TO REMOVE THE PORCH WALL

F0007

Just a few feet can make a world of difference in the use and furnishability of a room. Take the rear-facing family room shown. Its current 16′2″-×-12′4″ size was always tight, particularly because it serves as the access to the kitchen. When part of a room has to provide a circulation path to an adjoining room, it decreases the furnishing options. The addition shown adds six feet to this family room. A wood stove is suggested as a focal point, but you could substitute a fireplace. Sliding doors with transom lights above flank each side. The added space now provides enough depth to create a comfortable seating arrangement, including love seats and wrap-around couches. The kitchen could also be modernized, as shown, if desired.

ROOFLINES AVAILABLE:

GABLE (SHOWN), HIP, FLAT

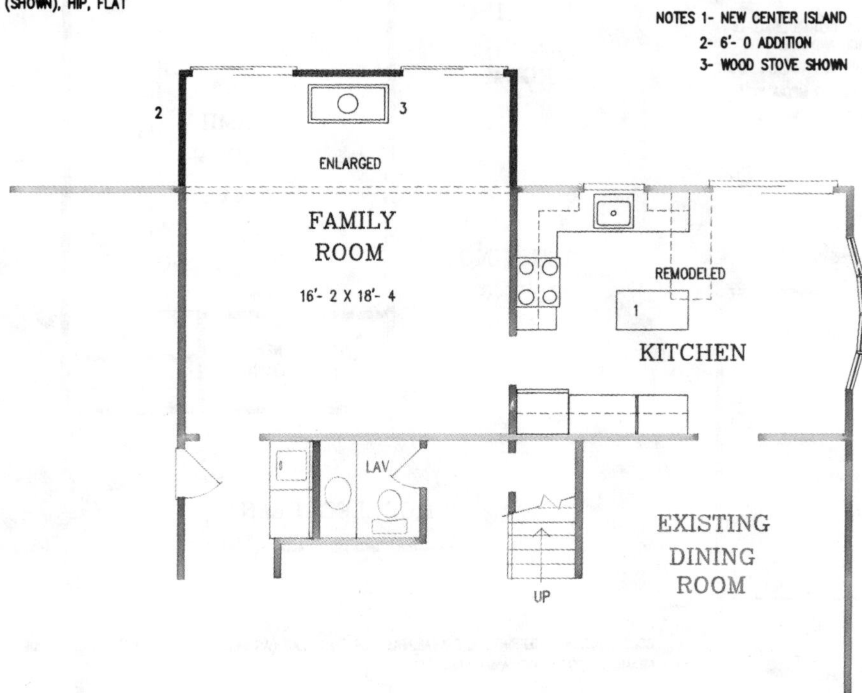

CONSTRUCTION BLUEPRINTS ARE AVAILABLE FOR THIS PLAN (AS WELL AS ALL OTHERS) AND INCLUDE THE DESIGN OF: THE HEADER FOR REMOVING THE REAR WALL AND A LIST OF MATERIALS.

F0009

There's a family room in your mind, but your budget is tight and you think you can live without that one-car garage. Beware! Many one-car garages are too narrow to create a workable family room. Twelve feet is minimum, especially for the length; 14 feet is better. If yours is a narrow garage but you have the side-yard space (albeit it needing a variance), give some thought to adding four feet as in the plan shown. It will cost more, but you won't be sorry. With the extra space, you could take some area for a laundry room or built-in media center, and even recoup some space for storage of auto or garden supplies. An extra benefit is the interesting roofline and a high clearstory window that the addition permits.

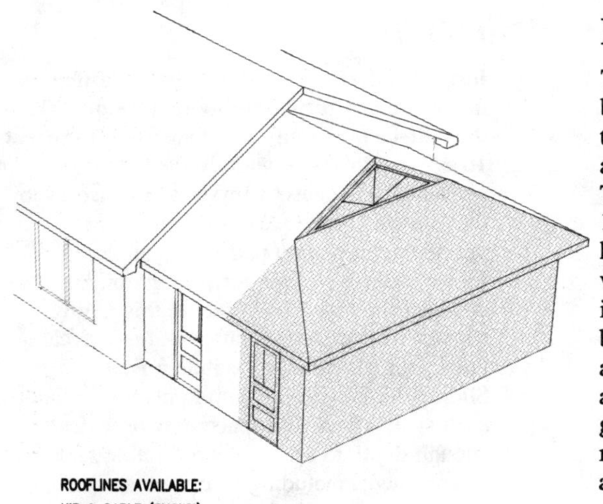

ROOFLINES AVAILABLE:
HIP & GABLE (SHOWN)

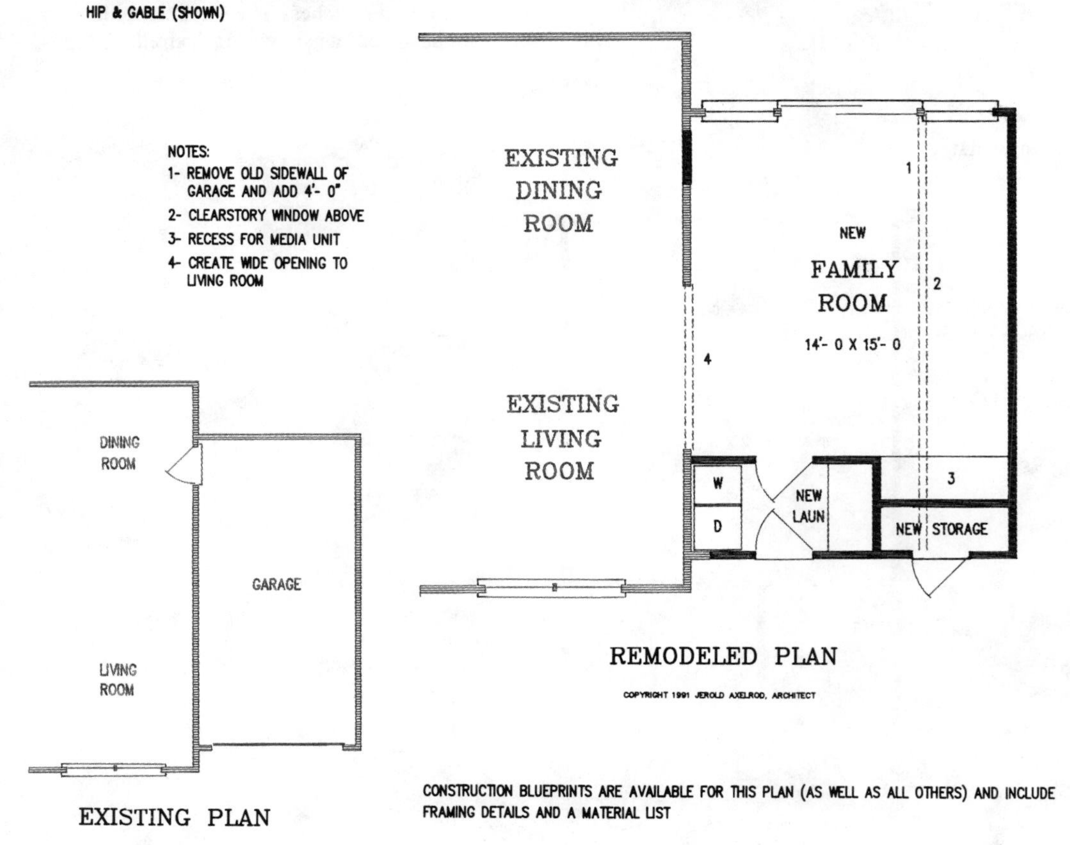

REMODELED PLAN

EXISTING PLAN

CONSTRUCTION BLUEPRINTS ARE AVAILABLE FOR THIS PLAN (AS WELL AS ALL OTHERS) AND INCLUDE FRAMING DETAILS AND A MATERIAL LIST

F0013

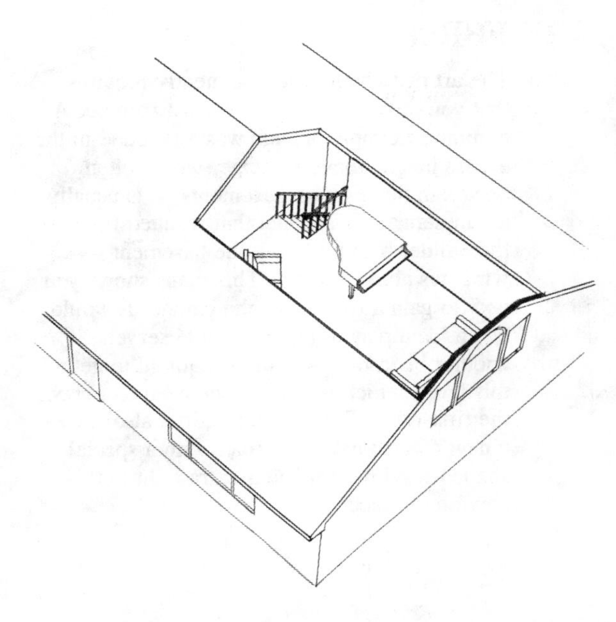

Have you ever crawled up into your attic, realized that you can stand straight for some sizeable distance, and wondered if it could be converted to living space? This design shows you how. It creates a room 11 feet wide, by whatever length you have available (19′6″ shown on the plan). No structural work is required, other than the placement of large attractive gable-end windows for light and the installation of a stair. The latter is no small task, as you do have to find an area approximately 3′×10′ for a stair, the top end of which should be fairly close to the middle of your home (under the ridge, where you have maximum headroom). Note that if your attic becomes a third floor above-grade, some codes might not permit it.

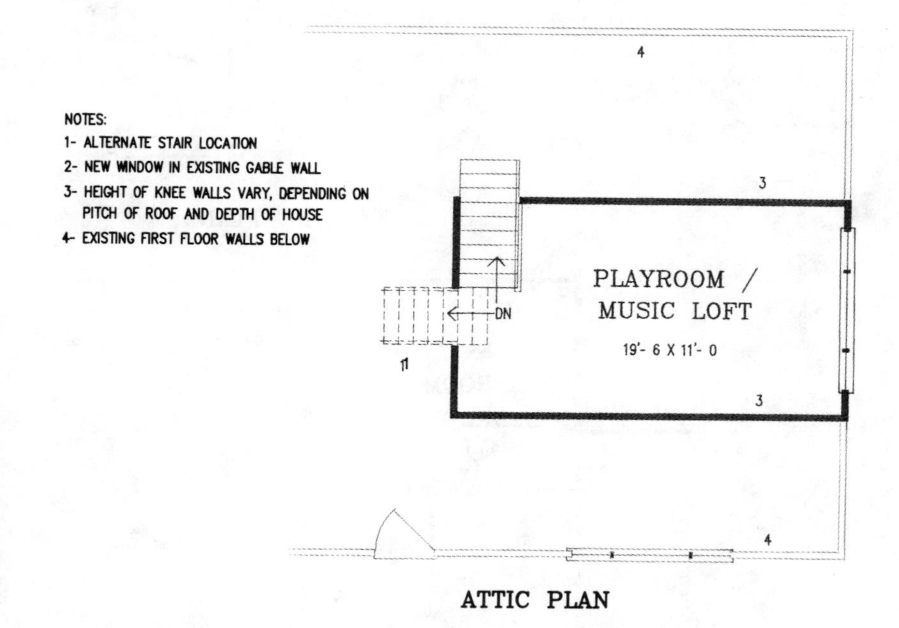

ATTIC PLAN

NOTE: CONSTRUCTION BLUEPRINTS CAN BE READILY MODIFIED TO ACCOMMODATE YOUR SPECIFIC HOUSE OR ROOM SIZES.

FBR08

The art of finding space frequently requires that you look overhead for wasted volume. A common example of such waste is found in the excess height above a garage, which often occurs in homes with basements. It is usually an unplanned by-product that resulted from the builder trying to keep the basement windows above ground. This plans shows you how to gain a room over the garage. It would be an ideal playroom, or it could serve as another bedroom. A stair is required to get to this level, which is likely to be 4 – 6 feet above the first floor. The design pictured also shows a new reverse-gabled dormer with a special shaped window that sheds extra light to this newfound space.

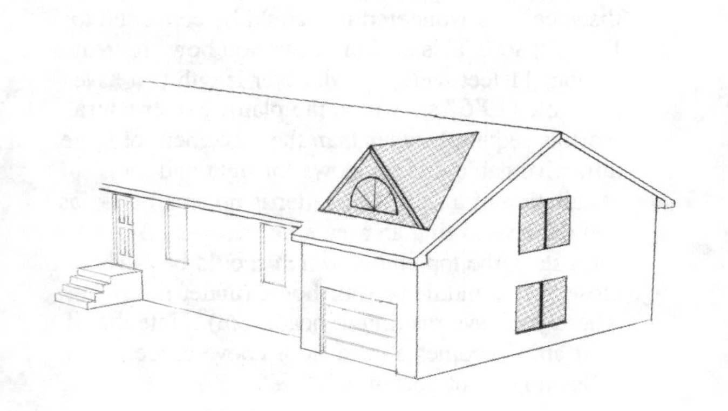

ROOFLINES AVAILABLE:

ONLY AS SHOWN

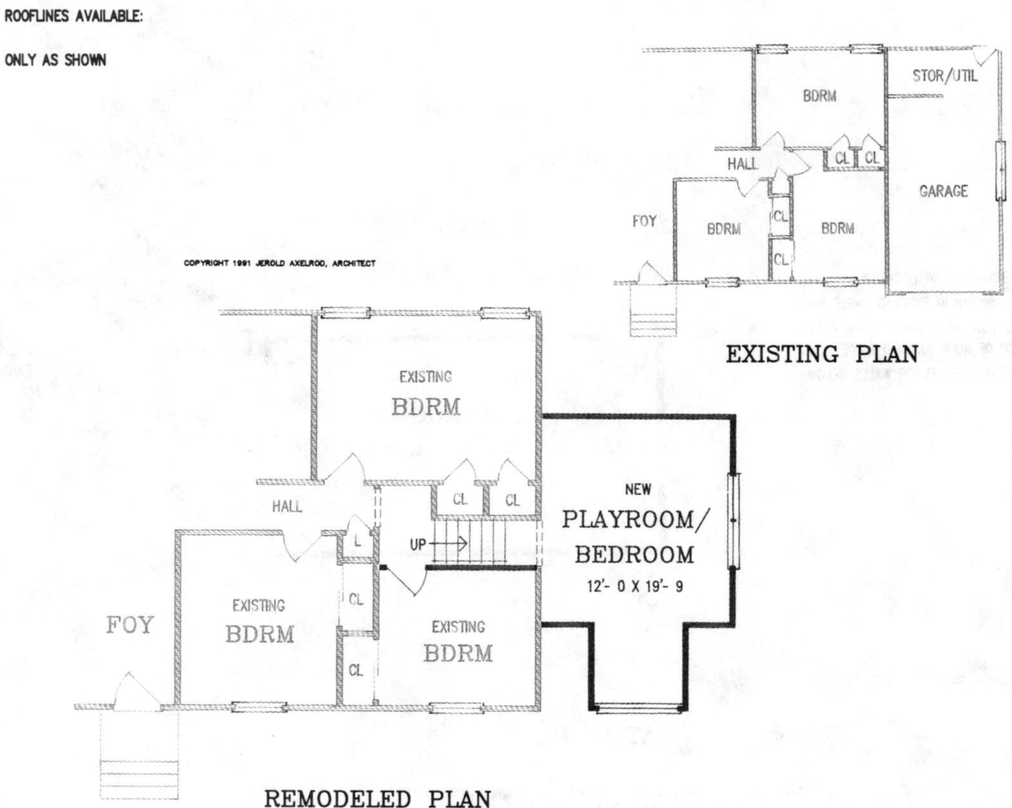

CONSTRUCTION BLUEPRINTS ARE AVAILABLE FOR THIS PLAN (AS WELL AS ALL OTHERS) AND INCLUDE INFORMATION ON HOW TO EVALUATE POTENTIAL ROOM SIZE

KF001

A common problem: a stately, older two-story home, but little room to expand out; the kitchen is dated and has no eating space, and you would love to have a large family room. Solution: Remove the typical old rear porch (5′6″ deep in this plan as drawn), and push out to that line with a beautifully balanced extension that results in a sensational kitchen and an expanded family room. Each room includes a wide bay window flanking the double center doors, which provide direct access to the patio. The kitchen features a large center island with a bar sink. The dramatic surrounding cabinet layout shows you how to work existing constraints, such as a chimney bump, to your advantage. Review the plan for several internal changes suggested to improve circulation.

ROOFLINES AVAILABLE:
FLAT, (SHOWN), HIP, SHED, GABLE

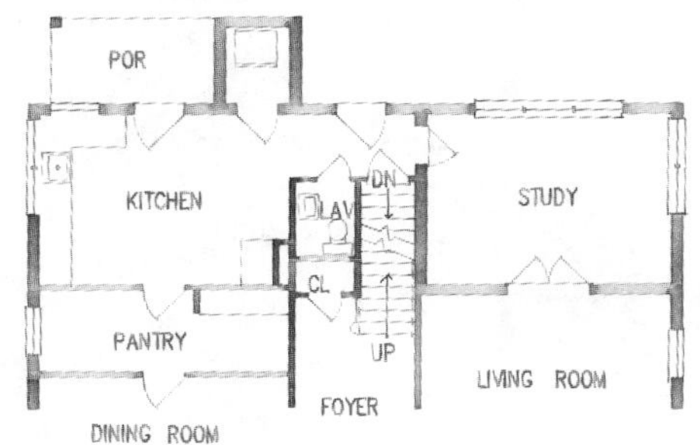

EXISTING PLAN

NOTES 1- REMOVE LAV AND CLOS AND OPEN UP TO CENTER HALL

2- REDUCE PANTRY AND RELOCATE LAV

3- REMOVE WALLS TO OPEN FAMILY ROOM

REMODELED PLAN

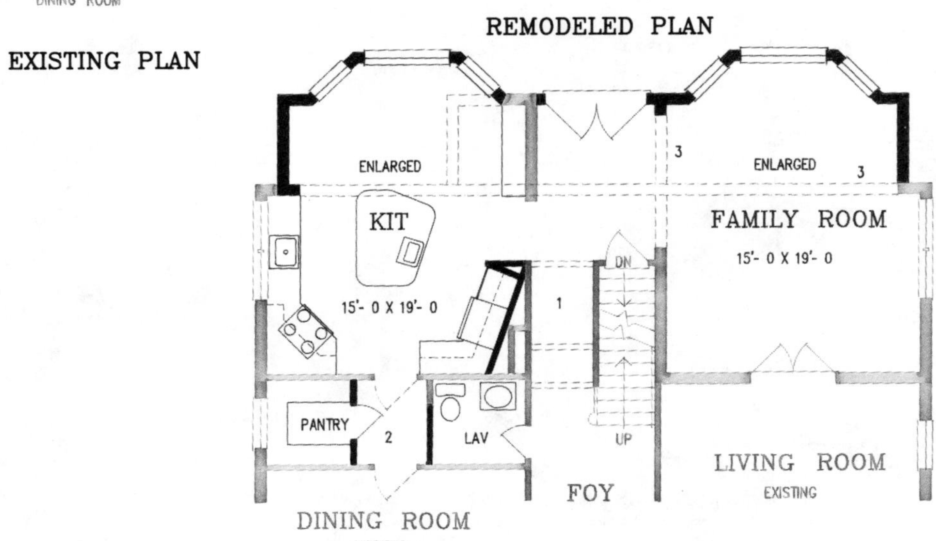

CONSTRUCTION BLUEPRINTS ARE AVAILABLE FOR THIS PLAN (AS WELL AS ALL OTHERS) AND INCLUDE THE DESIGN OF:
INFORMATION ON REMOVING THE REAR WALL

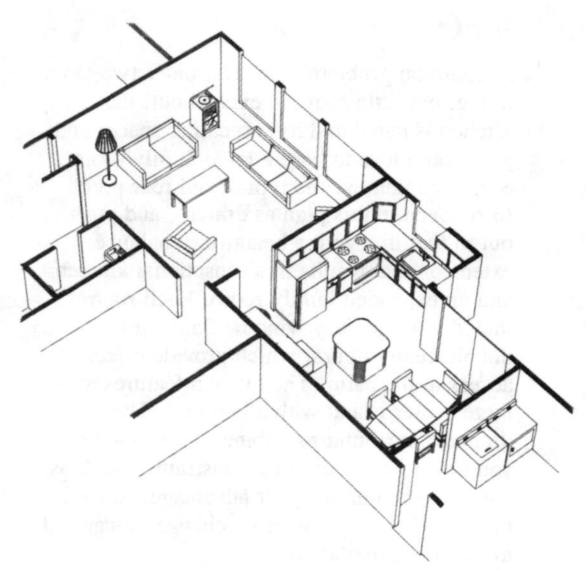

KF005

The rear-facing family room and the adjoining kitchen of this home are of adequate length, but their 11-foot width is tight. The 4′4″ addition proposed for both provides enough space to create a great new kitchen and a comfortable family room. You could add more if you want, the only limitation being a concern with the roof pitch of the extension so as to clear the bottom of second floor windows if yours is a two-story home. The kitchen, which had a peninsula return before, is now large enough to house a center island. Skylights are suggested in the new areas of both rooms.

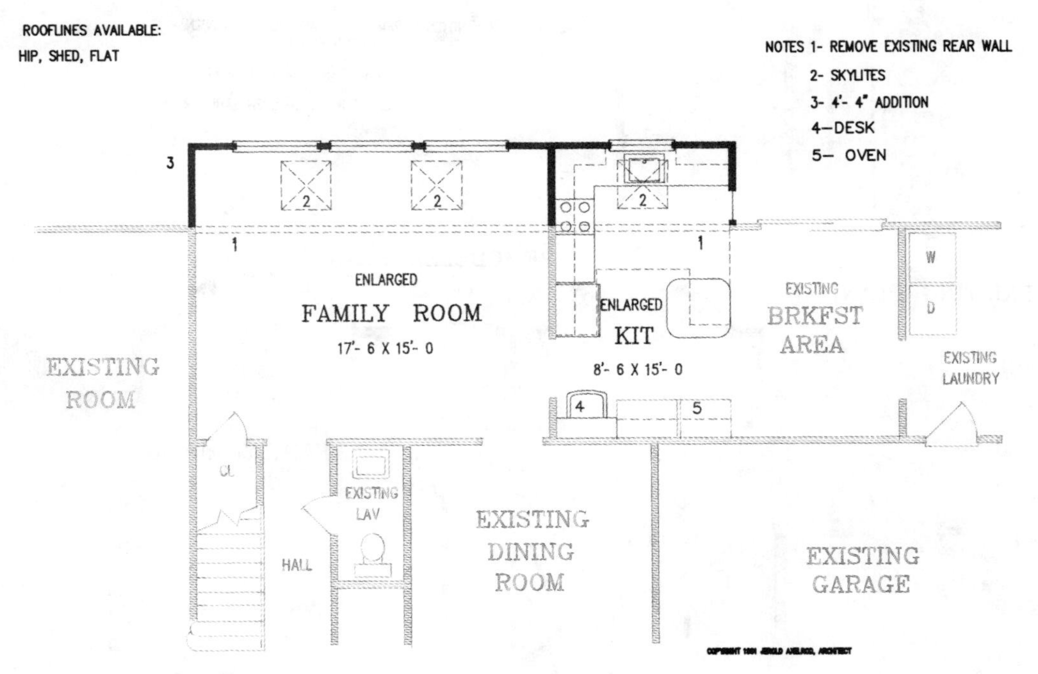

CONSTRUCTION BLUEPRINTS ARE AVAILABLE FOR THIS PLAN (AS WELL AS ALL OTHERS) AND INCLUDE THE HEADERS NECESSARY TO REMOVE THE REAR WALL

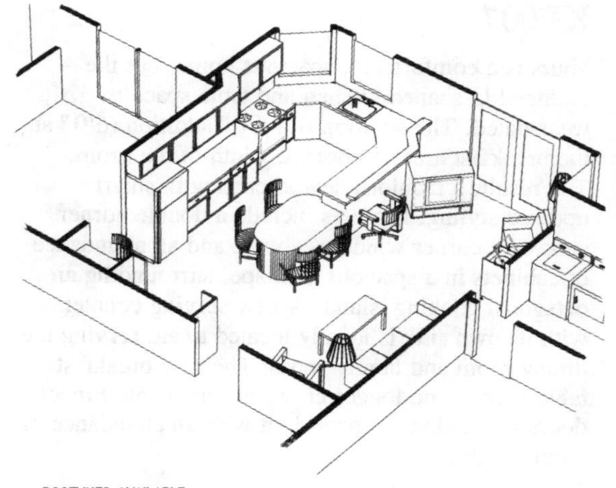

KF006

Is yours a narrow U-shaped kitchen, with a peninsula eating bar return that is forever crowded—especially so at mealtime? This 8′8″ addition to the kitchen (3′8″ at the family room side) provides all the space needed to create a wonderful country kitchen. The extra area permits the creation of a stylish working kitchen, and a lovely eating area that bridges the space between the kitchen and the family room. Angled corners serve to expand and enhance the kitchen. A desk and pantry are located at the inside wall, and a high counter serves to define the kitchen from the family room without actually separating them.

ROOFLINES AVAILABLE:

SHED, FLAT

NOTE: CONSTRUCTION BLUEPRINTS CAN BE READILY MODIFIED TO ACCOMMODATE YOUR SPECIFIC HOUSE OR ROOM SIZES.

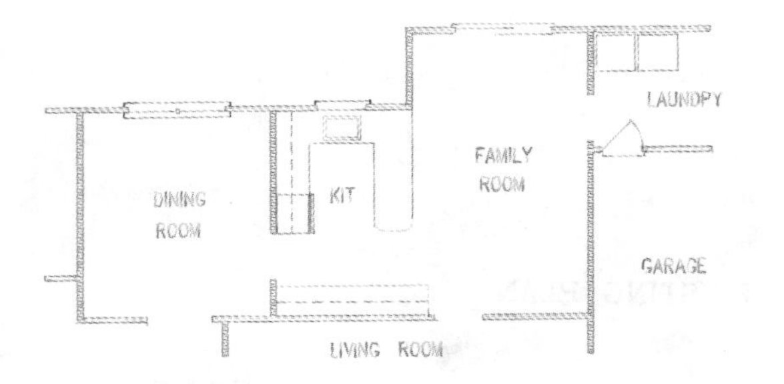

NOTES
1- BUILT-IN TABLE
2- 3'- 8 ADDITION
3- 8'- 8 ADDITION
4- SKYLITE

EXISTING PLAN

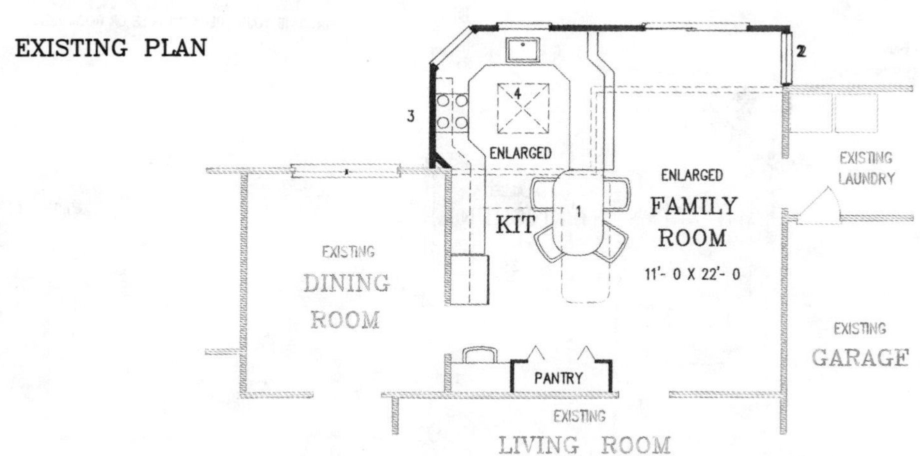

REMODELED PLAN

KF007

Yours is a comfortable, spacious home, but the cramped U-shaped kitchen and table space is wear-dated. The solution is a 5′0″ addition (6′0″ at the breakfast area) across the width of the room. The result: a fabulous new kitchen with smart, updated styling. Features include a double corner sink with corner windows above, and an abundance of cabinets in a spacious U-shape, surrounding an octagonal cooking island. A new serving counter with its own sink is ideally located to aid serving the dining room and family room. The new breakfast table space is no longer crowded, and triple French doors and a skylight provide it with an abundance of natural light.

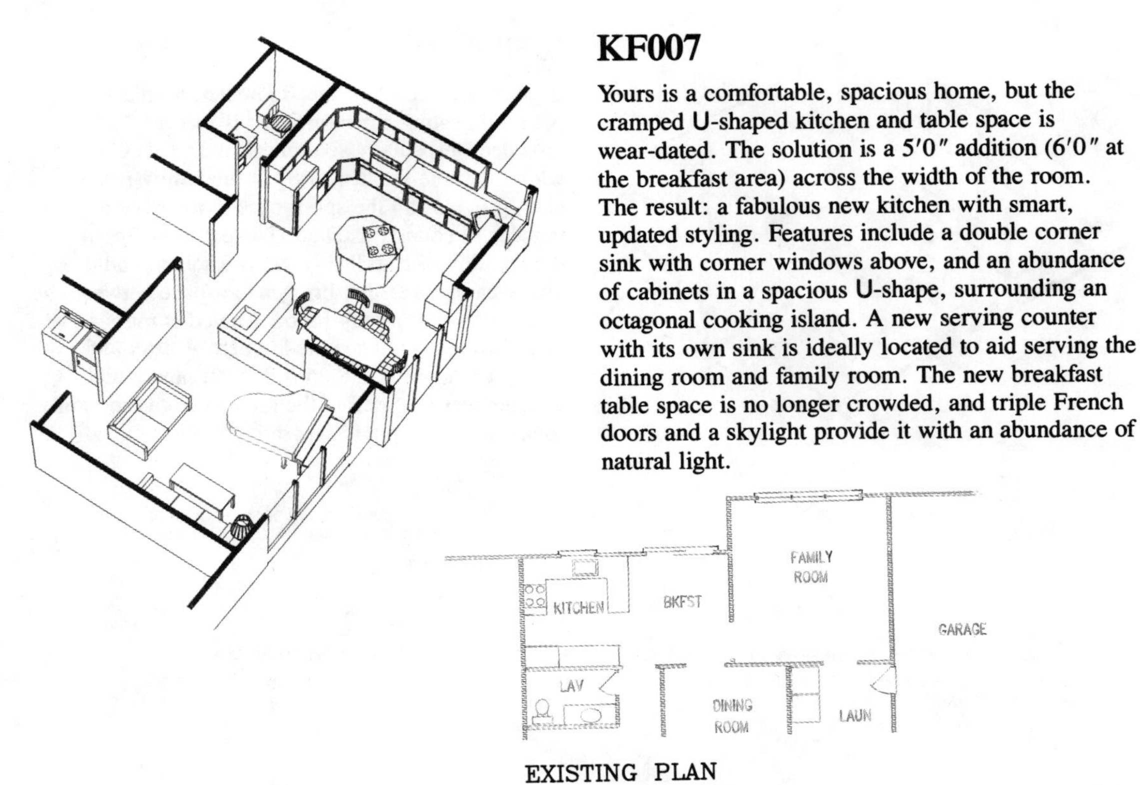

EXISTING PLAN

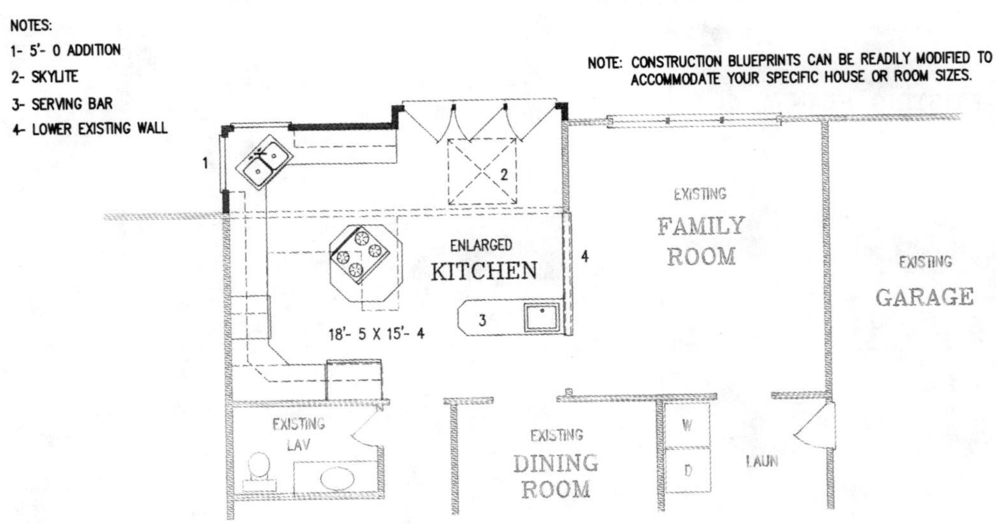

REMODELED PLAN

KFD01

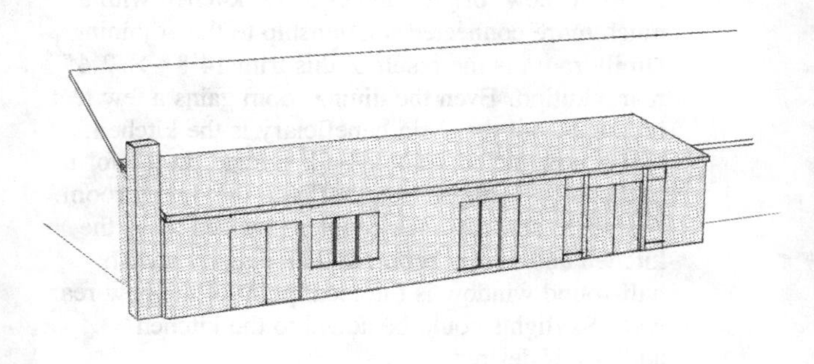

ROOFLINES AVAILABLE:
SHED (SHOWN), FLAT

Your kitchen, dining rooms, and old den are all just a little too tight for your current needs? This simple, easy-to-construct, 5′4″ wide shed-roof addition across the rear of all three rooms affords all the space necessary to solve the problem. Of course, the same could be accomplished on any one of the rooms, if that's all you need. The kitchen is now large enough to include a center island, and the skylit breakfast area is just delightful. The family room is now spacious enough for entertaining. A new corner fireplace, sliding doors, and skylights add to the room's newfound attraction. See plan BRB13, on page for a matching 5′4″ bedroom addition.

CONSTRUCTION BLUEPRINTS ARE AVAILABLE FOR THIS PLAN (AS WELL AS ALL OTHERS) AND INCLUDE THE DESIGN OF:
INFORMATION ON REMOVING THE REAR WALL

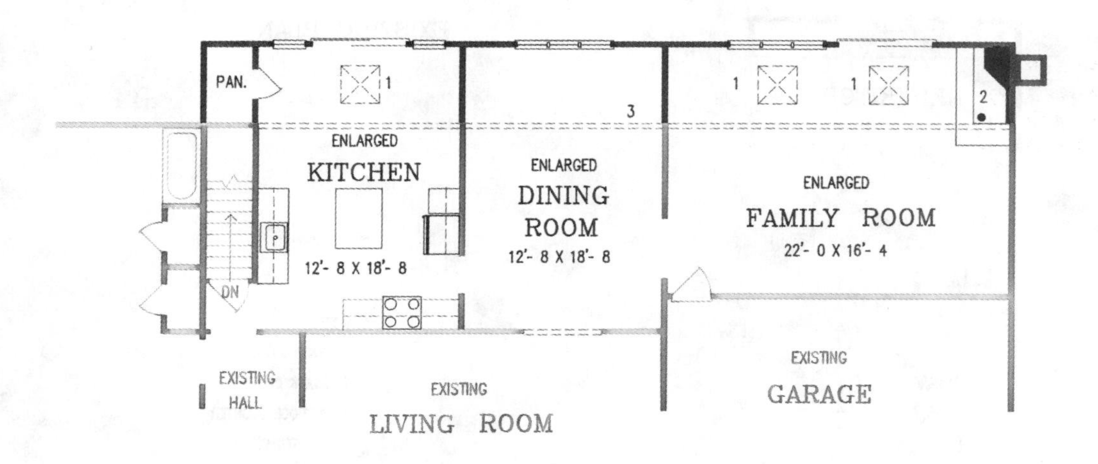

NOTES 1-SKYLITES
2-NEW CORNER FIREPLACE
3-REMOVE ENTIRE REAR WALL

KFD02

A smart, new, bright, and cheerful kitchen with a much more connected relationship to the adjoining family room is the result of this trim 14′8″-×-7′4″ rear addition. Even the dining room gains a few feet in length, but the main beneficiary is the kitchen, which now includes some bulk storage, in the form of a pantry, and a serving buffet to the dining room. A stylish, angled snack counter now separates the kitchen and family room, and an equally stylish half-round window is the focal point of the new rear wall. Skylights could be added to the kitchen addition if desired.

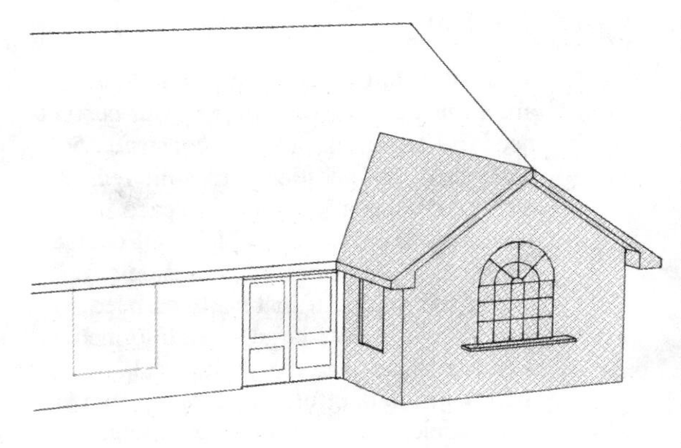

ROOFLINES AVAILABLE:
GABLE (SHOWN), HIP, SHED, FLAT

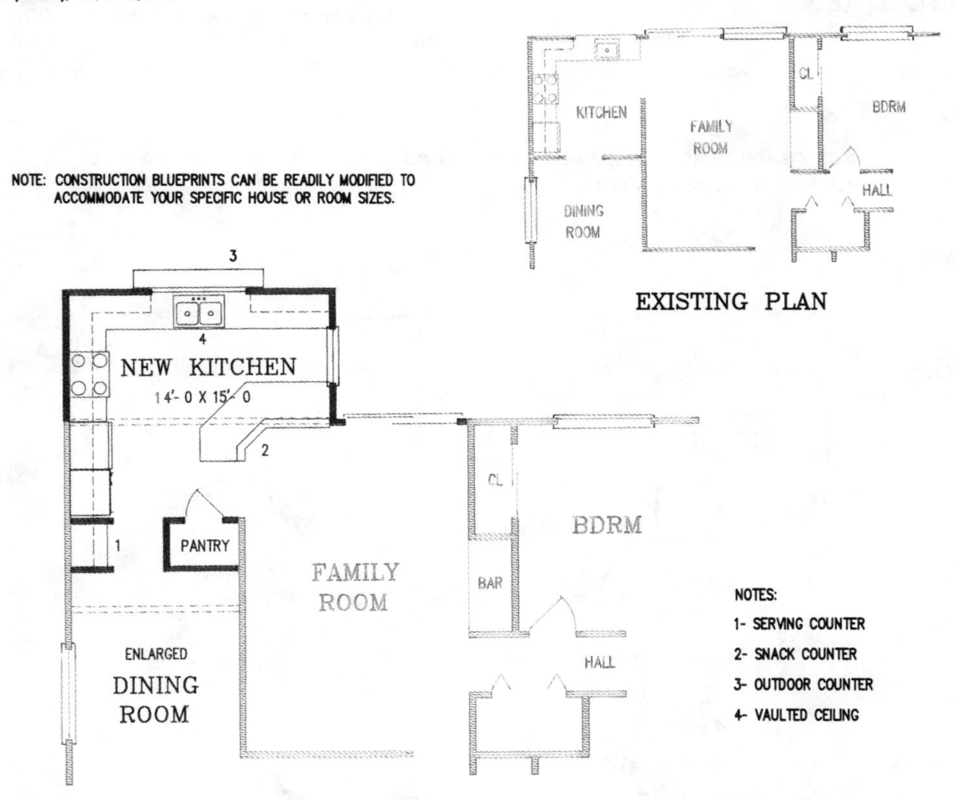

K0001

The first of three variations on remodeling a narrow kitchen with inadequate eating area. This version adds the most space in the form of a 9′2″-×-6′6″ addition in the shape of a half octagon. This delightfully sunny space is large enough to accommodate a breakfast table, enabling the introduction of a dramatic angled countertop to the working part of the kitchen. The angles are mirrored in the corner cabinets at the far end of the kitchen, thus creating an overall style and artistry to the remodeled kitchen. The kitchen window and doorway to the dining room can remain without change.

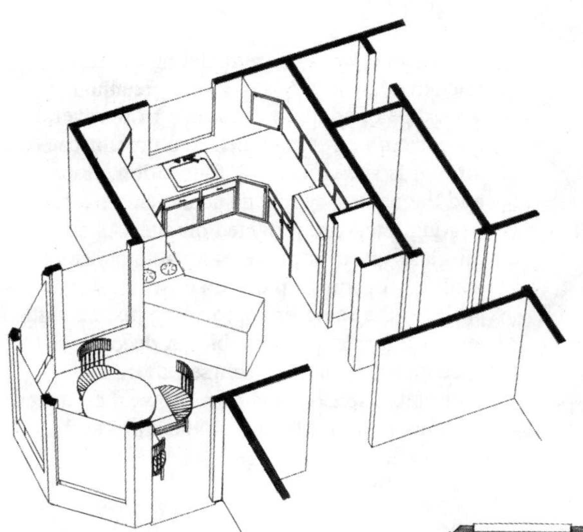

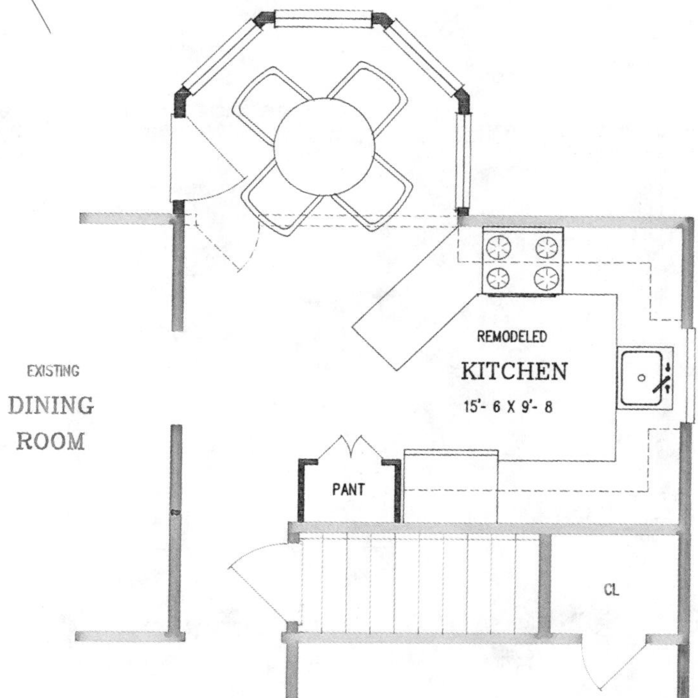

CONSTRUCTION BLUEPRINTS ARE AVAILABLE FOR THIS PLAN (AS WELL AS ALL OTHERS) AND INCLUDE THE DESIGN OF:
THE HEADER NECESSARY TO REMOVE THE REAR WALL

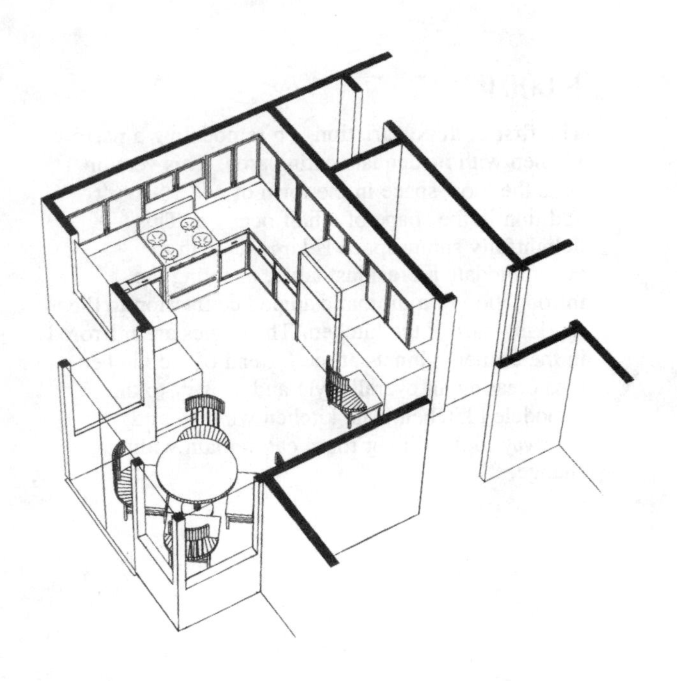

K0002

The second variation remodeling the same kitchen adds an 8′6″-×-3′6″ "greenhouse." This type of addition, available from several manufacturers, is provided in a prefabricated kit that is site-erected. A foundation, floor, and the necessary mechanicals (electrical, heating, etc.) are provided on-site. As an all-glass space, it provides an abundance of sunlight, opening up the smallish kitchen, while at the same time providing just enough space to locate a small table. A deeper greenhouse would, of course, afford even more table space. This plan shows the kitchen window moved, along with the sink, to the rear wall.

NOTE: CONSTRUCTION BLUEPRINTS CAN BE READILY MODIFIED TO ACCOMMODATE YOUR SPECIFIC HOUSE OR ROOM SIZES.

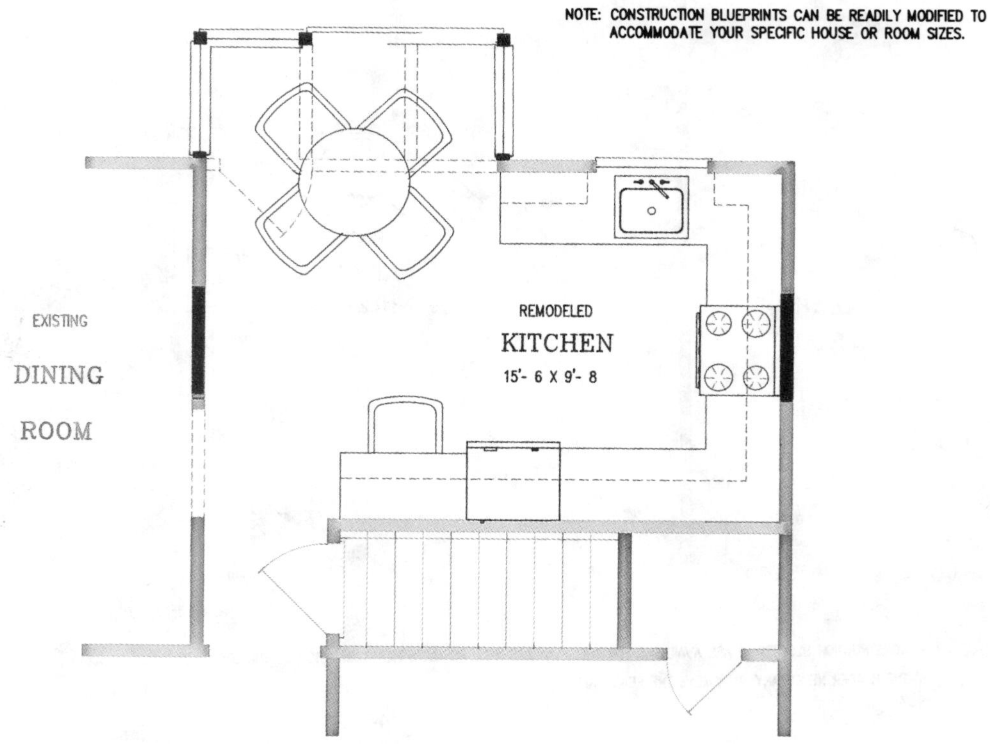

K0003

The third variation remodeling the same narrow, long kitchen, incorporates a bump in the form of an 8′4″-×-4′0″ angled bay. This is likely to be the least costly of the three additions, yet it accomplishes almost the same as the others. The bay provides plenty of light, space for a small breakfast table, and access to the yard, as do all three plans. The table space is a little tighter, but by moving the doorway to the dining room down, you can almost achieve the same table space as in the largest addition. The plan provides for a convenient desk and leaves the kitchen window on the side wall.

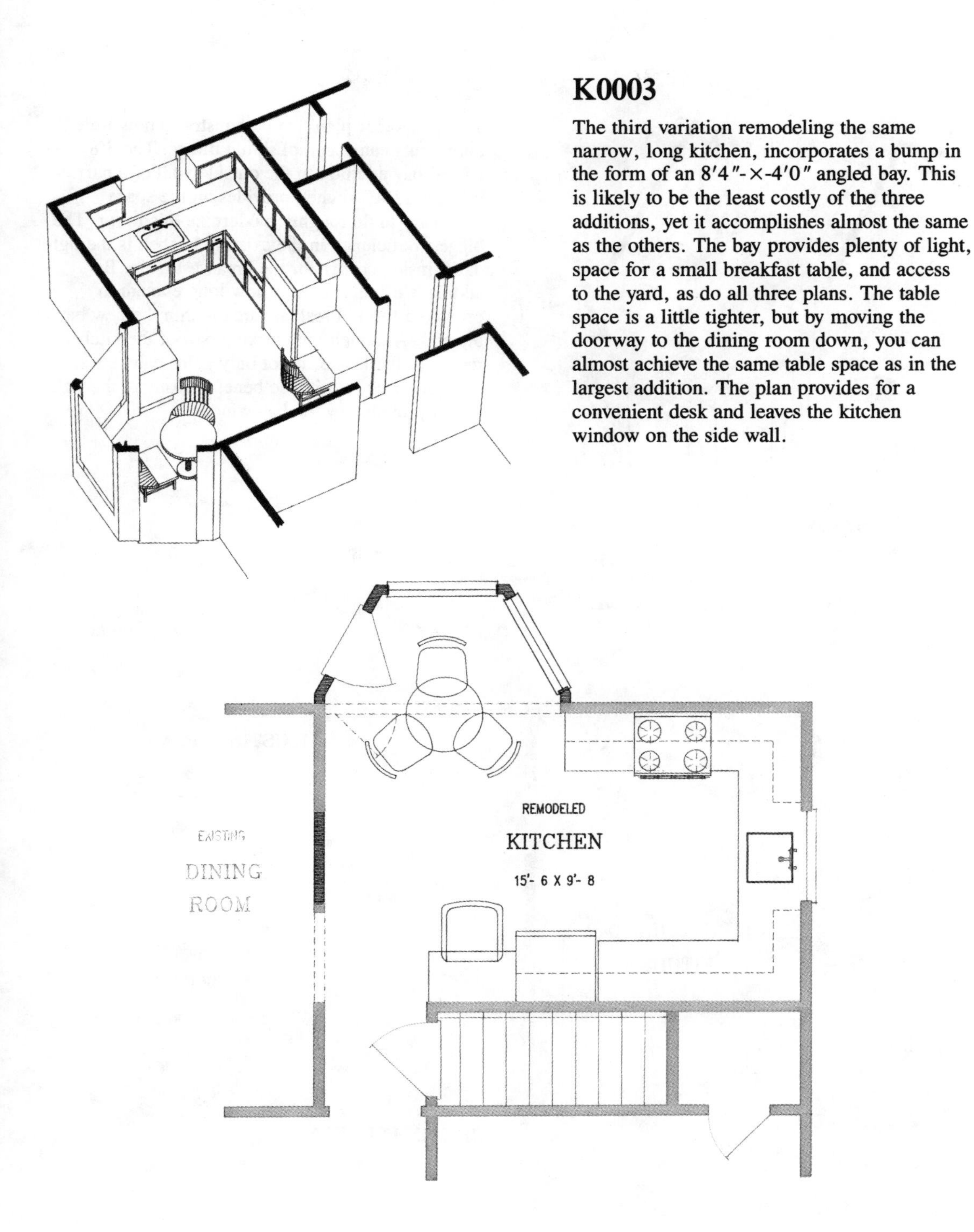

K0007

This is another plan that demonstrates how little bump-outs can accomplish big things. The 3′8″ angled bay addition to the outside wall of a narrow 9′0″-×-13′4″ kitchen provides all the space necessary to thoroughly modernize this room. The biggest deficiency in the existing kitchen is the tight, dark, inside corner for the breakfast table. By adding 3′8″, the room is now long enough to provide a lovely breakfast area within the new bay at the end. Although the working part of the kitchen moves to the inside, it not only gains cabinets and counter space, but it also benefits from all the light that is provided by the bay window.

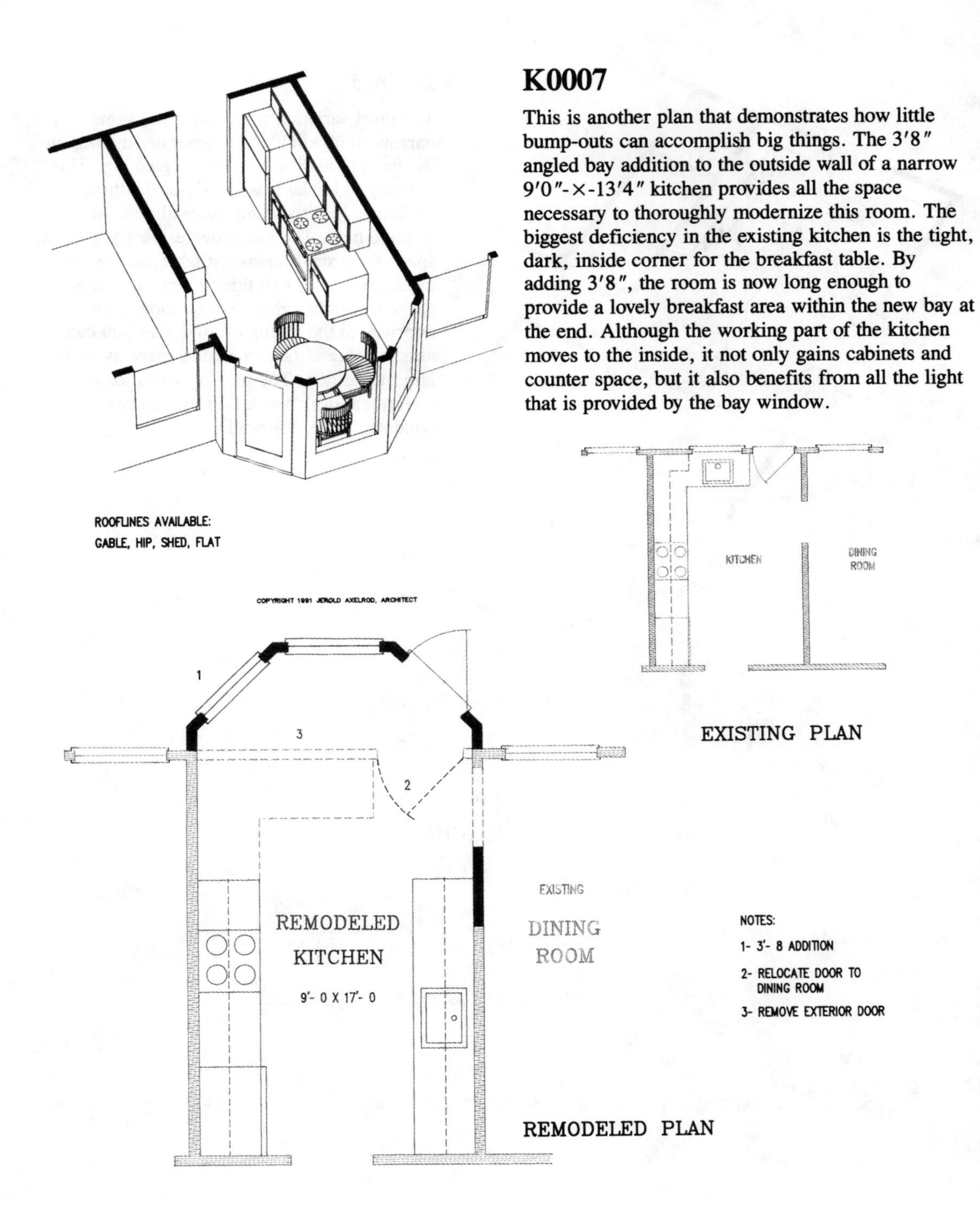

K0006

Modernizing the front-facing kitchen is potentially much more limiting than any other location; since it affects the front facade. Anything other than mere cabinet replacement should be undertaken with care. If your desire is also to expand the kitchen, the studied design of rooflines and windows becomes crucial. Setback limitations will likely govern, as in the home shown. The addition to the kitchen is designed to align with the forward bedroom wing. Although only 4′4″, it accomplishes both a sparkling new U-shaped kitchen and a smart new front facade. The dramatic window reflects the new vaulted ceiling inside.

ROOFLINES AVAILABLE:
GABLE (SHOWN), HIP, SHED, FLAT

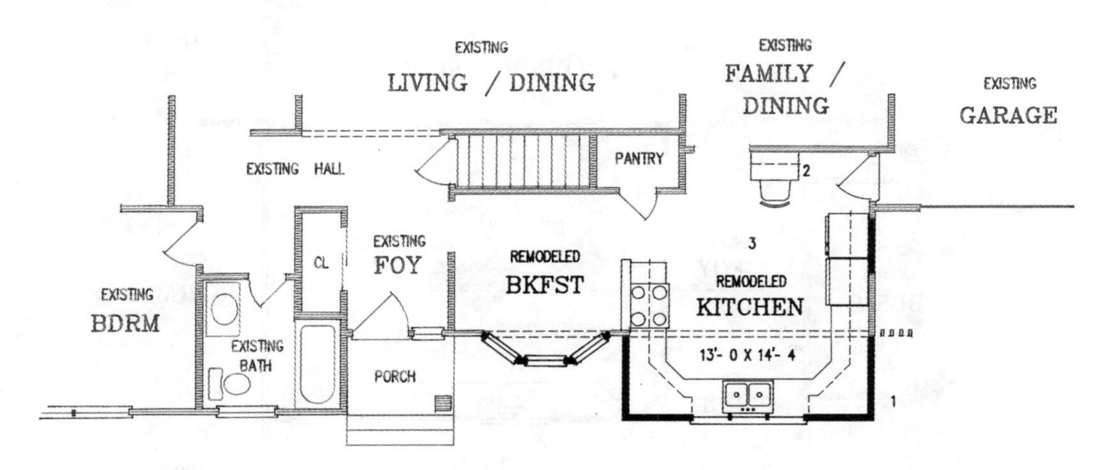

REMODELED PLAN

NOTES 1- 4'- 4 ADDITION
2- DESK
3- VAULTED CEILING

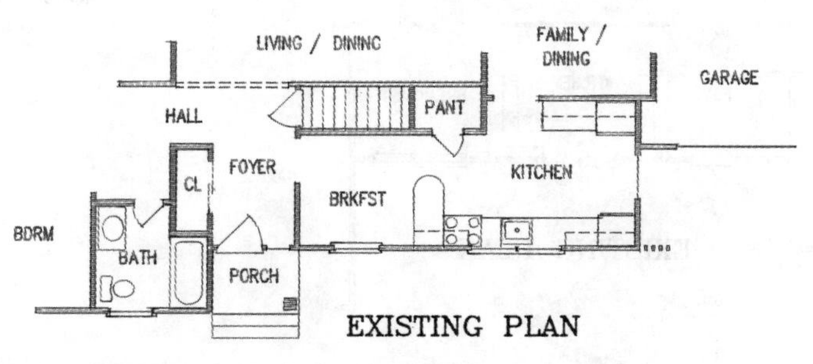

EXISTING PLAN

NOTE: CONSTRUCTION BLUEPRINTS CAN BE READILY MODIFIED TO ACCOMMODATE YOUR SPECIFIC HOUSE OR ROOM SIZES.

ROOFLINES AVAILABLE:
GABLE (SHOWN)

K0005

As indicated before, expanding a front-facing kitchen can be hazardous. Technical considerations aside, the new extension can redefine the character of your whole house. This home, as shown, takes on a new presence by infilling the recess between the existing bedroom wing and garage with a fashionable new reverse-gabled wing. A stylish new covered porch and expanded foyer are the "extras" to the main subject, the new kitchen. Features include a dramatic angular layout, an unusual island that defines the new breakfast table area, and a fabulous wall of built-ins.

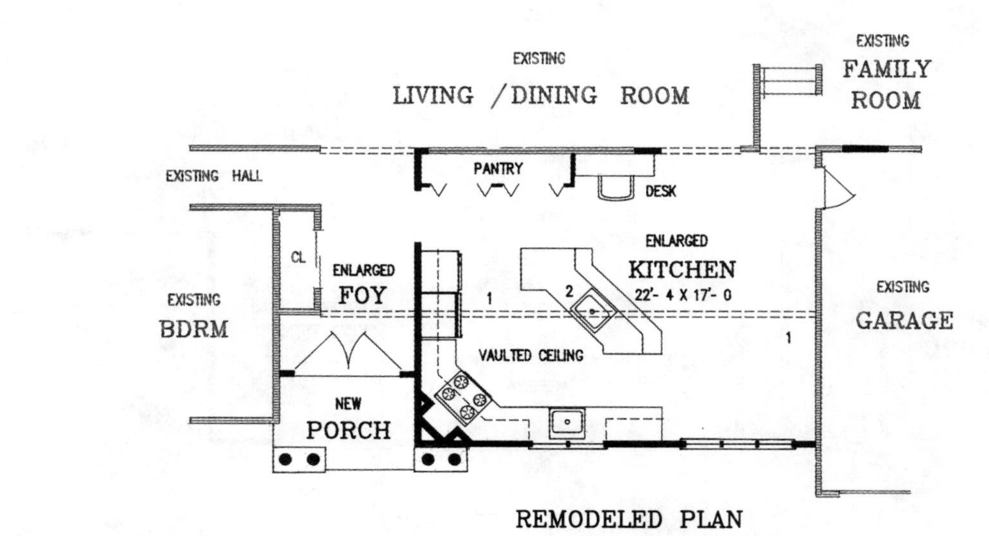

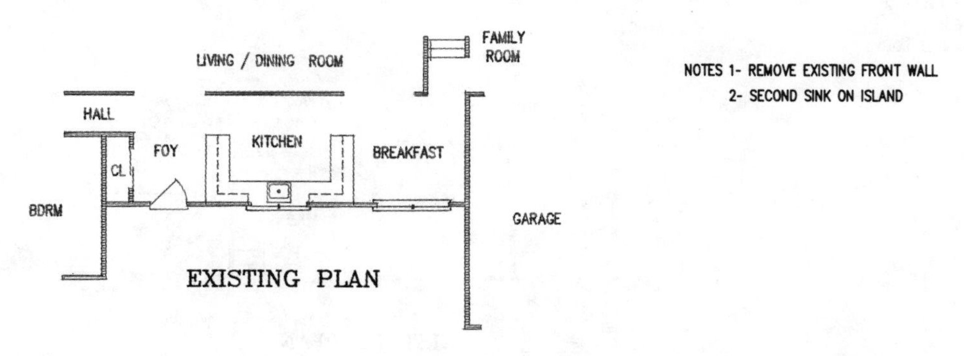

NOTES 1- REMOVE EXISTING FRONT WALL
2- SECOND SINK ON ISLAND

CONSTRUCTION BLUEPRINTS ARE AVAILABLE FOR THIS PLAN (AS WELL AS ALL OTHERS) AND INCLUDE FRAMING FOR NEW ROOFLINES, KITCHEN DETAILS, AND A MATERIAL LIST.

K0009

There are times when a remodeling project combines rooms in what might appear to be a loss of a room. Well, this new kitchen is such a plan. The existing 17′0″ width is somewhat cramped for both a kitchen and breakfast space. Assuming the goal is a dazzling expanded kitchen—eliminating the separate breakfast area is something to consider. The 4′0″ addition provides all the space needed to create a winning kitchen. The new plan is an expansive U-shape with an abundance of natural light. A large multilevel center island provides a comfortable table-height eating area on two sides, while leaving a large work surface for kitchen preparation on the opposite sides.

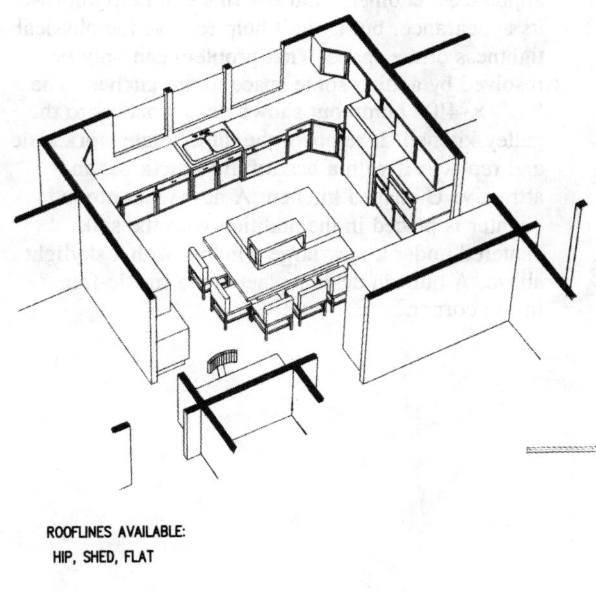

BRKFST

KITCHEN

17'- 0 X 12'- 0

EXISTING PLAN

CONSTRUCTION BLUEPRINTS ARE AVAILABLE FOR THIS PLAN (AS WELL AS ALL OTHERS) AND INCLUDE KITCHEN CABINET MANUFACTURERS INFORMATION AND A MATERIALS LIST.

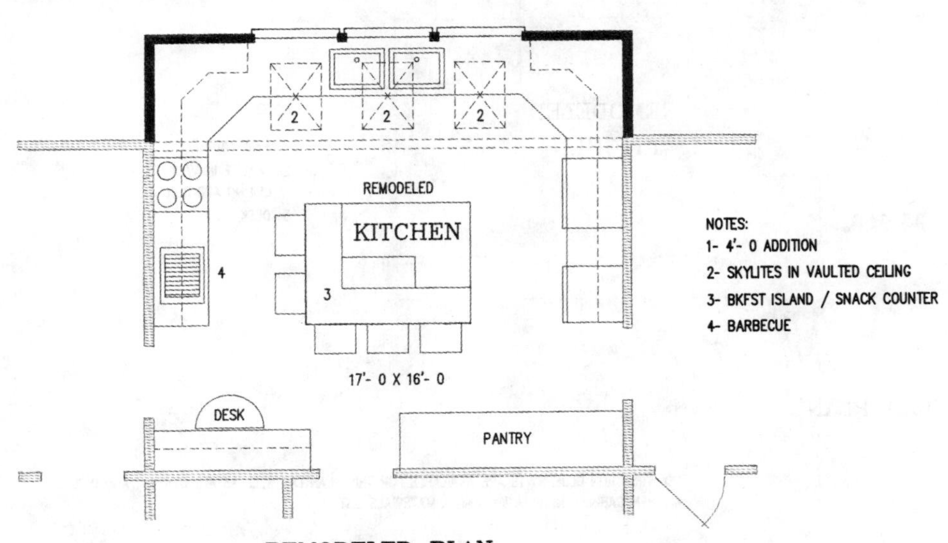

REMODELED PLAN

K0010

Remodeling a galley kitchen with new lighting, appliances, cabinets, and flooring can help improve its appearance, but it can't help remove the physical tightness of the space. That problem can only be resolved by adding some space to the kitchen. The 8′0″-×-4′0″ bump out shown does wonders to the galley kitchen. It removes the dull, single work aisle and replaces it with a beautifully functional and attractive U-shaped kitchen. A new wrap-around counter is placed in the addition with the sink centered under a new large window with a skylight above. A built-in desk is placed in a traffic-free inside corner.

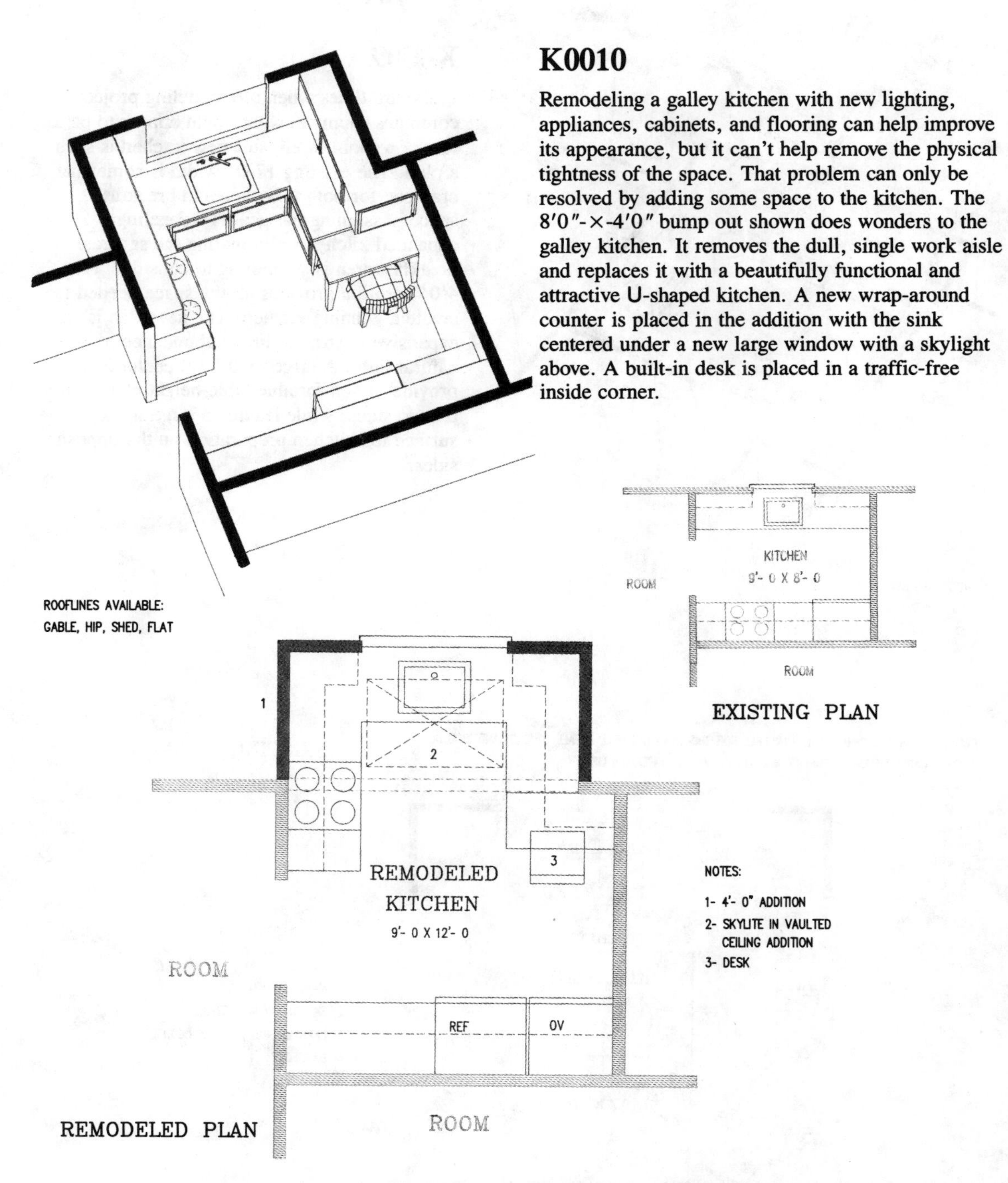

CONSTRUCTION BLUEPRINTS ARE AVAILABLE FOR THIS PLAN (AS WELL AS ALL OTHERS) AND INCLUDE KITCHEN CABINET INFORMATION AND A MATERIALS LIST

K0008

Given: a very tight kitchen that is really too small to allow you to set up a table, so you eat in the adjoining dining room all the time, but even that is an unattractive, cramped room. A solution: If formal dining is not your style anyway, consider creating a stunning new country kitchen from the two spaces. The result is an inviting room with a kitchen work area equal to what existed, but with a significant new storage component in the way of a pantry, a built-in desk, a corner for a built-in TV, and a center island eating counter. But beyond this, there is now space for a comfortable, informal, family area. To satisfy the occasional need for large dining you would move the couch and chair and set up a table, but since the space is now open, it will no longer feel cramped.

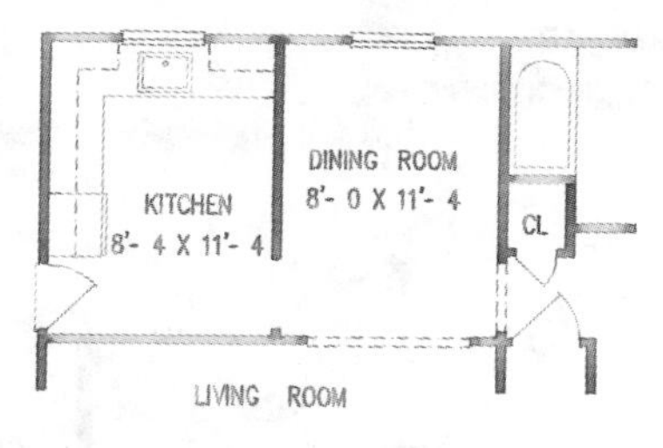

EXISTING PLAN

NOTES:
1- ISLAND SNACK COUNTER
2- PANTRY
3- DESK
4- BUILT-IN T.V.
5- REMOVE WALL

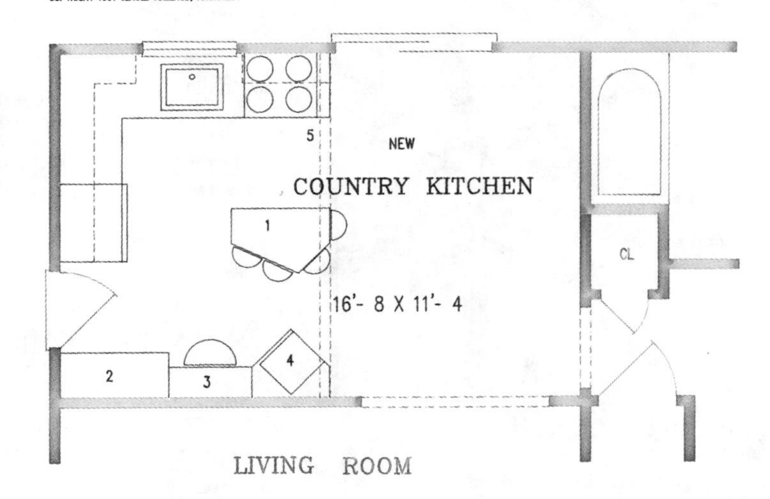

REMODELED PLAN

CONSTRUCTION BLUEPRINTS ARE AVAILABLE FOR THIS PLAN (AS WELL AS ALL OTHERS) AND INCLUDE KITCHEN CABINET MANUFACTURER'S INFORMATION AND A MATERIAL LIST

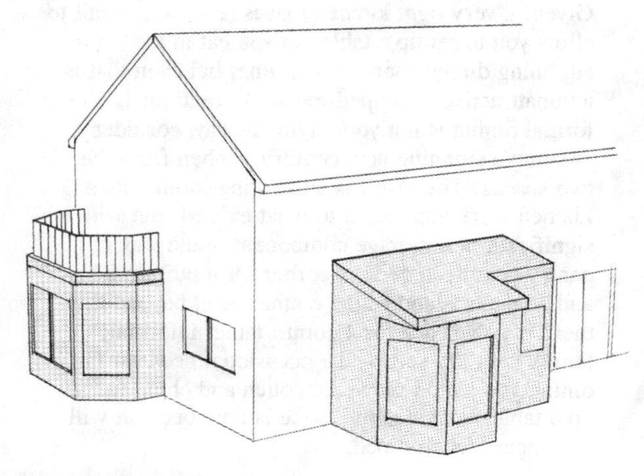

KD003

This plan is actually two separate additions that complement each other, but which could be undertaken independently. The home featured is typically an older colonial-styled two-story with a nice dining room and kitchen, but missing is a hall that would allow direct access to the kitchen. While the kitchen is ample, it's a little too tight to provide attractive table space. The two additions, shown with two different rooflines, provide not only the space necessary to accomplish the goals, but do so tastefully and in harmony with the home. Two columns and a dropped header define the new hall space without the confining use of walls. The dining room bump is 5′4″, and the rear extension is 14′4″×7′10″, including the lavatory as shown.

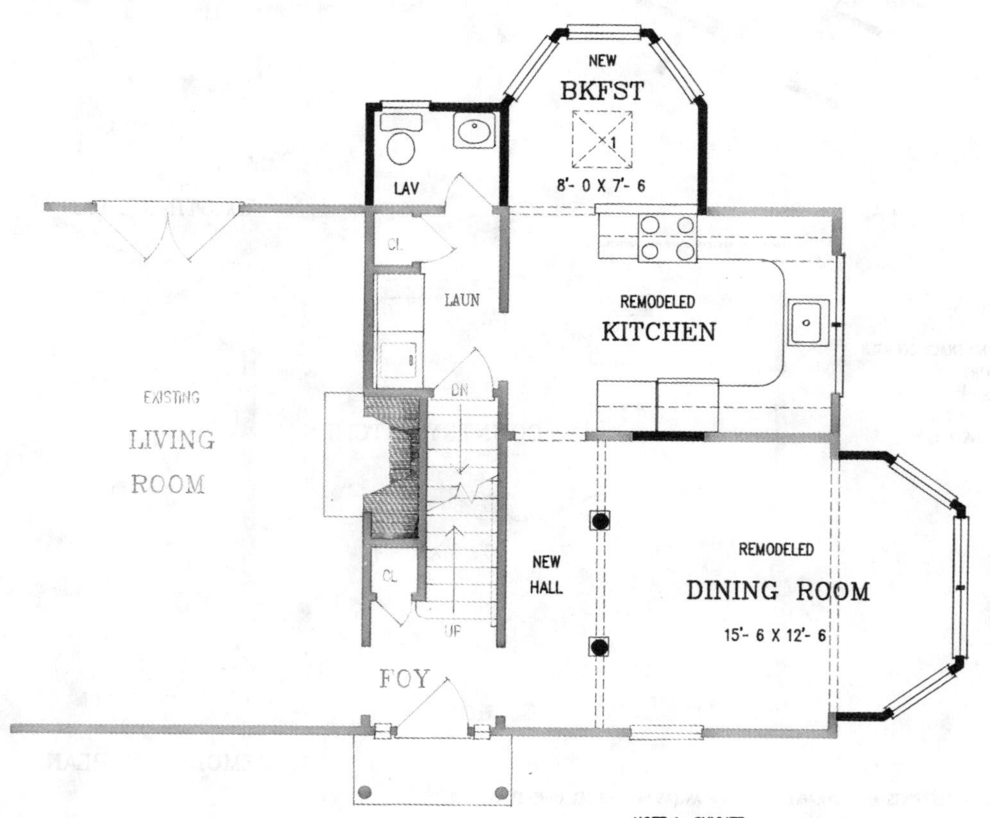

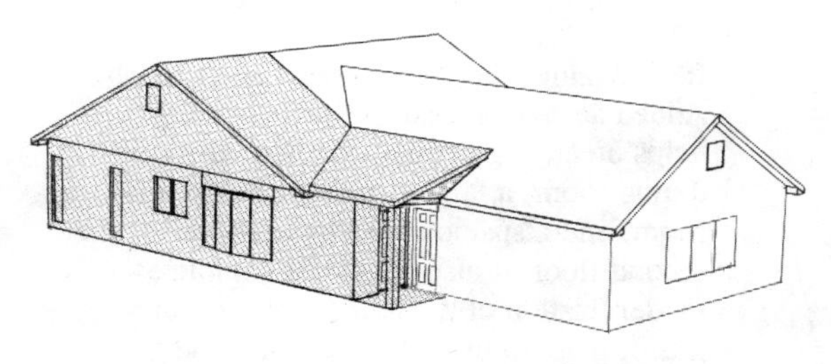

ROOFLINES AVAILABLE:
GABLE (SHOWN), HIP

KD005

A narrow, side-facing kitchen is common to many one-story homes. Enlarging one to create a fabulous, up-to-date space is the subject of this plan; it requires six feet of your side yard, although without a new laundry room it could be built in four feet. The kitchen is a sensational U, surrounding a dramatic center island. Built-ins, including a desk and pantry, line the inside wall. It is strongly suggested that the family room be remodeled and opened to the new kitchen, thereby creating a unified informal family living area. The dining room expansion, new laundry, and covered porch are the "bonuses" that go along with this plan.

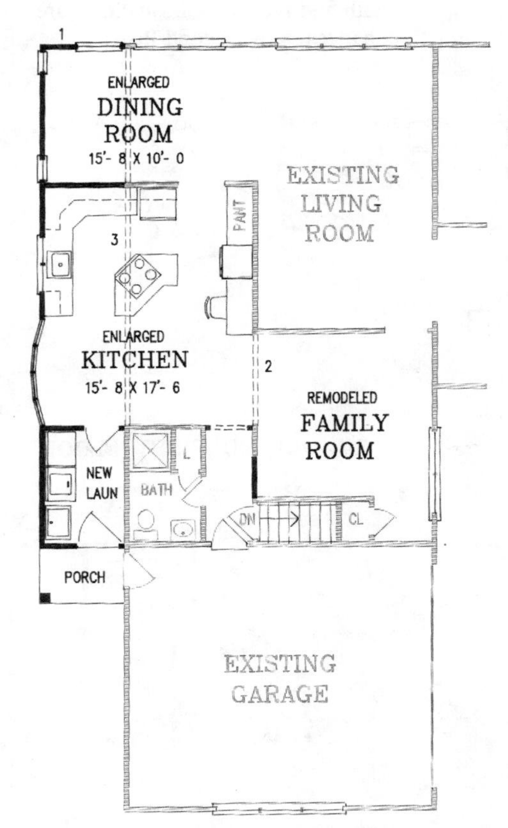

REMODELED PLAN

NOTES 1- 6'- 0 ADDITION
2- REMOVE WALL TO FAMILY ROOM
3- REMOVE EXISTING SIDE WALL

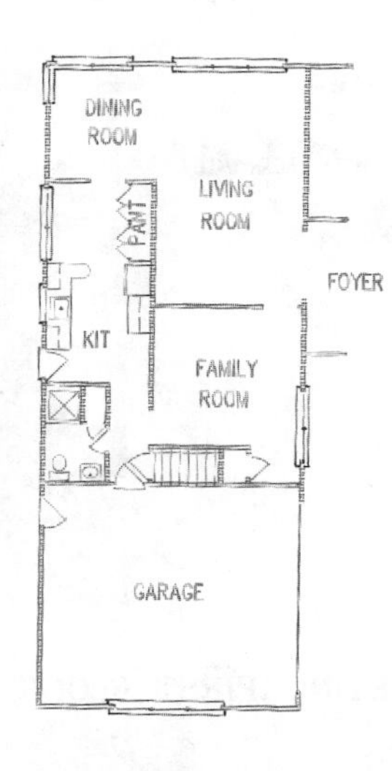

EXISTING PLAN

NOTE: CONSTRUCTION BLUEPRINTS CAN BE READILY MODIFIED TO ACCOMMODATE YOUR SPECIFIC HOUSE OR ROOM SIZES.

KDEB1

It's amazing what five feet can accomplish. Added across the rear of this two-story, it helps create a great new kitchen, an expanded dining room, a large new laundry/exercise room, and a spacious new master bath at the second floor. It also enables a complete modernization of the rear facade, permitting lots of glass to bring light inside. The remodeled kitchen is a visual delight; a corner sink, a center island, and skylights in the new addition create a dramatic new space. The expanded laundry provides an ideal space for exercise equipment or for sewing or a home office—maybe even all three. The new master bath and second walk-in closet are the notable changes to the second floor.

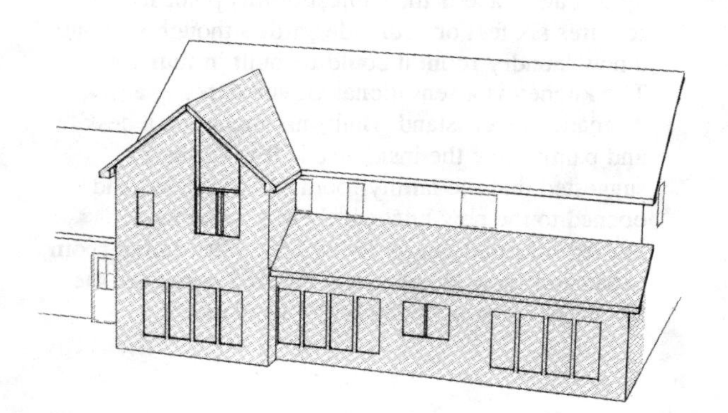

ROOFLINES AVAILABLE:
GABLE & SHED (SHOWN)

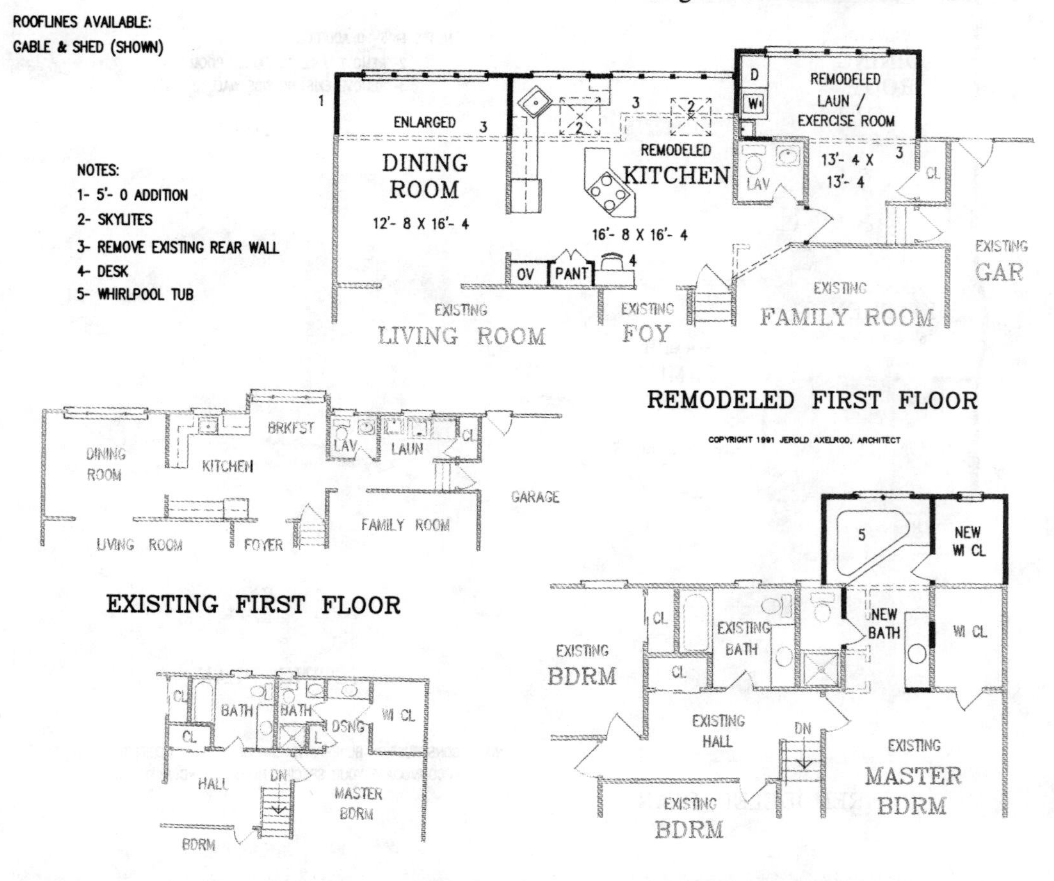

KDL01

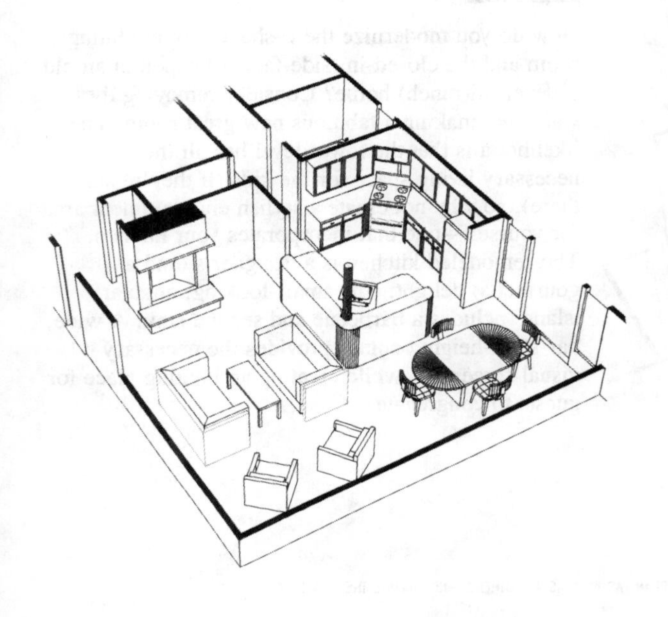

When there is no room to expand out, and your goal is to modernize a tight kitchen, consider the choice of opening the space to create a great room. This option, while not for everyone, is a technique frequently used to maximize limited space. The kitchen, while not increased in size, benefits from a visual integration with the adjoining living and dining rooms; interior walls are eliminated to the maximum extent possible, and a column is used as necessary. The old kitchen door and window are removed, since exterior access and an abundance of light are now provided from the rear of the great room. The kitchen layout is significantly enhanced.

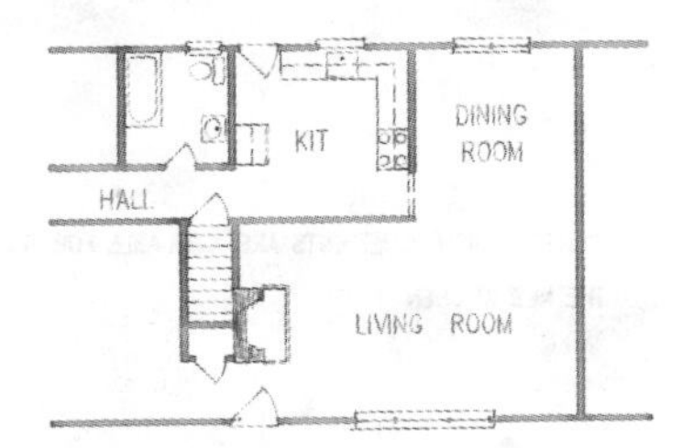

EXISTING PLAN

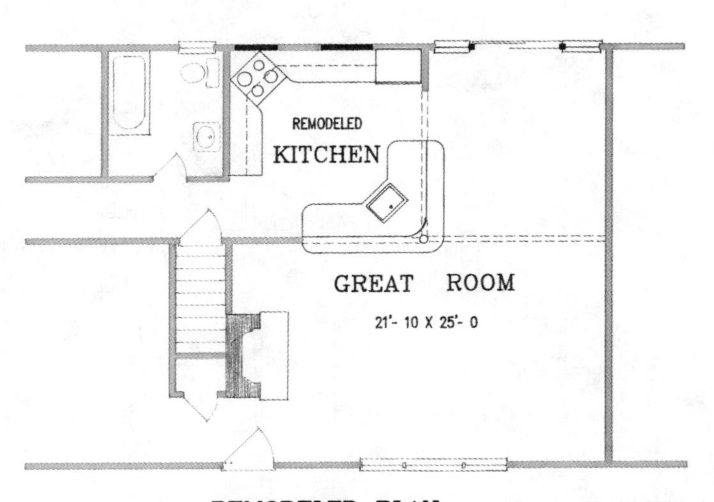

REMODELED PLAN

CONSTRUCTION BLUEPRINTS ARE AVAILABLE FOR THIS PLAN (AS WELL AS ALL OTHERS) AND INCLUDE THE DESIGN OF:
THE HEADERS REQUIRED TO OPEN UP AND REMOVE THE WALLS

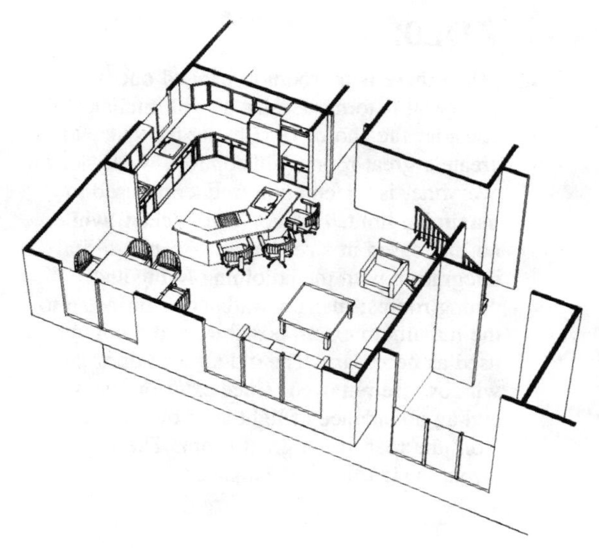

KDL02

How do you modernize the L-shaped living/dining room and the closed-in, side-facing kitchen in an old bi-level (hi-ranch) home? Consider removing the walls and making a fabulous new great room. The likelihood is that the lower level has all the necessary living space for the kids (if they're still there), so why not create a stylish entertainment area for yourself—one that incorporates your kitchen. The remodeled kitchen is a designer's (and a gourmet's) delight. The smart-looking, angular island includes a barbecue and second sink. A wide, bar-stool-height counter provides the necessary visual screening, while creating an inviting place for guests to congregate.

CONSTRUCTION BLUEPRINTS ARE AVAILABLE FOR THIS PLAN (AS WELL AS ALL OTHERS) AND INCLUDE THE DESIGN OF:
THE NEW KITCHEN

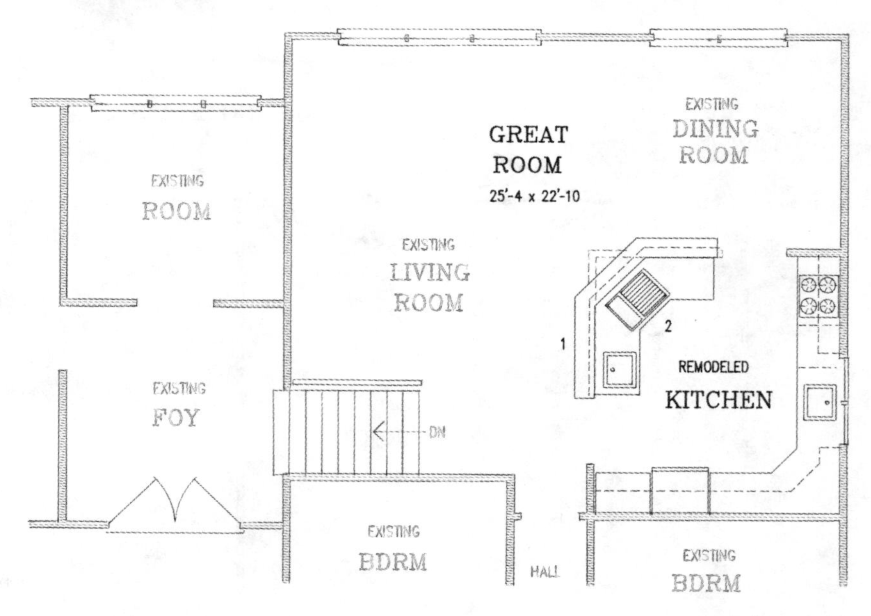

NOTES 1- RAISED COUNTER
2- BARBECUE UNIT

KDL03

It's not that you need more space that has you considering the remodeling of the entire living area of this bi-level (hi-ranch); it's the desire for a change in lifestyle. You're tired of chasing up and down the stairs to the lower-level family room, and, what's more, you are excited by those splashy new great rooms. This plan can help turn dreams to reality; walls are eliminated, replaced by decorative columns, and the entire area becomes a sensational living and entertaining space, that has a dramatic new kitchen as its focal point. The only expansion is the rear deck, which has been enlarged in the form of a half-circle.

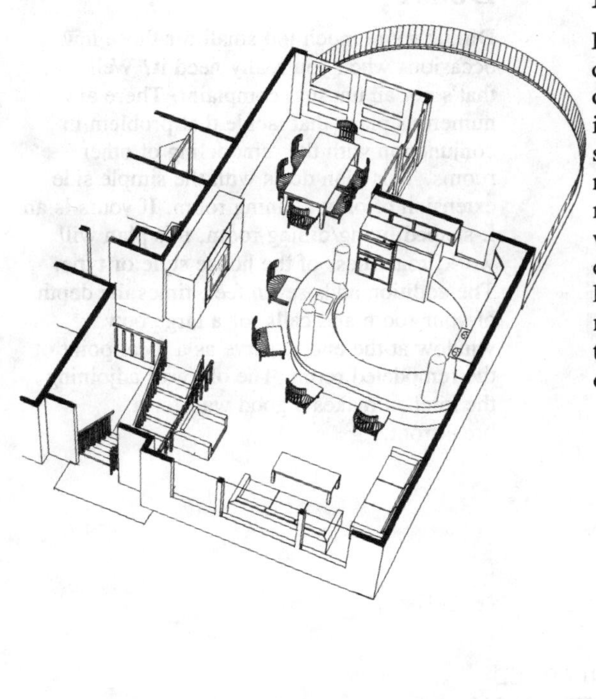

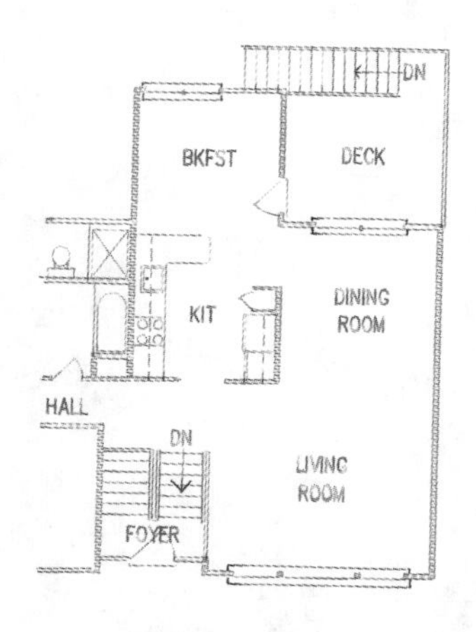

EXISTING PLAN

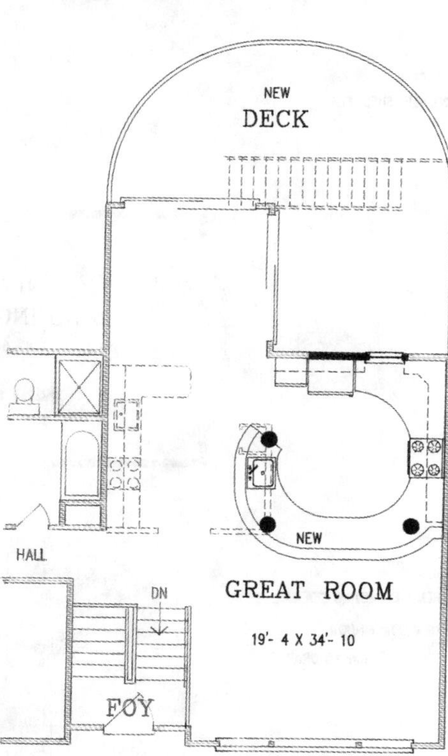

REMODELED PLAN

CONSTRUCTION BLUEPRINTS ARE AVAILABLE FOR THIS PLAN (AS WELL AS ALL OTHERS) AND INCLUDE THE DESIGN OF THE KITCHEN AND STRUCTURAL HEADERS

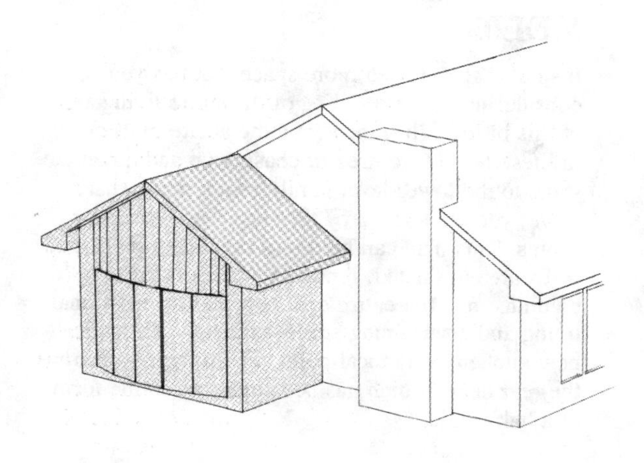

ROOFLINES AVAILABLE:
GABLE (SHOWN), HIP, SHED, FLAT

D0001

Dining room much too small for those few occasions when you really need it? Well, that's not an unusual complaint. There are numerous plans that tackle that problem in conjunction with the remodeling of other rooms. This plan deals with the simple side extension of only a dining room. If yours is an L-shaped living/dining room, this plan will work, regardless of the house style or type. The addition adds seven feet, times the depth of your room and calls for a large bow window at the end to serve as a focal point of the remodeled room. The old wall adjoining the kitchen makes a good wall for a breakfront.

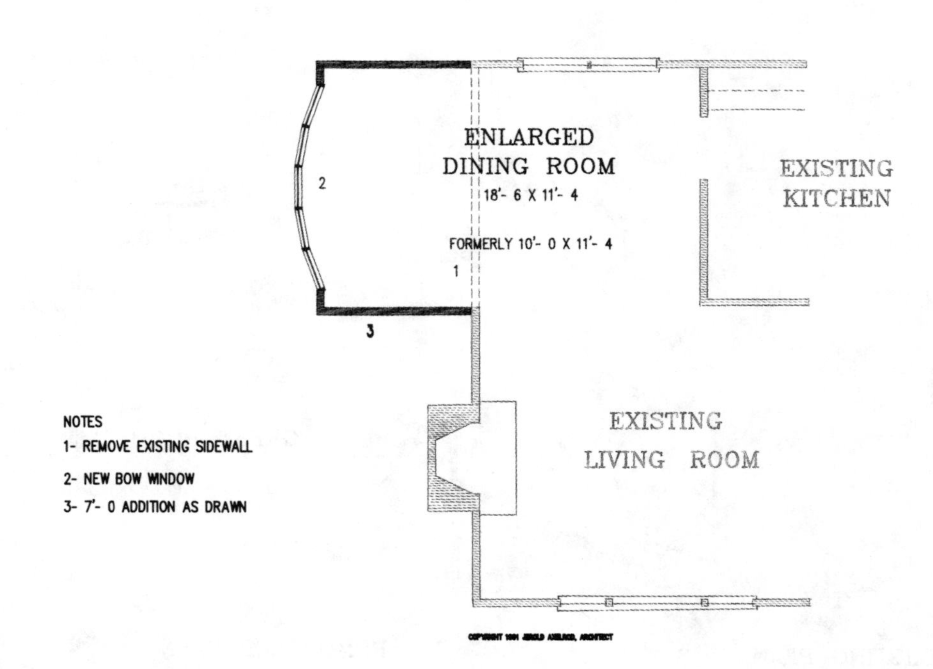

NOTE: CONSTRUCTION BLUEPRINTS CAN BE READILY MODIFIED TO ACCOMMODATE YOUR SPECIFIC HOUSE OR ROOM SIZES.

D0002

Expanding a front-facing dining room in a center-hall two-story design can be risky business. Most floor plans will only allow you to extend forward, which becomes a minor structural concern and a major aesthetic one. Trying to emulate and design an addition within the style of the existing home frequently presents an unworkable solution, particularly because the new element is too small. Although maybe not for everyone, this glass-enclosed greenhouse-style addition could be the answer. While it creates a major departure from the style of the front, it does so dramatically and tastefully. The color of the structure should blend with your windows, and the use of traditional draperies could tie it in beautifully.

NOTE: CONSTRUCTION BLUEPRINTS CAN BE READILY MODIFIED TO ACCOMMODATE YOUR SPECIFIC HOUSE OR ROOM SIZES.

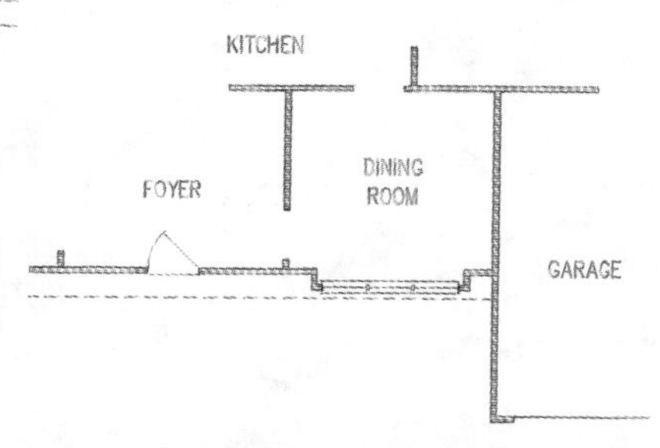

EXISTING FIRST FLOOR

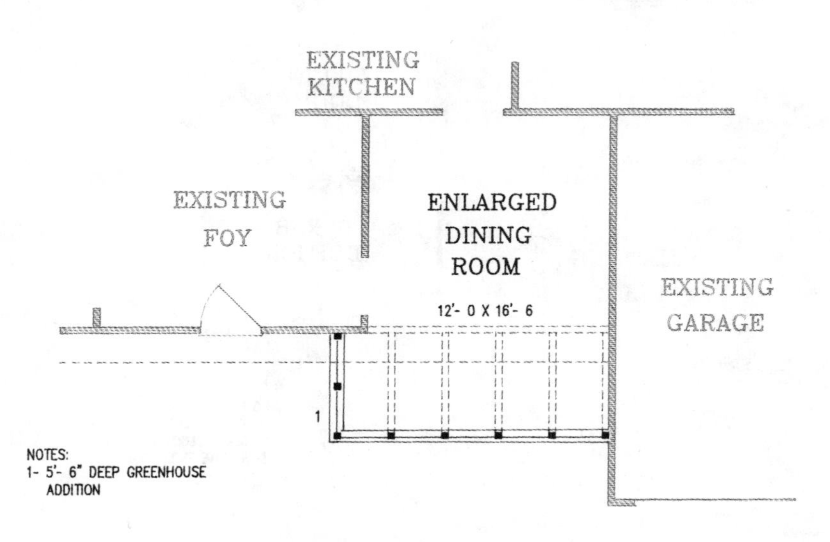

REMODELED FIRST FLOOR

L0003

The inherent flexibility for change that most split-level homes enjoy is demonstrated in this living room addition. The split-level also lends itself to contemporary adaptations, and if you will be adding to the front, as in this design, it is a marvelous opportunity to create a striking change. This reverse-gabled addition features a dramatic curved glass-block wall that admits lights, provides privacy, and is handsome to look at. The space that is added is ideal as a library or music alcove, and is a perfect enhancement for the entertaining capabilities of the living and dining rooms.

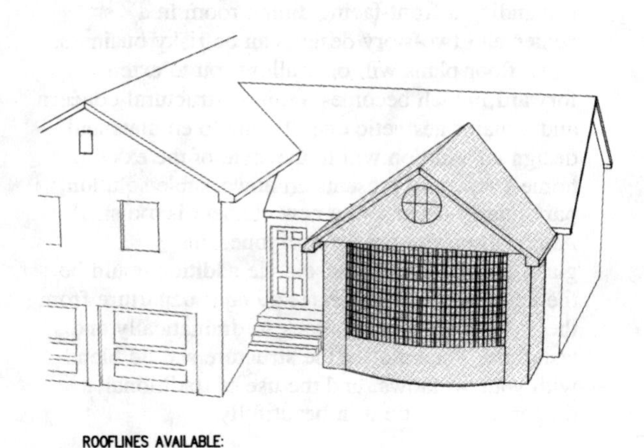

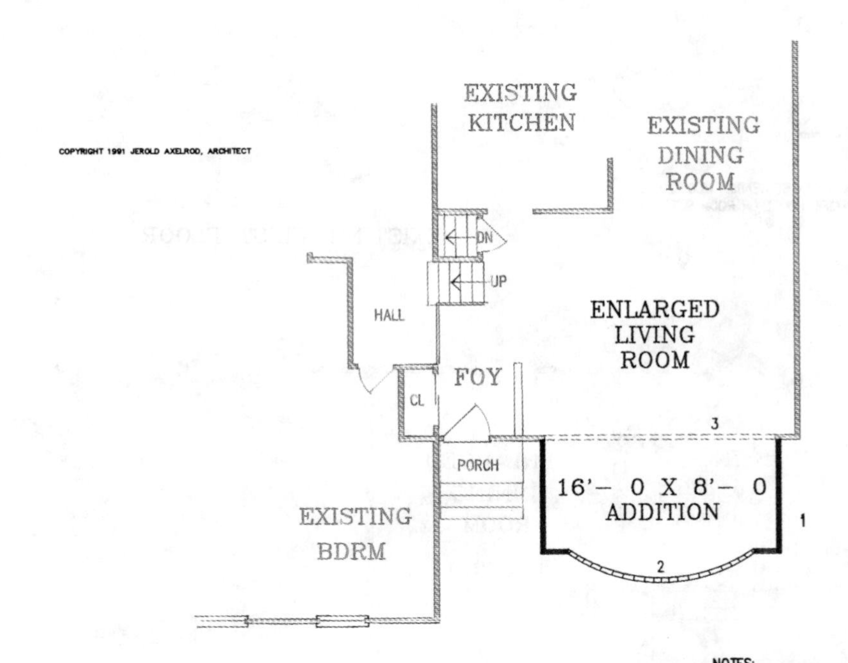

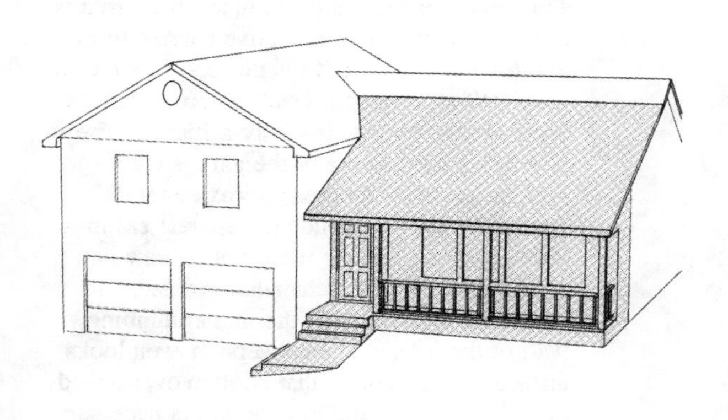

LP001

Just as the open site area formed by the wings of an L-shaped one-story offers potential for new space, the area in front of the living room on an L-shaped split-level is also space waiting to be found. In the design shown, the living room has been graciously increased by five feet, and a new front porch is also added. Bonuses of this addition include a new entrance foyer and a newfound exterior appeal. In the next chapter, you will find plans showing the complete renovation of the facade of several split-level homes. When planning any front addition, remember to investigate the location of all underground utility connections.

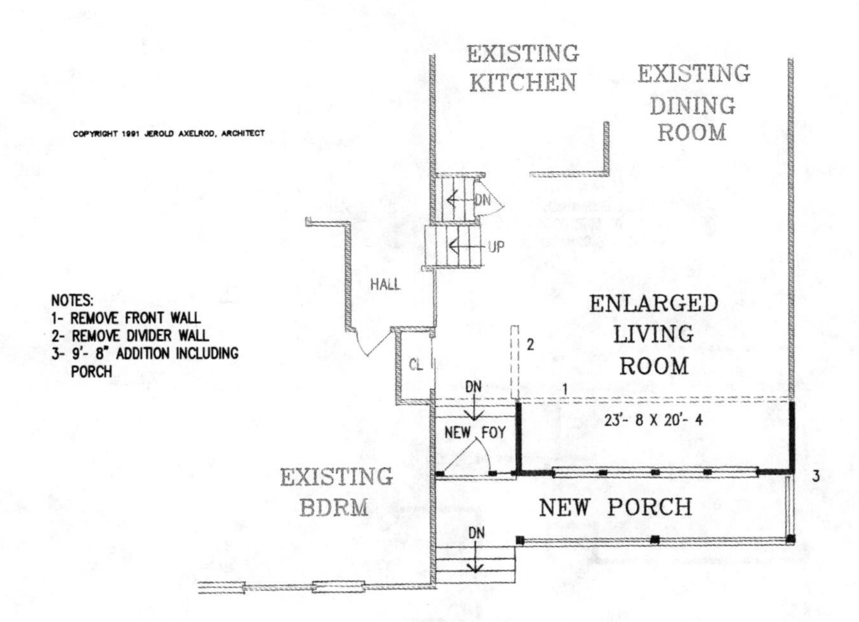

NOTE: CONSTRUCTION BLUEPRINTS CAN BE READILY MODIFIED TO ACCOMMODATE YOUR SPECIFIC HOUSE OR ROOM SIZES.

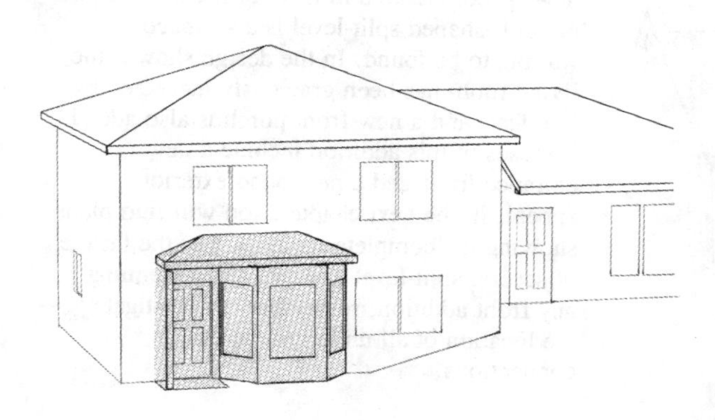

HAPT4

This is one of several plans in the book that is designed to create an attractive barrier-free apartment. The location is perfect: It is the ground floor of a split-level, where direct access to the outside is easily achieved. The plan takes up the area of the garage and recreation room, an area approximately 27 feet × 26 feet. It includes a separate eat-in kitchen, a large living room with a new bay window, and a wheelchair-accessible bathroom located along the main plumbing wall of the home. The conversion even looks attractive—something that is often overlooked, despite the new windows and doors necessary. Plans that add back a family room and garage are shown in the preceding chapter.

ROOFLINES AVAILABLE:
HIP (SHOWN), GABLE

NOTES:
1- 4'- 0" WIDE HALLS
2- ROLL IN SHOWER
3- 3'- 0" WIDE DOORS
4- REMOVE GARAGE WALL

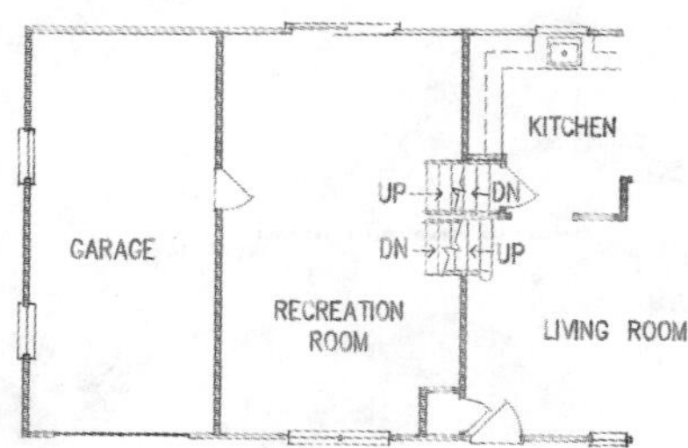

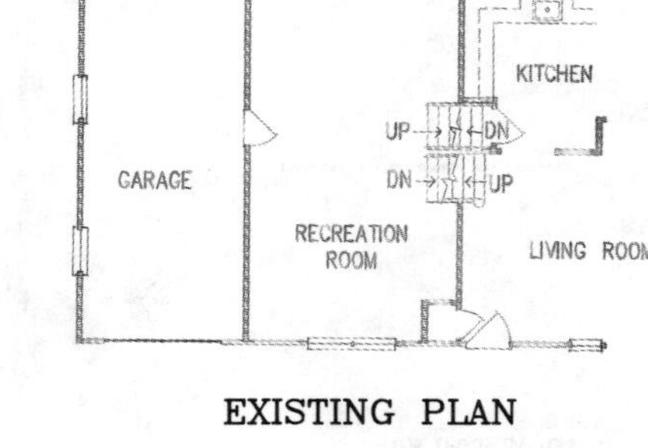

EXISTING PLAN

NEW BDRM 13'- 0 X 11'- 0
CL
NEW BATH
EXISTING KIT
UP DN
NEW LIVING ROOM 15'- 6 X 13'- 8
NEW KIT 10'- 6 X 9'- 0
EXISTING LIVING ROOM
CL
PORCH

REMODELED PLAN

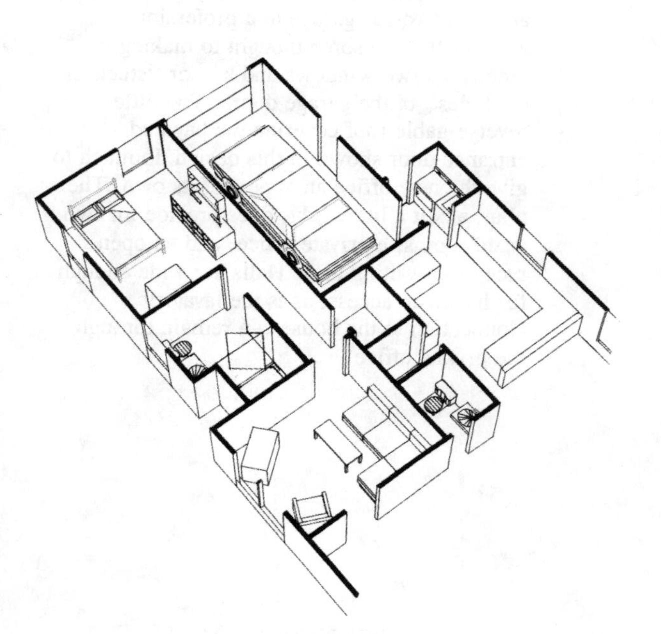

HAPT1

This is another barrier-free conversion. The plan creates an attractive one-bedroom, fully handicapped-accessible apartment from an existing garage and adjacent room—most likely a dining room, as shown. The plan assumes a two-car side-entry garage, and still leaves a one-car garage with direct interior access. The half garage utilized is converted into a bedroom and a fully wheelchair-accessible bathroom with a roll-in shower. The dining room makes a fine living room, but need not be converted if only a bedroom is necessary. Other than the closure of one garage door, there are no other changes required to the exterior of the home. Other barrier-free apartments are shown in chapter 7.

NOTES 1- POCKET DOOR
2- ROLL-IN SHOWER
3- 4' WIDE HALL
4- 3' WIDE DOORS

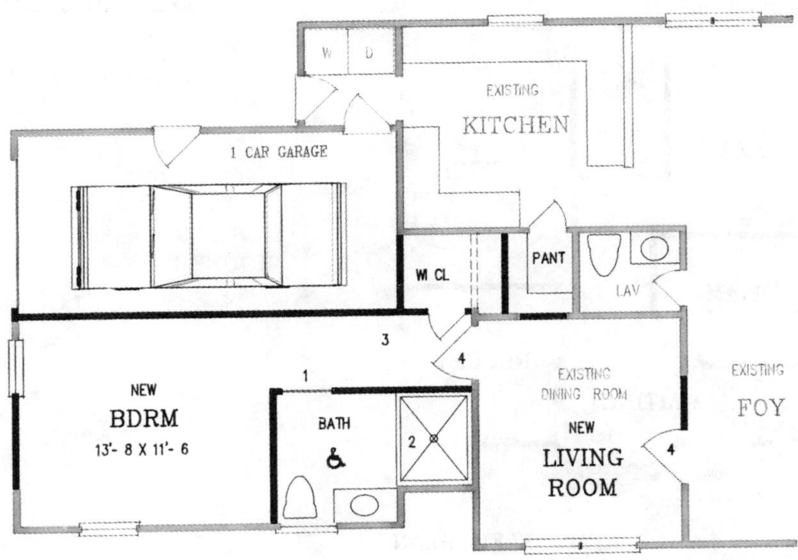

CONSTRUCTION BLUEPRINTS ARE AVAILABLE FOR THIS PLAN (AS WELL AS ALL OTHERS) AND INCLUDE THE DESIGN OF:
THE WHEEL CHAIR ACCESSIBLE BATHROOM

APT05

If you are considering converting your attached two-car garage to a professional apartment, give some thought to making it other than two windows and a door "stuck in the holes" of the garage doors. The little reverse-gable roof covering the bay and entrance door shown in this design do much to give the new office an appeal of its own. The plan, albeit a little tight, does provide for two exam rooms, a private office, and an open reception/waiting area. Halls are wide enough for handicap access, as is the lavatory. Connection to the house can remain through the private office.

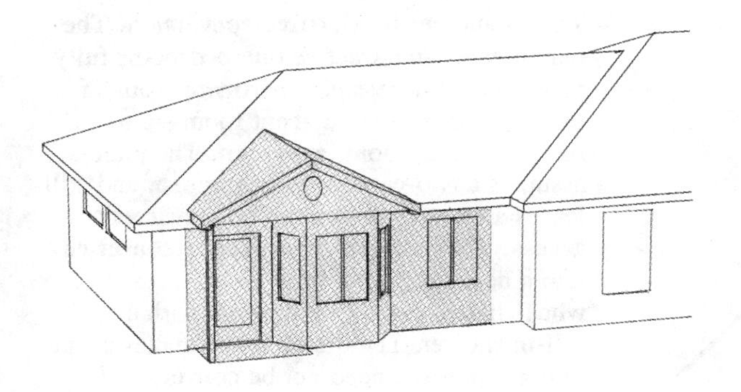

ROOFLINES AVAILABLE:
GABLE & HIP (AS SHOWN)

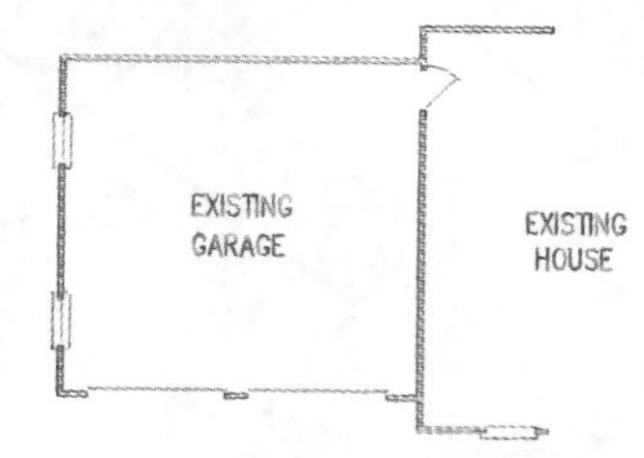

EXISTING PLAN

NOTES:
1- DOOR TO HOUSE TO REMAIN
2- COUNTER
3- NEW BAY WINDOW

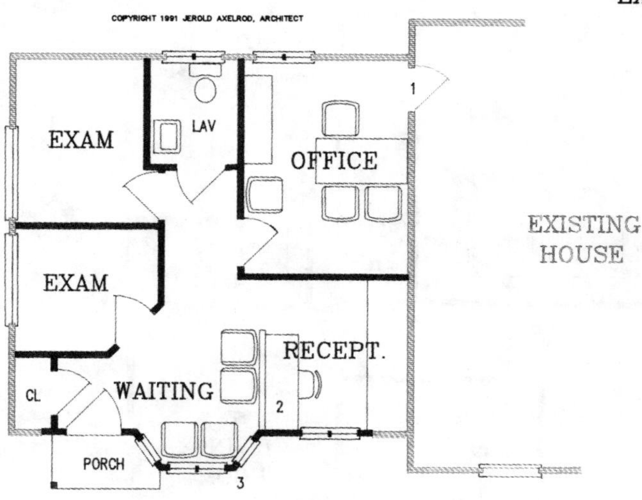

REMODELED PLAN

CONSTRUCTION BLUEPRINTS ARE AVAILABLE FOR THIS PLAN (AS WELL AS ALL OTHERS) AND INCLUDE DIMENSIONED PLAN FOR CONVERSION TO OFFICE AND A MATERIALS LIST

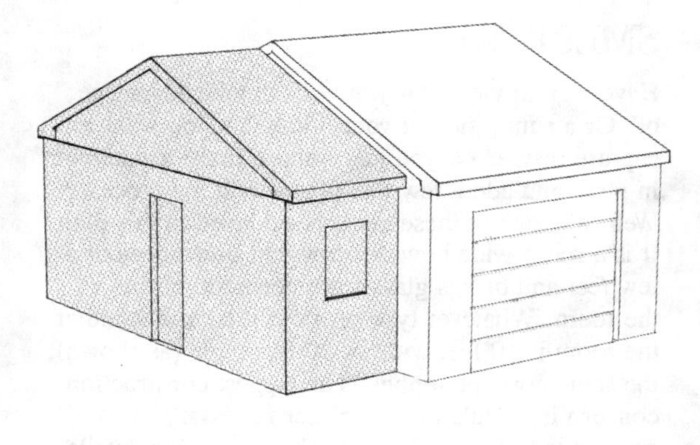

ROOFLINES AVAILABLE:

GABLE (SHOWN), HIP, SHED, FLAT

STR01

If you would love to have some storage or workshop space, and your lot gives you five or six feet to expand alongside your garage, you could consider the addition shown here. There are some pitfalls of which to be aware. Even though you could use all the space, the small set-back at the front is advisable. It creates an attractive form and roofline break, and with some accents (such as a specialty window or some stone veneer), it could become an attractive addition. If you have seven or eight feet that can be added, consider making the addition flush with the garage and reconstructing the garage front with one or two overhead doors. Bear in mind, however, that a two-car garage should have a minimum of 17 or 18 feet in width.

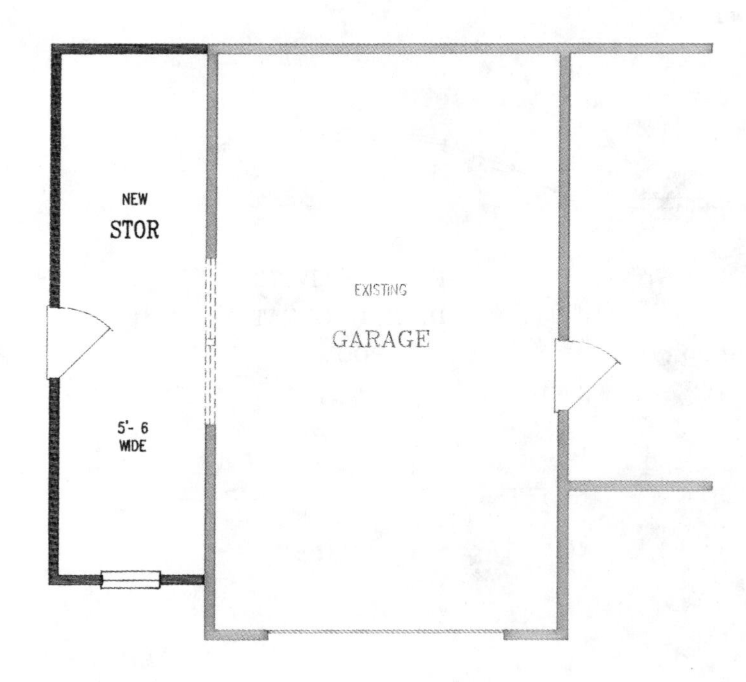

NOTE: CONSTRUCTION BLUEPRINTS CAN BE READILY MODIFIED TO ACCOMMODATE YOUR SPECIFIC HOUSE OR ROOM SIZES.

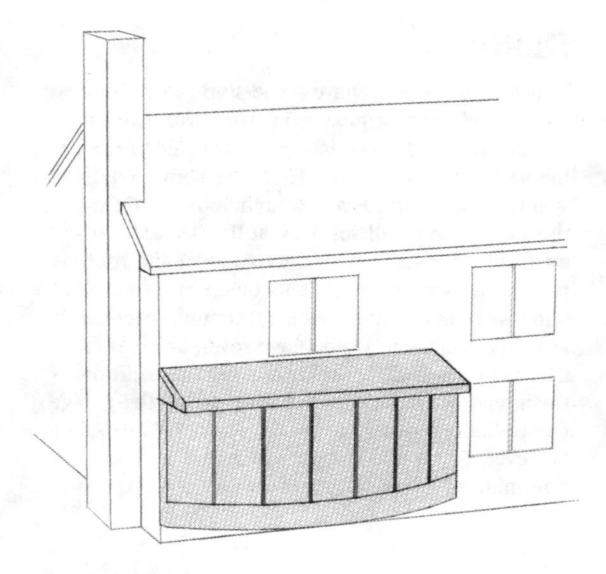

SML01

Have a great view that you want to take advantage of? Or a sunny side of your home that you want to capture inside? Or you just want to make a statement in glass and add a few feet of space in the process? Well, any one of these goals is achieved in this plan. It is a room-wide bow window that protrudes out a few feet and brings glass from corner to corner of the room. Whatever type of room it is, and whether the room is 10 feet wide or 20 feet wide (as shown), the same concept applies. The biggest construction concern is calculating the girder necessary to support the roof (and floors) above, and the details necessary to install this girder; see chapter 11 on how to accomplish this.

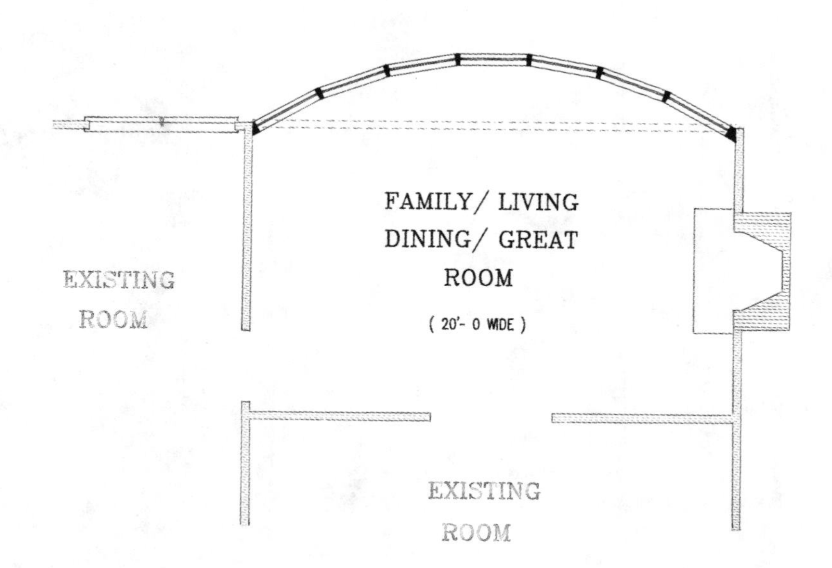

NOTE: CONSTRUCTION BLUEPRINTS CAN BE READILY MODIFIED TO ACCOMMODATE YOUR SPECIFIC HOUSE OR ROOM SIZES.

SML04/05

The two roof clearstory plans pictured here are amongst the many plans in this book that have as their focus the addition of light. This propensity for light is a driving force in contemporary design. It has also become a yearning of many homeowners when undertaking a remodeling project. We remove walls, open outside walls, and install new larger windows; we also install skylights. To many people, though, the skylight provides too direct a source of light. Skylights are also difficult to shade. The clearstory (or *roof monitor*, as some call it) brings in light through its walls. Shown on this page is a shed roof clearstory and a reverse-gable clearstory. Two other designs follow.

CLEARSTORY PLAN

PLAN# SML04

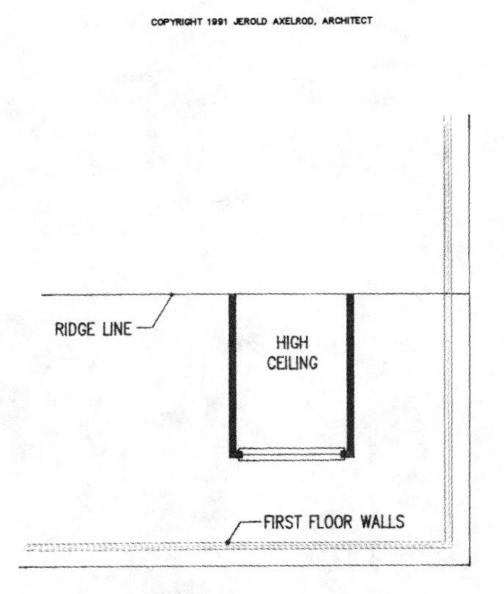

CLEARSTORY PLAN

PLAN# SML05

CONSTRUCTION BLUEPRINTS ARE AVAILABLE FOR THIS PLAN (AS WELL AS ALL OTHERS) AND INCLUDE FRAMING DATA AND A MATERIALS LIST

SML06/07

This page presents a hip-roofed clearstory and a multisided conical style clearstory or monitor. With roof overhangs any of these clearstories can be designed to eliminate direct sun, if that is desired. The clearstory also provides light more directly to walls below it and can be used as an effective design tool to light specific areas of the floor below. The installation of a clearstory requires the removal of some ceiling below, and maybe even the entire ceiling of a room. As such, a clearstory also has the effect of elevating the volume of a room and can visually enhance the appearance of the space. Clearstories are also exterior forms that add interest and appeal to the roofline of your home.

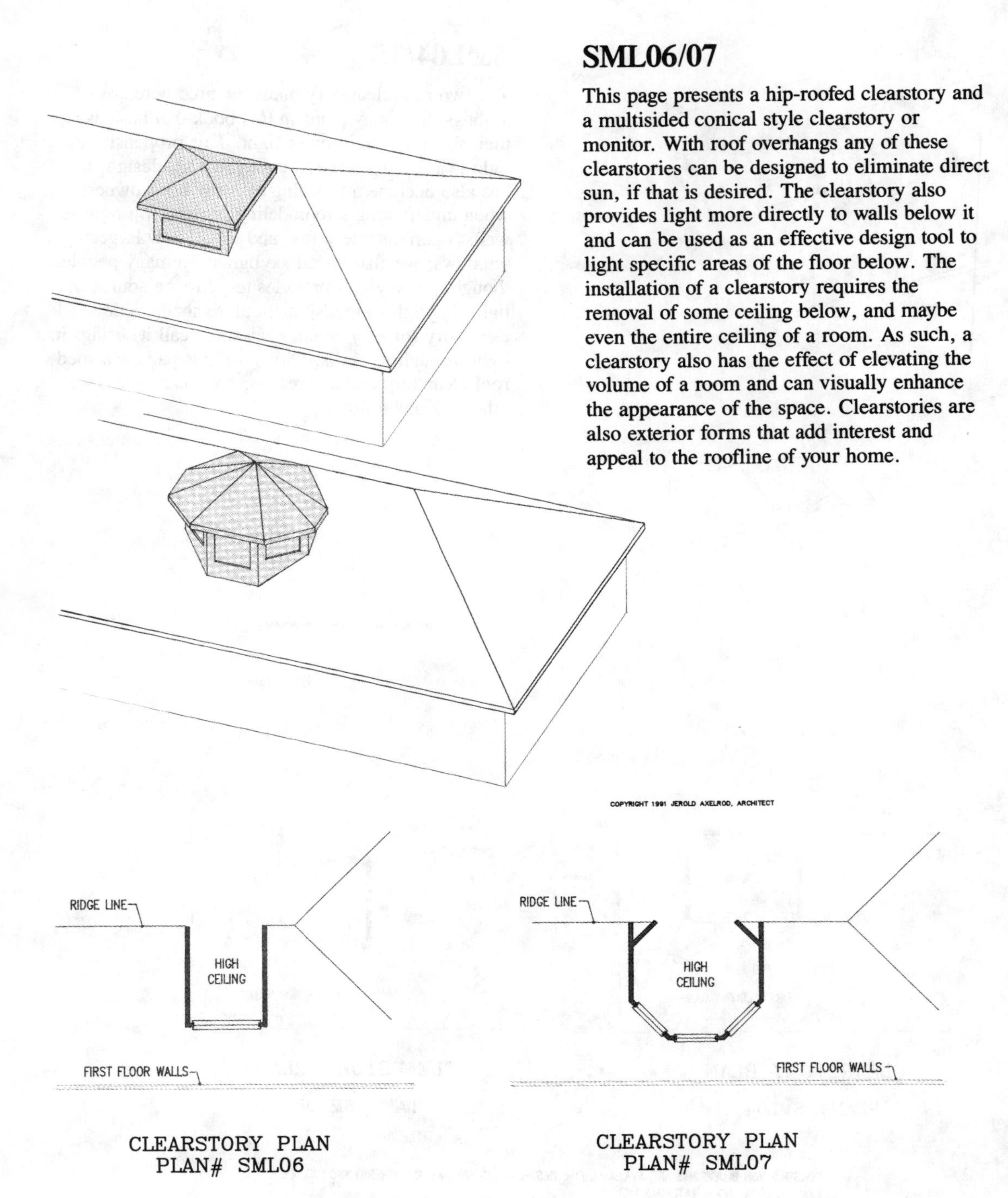

CLEARSTORY PLAN
PLAN# SML06

CLEARSTORY PLAN
PLAN# SML07

NOTE: CONSTRUCTION BLUEPRINTS CAN BE READILY MODIFIED TO ACCOMMODATE YOUR SPECIFIC HOUSE OR ROOM SIZES.

SML08

Here is a fabulous way to get some natural daylight into a basement room. The search for natural light is certainly one of today's prevalent design themes. Removing ceilings and installing skylights is a given to most remodeling projects, as is the installation of larger expanses of new windows. But how do you achieve that in a basement? This plan shows how. Pick an area where the ground is below the basement windows. A wide new areaway wall is built, and above that, a glass-enclosed greenhouse structure is installed.

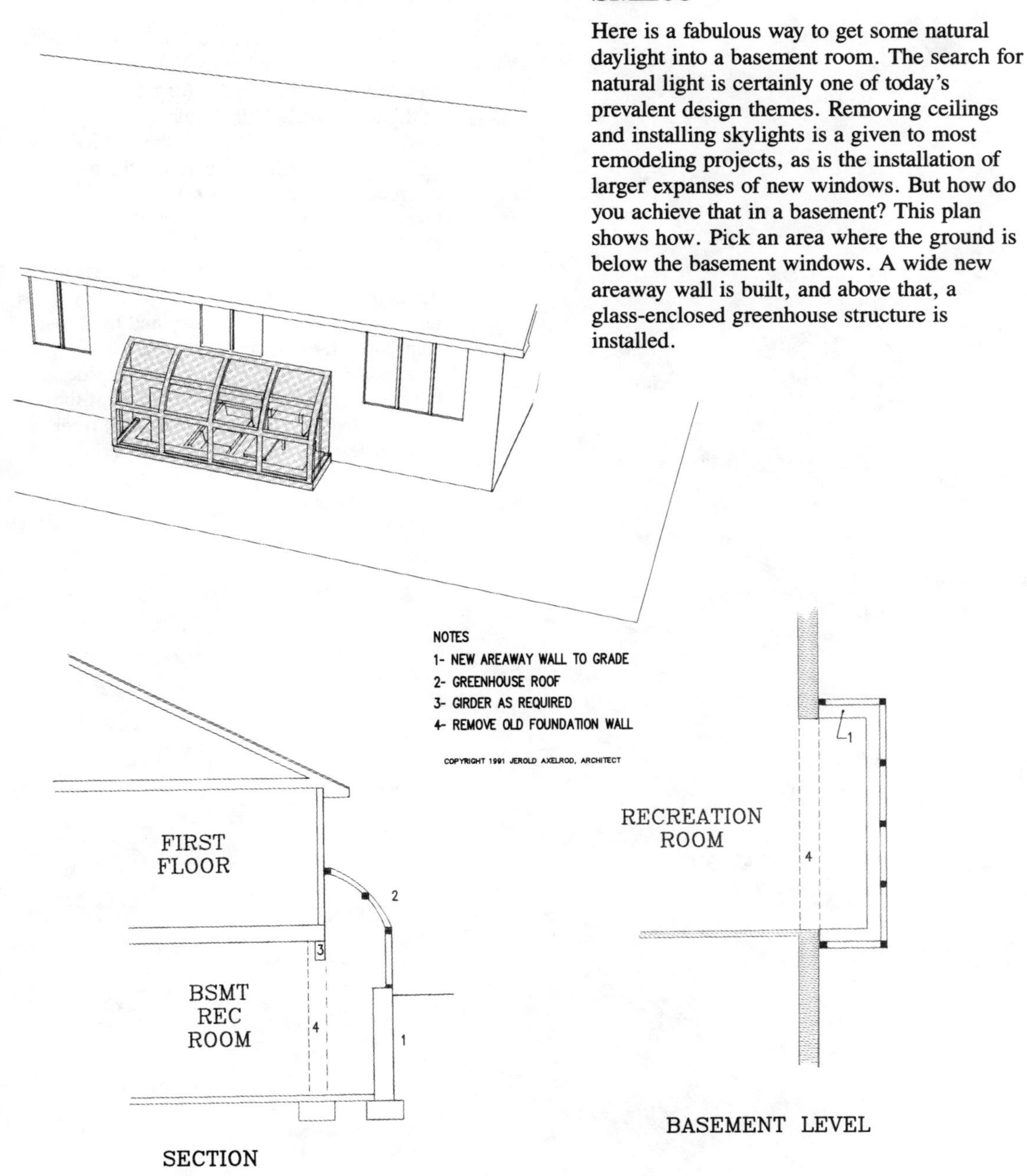

CONSTRUCTION BLUEPRINTS ARE AVAILABLE FOR THIS PLAN (AS WELL AS ALL OTHERS) AND INCLUDE GREENHOUSE MANUFACTURER'S DATA AND INFORMATION ON REMOVING THE EXISTING WALL

SML09

The design presented here is another idea for you to consider that brings light into a basement recreation room (or any other room in the basement). This design goes one step further than plan SML08 on the last page by also providing exterior access to the basement. A broad, stylish, angular set of steps, built from either wood ties or concrete, goes down from the exterior grade to a wide, comfortable, sunken entry. Triple French doors provide light and access. The key is the large size of the sunken entry and the broad steps, which eliminates the negative impression of the typical narrow outside basement stair. This solution is best if the distance from grade to the basement floor is 4′6″ or less.

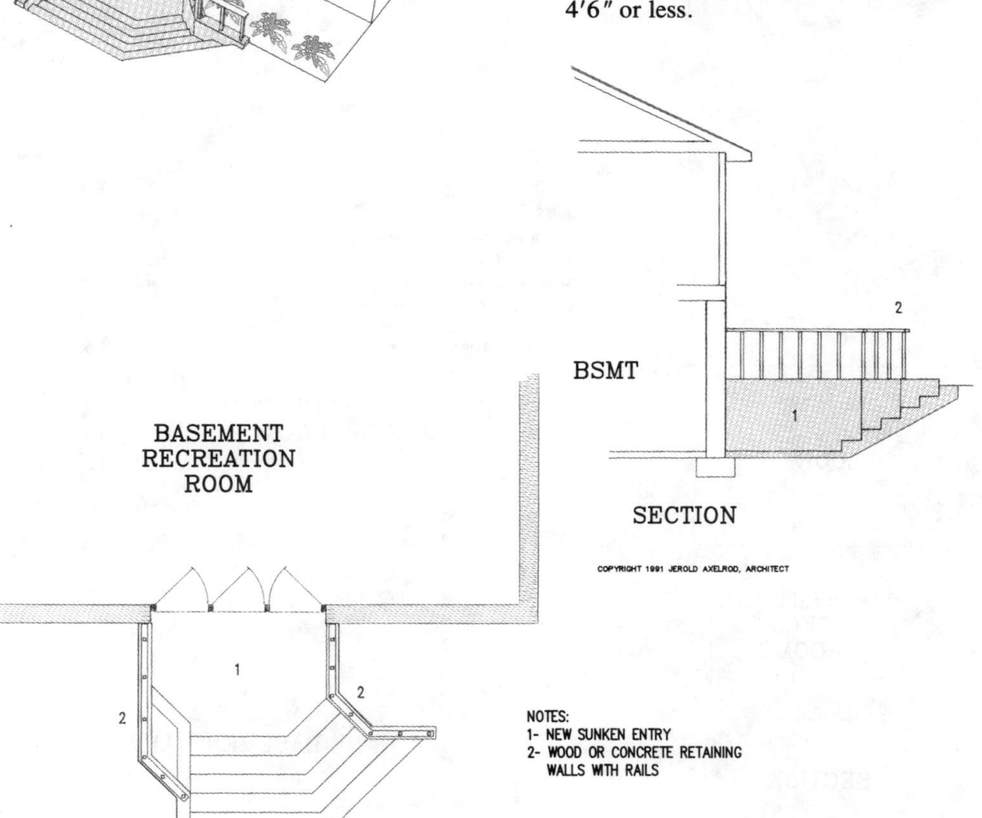

CONSTRUCTION BLUEPRINTS ARE AVAILABLE FOR THIS PLAN (AS WELL AS ALL OTHERS) AND INCLUDE A MATERIALS LIST & INFORMATION ON HOW TO MAKE CHANGES TO PLANS

9 CHAPTER

New faces

As you likely have noticed by now, the exterior designs pictured so far have been devoid of materials and details. This should not at all suggest that this is the way to design. On the contrary, when preparing a custom design, I am always studying facade details simultaneously as the plan develops. However, for the purposes of preparing these prototype, ready-to-use plans, it is helpful to deal with the choice of facades as a separate topic. There are three different scenarios in which the following drawings will prove helpful.

New faces as a planned product

Almost every remodeling project simultaneously produces some exterior redesign. Even if you are just remodeling a bathroom or a kitchen, you are likely to change a window. What impact does this have on the outside of the home? One window change could be major, particularly if it's on a front wall.

Changing a window might require some fixing of siding or brickwork. What if that siding is no longer available, or what if the brick can't be matched? Will you repaint the whole house? Well, if you do, you might as well change some other windows, too. Such questions multiply significantly the more complex the remodeling. You've likely read my comments on certain plans, where I cautioned that making an addition flush with existing walls requires the perfect match of siding and roofing.

Yet this is the opportune time to consider a new face. Additions, especially those that affect the forms and massing of the home, might even trigger the question of changing style. As you undertake a major addition, if yours is one of those homes built over the last 40 years that lacked any real discernible character, I strongly recommend that you now give it some character—budget permitting, of course.

Virtually all of the additions shown in the previous chapters have already been designed from the point of view of forms and mass. All they need are materials and trim to complete the picture, and that you can do on your own, or you could pick from the drawings in this chapter. If you prefer one style over another, and that suggests changing window designs, that is perfectly acceptable. A custom sketch could be prepared if you are unsure.

New faces as an unplanned product

An unplanned change can be a most unfortunate experience, and it usually happens when there has been a lack of proper design. Just as I sketched some "don'ts" in chapter 4 on forms, I could do the same here; but that would take a great deal of space, and I would rather show you what you should do. The best advice is to hire a professional or to build from the ready-made designs in this book. If you insist on designing the whole project yourself, make sure you study all sides of the project, and even build a scale model.

New faces per se

Many remodeling projects are only skin deep. The project that only replaces windows, siding, and trim is quite common. The motivation is generally a practical consideration, stemming from rotted windows, broken siding, or peeling trim. Yet all too often the homeowner fails to seize upon this as a golden opportunity to give their home some character.

If you are considering replacing windows, also give some thought to changing their location, size, and style. As many have learned, the mere changing of the windows can have a profound effect on a home's appearance. The enormous variety of window shapes, sizes, and styles that are readily available today provide today's remodeler with a wonderful opportunity to create a new look, whatever its style. Years ago, an arched head window with intricate, lacy, divided lights was a custom order that could blow your budget. Today, you can pick it up at your lumber yard.

If you are planning to redo the exterior, look at the 34 prototypes that follow. Hopefully they will give you some direction that can be developed and embellished further as warranted.

New faces as motivation for remodeling

As a society, we have become more involved with our homes. We care how they look, how they appear to ourselves and to others. Forty years ago, the obsession was pure shelter; today it is style. As a consequence, some remodeling projects start with the exterior. Typically such a renovation starts out where the owners have a strong feeling for the way they would like their home to appear. All other aspects of the program, then, take a back seat to that direction. It is not wrong. In fact, it can lead to a very successful renovation because there is a strong cohesive direction to the project.

Frequently, this type of program leads to a veritable gutting of the existing home, although this need not be the rule. Several of the new facades shown on the following pages are of this nature, whereas any of the facades presented could serve as the prototype for an exterior-directed remodeling project.

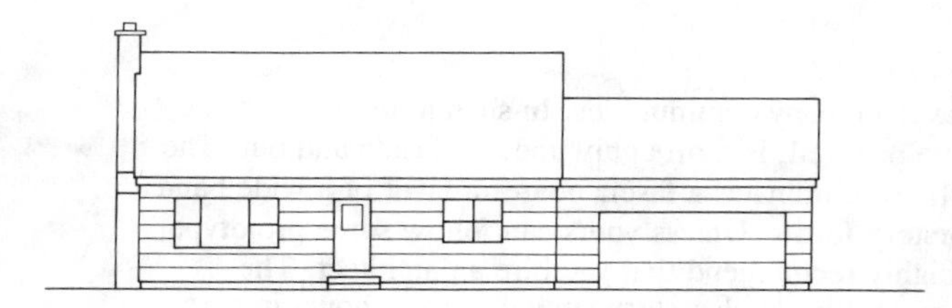

FAC01

This simple one-and-one-half-story (a.k.a., a cape cod, farm ranch, cottage, etc.) can transform into a seductive country-style home with the mere addition of a front porch. Two plans showing such porches are presented in the prior chapter, as plan P0008 on page 161 and plan P0013 on page 159. Siding and windows must be replaced of course, as well as trim, but chances are they were due anyway.

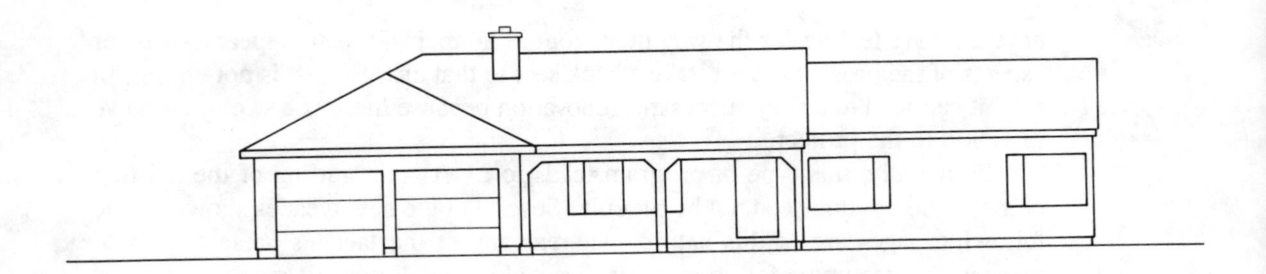

FAC02

Modern renovations are still much in favor in many communities. In such a project, a conventional one-story, such as the home pictured, is thoroughly updated inside and out. The eclectic facade shown raises some rooflines, eliminates a fascia board in favor of a wide band and introduces large windows and clearstory forms. Unless yours can follow some prototype, like the few pictured in this section, I highly recommend that you hire an architect. The striking forms of this facade include a reverse-gable clearstory similar to that shown in plan SML05 on page 233.

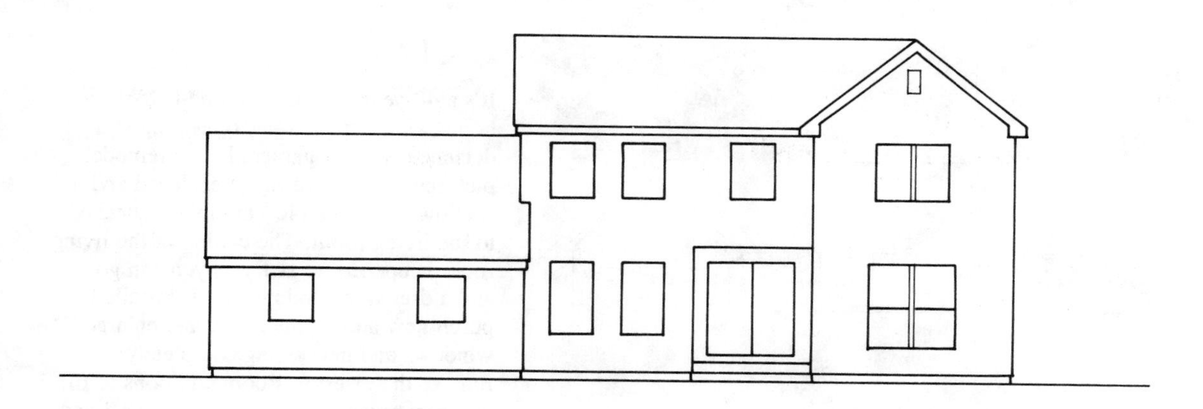

FAC03

A decorative balcony over the entry is the only structural change of substance necessary to achieve the new facade pictured. The new look is best classified as post-modern contemporary. The elements that distinguish it are odd-shaped windows, the use of glass block, and the introduction of roof dormers. However, the use of stucco as a siding adds a Spanish influence that affords this remodel its distinctive appearance.

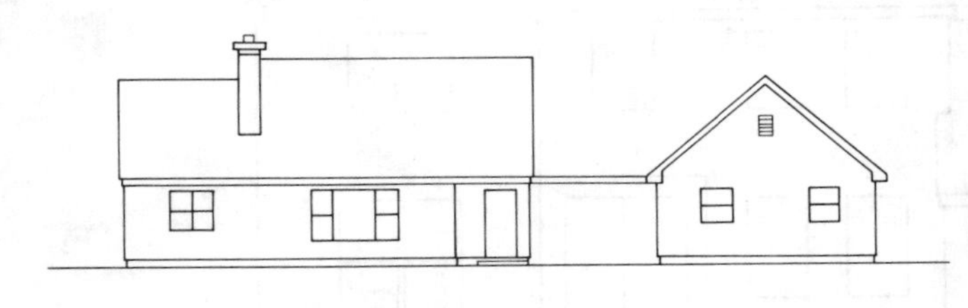

FAC04

It's a 40-year-old one-and-one-half-story home with a breezeway connecting to a detached two-car garage. In the remodel pictured, the breezeway is enclosed and a shallow reverse-gable "bump" is attached to the living room. The ceiling of the living room is opened up as far as you can go, and a dramatic window wall is installed. A porch now adorns the entry, and enlarged windows and new siding completely remake the exterior. Room additions to fill in a breezeway are shown in chapter 7 and "bumps" are pictured in chapter 8.

FAC05

Even a plain old split-level can be dressed up in country gear, if that's your leaning. The introduction of a covered porch at the entrance, with an open balcony over, and a wrap-around porch at the mid-level are the main structural changes required. The wrap-around porch can be as shown, with a new, higher roofline or it can be similar to plan P0013 on page 159. New windows, siding, and a false roof dormer complete the redesign.

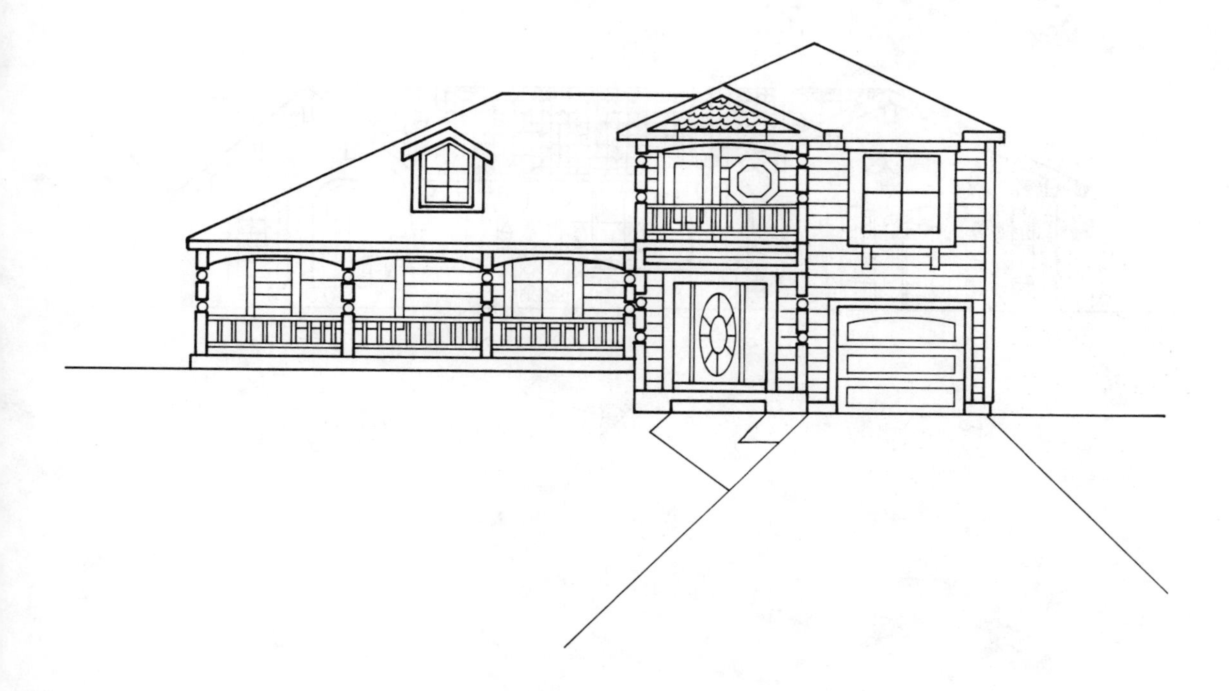

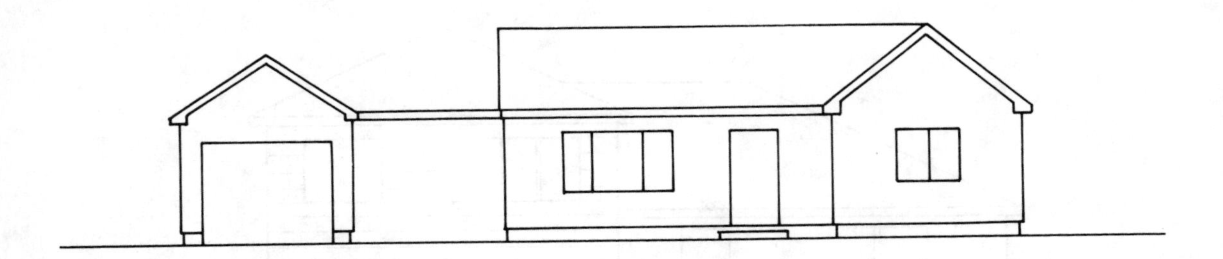

FAC06

The warm, inviting styling of a southwestern or Spanish-inspired new facade was a likely choice for this L-shaped one-story home. A new two-car garage, forward of the bedroom wing, will create a courtyard which is shown with a low stucco wall and iron gate enclosing its open end. Plan GAR03 on page 147 shows such a two-car garage and an arched enclosing wall. A Spanish tile roof, wood beams, and coach lanterns add to the appeal of this new facade.

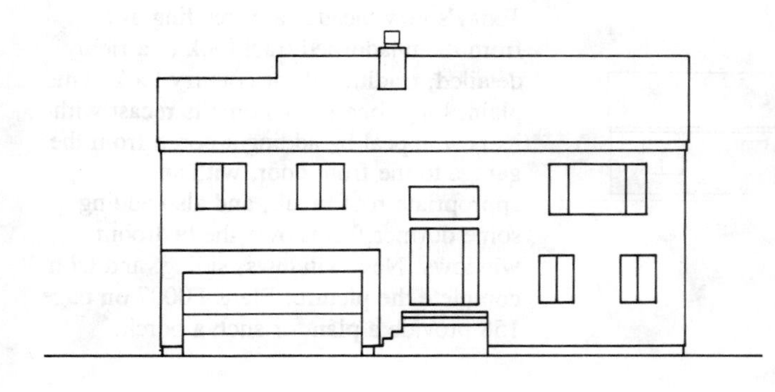

FAC07

The hi-ranch (a.k.a., a bi-level or split-foyer) is a common target for facade remakes. Because of its different door and window heights and the absence of symmetry, it is frequently refaced in a contemporary mode. In the plan shown, not only are windows and doors changed and the house resided, but a reverse gable is proposed for the living room. This would allow you to raise the ceiling and install a tall window. Finally, the roof over the entry has been pushed out a few feet to provide some cover to the front door.

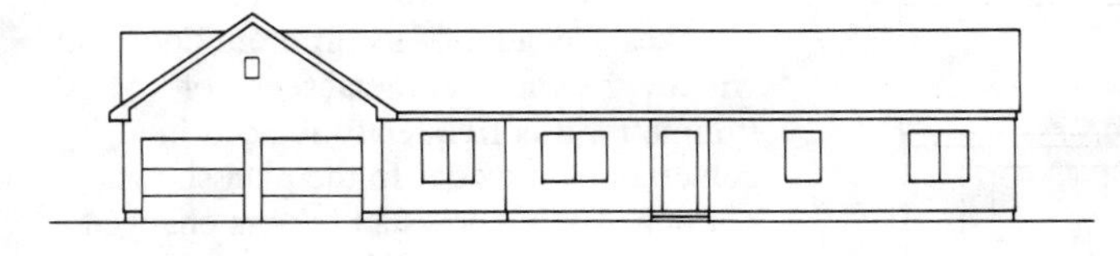

FAC08

Today's new facades are trending away from the unadorned tract look to a richly detailed, traditional, or country look. This plain, long, one-story home is recast with its new appeal by adding a porch from the garage to the front door, with an appropriate roof break, and also adding some dormer forms over the bedroom windows. New windows, siding, and trim complete the picture. Plans P0007 on page 158 provide a plan for such a porch.

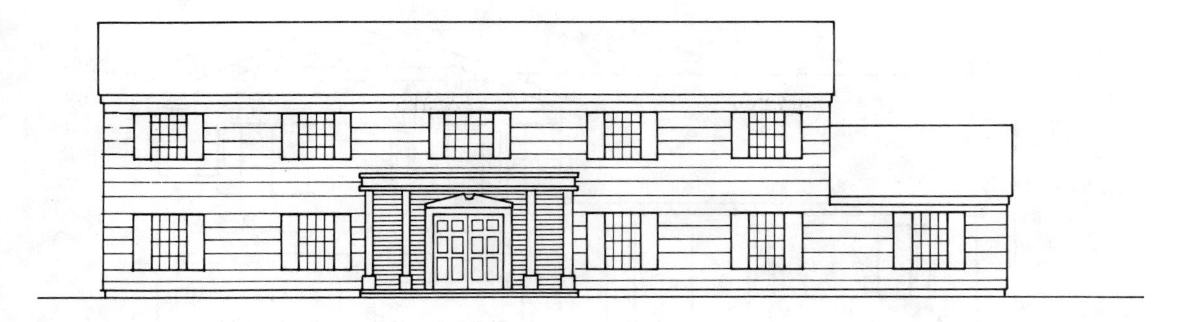

FAC09

You've stared at this big hulk of a two-story for years. For most of this time it looked like the others on the street, but that's changing because everyone seems to be remodeling. It's that time—new baths and a new kitchen—but it is also time to rethink the exterior. An eclectic, contemporary approach, such as the one shown, might be just perfect. It suits you, it is accepted in the community, and you want to be back in style. The facade shown mixes stucco, glass block, and wood siding; it proposes a new entry porch and lots of glass.

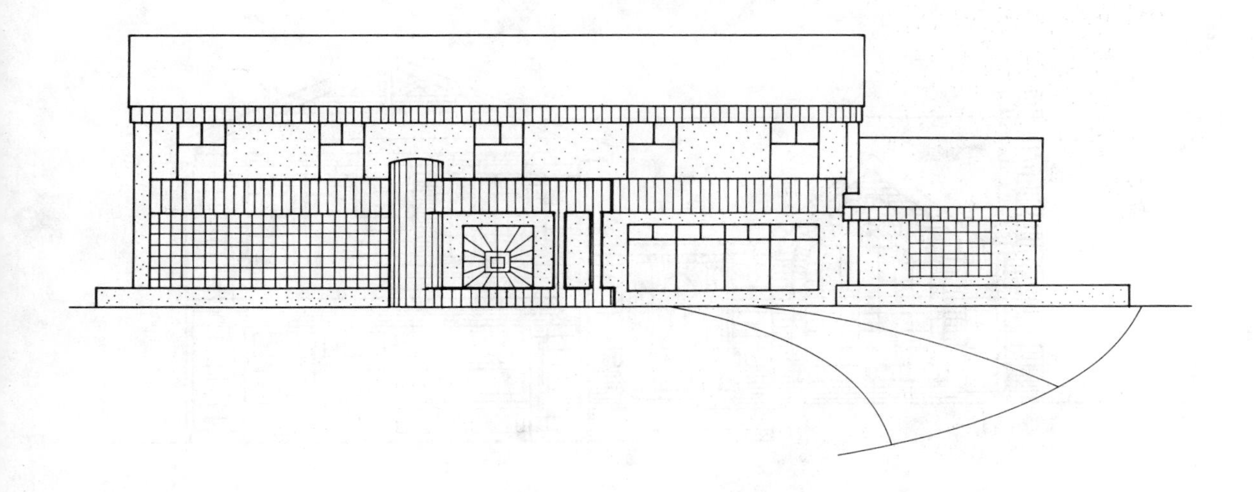

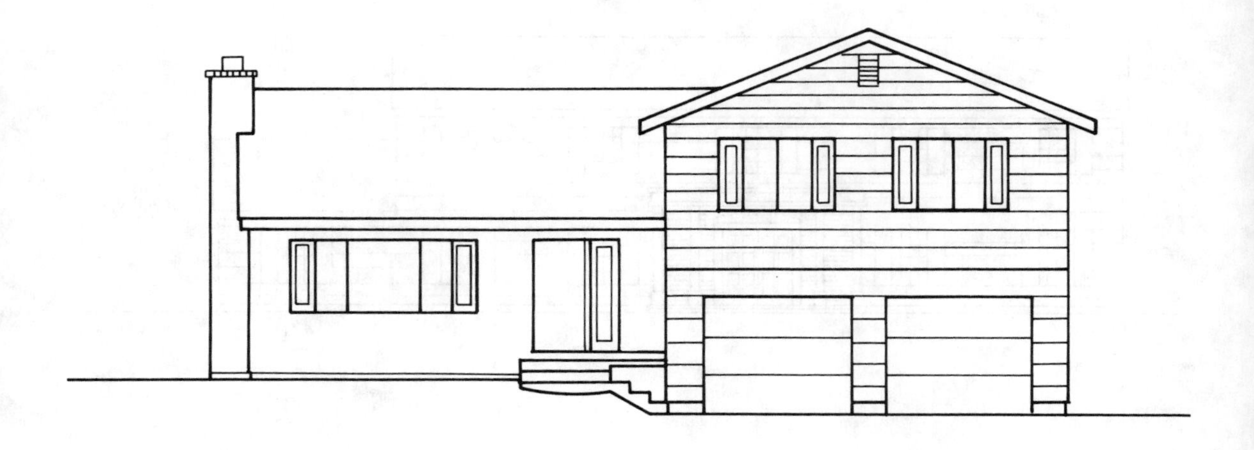

FAC10

The problem with the facades of so many post-World War II homes was that they were built without any attention to exterior detail. This split-level was no exception. Fortunately, it does not take much to give it some style and some flair. By installing a small reverse-gable roof and raising the ceiling in the living room, it is now possible to install one of those great-looking new windows with rounded tops. New front steps, a new front door, and some trim around new windows and doors complete the refurbishing.

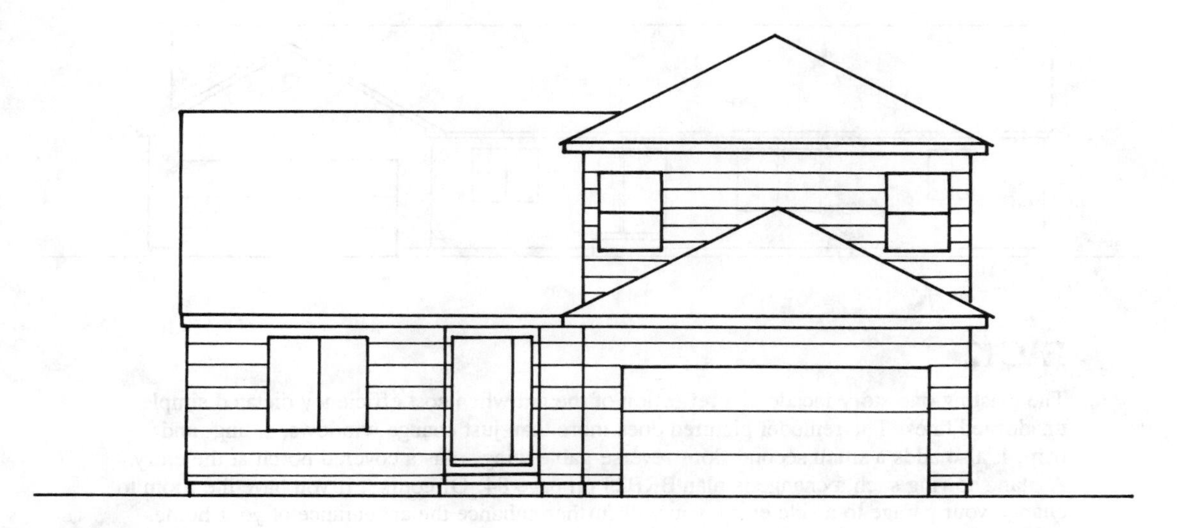

FAC11

This L-shaped two-story is a good choice for a remodeling in the trendy contemporary Spanish mode. Certainly not a pure style, but one that has many adherents, it uses stucco to create shapes and forms of interest, whether relevant or not. The structural changes pictured here include a shed roof clearstory window that brings light deep to the interior of the first floor and a prow-shaped addition to the living room, similar to plan KFLD2 on page 126.

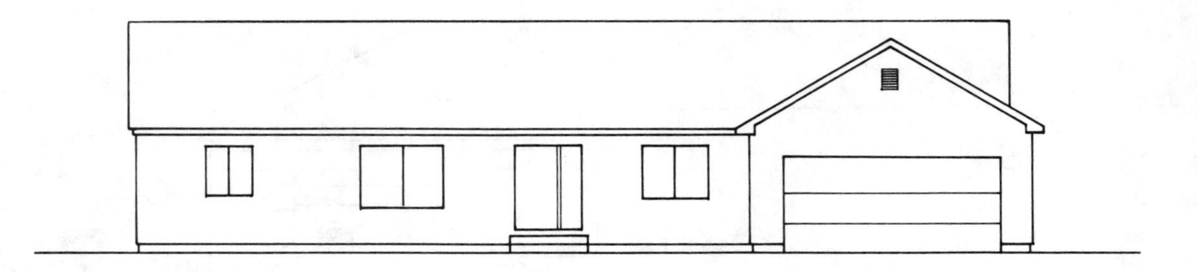

FAC12

The existing one-story facade is a reflection of the era when cost efficiency dictated simple, unadorned lines. The remodel pictured does more than just change windows, siding, and trim; it also adds a small second floor reverse-gable wing, plus a covered porch at the entry. A plan showing such a change is plan BRB23 on page 84. Of course, if you have the room to change your garage to a side entry, you will further enhance the appearance of your home.

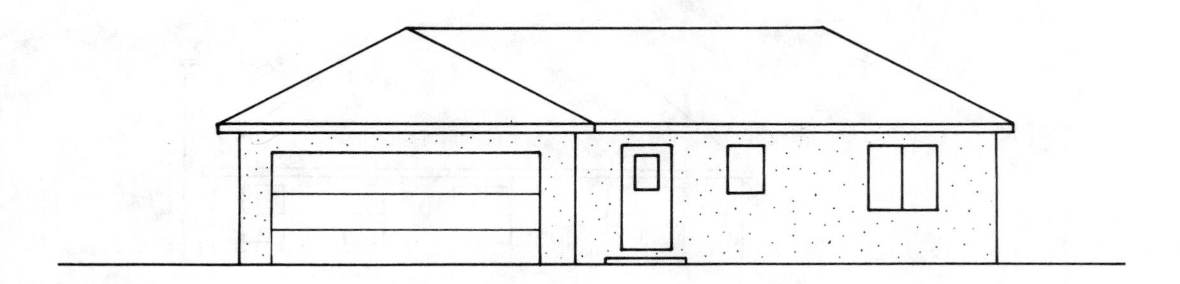

FAC13

The modest L-shaped stucco one-story takes on a smart new appearance in the new facade pictured. The styling: an eclectic mix of traditional and contemporary elements that might be called post-modern. Traditional quoins and traditional windows are freely mixed with contemporary stucco shapes and contemporary fenestration to create this look. The entrance is now accented, as is each element of the small facade. The bottom line: The original box shown above now has presence.

FAC14

No big addition is needed to turn this flush-front center-hall two-story into the country Victorian pictured below. A two-foot reverse-gabled addition is all that is needed at the left side of the entrance. Then add a wrap-around country porch like the one shown in plan P0013 on page 159, and you have achieved the new look. The dormer on the garage roof is nice, but not necessary, and it doesn't have to be real space anyway.

FAC15

I would be remiss if I didn't present at least one Tudor facade option. Although there appears to be very little call for new Tudor homes these days, it is not absent. More importantly, though, if you are looking to make a change from "builder's basic," Tudor is a readily achievable style that can be cost-effective. The remodel pictured is accomplished by window replacement, minimal framing for the roof dormers, and a refacing with thin brick veneer and stucco.

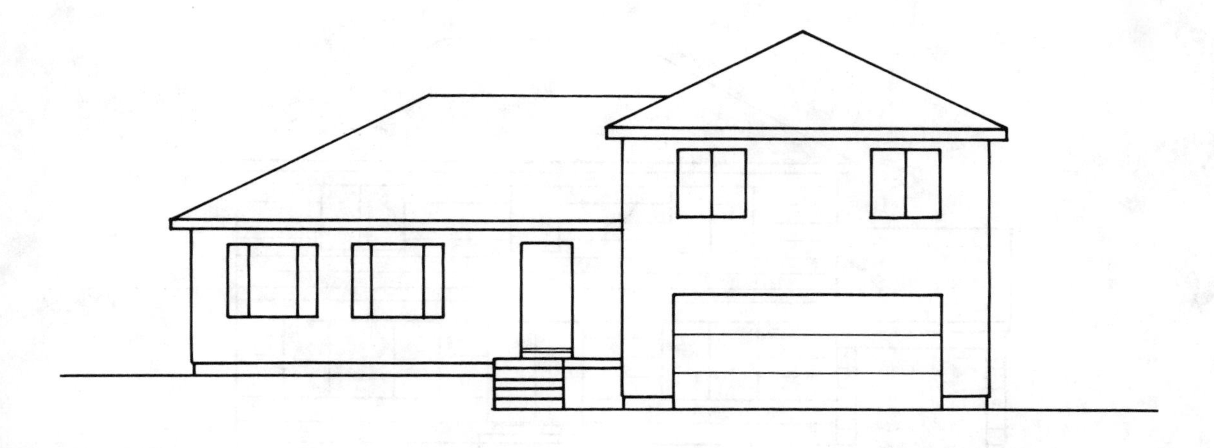

FAC16

Because of its change in rooflines and its asymmetrical forms, the split-level makes an ideal choice for remodeling contemporary. The options are boundless, enabling a skilled designer to create a very striking new facade. The plan pictured here incorporates a curved glass-block addition at the living room, similar to plan L0003 on page 226, and a new set of entrance steps. Redoing the steps in itself is a major consideration, and designs for two new styles of steps are shown in plans SML02/SML03 on page 168.

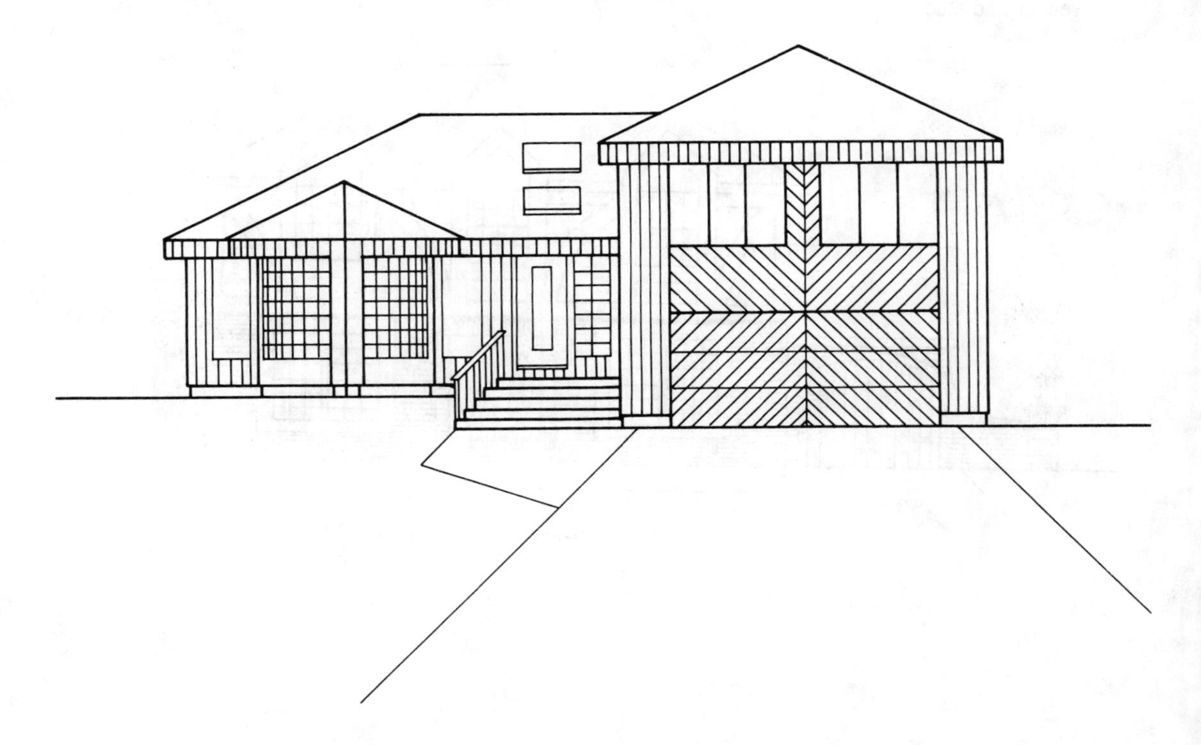

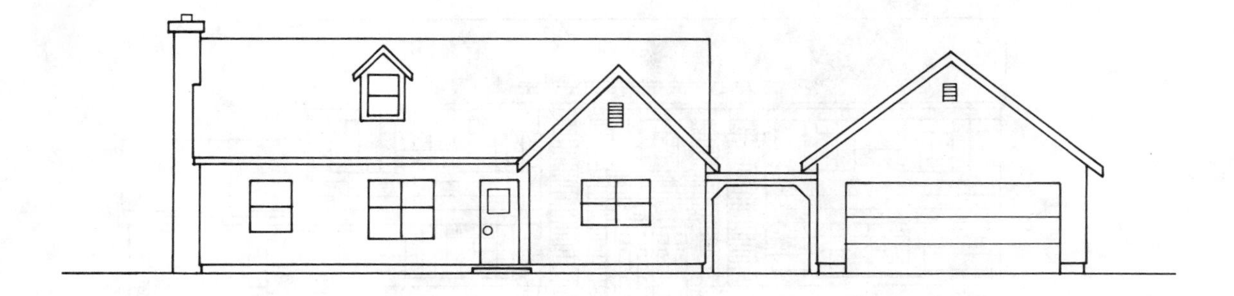

FAC17

This one-and-one-half-story is 40 years old, and it looks it. It's time for a change, and your taste leans toward country. New windows are certain, especially the large ones in the remodeled gable wall. A new country porch, similar to plan P0013 on page 159, is part of the picture, as are two more roof dormers. The breezeway could be enclosed (see plan F0002 on page 108), and you could consider raising the roof of the garage, but that is least necessary to achieve the look.

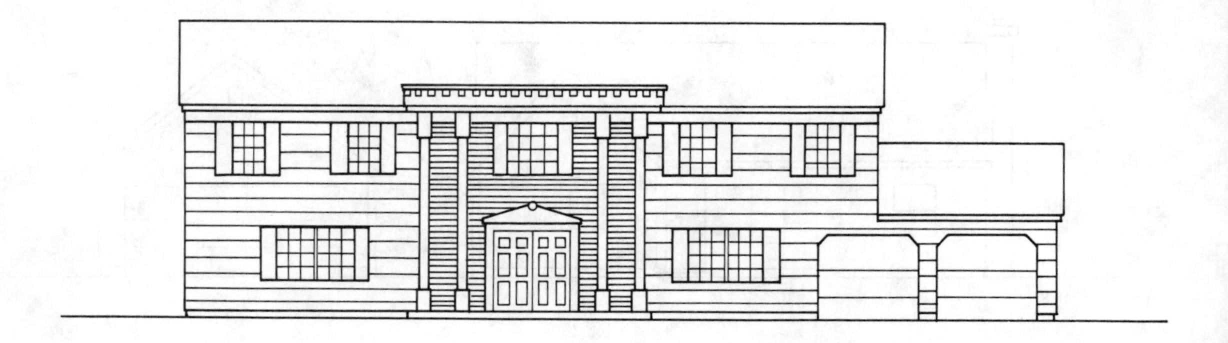

FAC18

There was a time that visitors raved about that expansive facade of yours, with its elegant tall columns. Nobody has said a word to that effect in years, but they do talk about the new modern facade on your neighbor's house. Maybe it is time for a change such as the one shown. A one-story porch is proposed, as is a false-front wall at the left to force a roofline break. Both help to differentiate the facade into smaller contrasting shapes. Large windows and a wide fascia band are also suggested.

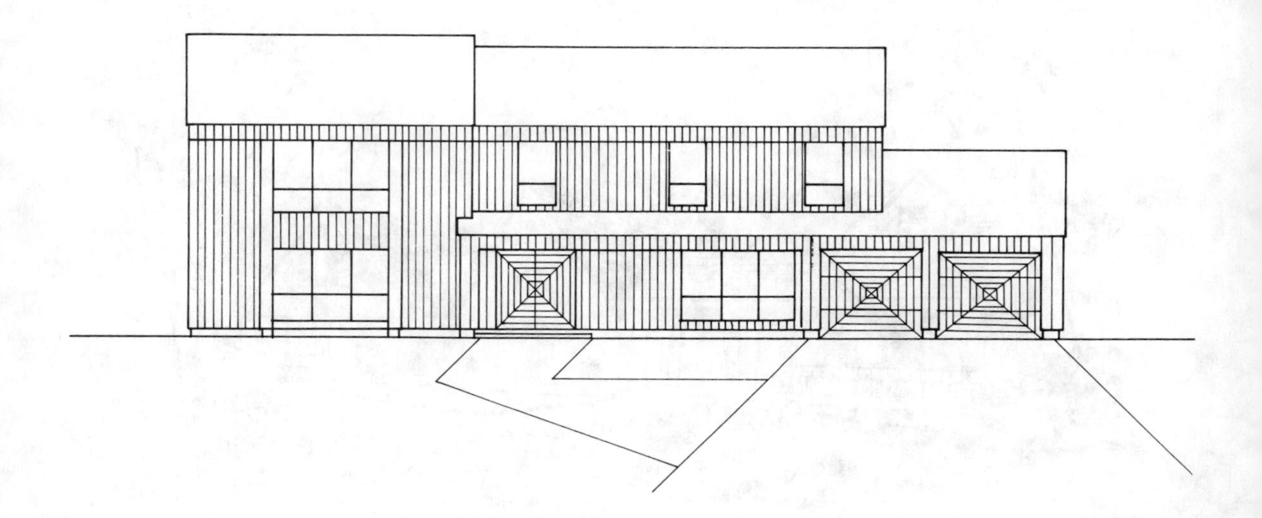

FAC19

The changes wrought in its new facade have a profound effect on the character of this center-hall home, despite the fact that the changes are very minimal. A small bump-out is created surrounding the main entrance. This extends two stories in height, and, above it, the shed roof simply continues. Two contrasting reverse gables at each side emphasize the formality of the plan and, together with the richly detailed trim, give the facade its distinguishing features. See plans PF002 and PFB02 on pages 172 and 174 respectively for the entrance foyer changes.

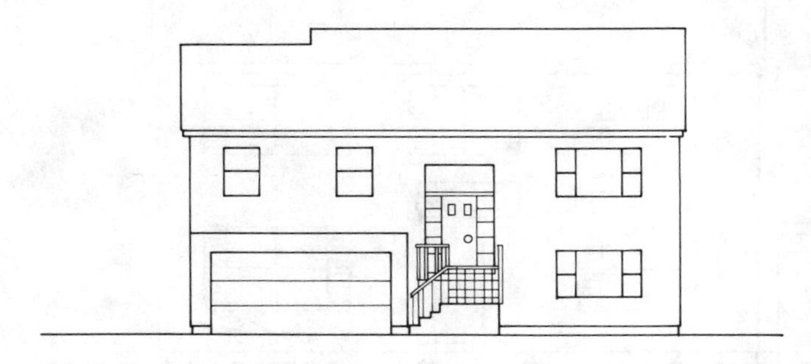

FAC20

Oh, that ugly big stoop! Owners of hi-ranches (a.k.a., bi-level, split-entry or split-foyer) have covered them in wood, marble, brick—you name it. If you have wished there was a way to get rid of it, this remodel shows how. It does require removing the stoop, but if you're up to that, you can recast your home as a two-story, as shown. A new porch and foyer are created, and the design for these is plan P0019 on page 166. You can also reverse gable the living room, as pictured, and raise the volume to create a high vaulted ceiling.

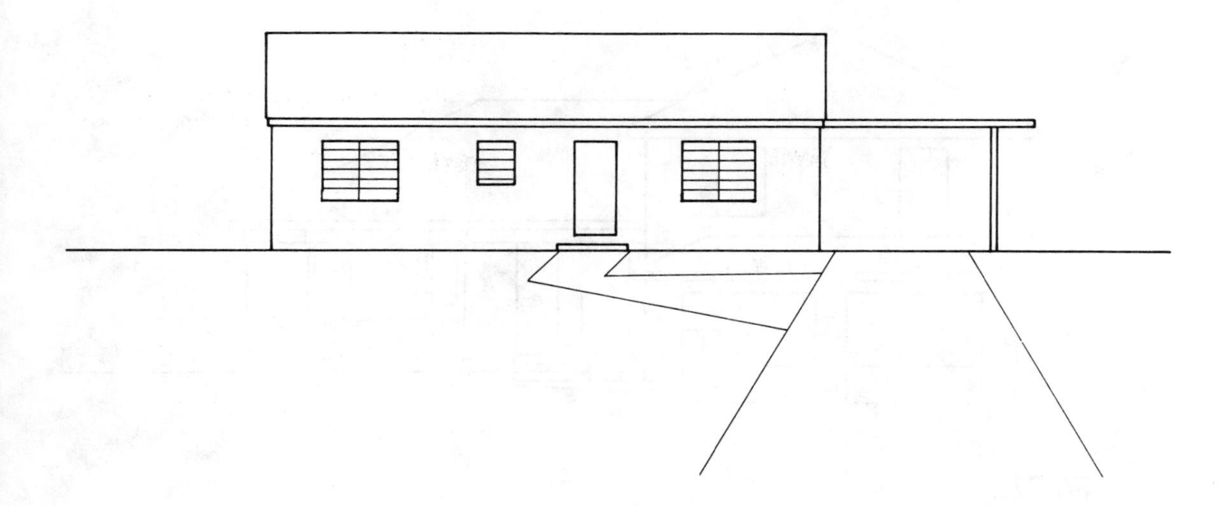

FAC21

The existing stucco box, with its attached carport, does little to excite anyone's aesthetic senses. The remodel pictured here begins to suggest that this might be a home for someone with a sense of pride. The styling is essentially post-modern with some individual nuances. Focus is placed on the newly developed main entrance with its high arched recess. The carport is offered some mass with the introduction of a storage unit and a circular stair to the roof—maybe there's a view.

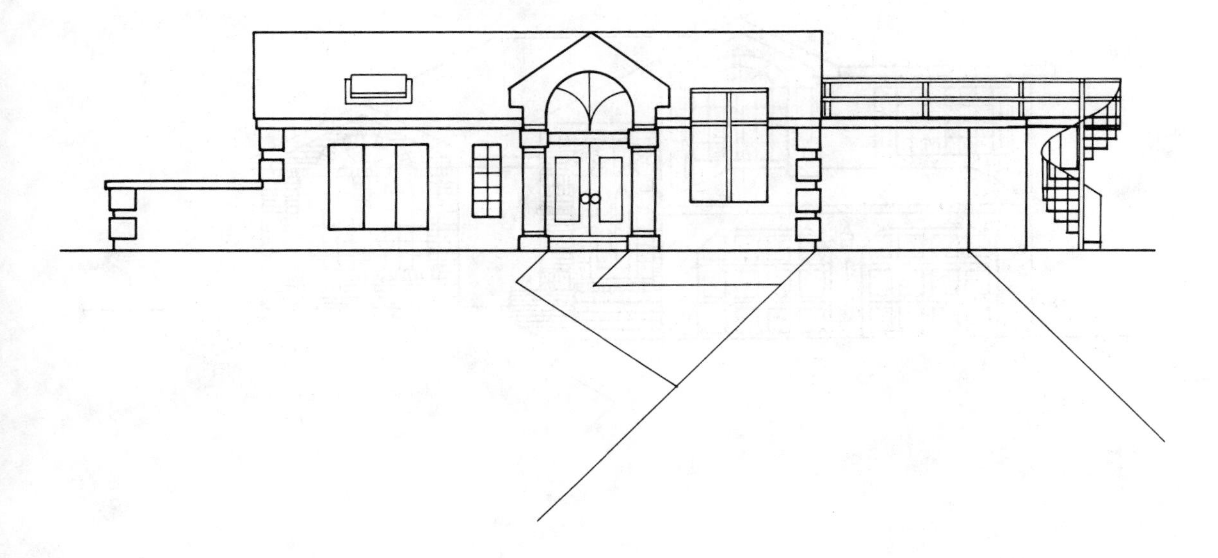

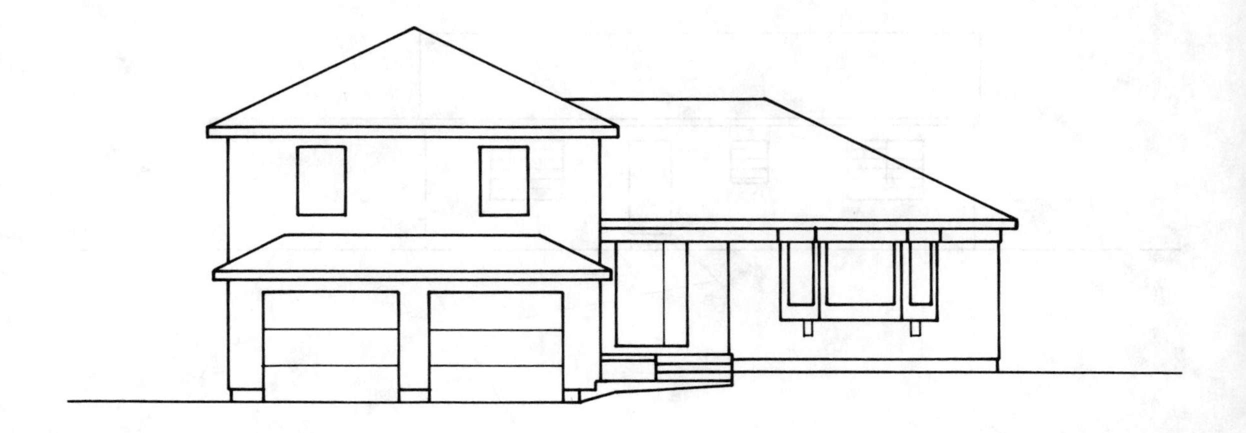

FAC22

This split-level has likely served its owners well, but it's time for a change—not a big one, but just enough to freshen up the home's appearance and put it in sync with today. New windows, especially some half-rounds, do that with flair. The small reverse-gabled bump (18 inches is OK) at the living room, with larger windows, is another element pictured. And then there's a new entrance foyer that was always missing. This can be seen in greater detail in plan P0016 on page 167.

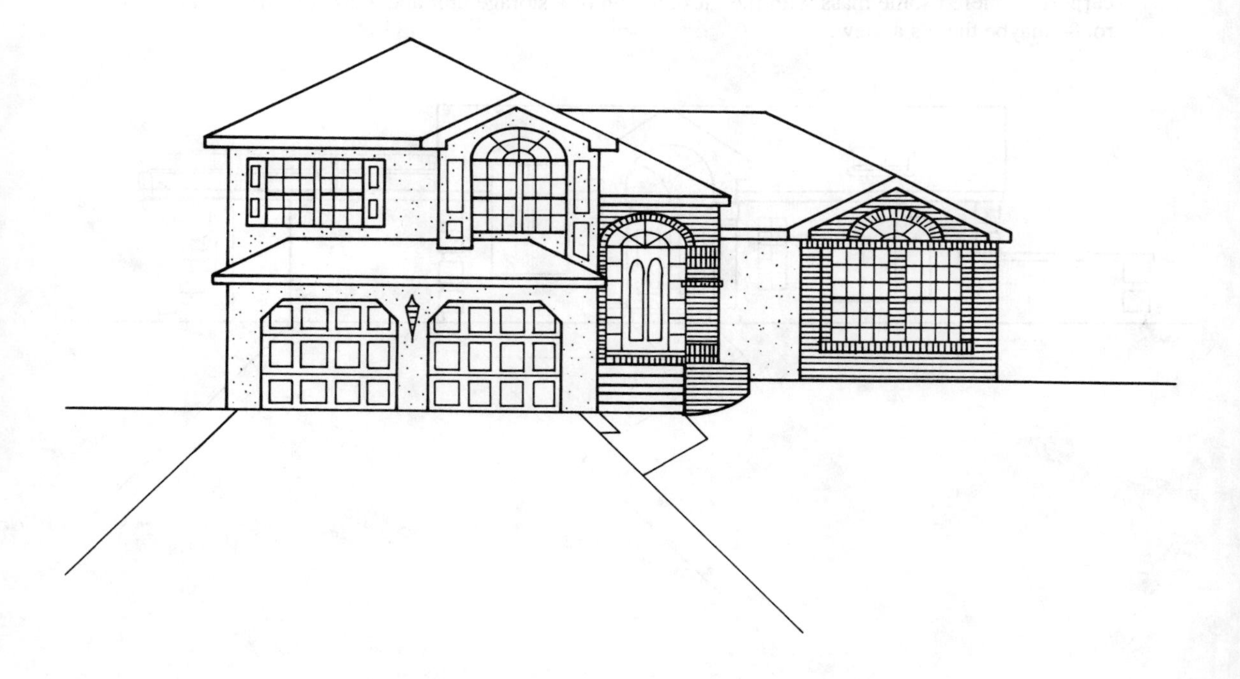

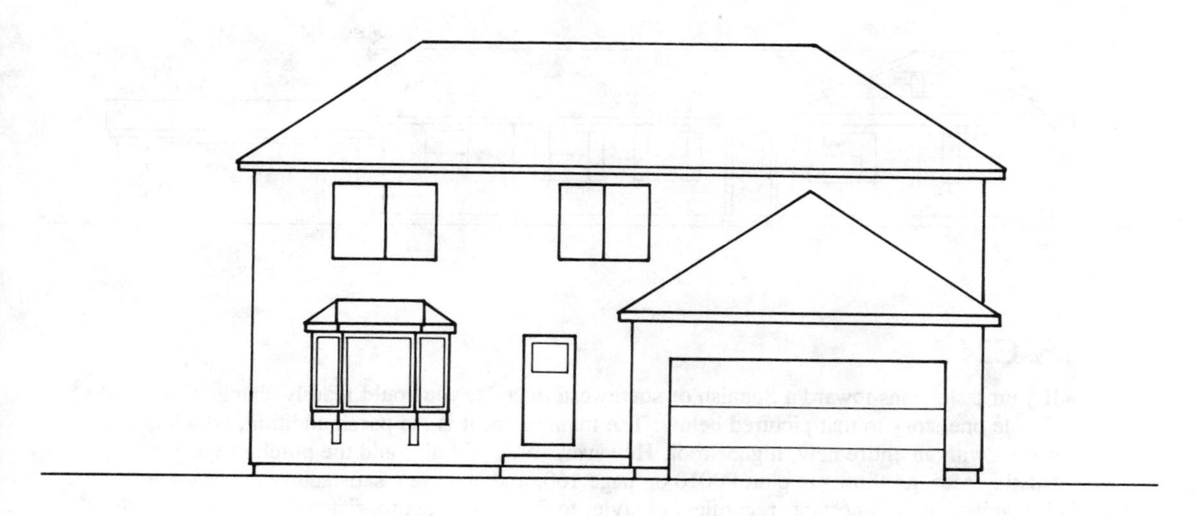

FAC23

A wrap-around contemporary brick porch with brick piers and protruding second floor window bays are the key elements that relieve the starkness of the existing facade of this two-story home. The porch serves to broaden the appearance of the home, while focusing on and sheltering the main entrance. Other nuances include two decorative windows on the wall above the garage and very decorative doors within a new double-door entry.

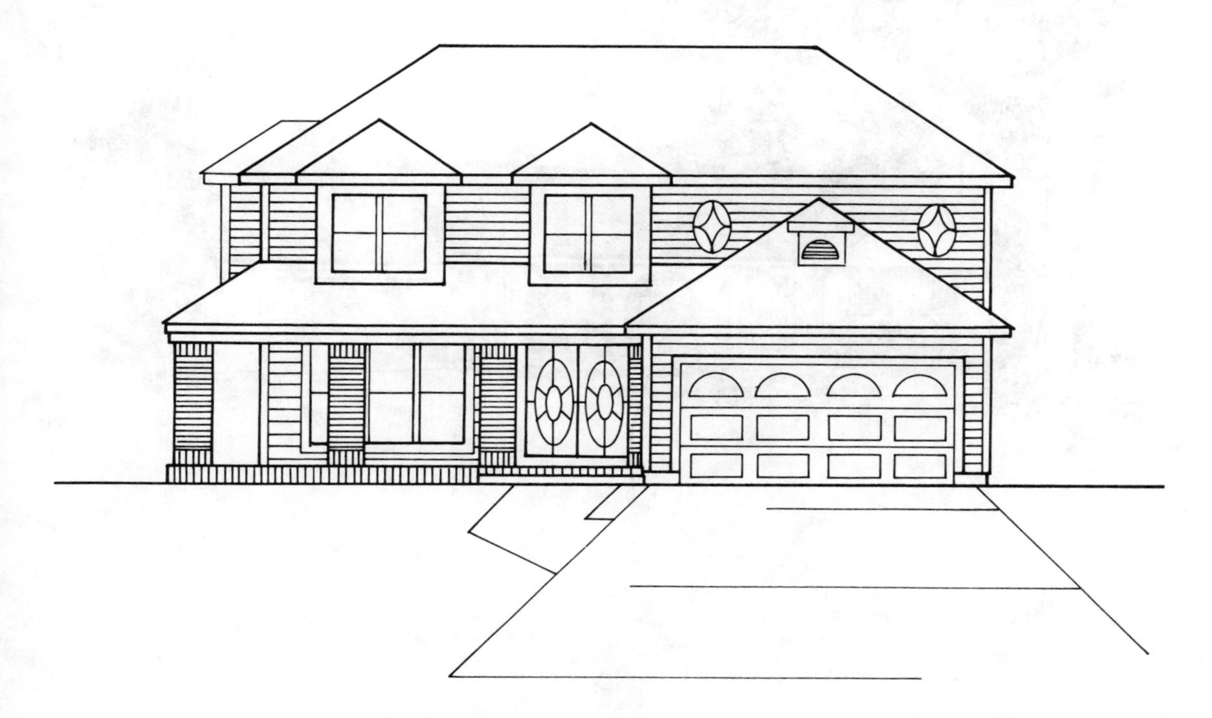

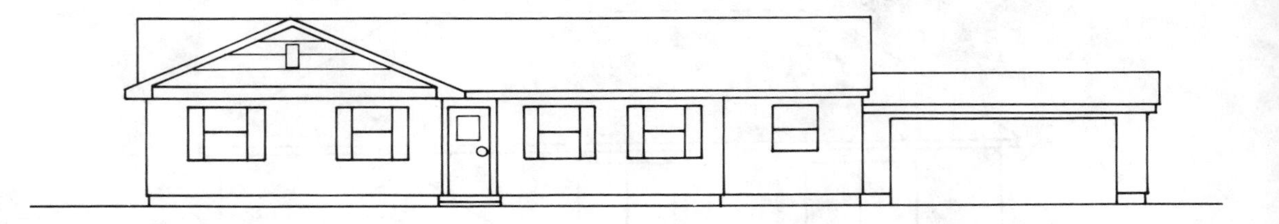

FAC24

If your taste leans toward a Spanish or southwestern style, you could readily change this simple one-story to that pictured below. The main element is the porch addition, which is shown with an entire new, higher roof. However, you could also add the porch to your existing roof, as shown in plan P0010 on page 160, and achieve a satisfactory appearance. Remember, it is important, regardless of style, to finish with appropriate trim. In this instance it should be rough-sawn, dark-stained wood.

FAC25

There is absolutely nothing wrong with this existing facade—that is, unless you want to change it. While I was taught countless aesthetic rules, that practice confirmed or denied and that I have taught to others, there is one rule that only practice taught. It is that the home is each owner's personal statement, and I should only advise and suggest—not dictate—taste. I never have and never will. In that regard, if you want to turn this lovely center hall into a modified Early American, and you do it tastefully, as pictured, go ahead.

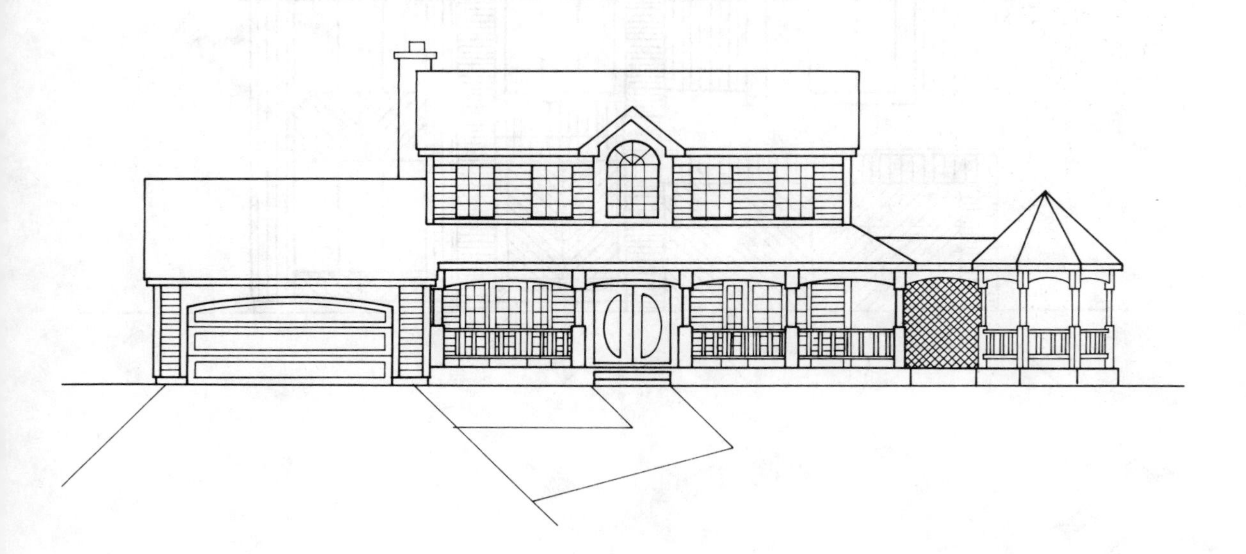

FAC26

Because of its inherent asymmetry, a narrow two-story home with a protruding front garage also makes an ideal candidate for a contemporary remodeling. Brick piers are important to the facade pictured. They help define the entrance and afford it greater prominence by extending vertically. A wide fascia band of vertical siding defines a contrasting horizontal plane, and diagonal siding is used in several areas as an accent. Windows are enlarged and simplified into repetitive elements.

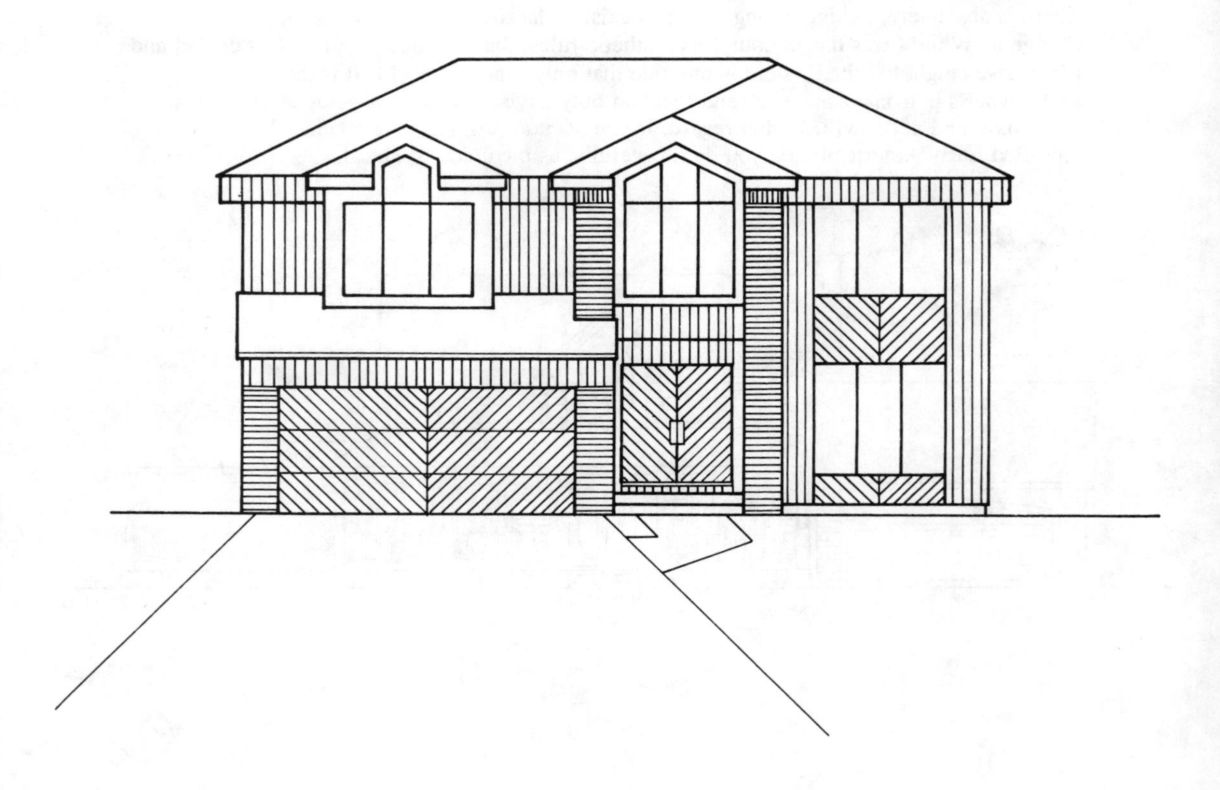

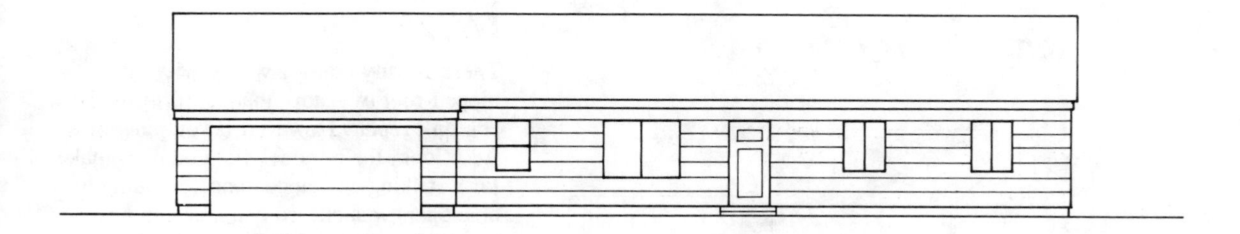

FAC27

The existing facade of this long one-story home has one distinguishing feature—it is long. It is devoid of any character because that was not the concern of its builder. Your concerns are personal. While you can develop almost any stylistic leaning on such a home, I have shown it with an Early American theme. A porch similar to those shown in chapter 8, plus new, larger windows, paneled shutters, flower boxes, friezes, and corner boards are the elements that create this new facade.

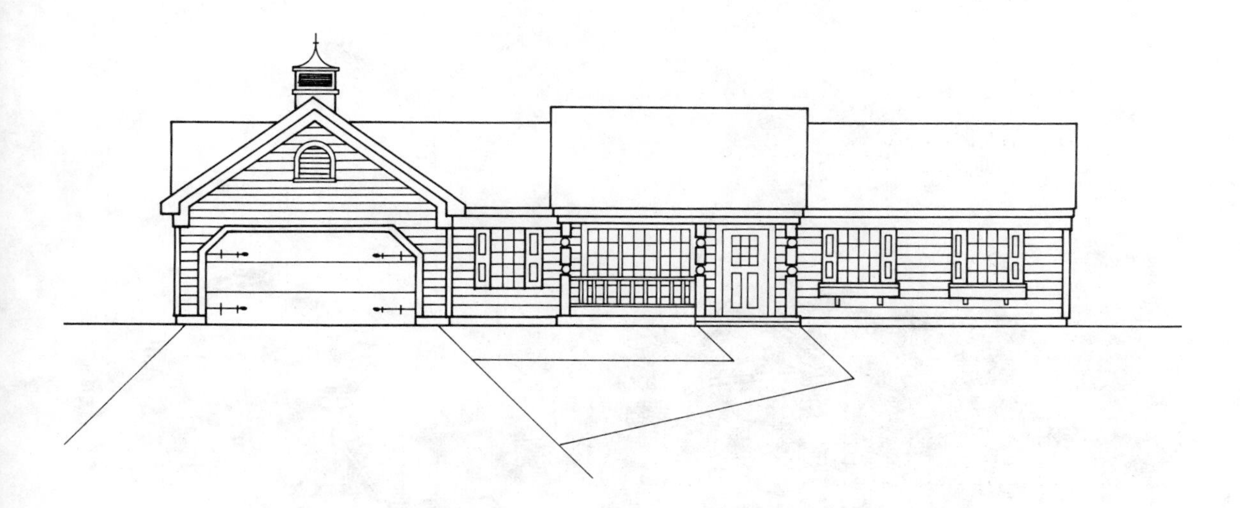

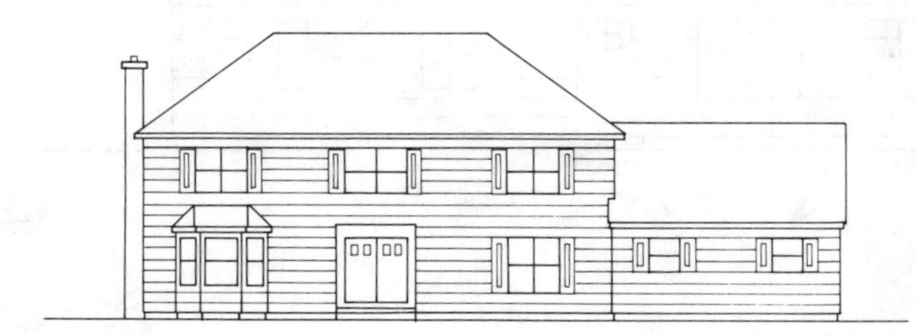

FAC28

There is truly nothing wrong with this hipped-roof two-story that some cosmetics couldn't repair. However, if you pine for a fresh look, the remodel pictured does make for a striking new appearance. A bump to create a new sheltered entrance can be found in plans PF002, PFB01 & PFB02 within chapter 8, and the new reverse gable over the garage represents an apartment (or it could be a second floor playroom), which can be seen in plan APT03 on page 155. A new brick front could be achieved with the use of thin brick that does not require a foundation.

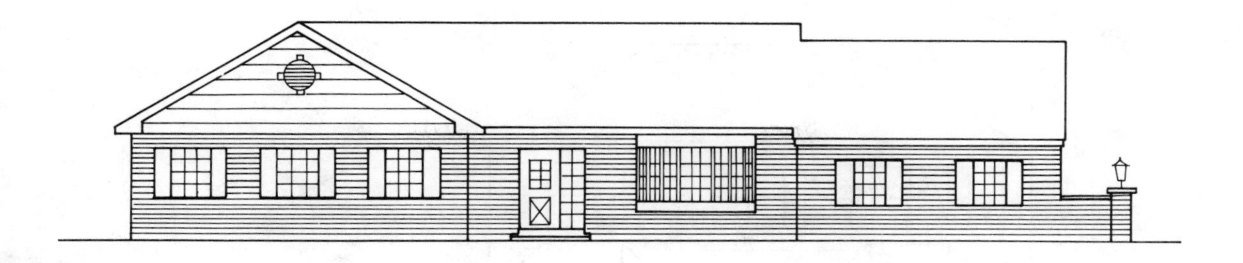

FAC29

There will likely be many readers who will be at a loss to understand why someone would redo the facade on this large, obviously expensive home. The answer is likely twofold: such is the personal wishes of the owner, and such renovations are commonplace in the community in which the owner resides. Modern is not out. In fact, in certain such communities, it is still the driving force. In such a renovation it is likely that the whole house will be updated and ceilings raised. The renovation shown proposes no roofline changes other than the clearstory. (See plan SML09 on page 236.)

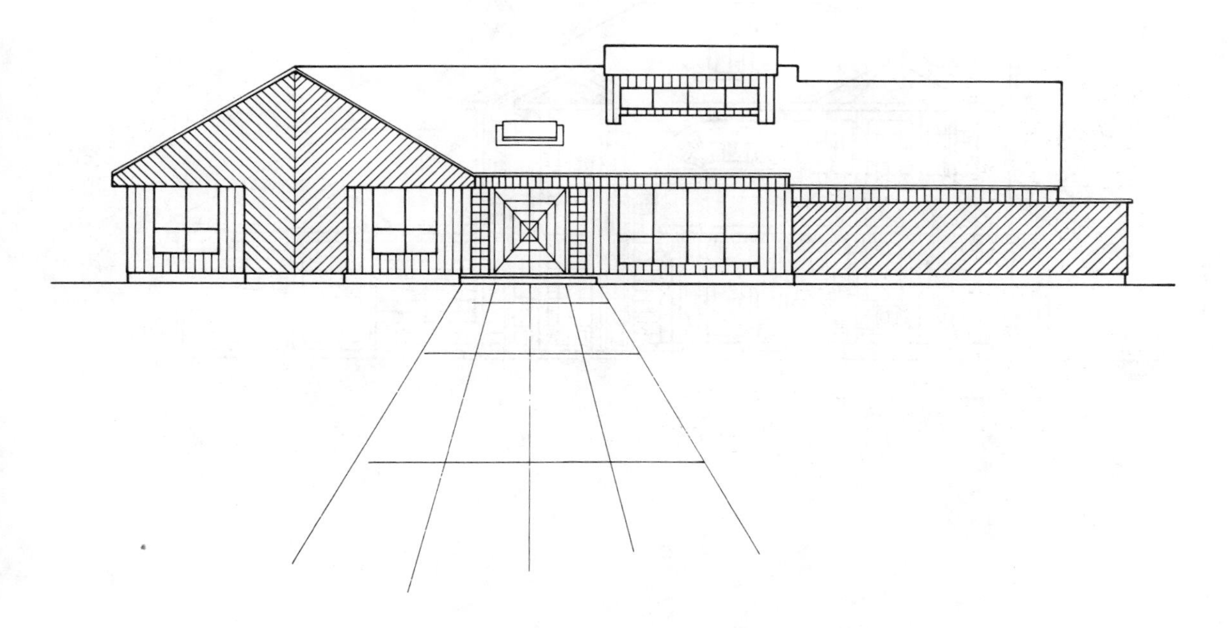

FAC30

If you're tired of the formal balance of your center-hall two-story, it is possible to unbalance it if you lean toward a whimsical style, such as the pseudo-Victorian pictured. The bay on the right side becomes vertical by extending it two stories high, whereas the opposite side develops a horizontal theme by virtue of the introduction of the wrap-around porch. Odd-shaped windows and scalloped trim add to the whimsy of this style.

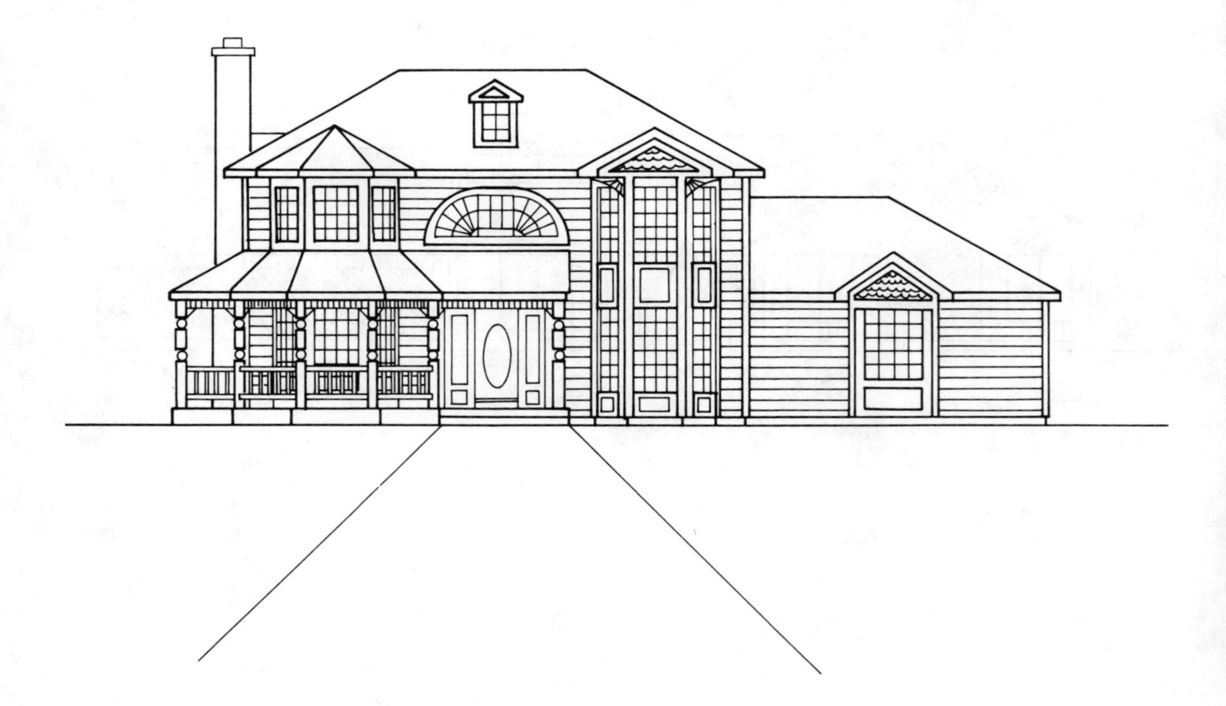

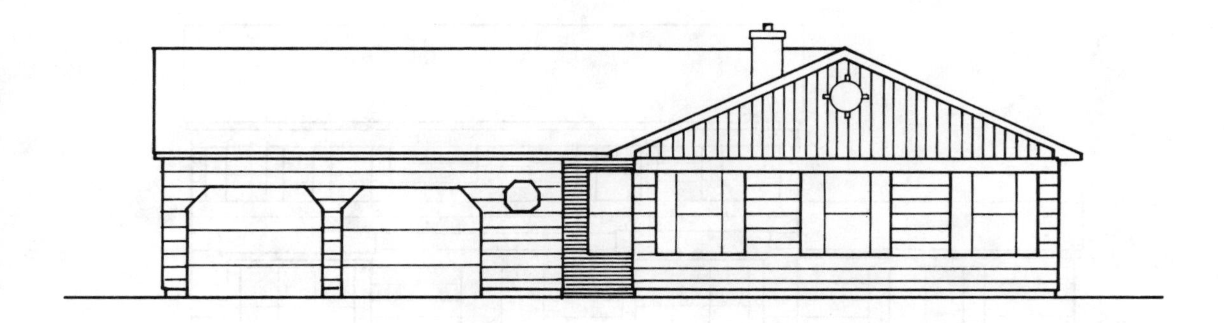

FAC31

Another in the series of contemporary choices, this one is a post-modern adaptation of a standard L-shaped one-story home. In this design, the small front wall is elongated to give the home a greater presence at the setback line. The wall becomes a screen wall that extends on both sides of the home, partially shielding the garage from view. A narrow, shiplap-joint horizontal siding is proposed, along with stucco and glass block. Post-modern shapes in the stucco provide interest and a personality to the new facade.

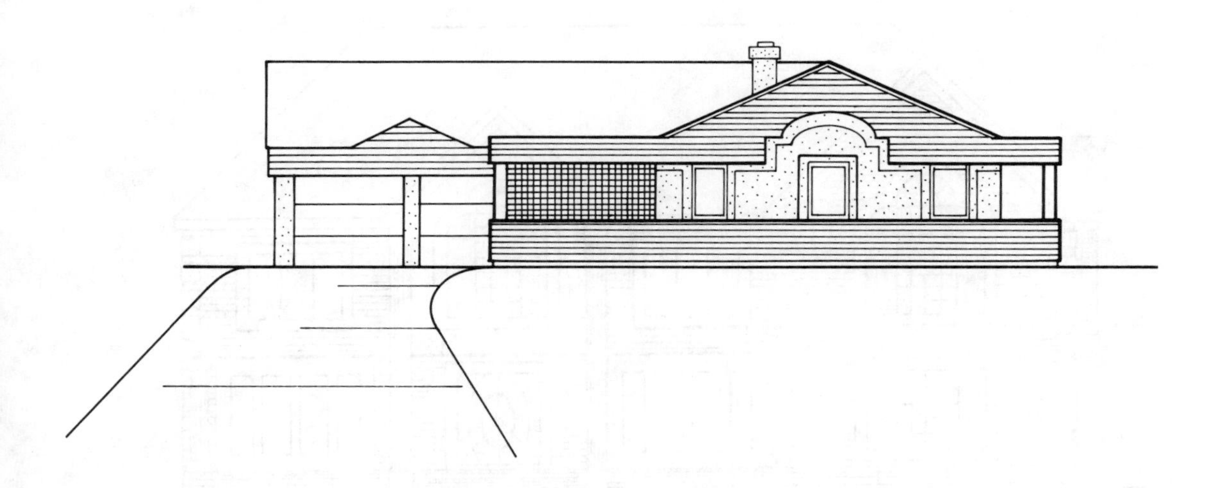

FAC32

Other than being dated, the existing two-story facade is nothing to be ashamed of. However, for the restless, we can always improve and update. This facade remake does involve several additions, as well, including an entire room over the garage and an extension to the front rooms. The new facade is an elegant post-modern design that is certain to please. One of the main design elements needed to effectively create such a new facade is the new, steeper roofline—an item of significant expense, but without which the new facade would not work.

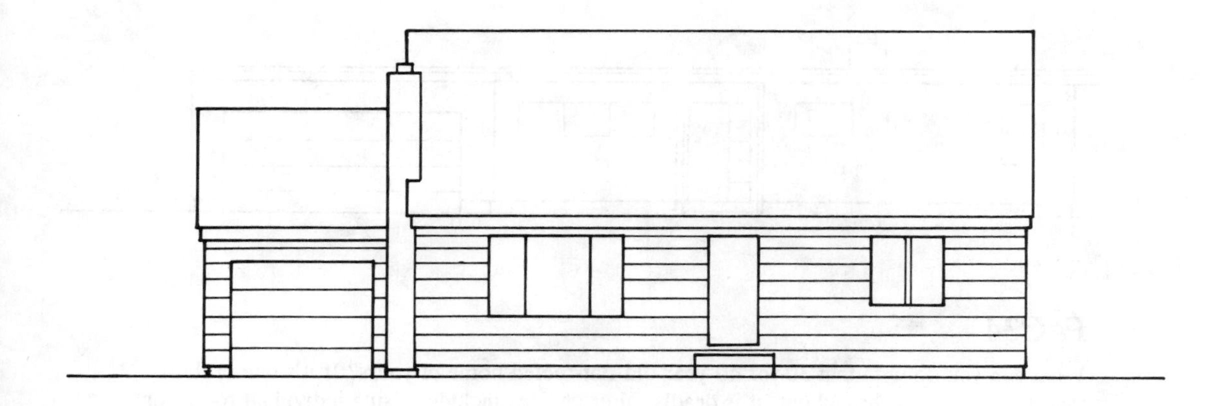

FAC33

This is not a mistake. The new facade was built on the same home as shown above. Well, don't feel bad if you can't figure it out. The small one-and-one-half-story was completely redesigned, added to, ripped apart, you name it, and the result is what is pictured. The point to be made is that while most of this section has dealt with facade changes per se, one of the other more common ways we change a home's facade is the product of a custom design undertaken by a qualified design professional.

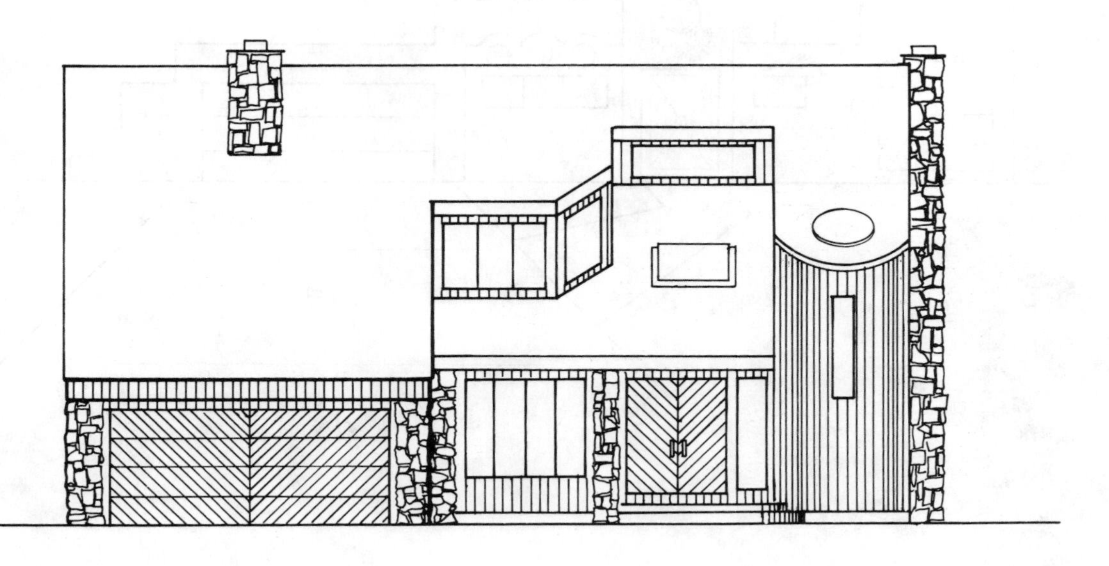

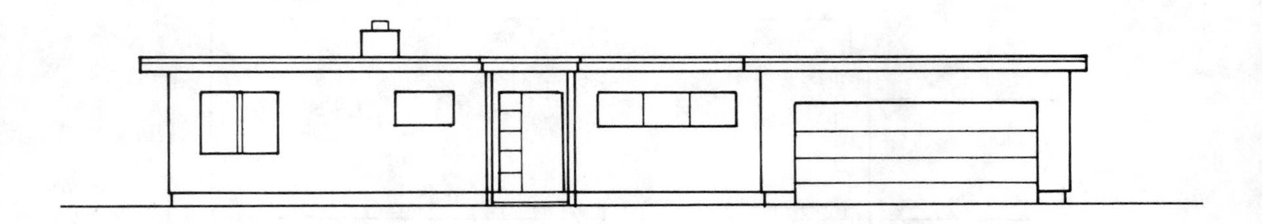

FAC34

What can you do with that 30- or 40-year-old flat-roofed home of yours? If it's one of those that has one height throughout, it is deadly. Your choices include raising individual rooms or wings to create some interest and some interior volume. You might also be able to put a pitched roof over it. The proposal shown opts for a variation of the first idea. It proposes moving the roofline—or the apparent rooflines—up and down in a southwestern-inspired stucco contemporary motif. Only some rooms need to have their actual height changed, while most of the height changes need only be in the exterior wall.

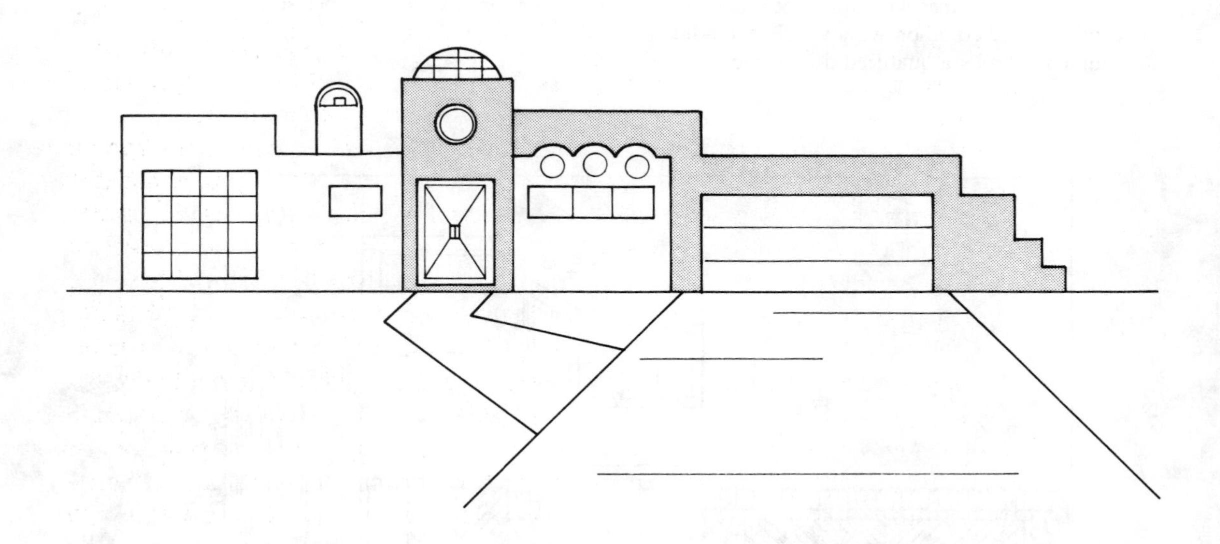

10 CHAPTER

Whole-house remakes

There are times when a remodeling project involves such a large addition, or additions, or so many bumps, bays, and remodelings of interior space that it takes on the scope of a virtual remodeling of the entire home. There is no precise definition of when that threshold occurs, but I like to think that it occurs when your project begins to involve 40% or more of the home. Remodeling projects of this nature are expensive and frightening to many, but they are also very exciting. A renovation of this nature can completely transform an inadequate home into a palace.

Such projects are frequently featured in home journals because they can be beautiful, but they also can be more realistic for you to consider if you realize that an undertaking of this nature need not be done all at once. I will deal in greater detail on the subject of staging in the next chapter. Briefly stated, *staging* involves some preplanning on how you will accomplish the project while living there. In a whole-house renovation, proper staging can frequently make the project palatable.

The plans that follow are like all the prior plans readily duplicated by you on your home—presuming of course, that there's a match. Let's talk about that a bit. There are two types of whole-house remakes presented. The first type involves the complete renovation of a small home that is likely to be at least forty years old or more. These are typically old bungalows, cottages, split-levels, or one-and-one-half stories (cape cods, expansion attic homes, etc.), that are so poorly suited to contemporary living needs that the only solution to many of their deficiencies frequently involves a complete remodeling. You will find a number of these to peruse. Although your home may not precisely match the existing floor plan, it could be similar enough for these plans to work, or it is very possible that you can modify the plans to suit your home. The end result of all these is to create an exciting, new, beautiful home from one that is somewhat sour.

While this first type of plan involves a remodeling project frequently motivated by the need to correct a home with grossly inadequate features, there are other motivations that lead to a whole-house renovation. These are usually not as corrective or mandatory in nature, but more selective, representing your individual desires or needs. Whole-house projects of this nature frequently add more space for special needs and may significantly renovate the exterior design of the home by choice. Projects of this nature usually do not have to correct circulation problems or reutilize poorly functioning space, and are inherently more flexible.

The potential number of variations on such plans is limitless. But the plans have some very common ground that we have already explored in the plans of the prior chapters. Whether your home is a one-story, two-story, or a split-level, if you want a new family room, country kitchen, master suite, guest apartment, or whatever, these have been shown. So if you have a large wish list, it is possible to combine many of the plans shown to create your own personal whole-house renovation.

I have presented a number of examples of such potential whole-house renovations. These take various plans from the prior chapters and combine them to create several wonderful new homes, which are each worth studying in detail. The numbers of the plans utilized are referenced on each page, so you can see the small modifications that might have been required to incorporate them into the grand scheme. It is my hope that with these examples, you will be able to see your way through the process of creating your own dream home. Remember that if you have a long wish list and it is not all achievable at once, you can easily stage projects of this nature.

CAP04

Almost all the space needed to design this whole-house renovation is found on the second floor. By carefully pushing out the second floor walls, a very appealing two-story home is created from this 34′0″-×-26′0″ very ordinary one-and-one-half-story plan. A very small side bay is all that is added to the first floor. It expands the fabulous new great room just enough to provide extra light and a built-in media center. This open concept kitchen and great room is a trendy way to deal with limited space that is also very much in keeping with today's less-formal lifestyle. The new center hall, created from part of the former front bedroom is a modest, but important, contribution to the great room.

ROOFLINES AVAILABLE:
GABLE (SHOWN), HIP, SHED

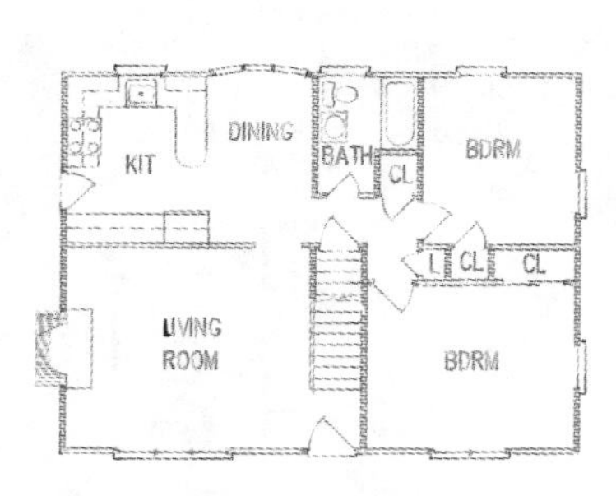

EXISTING FIRST FLOOR

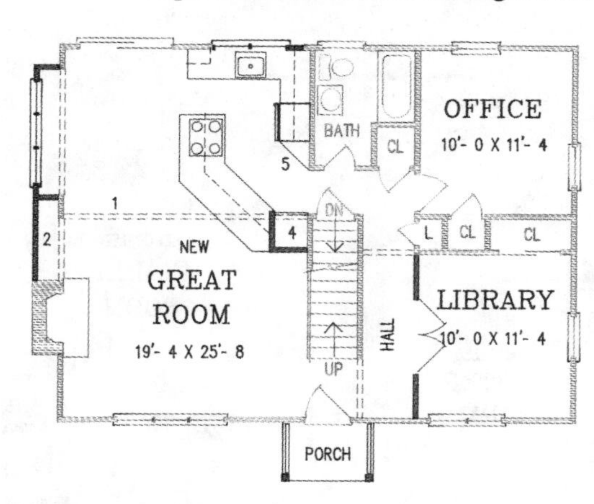

REMODELED FIRST FLOOR

NOTES:
1- REMOVE CENTER BEARING PARTITION
2- MEDIA UNIT RECESS
3- REMOVE EXISTING KNEE WALL
4- WALL OVEN
5- CORNER DESK

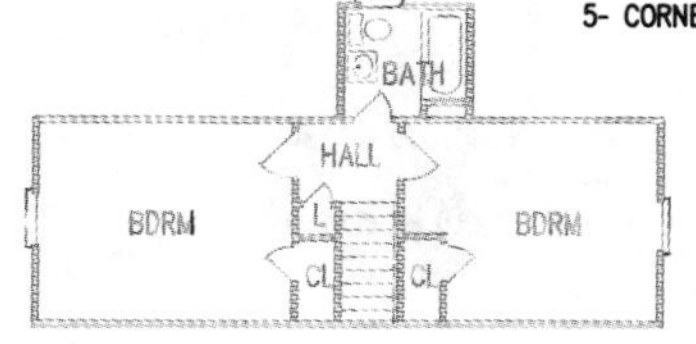

EXISTING SECOND FLOOR

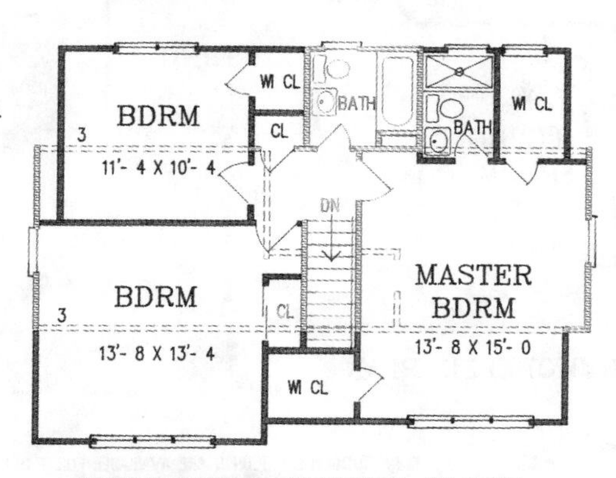

REMODELED SECOND FLOOR

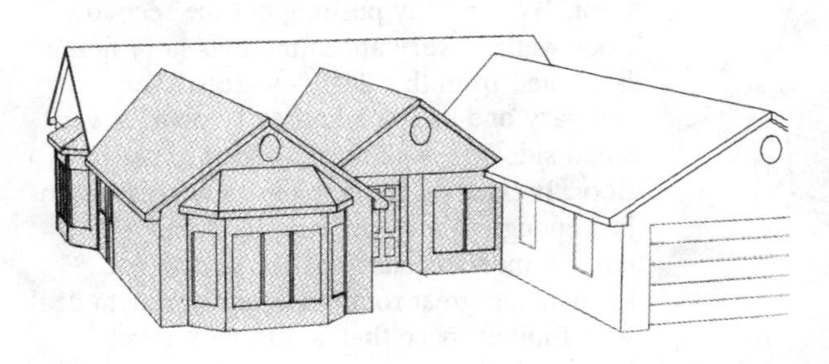

RAN01

In chapter 3, in reviewing the art of finding space, I discussed the L-shaped ranch. This whole-house renovation is one that adds on by filling in the site area of the L, while remodeling the existing plan to create an impressive, up-to-date, one-story home. The area of the handsome front addition provides for a new living room, entrance foyer, part of the new dining room, and a new master bath and dressing room. The changes to the existing part are designed to allow you to stage this remodeling while living here. If the front of your home is turned 90 degrees (bedroom wing at front), a similar plan is possible by simply rotating the new living room to the garage wall.

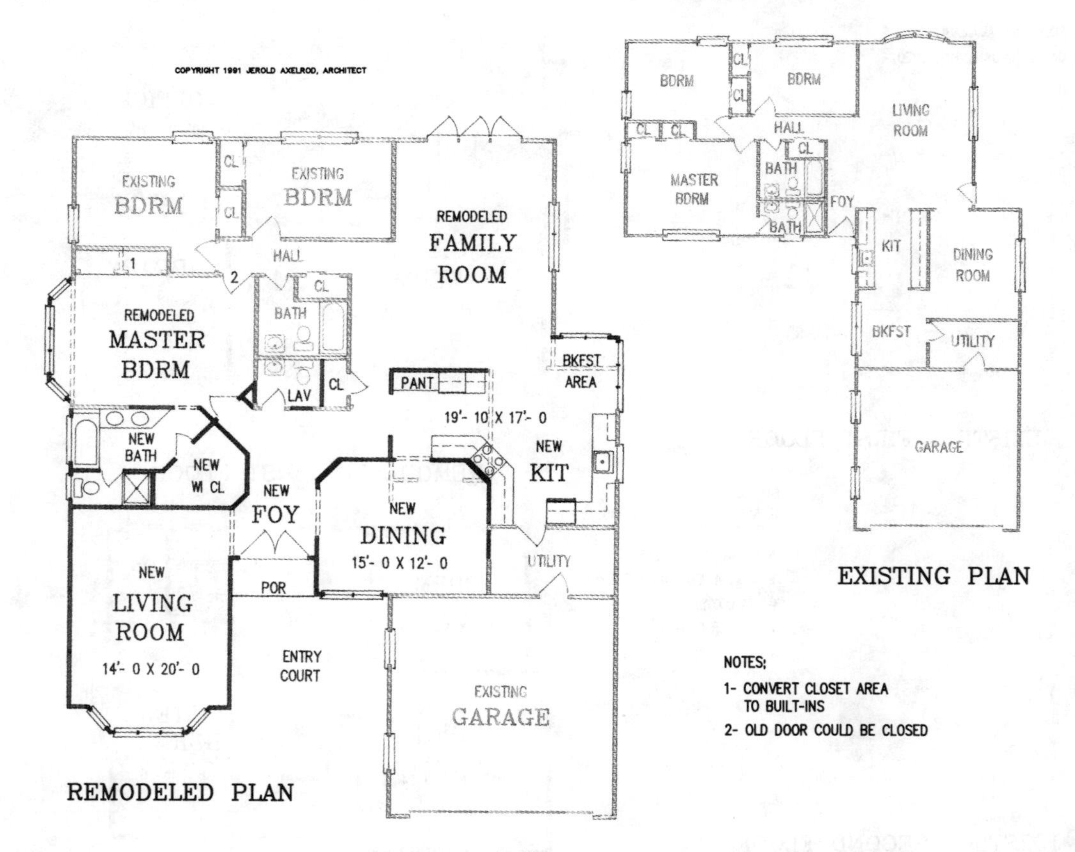

CONSTRUCTION BLUEPRINTS ARE AVAILABLE FOR THIS PLAN (AS WELL AS ALL OTHERS) AND INCLUDE A MATERIAL LIST

CTG01

It's a lovely old three-bedroom bungalow or narrow one-and-one-half-story home, but you need more space, and there's only a few feet to expand in each direction. Then take a hard look here! A major renovation of this 28′0″-×-39′6″ plan adds just a few feet but goes up, too, and makes a stunning overall improvement. The resulting renovation to the original first floor provides a remodeled eat-in kitchen, a large family room, and a generous sized master suite. The space for all this is achieved by locating two children's bedrooms (and a full bath) on a new partial second floor. The front end of the home remains totally unaffected in this remodel, allowing you to stage the renovation. See the discussion and chart on staging in chapter 11.

ROOFLINES AVAILABLE:
ONLY AS SHOWN

NEW MASTER BDRM 17'- 0 X 12'- 0
NEW FAMILY ROOM 12'- 0 X 17'- 4 HIGH CEILING
CL CL CL UP DN 1 DN HALL
BATH LAV REMODELED KITCHEN
LIVING ROOM DINING ROOM
PORCH

FIRST FLOOR
REMODELED

NOTE 1- REMOVE FLOOR AND SET FAMILY ROOM LEVEL WITH STAIR PLATFORM AND SIDE YARD GRADE.

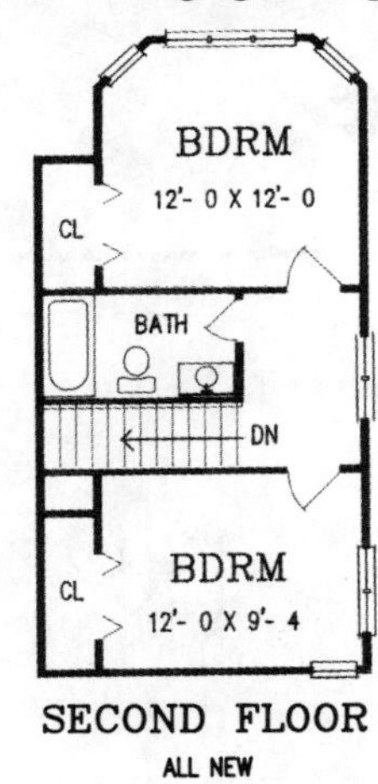

SECOND FLOOR
ALL NEW

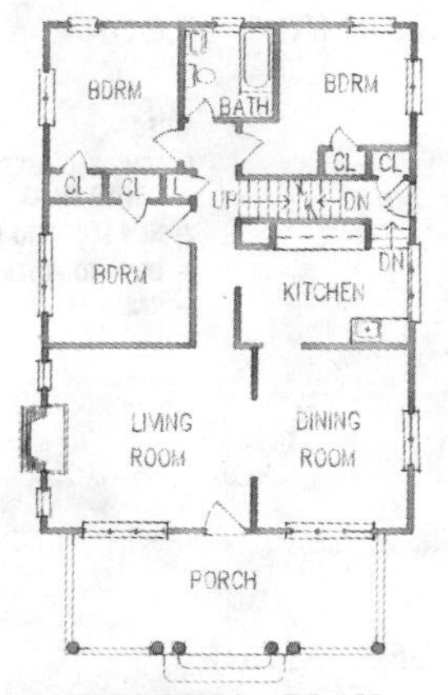

EXISTING PLAN

TWS03

The lot is lovely, the existing one-and-one-half-story cottage is charming, but it is out of date and cramped. Furthermore, there is no room to expand out, so your only solution is to go up. But in order to do it with style, building upon and reinforcing the charm of the existing home is a challenge. This whole-house renovation meets that task with flair. The resulting home is a wonderful three-bedroom two-story that retains and enhances the appeal of the old cottage, while providing a stylish, contemporary living environment. The former first floor bedrooms are remodeled to create a center-hall circulation pattern and space for a breakfast room and family/media room. The kitchen is enlarged and remodeled, and the second floor is completely rebuilt.

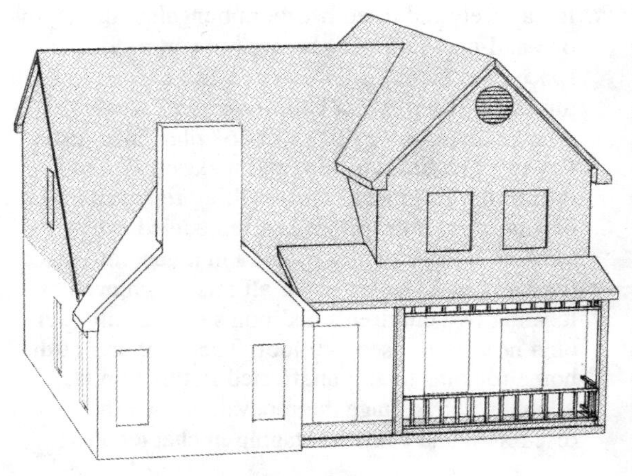

ROOFLINES AVAILABLE:
GABLE (SHOWN), HIP

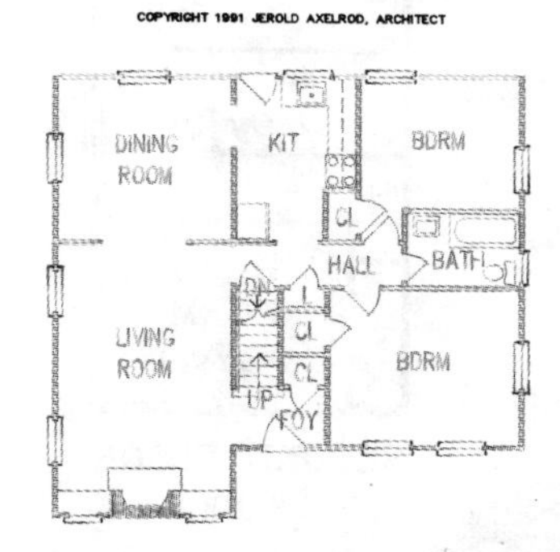

EXISTING FIRST FLOOR

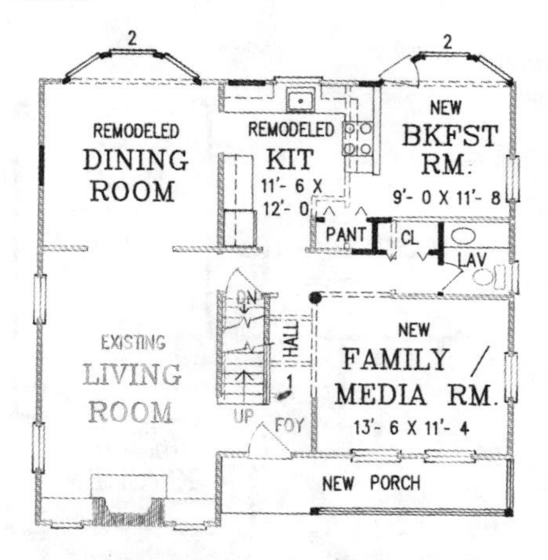

REMODELED FIRST FLOOR

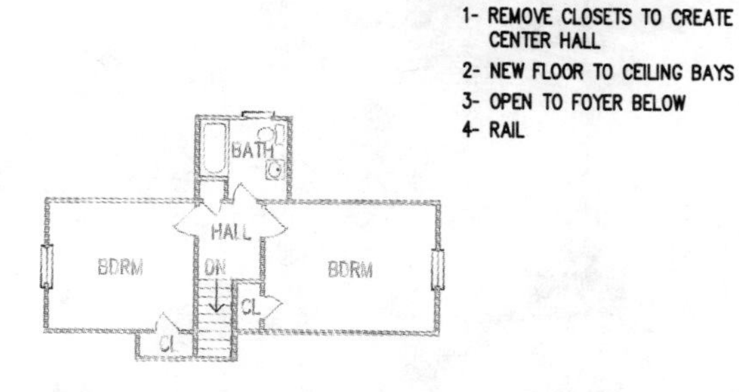

EXISTING SECOND FLOOR

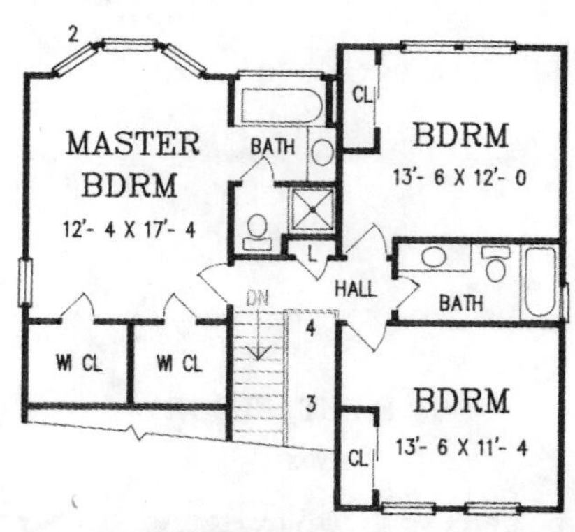

REMODELED SECOND FLOOR

CAP01

A great neighborhood, a nice lot, and the existing 32-×-26-foot cottage is much too small for your needs, but there is no room to expand out. You can go up! This whole-house renovation doubles the square footage by turning the modest cottage into a full three- or four-bedroom center-hall-style two-story. The old basement stair is relocated to the front center where it becomes part of a dramatic two-story-high entrance foyer. The expanded kitchen is now a delight, with its large center island and convenient laundry. The new second floor master suite is spacious and up-to-date. It includes a large walk-in closet and a five-fixture private bath.

ROOFLINES AVAILABLE:

HIP (SHOWN), GABLE, FLAT

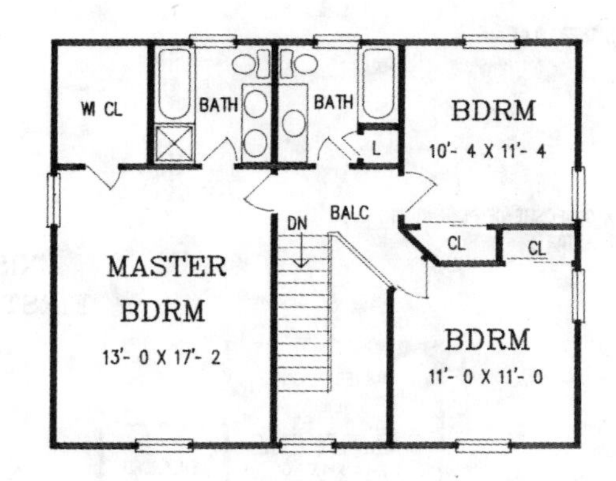

NOTES 1- REMOVE EXISTING BASEMENT STAIR

2- FAMILY ROOM COULD BE ADDED HERE

3- LIVING ROOM COULD EXTEND TO REAR

SECOND FLOOR

ALL NEW

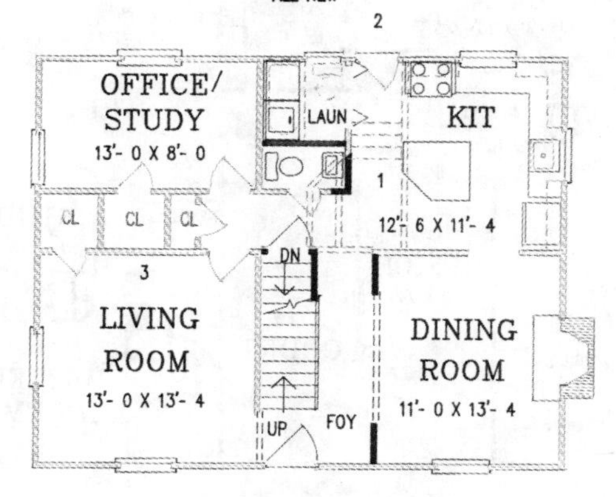

EXISTING PLAN

FIRST FLOOR

REMODELED

RAN02

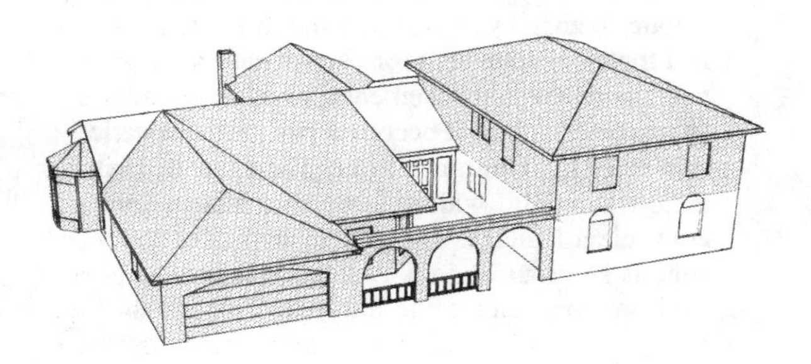

By combining four of the plans shown in prior chapters, this ordinary, L-shaped one-story has been refashioned as a sensational Spanish courtyard residence of impressive proportions. The design expands into the open site area created by the L and also utilizes a rear yard addition. The children's bedrooms move to their own second floor wing with the master bedroom now encompassing the area vacated by them. There is a dramatic new foyer and a fabulous new kitchen in an open arrangement with the new family room. The former garage is utilized to create a lovely apartment, with its own side entrance, and a spacious office is created alongside the new garage.

ROOFLINES AVAILABLE:
GABLE & HIP (SHOWN), SHED, FLAT

NOTE: THIS PLAN IS A COMPOSITE OF PLAN NUMBERS:
GAR03, BRB02, FD001, APT04

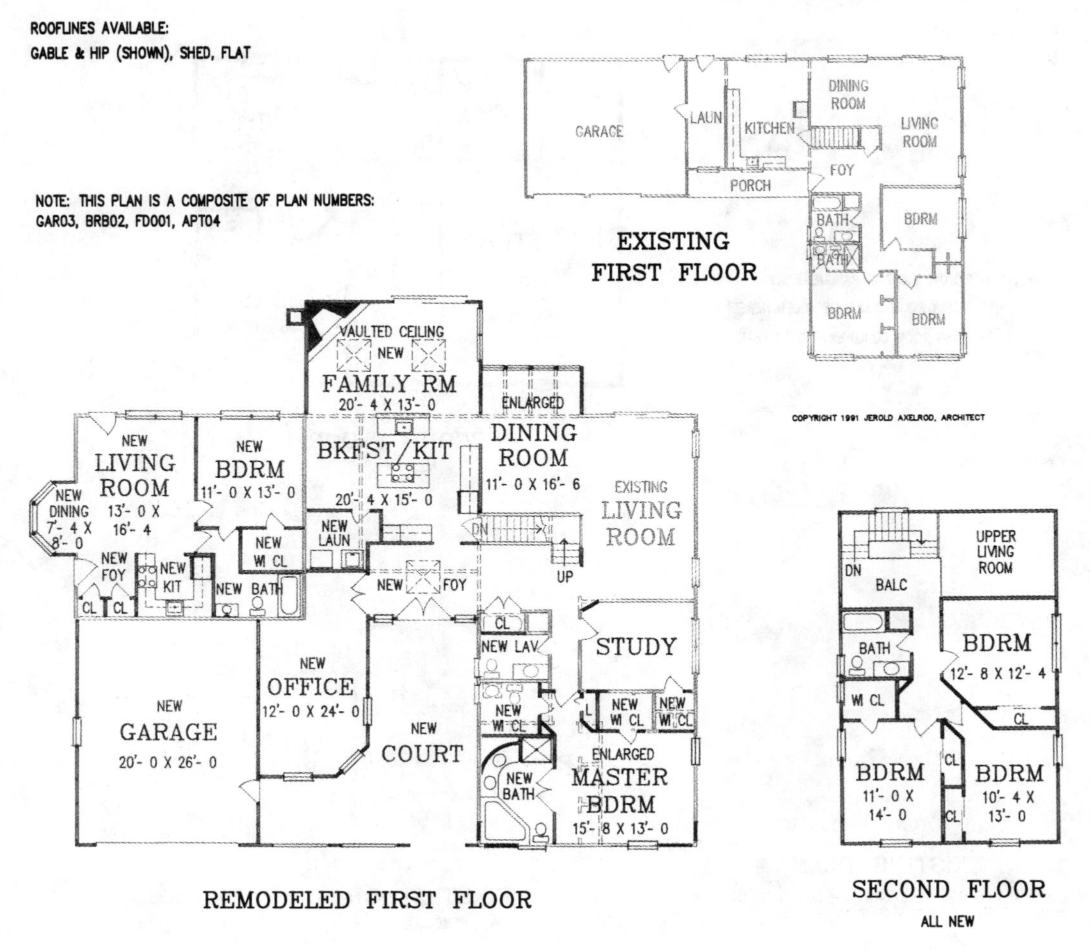

SPL01

The front-to-rear split-level is one of those few homes that I find great difficulty in providing partial help for. The tidy solution presented here is a whole-house renovation that goes up. It adds a full floor over the first floor and relocates rooms from top to bottom. The new second floor is now a self-contained children's wing, complete with its own study area and recreation room. The former bedrooms are now occupied by a living room and a master suite. The ground floor now has ample space for a large kitchen and breakfast room, wider staircases, and a new dining room. It isn't an easy project, but it is readily constructed in 3 or 4 stages.

ROOFLINES AVAILABLE:
GABLE (SHOWN)

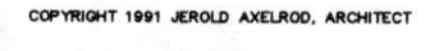

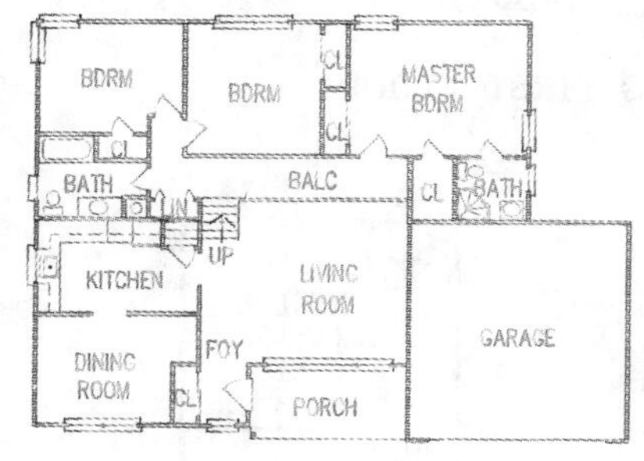

EXISTING
FIRST & UPPER FLOORS

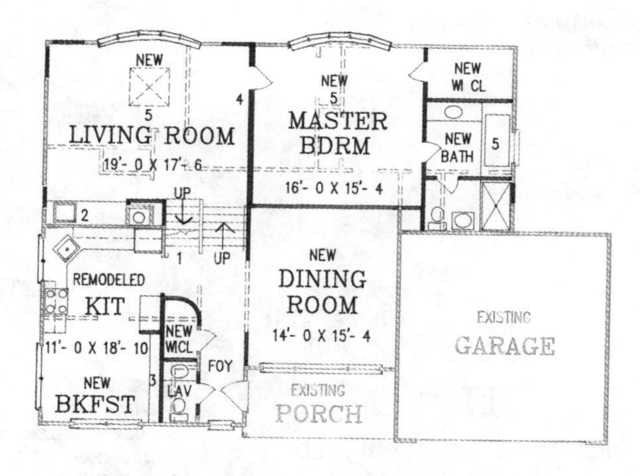

REMODELED FIRST & UPPER FLOORS

NOTES:
1- STAIR TO LOWER LEVEL RECREATION ROOM AND BASEMENT
2- BUILT-IN WET BAR AND MEDIA CENTER
3- PANTRY WALL
4- SOUND INSULATED WALL
5- VAULTED CEILINGS

NOTE: CONSTRUCTION BLUEPRINTS CAN BE READILY MODIFIED TO ACCOMMODATE YOUR SPECIFIC HOUSE OR ROOM SIZES.

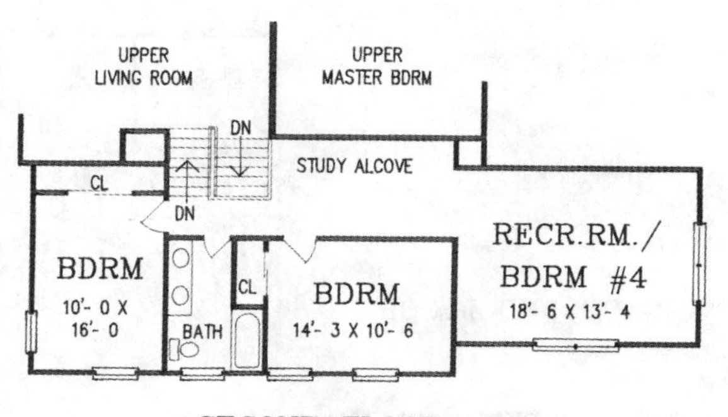

SECOND FLOOR
ALL NEW

TWS05

This is another whole-house project drawn from the hundreds of plans presented. It is a remake of a 25–35-year-old two-story that provides a new facelift amongst its improvements. This new facade can be seen in plan PF002 on page 172. The rear view pictured here shows an enlarged first floor rear, comprising a fabulous new kitchen and expanded dining room, a new master bath and dressing area, and an apartment on the first floor, which is located behind the garage. The apartment, designed to house elderly parents, is a barrier-free design. The connection to the apartment is through an enlarged laundry/exercise room.

ROOFLINES AVAILABLE:
GABLE (SHOWN), HIP

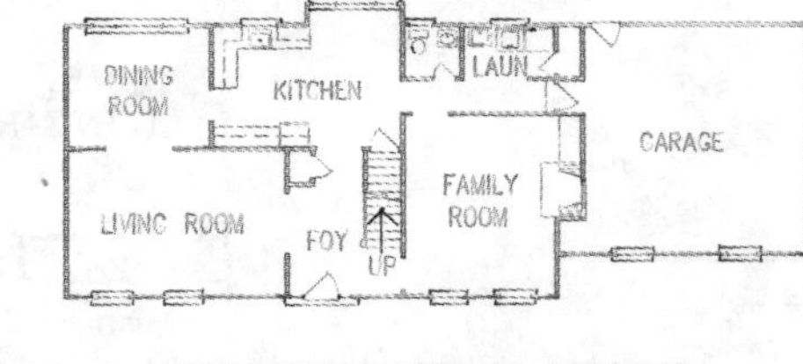

EXISTING FIRST FLOOR

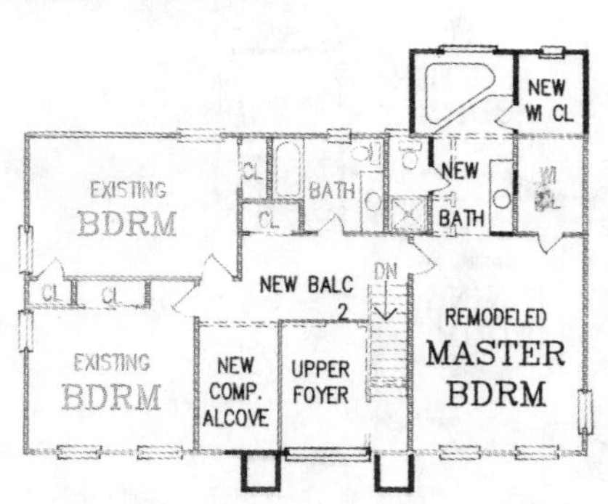

REMODELED SECOND FLOOR

NOTES:
1- TWO STORY HIGH FOYER
2- BALCONY RAILING
3- HANDICAP ACCESSIBLE APT.& BATH
4- WHEEL CHAIR STORAGE
5- 27' 8" DEEP ADDITION

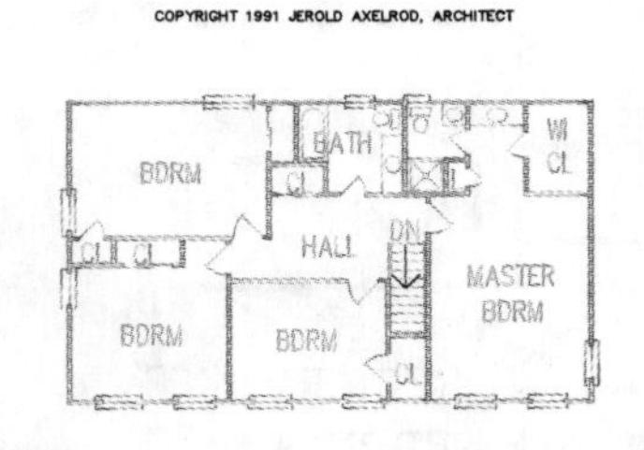

EXISTING SECOND FLOOR

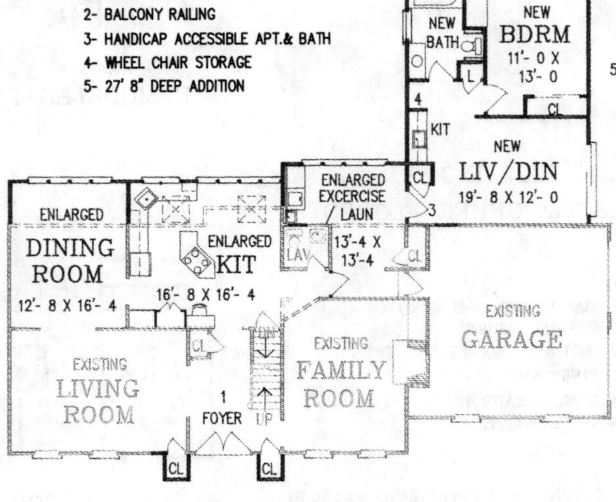

REMODELED FIRST FLOOR

NOTE: THIS PLAN IS A COMPOSITE OF PLAN NUMBERS:
KDEB1, PF002, HAPT2

CAP02

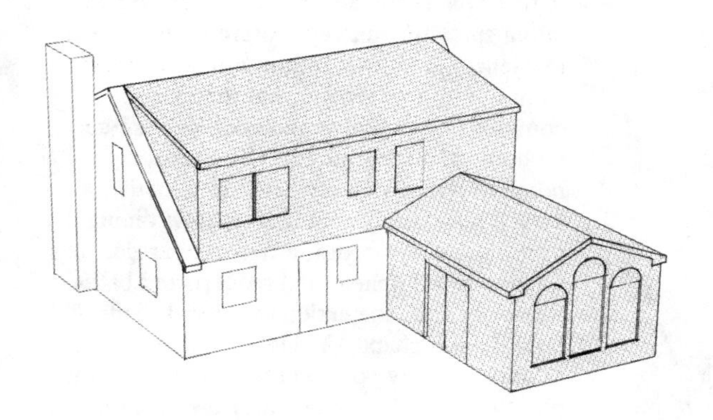

A comfortable, delightful, three-bedroom, home is the result of the whole-house renovation of the same basic 32-×-26-foot two-bedroom cottage. The plan features a one-story rear addition that provides the space for a lovely new master suite with its own bath that is carved from an existing small bedroom. The basement stair is relocated to the front, by enlarging the existing rear kitchen. The former master bedroom is enlarged to create a living room, and a new dining room results from the decreased former living room. It is likely that a new, steeper roof will have to be framed to enclose the two bedrooms and bath at the new second floor.

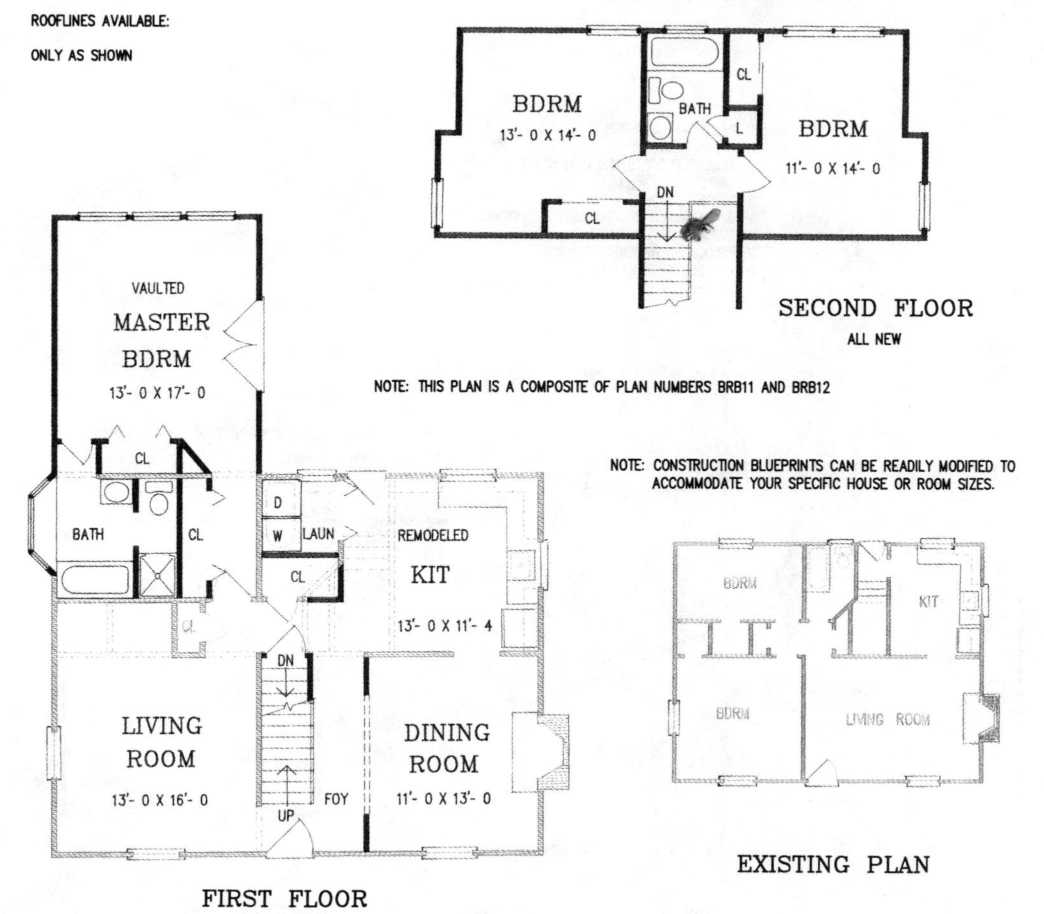

RAN03

A tiny four-room starter home is converted into a sprawling and comfortable seven-room residence in this whole-house renovation. The front two rooms remain, but the rear is completely remodeled and additions are added on both sides plus the rear. The resulting home provides a lovely new master suite, a stylish new family room, a separate dining room, a laundry room, a two-car garage, plus a brand new kitchen, and modernized baths. The plan is an adaptation of three designs presented in chapter 7, and it is another example of how you can successfully use the plans presented to create your own dream home. This is a project that can be readily staged over a long term.

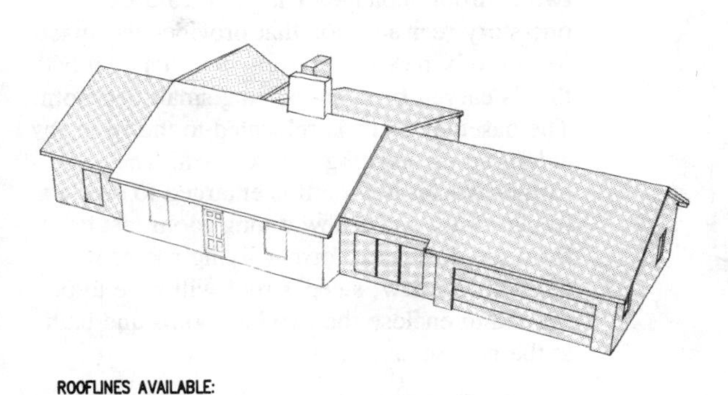

ROOFLINES AVAILABLE:
GABLE (SHOWN), HIP

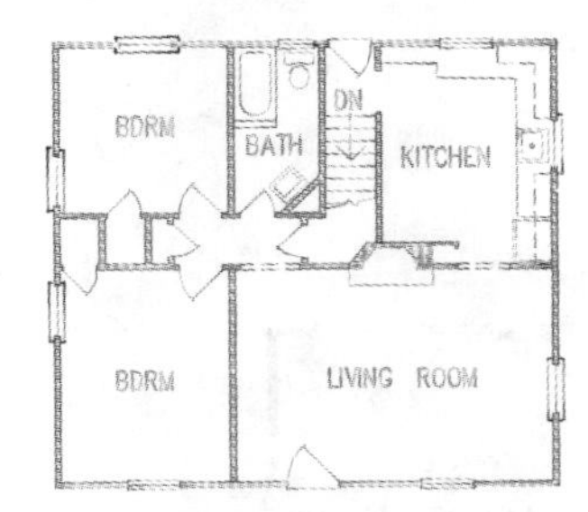

EXISTING PLAN

NOTES:
1- PASS-THRU COUNTER
2- SNACK COUNTER
3- REMOVE CLOSETS AND EXTEND HALL
4- SKYLITE
5- TRAYED CEILING
6- BATHS AND BEDROOM COULD BE OPENED TO LANDSCAPED COURT
7- RELOCATED BASEMENT STAIR
8- PORCH COULD BE ADDED

NOTE: THIS PLAN IS A COMPOSITE OF PLAN NUMBERS: DMG02, BRB09, & F0003

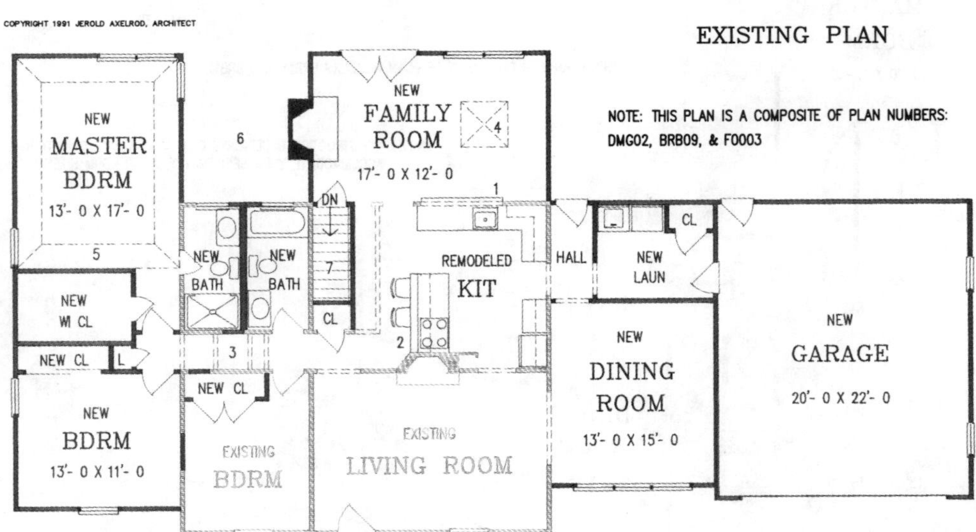

REMODELED PLAN

CTG02

This 23′0″-×-40′6″ two-bedroom cottage has been given a refreshingly new inside, while retaining most of its charming exterior character. Other than the area of the rear bedroom and bath, the balance of the first floor has been blown wide-open, which establishes a smart, creative interior with lots of nooks, corners, and new windows and skylights for light. A great new kitchen results, as does a beautifully flowing living and dining room, and there's a little corner for an office. A dramatic circular stair provides access to a new second floor wing that houses a very chic new master suite.

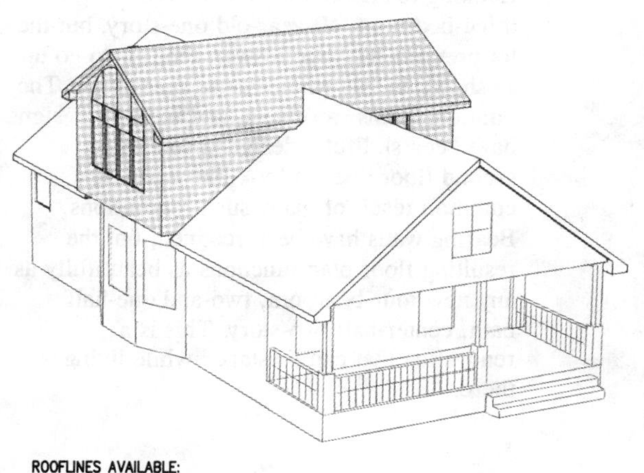

ROOFLINES AVAILABLE:
GABLE (SHOWN), HIP

NOTES 1- REMOVE EXISTING WALL
2- SKYLIGHTS
3- OPEN TO BELOW
4- RADIUS GLASS BLOCK

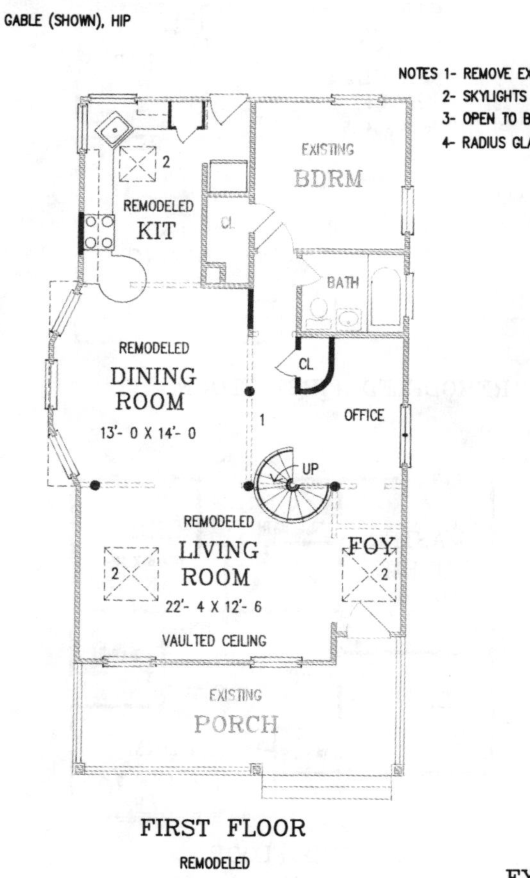

FIRST FLOOR
REMODELED

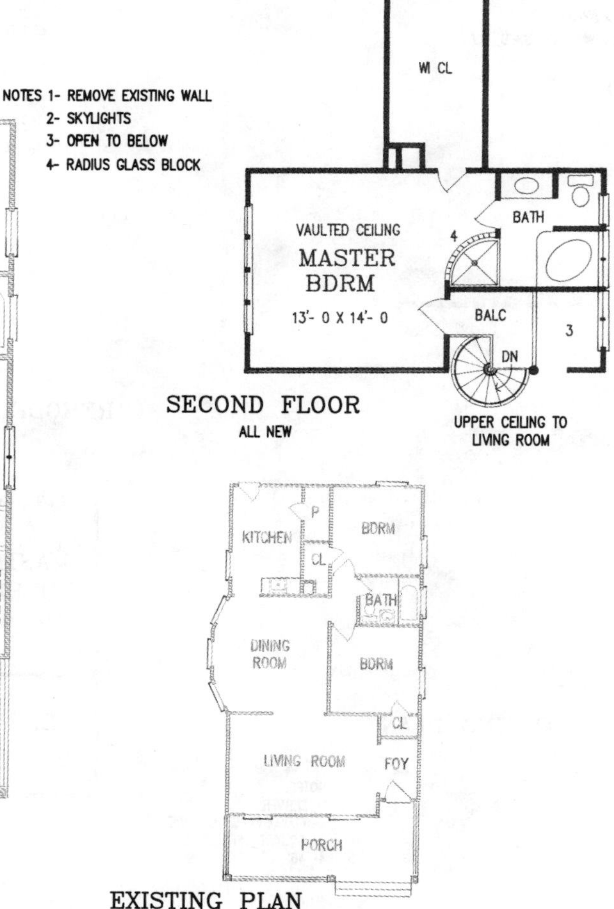

TWS06

Looking to double the area of your three-bedroom, 40-year-old one-story, but the lot prevents it? Your answer, then, is to go up as shown in this whole-house renovation. The building forms, rooflines, and window designs have been skillfully designed so that the second floor doesn't look like a hat—a common result of many such renovations. Bearing walls have been retained, but the resulting floor plan functions as beautifully as any new four-bedroom, two-and one-half bath, center-hall two-story. This is a renovation that can be staged while living there.

ROOFLINES AVAILABLE:
GABLE (SHOWN), HIP, SHED, FLAT

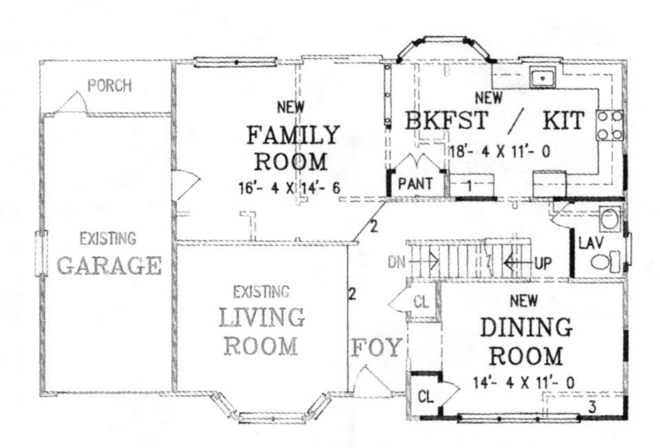

REMODELED FIRST FLOOR

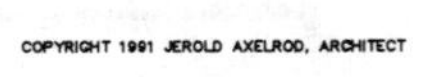

4

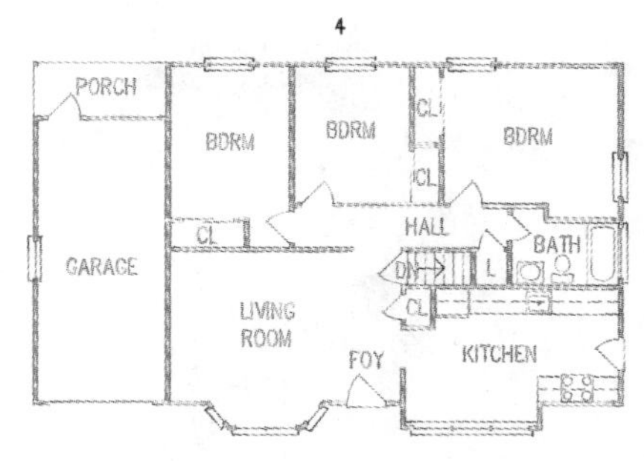

EXISTING PLAN

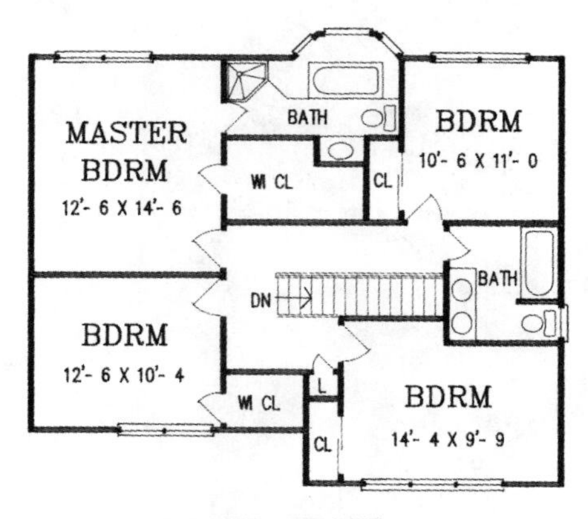

SECOND FLOOR
ALL NEW

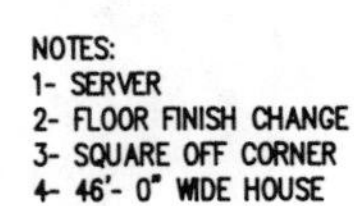

CAP03

An eight-foot-deep, two-story-high, addition across the rear provides all the space needed to transform this 36-foot-wide, modest-sized, one-and-one-half-story home into a spacious and comfortable two-story. As viewed from the rear, the forms of the old steeper rooflines are retained at each side—an important design consideration to keep in mind. The remodeled first floor now includes a tidy new master suite with its own tub bath and a charming new country kitchen. Since the second floor now provides three bedrooms and a play loft, you could consider multiple other uses for the old front bedroom on the first floor, such as a media room, office, or even a dining room.

ROOFLINES AVAILABLE:
SHED (SHOWN)

NOTE: CONSTRUCTION BLUEPRINTS CAN BE READILY MODIFIED TO ACCOMMODATE YOUR SPECIFIC HOUSE OR ROOM SIZES.

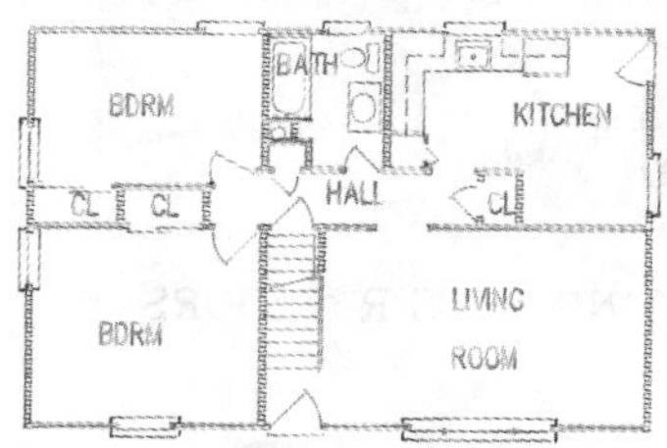

EXISTING FIRST FLOOR

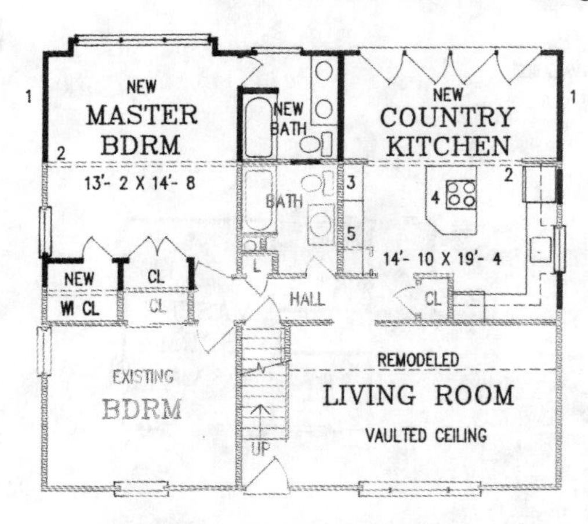

REMODELED FIRST FLOOR

NOTES:
1- 8'- 0 ADDITION
2- REMOVE REAR WALL
3- DESK
4- SNACK COUNTER
5- PANTRY CABINETS
6- RAIL
7- UPPER PART OF LIVING ROOM

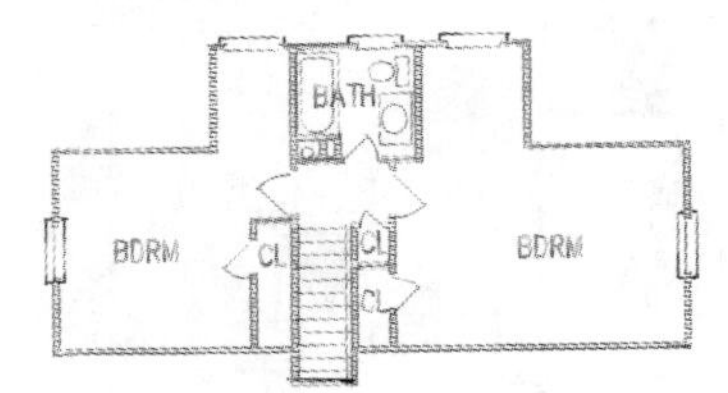

EXISTING SECOND FLOOR

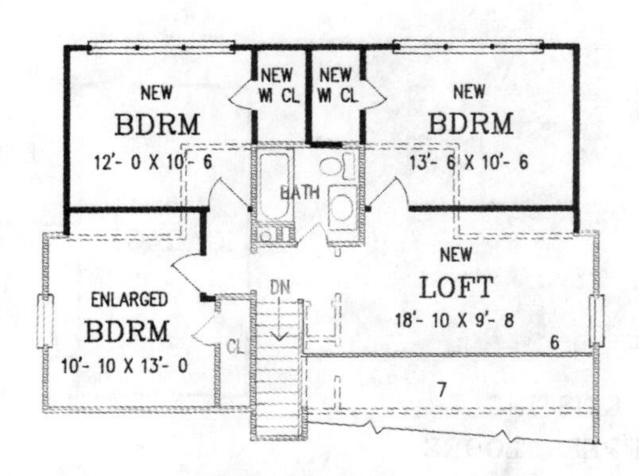

REMODELED SECOND FLOOR

SPL02

Everything you might need to add, or remodel, has been incorporated into the remake of this split-level. The architectural design takes four plans pictured in the previous sections and modifies them, as necessary, to achieve this winning new home. A beautiful main-level family room, located off the remodeled kitchen, is one of the striking highlights. A fabulous new master suite over the living room is another. The former recreation room is now the location for a ground-level apartment that is planned for handicap accessibility. A one-car garage is also added to compensate for the garage area that was included in the apartment.

ROOFLINES AVAILABLE:
HIP (SHOWN), GABLE

NOTE: THIS PLAN IS A COMPOSITE OF PLAN NUMBERS:
BRB18, F0006, HAPT4, GAR05

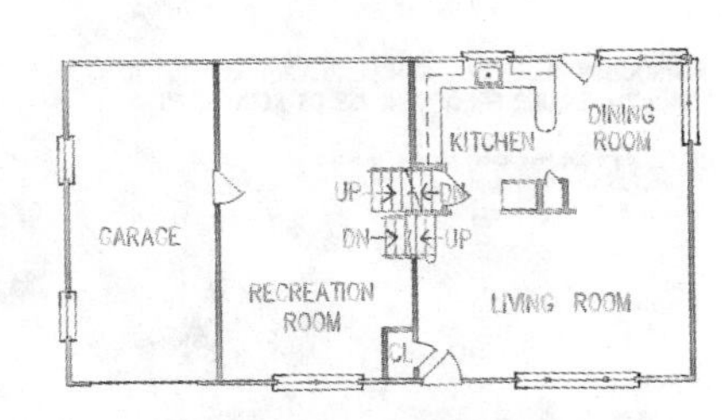

EXISTING LOWER FLOORS

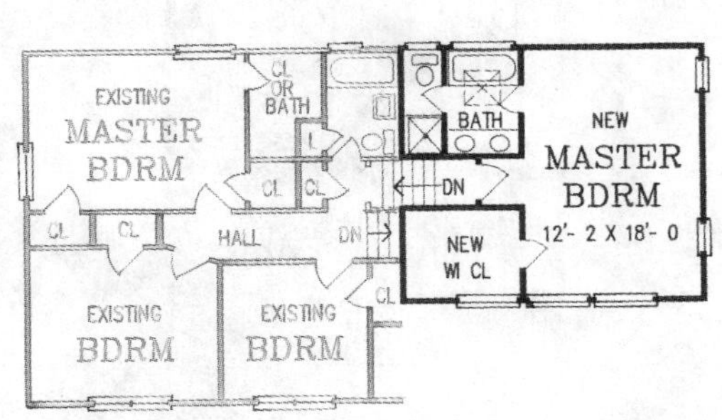

REMODELED UPPER FLOORS

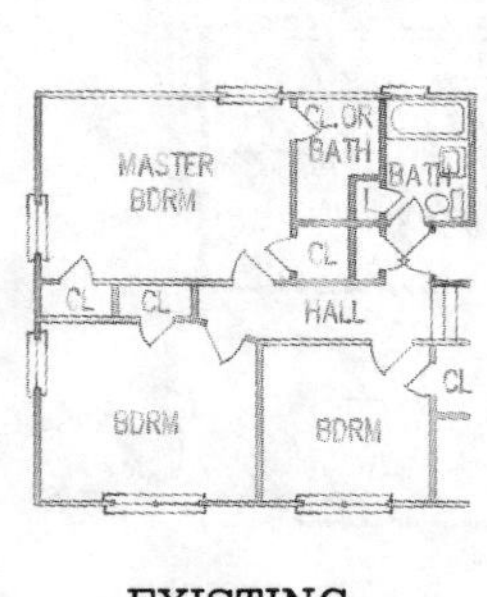

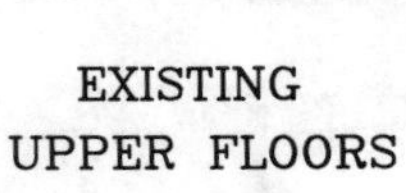

EXISTING
UPPER FLOORS

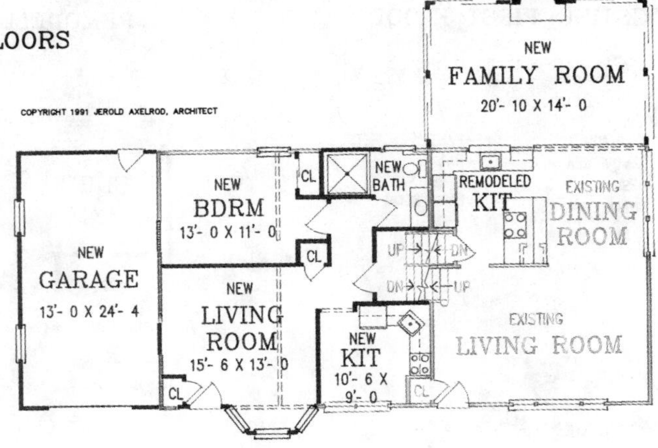

REMODELED LOWER FLOORS

TWS01

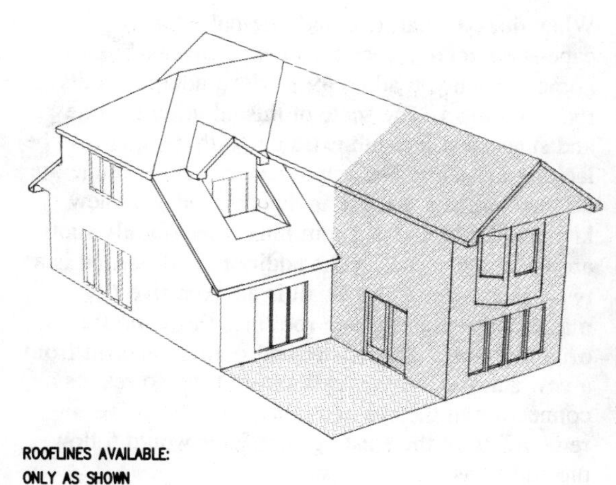

ROOFLINES AVAILABLE:
ONLY AS SHOWN

Many homes 60 or more years old suffer badly in terms of today's amenities, particularly as concerns their baths and kitchens. They also are likely to be devoid of good circulation patterns and are surely missing a family room. But despite these deficiencies, many enjoy such a wonderful character that they are worth the investment to modernize. This two-story is such a home. The renovation shown adds a family room and a new master suite. It converts a covered rear porch to a dining room. The old dining room now becomes a sensational new kitchen, ideally situated in respect to the new family room. Finally, the old kitchen now houses a large laundry room, powder room, and center hall.

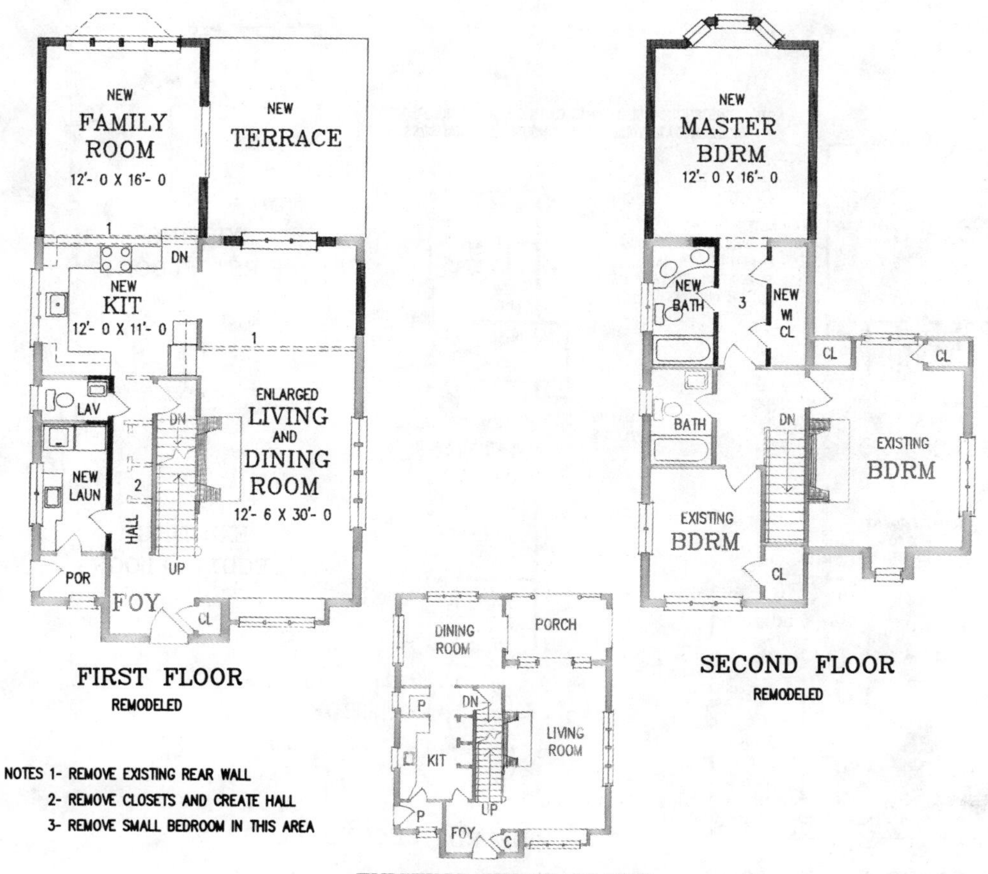

CAP05

When does a small one-and-one-half-story cape-style cottage become an expansive, fabulous home? When you add a 49′8″ deep addition to its rear, as shown. The scale of this addition is large and suggests a lot with extra depth that might also be lacking in width. The improvements to the home are all the elements that are in demand today. A new kitchen and breakfast room plus a new family room are the centerpieces of the addition, and beyond is a two-car garage. Above it all is an extensive new master suite, built within rooflines that echo the original house. The master bedroom is reached from a new staircase in the addition, but it also retains a connection to the old bedroom hall. Extensive remodeling of the existing first floor would follow the additions.

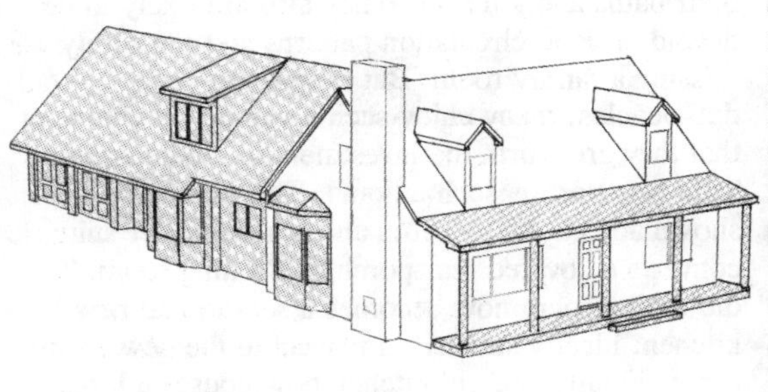

ROOFLINES AVAILABLE:
GABLE & SHED (SHOWN)

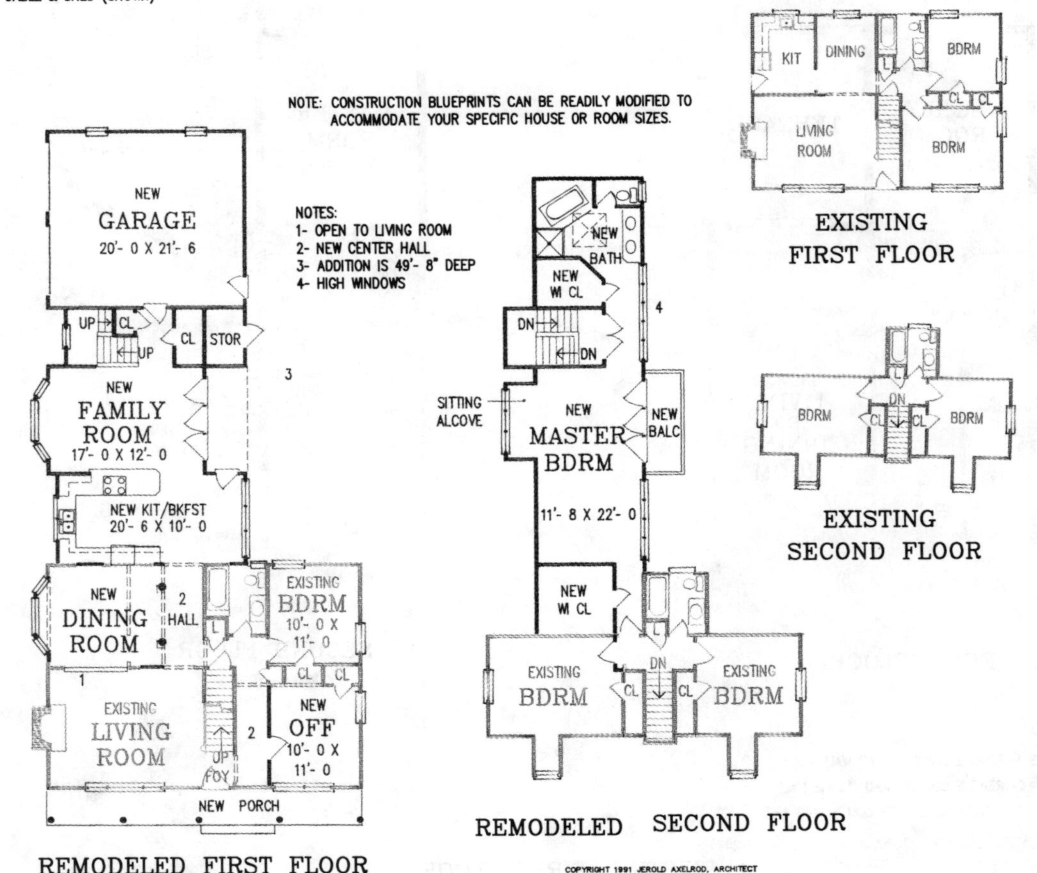

CTG03

A charming 24′0″-×-32′8″ cottage becomes a stylish new three- or four-bedroom winner in this whole-house remake. By adding 5′4″ to one side and going up in the rear, ample space is gained to enable a complete transformation, while still retaining the integrity and charm of the original home. The front porch is wrapped around one side, moving the entrance and creating a foyer and center hall. A private first floor master suite is provided, and two children's bedrooms are placed in the new second floor. The kitchen is remodeled in place, and a small greenhouse "bump" is suggested as a breakfast nook.

ROOFLINES AVAILABLE:
ONLY AS SHOWN

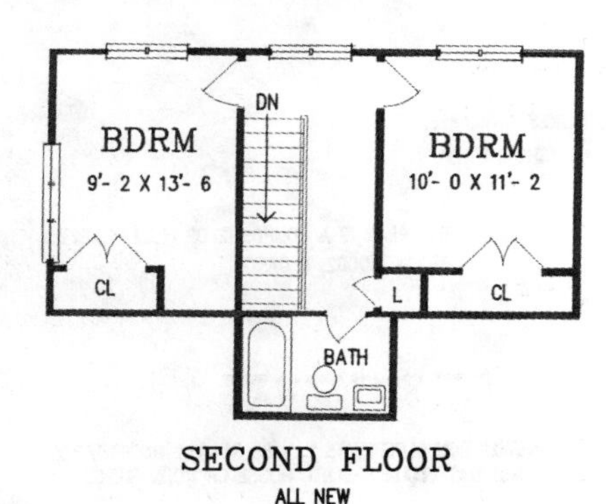

SECOND FLOOR
ALL NEW

NOTES 1- WIDEN DINING ROOM WITH BAY WINDOW
2- OPT. GREENHOUSE UNIT SERVES AS ATTRACTIVE BREAKFAST NOOK
3- REMOVE ENTRY DOOR AND REPLACE WITH FRENCH DOORS

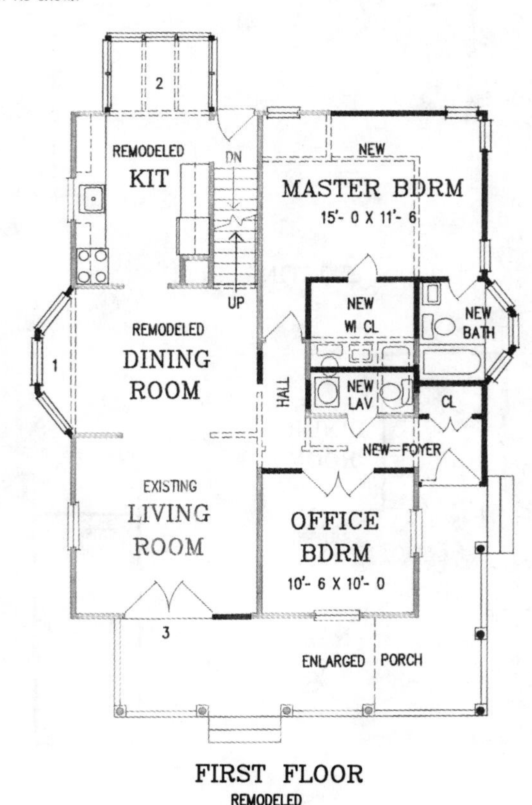

FIRST FLOOR
REMODELED

NOTE: CONSTRUCTION BLUEPRINTS CAN BE READILY MODIFIED TO ACCOMMODATE YOUR SPECIFIC HOUSE OR ROOM SIZES.

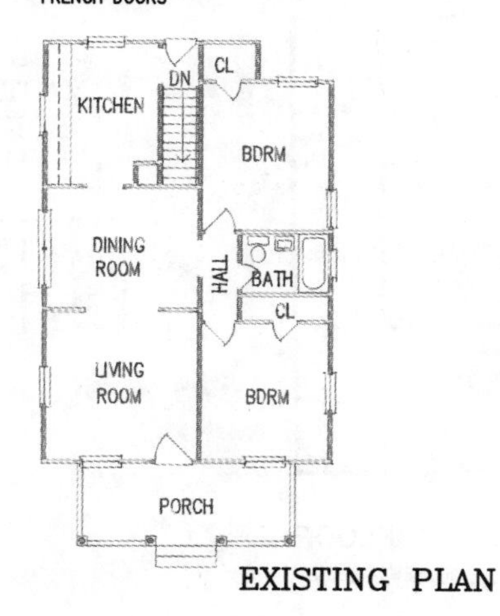

EXISTING PLAN

RAN04

The whole-house renovation of this dated U-shaped one-story expands out in almost every direction to create a lovely new home that is stylish and up-to-date throughout. Major remodeling includes a new children's wing on the second floor, a beautiful new first floor master suite, an expanded kitchen with a lovely new breakfast room, an enlarged dining room, and a new garage addition. Other benefits of the remodeling include an improved access to the den and a charming new front porch. This design is one of those culled from other plans in the book, and it demonstrates how easily you can create a whole-house scheme for yourself.

ROOFLINES AVAILABLE:
GABLE (SHOWN), HIP

NOTE: THIS PLAN IS A COMPOSITE OF PLAN NUMBERS:
BRB04, KD002, & GAR01

NOTE: CONSTRUCTION BLUEPRINTS CAN BE READILY MODIFIED TO ACCOMMODATE YOUR SPECIFIC HOUSE OR ROOM SIZES.

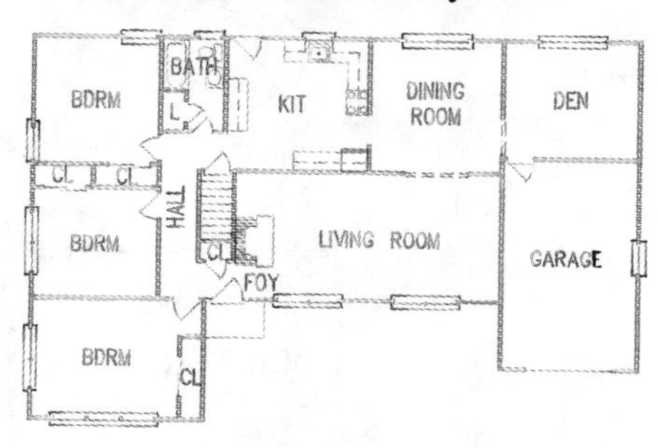

EXISTING PLAN

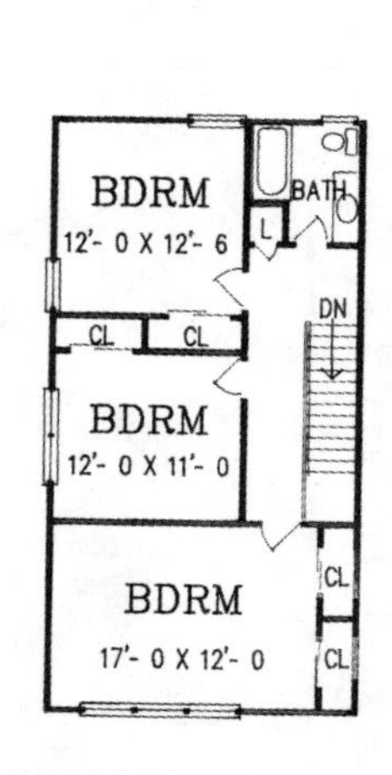

SECOND FLOOR
ALL NEW

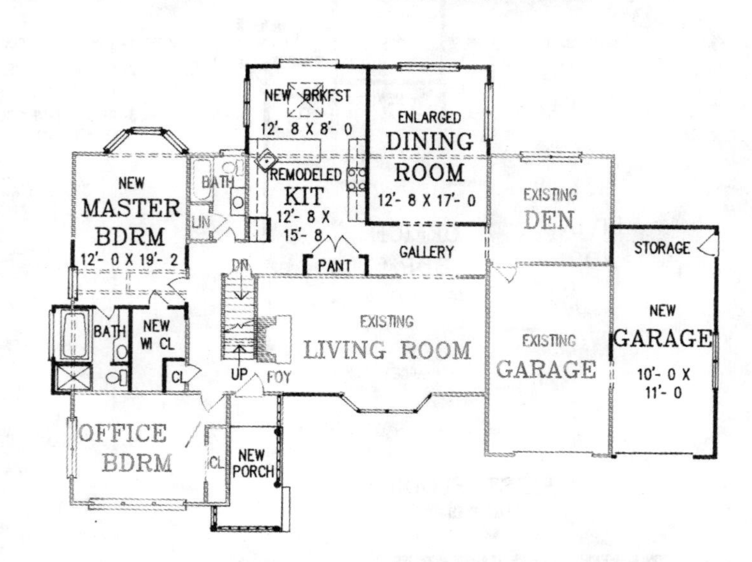

REMODELED FIRST FLOOR

TWS02

When does a 24′0″-×-40′0″ one-story bungalow or cape transform into a 24′0″-×-40′0″ thoroughly updated two-story? This is likely to occur when you love the location of the old charmer and there's no room to expand out, yet your program demands no less than you would achieve in a brand new home. That's the subject of this whole-house remake. The remodeled first floor includes a dramatic new two-story-high reception foyer, a true center-hall layout, a laundry adjacent to a new side-door entry, a great new eat-in kitchen, and an adjoining new family room. The all-new second floor has a dramatic balcony and three bedrooms; the master suite features a compartmented private bath, two walk-in closets, and a lovely little lounge or office alcove.

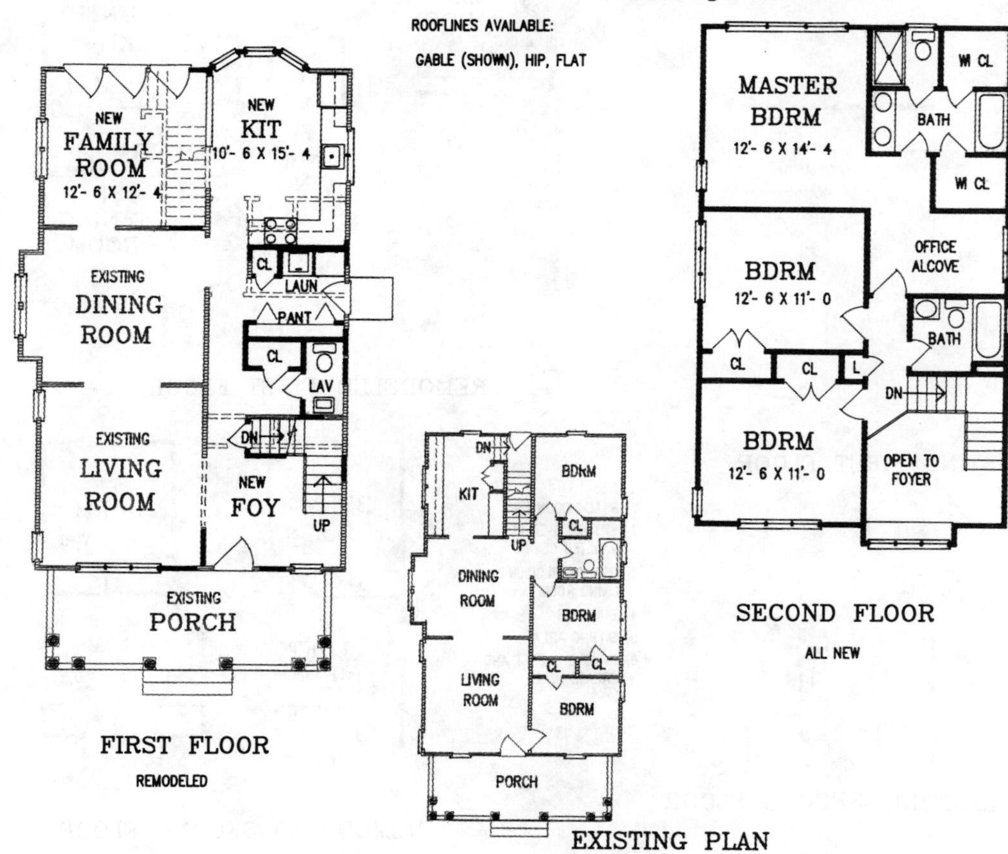

CAP06

The most unfortunate aspects of this existing one-and-one-half-story home are its lack of relationship to the rear yard and its poor circulation. The whole-house redesign pictured eliminates these deficiencies. A new first floor master suite is created by utilizing one rear bedroom and adding the unused space behind the garage. The other rear bedroom is now incorporated into a country kitchen. A small front vestibule is added to expand the circulation space, as well as to help define a fresh new look for the house. The new second floor now houses three bedrooms and a small loft. Removal of the old dormers helps complete the new look.

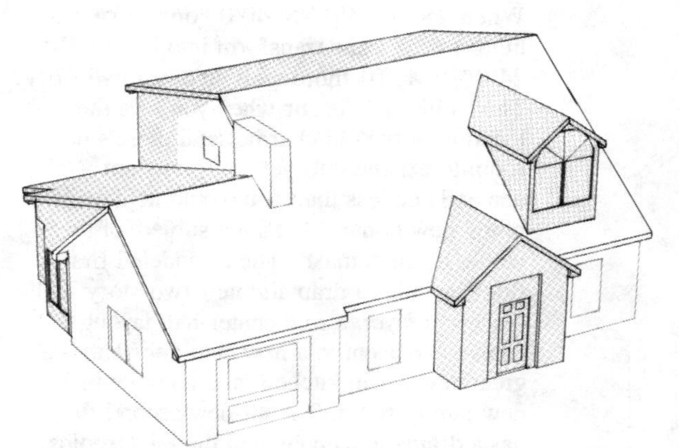

ROOFLINES AVAILABLE:

ONLY AS SHOWN

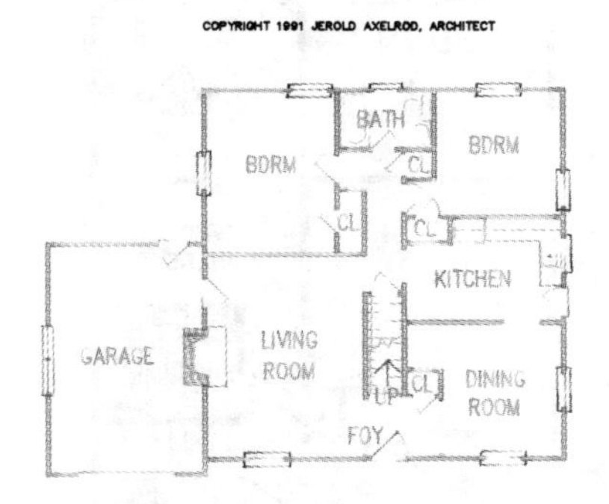

EXISTING FIRST FLOOR

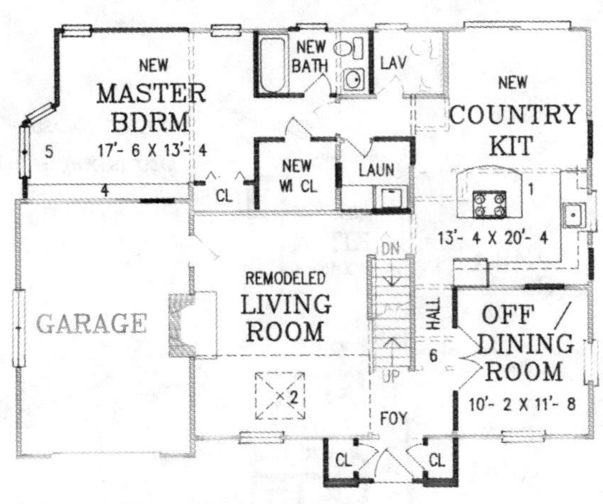

REMODELED FIRST FLOOR

NOTES:
1- SNACK COUNTER
2- SKYLITE OVER IN VAULTED CEILING
3- RAIL OVERLOOKING LIVING ROOM
4- BUILT-INS
5- SITTING AREA
6- REMOVE CLOSET AND CREATE HALL

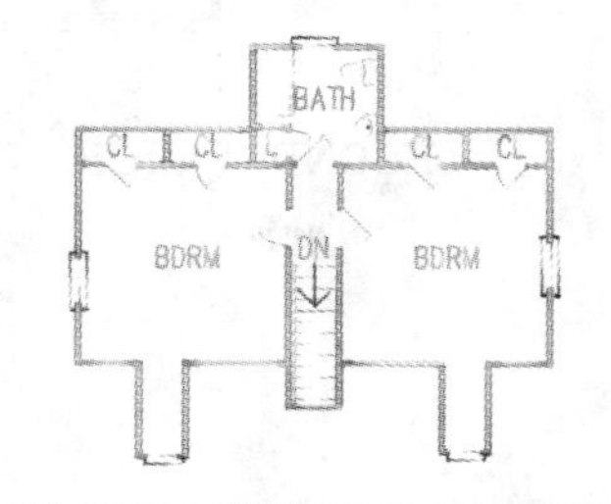

EXISTING SECOND FLOOR

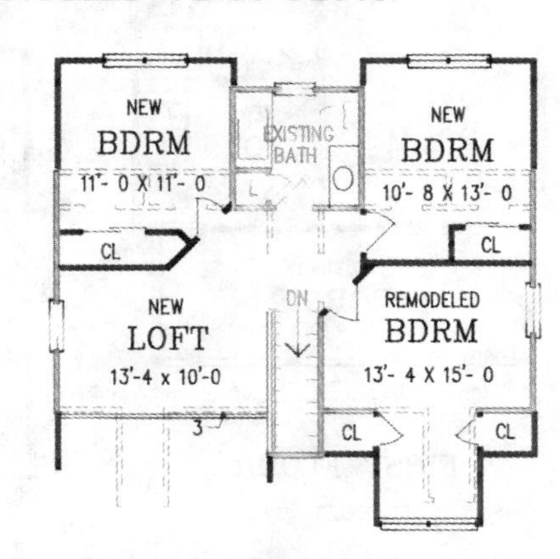

REMODELED SECOND FLOOR

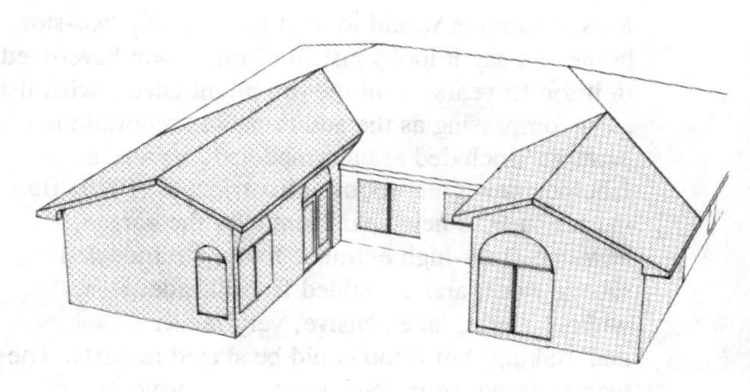

ROOFLINES AVAILABLE:
GABLE (SHOWN), HIP

RAN05

A wholesale change is necessary if your goal is to properly bring the existing hip-roofed one-story home pictured to today's standards. One serious problem with the home is its lack of relationship to the rear yard, and the solution requires a relocation of the kitchen. Two new wings are added, one for a new family room, the other for a new master bedroom suite. Both feature vaulted ceilings and lots of glass focusing about the beautiful new center court that they create. As complex as these changes may appear, the subject of planning a project in stages uses this home as an example. For an elegant two-story whole-house renovation from this same home, see plan TWS06 on page 286.

NOTE: CONSTRUCTION BLUEPRINTS CAN BE READILY MODIFIED TO ACCOMMODATE YOUR SPECIFIC HOUSE OR ROOM SIZES.

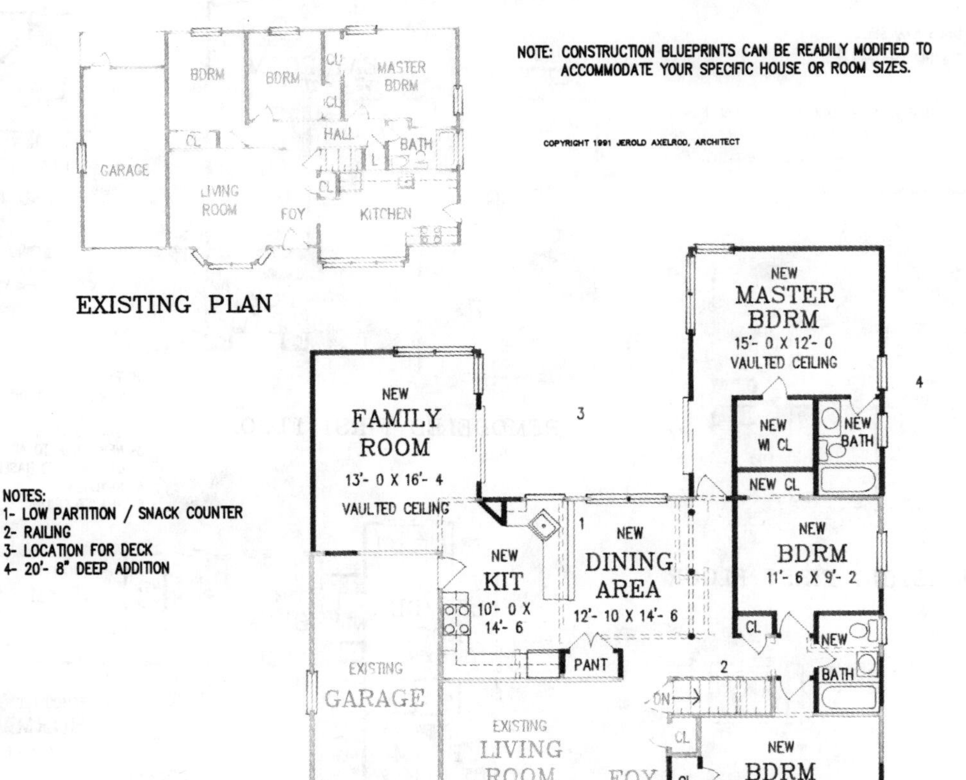

EXISTING PLAN

NOTES:
1- LOW PARTITION / SNACK COUNTER
2- RAILING
3- LOCATION FOR DECK
4- 20'- 8" DEEP ADDITION

REMODELED PLAN

TWS04

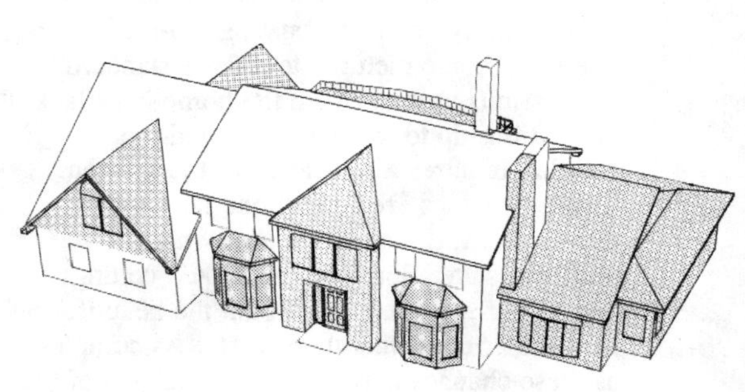

Most observers would look at this existing two-story home and say it looks just fine. But if you have lived in it for 15 years or more, you might have a wish list as encompassing as the additions and renovations pictured. Included are a remodeled kitchen, a fabulous new great room, a luxurious new first floor master suite, a new apartment over the garage, a new two-story-high entrance foyer, a remodeled second floor, and a restyled front facade. It is, without doubt, an extensive, very lavish undertaking, but it too could be staged in parts. The plan is drawn from four partial plan projects presented in the prior chapters and is another example that shows you how to adapt the ideas in this book to your home.

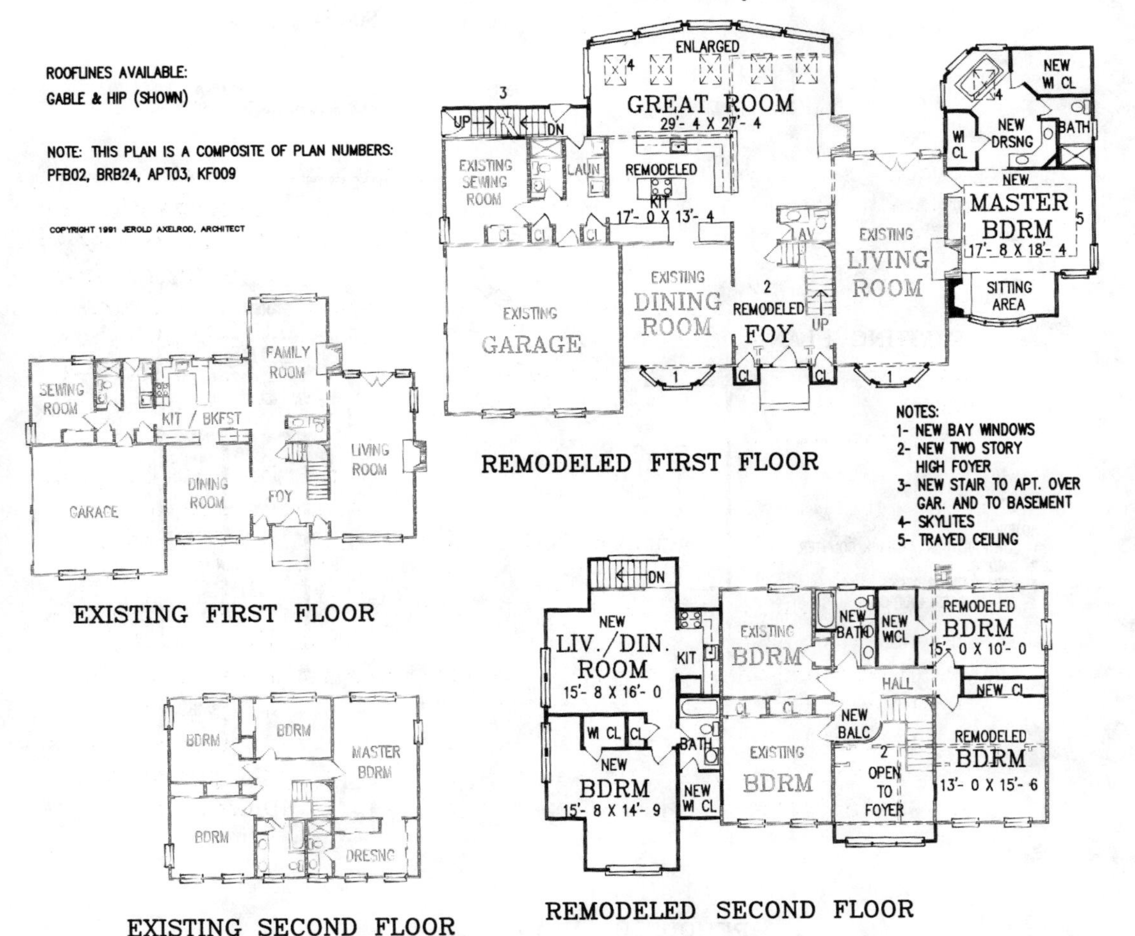

11
CHAPTER

Construction details & staging

I have said that this is not a how-to book, but a what-to book . . . so why is there a chapter on construction details? Because the construction subjects covered here are generally design-related, and may be missing from the how-to books you will consult. I also might have referred to a number of these in the text matter associated with individual designs. Furthermore, a few details shown are frequently misunderstood or overlooked by the do-it-yourselfer, and could cause severe problems. I've also included some favorite details of mine that lead to successful remodelings. As necessary, there is a brief discussion on each page, but let's talk about two of the subjects here.

An inordinate amount of time has gone into our preparation of the girder charts. Because I emphasize the opening up of walls to visually enhance or combine space, I believe you will find this to be extremely valuable in your pursuit of the open plan. It also will prove helpful during construction if you suddenly realize a wall is bearing, or if you decide to open a wall more than originally planned. A discussion on how to use the charts is included therein. These are obviously generalized conditions, but hopefully you should find that a condition of yours will fall somewhere within those tabulated.

The details on staging present proposed staging plans for two plans shown in prior chapters. These clearly illustrate how each remodeling can proceed in several predefined steps that will enable you to cope with the construction.

FOUNDATION DETAILS

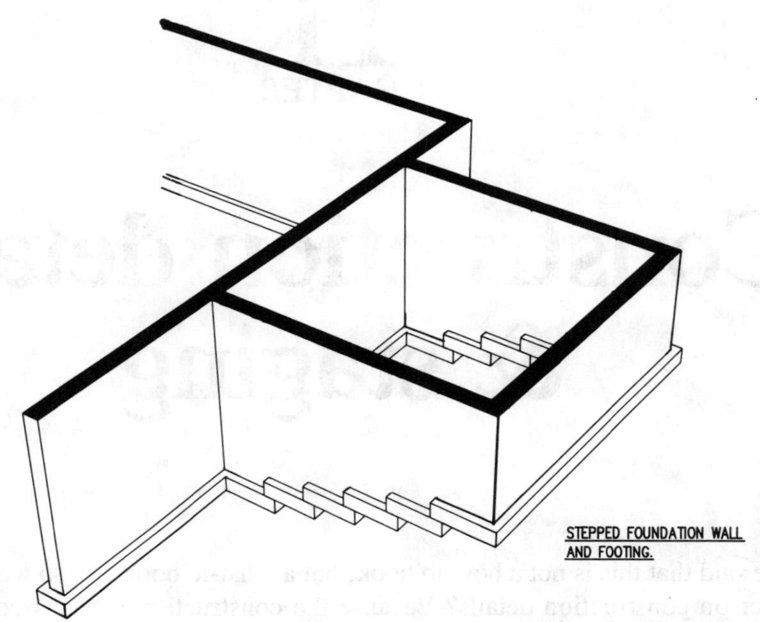

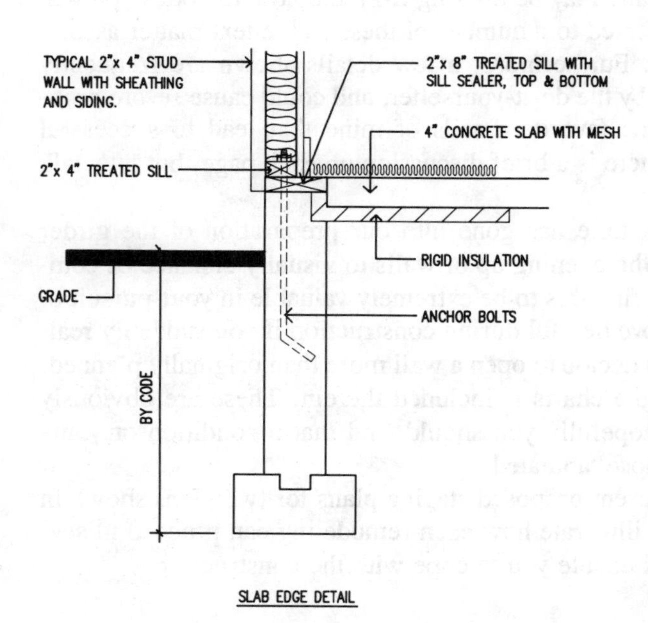

A common error in building an addition is the failure to step the footing and foundation wall to the depth of the existing footing and foundation. Why? To prevent the addition from settling independently of the existing house. The process is always necessary when adding an addition on a crawlspace adjacent to a basement. The process requires that you excavate to the top of the existing footing and then reduce the excavation to the minimum required by your code, or that required for the crawlspace.

If you intend to build with slab construction, I have developed a wall detail that I recommend you consider. The main purpose of the detail is to provide a positive method for insulating the edge of the slab from the exterior. It may not be necessary in southern climates, but is important in colder areas. The detail adds 2 inches extra in height to the exterior wood wall and does leave a strip around the perimeter in wood, rather than concrete. However, this strip is perfect for nailing your carpet into, and will not be noticed under ceramic tile.

DOUBLE FLOOR CONSTRUCTION

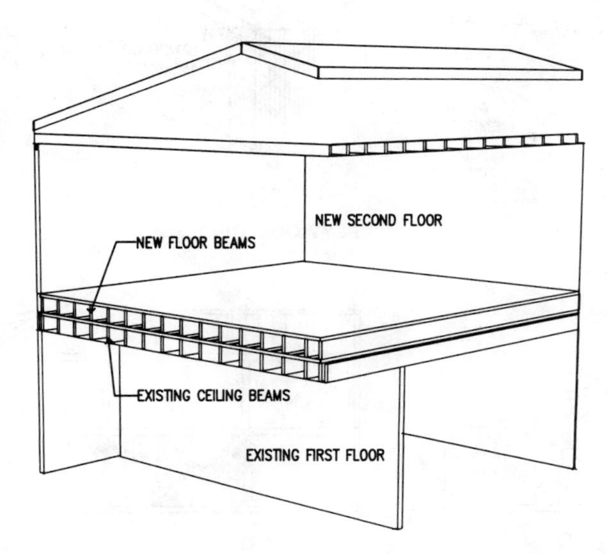

CUT-AWAY VIEW

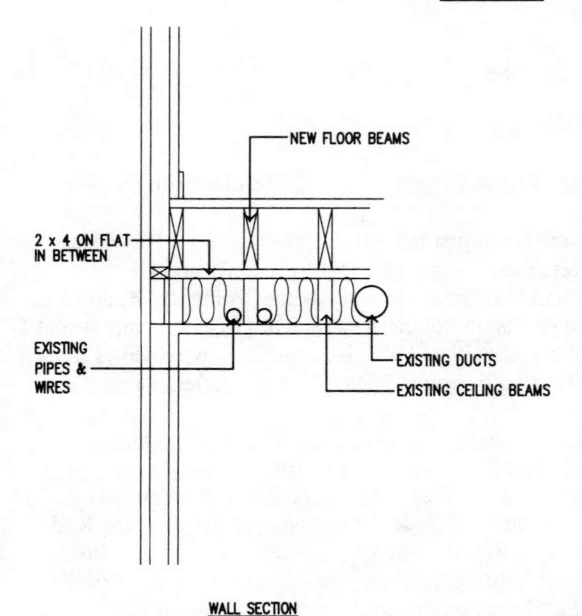

WALL SECTION

When adding a new addition up and over existing rooms it could be very advantageous to consider using a double floor. Why? It is likely that your existing ceiling beams are inadequate to support a floor, so new beams will be needed. If that is so, these new beams could be set with their bottom flush with the existing ceiling beams and extend higher as required. However, this frequently requires the cutting of wires and the relocation of pipes and ducts. It is also very messy and does pose a greater disruption factor to the rooms below.

By setting the new beams starting 2 inches above the existing ceiling beams, there is no need to relocate any wiring, light fixtures, pipes, etc. It also creates an excellent sound insulating floor. There are two potential problems you must consider though. The stair to this new floor will be longer; it will require at least one extra step, maybe two; and you will need the floor space to accommodate this stair. The second factor to consider is the extra height added to the house, which could create an aesthetic concern, especially if the addition is a small element on a very large roof, something I cautioned about back in chapter 3.

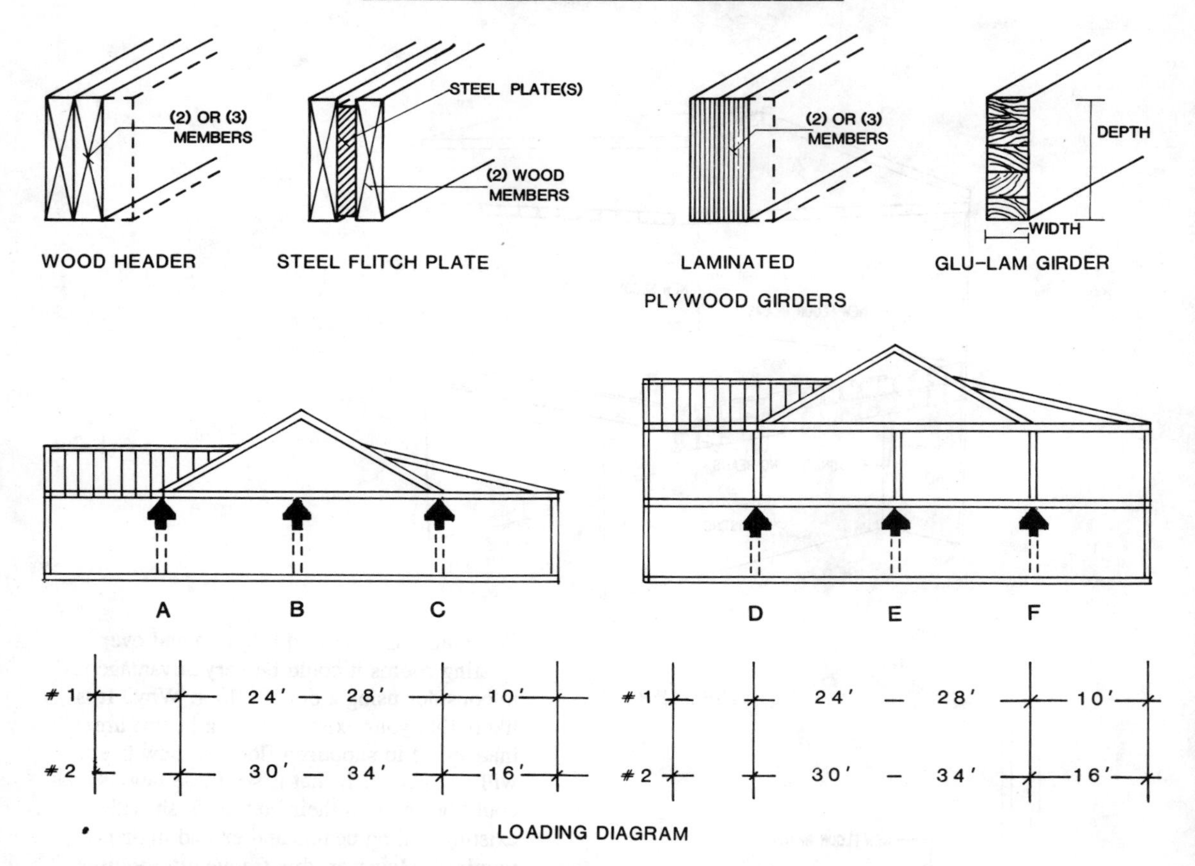

The girder details pictured above illustrate the most common types of wood girders. The charts on the next page are based on these types.

Various loading conditions found in a typical home are shown just below the girder details;. It is necessary for you to pick the correct location. Points A, B, and C represent ceiling and roof loads only and are for a one-story home without any second floor, and for the second floor of a two-story home. Points D, E, and F reflect conditions that also carry a second floor load. The charts do not cover basement girders. Points A and D reflect an outside wall with no loads from any new addition (these would also be used for new windows in an existing outside wall). Points B and E represent a center bearing partition. Points C and F reflect an existing outside wall that will also carry loads from a new addition.

Condition #1 reflects a home 24 to 26 feet wide and an addition no deeper than 10 feet. Condition #2 reflects a home 30 to 34 feet wide and an addition up to 16 feet deep. The three charts on the following page provide girder size information based on the loading diagrams and the span of the girder. The snow load 0-30-60 is the only variable you must obtain from your local building code. Enter the appropriate snow load chart once you have selected the loading condition that corresponds to your specific location. Cross-reference your loading condition A, B, C, etc., with the width of your house condition indicated at positions #1 or #2 in the diagrams. Then select a girder from the snow load chart based on the approximate opening (8 ft., 12 ft., or 16 ft.) that your girder will span. These girder openings reflect typical conditions and may vary slightly. If any of your actual conditions fall in the middle, opt for safety and go to the higher requirement.

0 SNOW LOAD

	8 FT. OPENING				12 FT. OPENING				16 FT. OPENING			
	WOOD HEADER	STEEL * FLITCH PL	LAM-PLYWOOD	GLU-LAM	WOOD HEADER	STEEL * FLITCH PL	LAM-PLYWOOD	GLU-LAM	WOOD HEADER	STEEL * FLITCH PL	LAM-PLYWOOD	GLU-LAM
A-1	(2) 2x 8	—	—	—	(3) 2x10	1/4"x 9"	(2) 1 3/4"x 9 1/2"	5 1/8"x 9"	(3) 2x12	1/4"x 11"	(2) 1 3/4"x 11 7/8"	3 1/8"x 12"
B-1	(2) 2x10	—	(2) 1 3/4"x 7 1/4"	5 1/8"x 6"	(3) 2x12	1/4"x 11"	(3) 1 3/4"x 9 1/2"	3 1/8"x 10 1/2"	—	1/2"x 11"	(3) 1 3/4"x 11 7/8"	3 1/8"x 13 1/2"
C-1	(2) 2x10	—	—	3 1/8"x 6"	(3) 2x10	1/4"x 9"	(2) 1 3/4"x 9 1/2"	3 1/8"x 9"	—	1/2"x 9"	(2) 1 3/4"x 11 7/8"	3 1/8"x 12"
D-1	(3) 2x10	1/4"x 9"	(3) 1 3/4"x 7 1/4"	3 1/8"x 7 1/2"	—	1/2"x 9"	(2) 1 3/4"x 11 7/8"	5 1/8"x 9"	—	(2) 1/2"x 9"	(3) 1 3/4"x 14"	6 3/4"x 12"
E-1	(3) 2x12	1/4"x 11"	(2) 1 3/4"x 9 1/2"	5 1/8"x 7 1/2"	—	(2)1/2"x 9"	(3) 1 3/4"x 11 7/8"	6 3/4"x 10 1/2"	—	t	(3) 1 3/4"x 16"	5 1/8"x 15"
F-1	(3) 2x12	1/4"x 11"	(2) 1 3/4"x 9 1/2"	5 1/8"x 7 1/2"	—	1/2"x 11"	(3) 1 3/4"x 11 7/8"	5 1/8"x 9"	—	(2) 1/2"x 11"	(3) 1 3/4"x 16"	6 3/4"x 13 1/2"
A-2	(3) 2x 8	—	(2) 1 3/4"x 7 1/4"	3 1/8"x 6	(3) 2x10	1/4"x 9"	(2) 1 3/4"x 9 1/2"	3 1/8"x 9"	—	1/2"x 9"	(3) 1 3/4"x 11 7/8"	5 1/8"x 10 1/2"
B-2	(2) 2x10	1/4"x 7"	(2) 1 3/4"x 7 1/4"	3 1/8"x 7 1/2"	(3) 2x12	1/4"x 11"	(2) 1 3/4"r 11 7/8"	5 1/8"x 9"	—	3/4"x 9"	(2) 1 3/4"x 14"	5 1/8"x 12"
C-2	(2) 2x10	1/4"x 7"	(2) 1 3/4"x 7 1/4"	5 1/8"x 6	(3) 2x12	1/4"x 11"	(2) 1 3/4"x 11 7/8"	5 1/8"x 9"	—	3/4"x 9"	(3) 1 3/4"x 11 7/8"	5 1/8"x 12"
D-2	(3) 2x12	1/4"x 9"	(2) 1 3/4"x 9 1/2"	3 1/8"x 9"	—	1/2"x 11"	(3) 1 3/4"x 11 7/8"	5 1/8"x 10 1/2"	—	3/4"x 11"	(3) 1 3/4"x 14"	5 1/8"x 13 1/2"
E-2	—	1/2"x 9"	(2) 1 3/4"x 11 7/8"	5 1/8"x 9"	—	3/4"x 11"	(3) 1 3/4"x 14"	5 1/8"x 13 1/2"	—	t	—	6 3/4"x 15"
F-2	—	1/2"x 9"	(3) 1 3/4"x 9 1/2"	5 1/8"x 9"	—	3/4"x 11"	(3) 1 3/4"x 14"	5 1/8"x 12"	—	t	—	5 1/8"x 16 1/2"

t Indicates use of a steel "I" beam

30 SNOW LOAD

	8 FT. OPENING				12 FT. OPENING				16 FT. OPENING			
	WOOD HEADER	STEEL * FLITCH PL	LAM-PLYWOOD	GLU-LAM	WOOD HEADER	STEEL * FLITCH PL	LAM-PLYWOOD	GLU-LAM	WOOD HEADER	STEEL * FLITCH PL	LAM-PLYWOOD	GLU-LAM
A-1	(2) 2x12	1/4"x 9"	(2) 1 3/4"x 9 1/2"	3 1/8"x 7 1/2"	—	1/2"x 9"	(2) 1 3/4"x 11 7/8"	5 1/8"x 10 1/2"	—	(2) 1/2"x 9"	—	5 1/8"x 13 1/2"
B-1	(2) 2x10	—	(2) 1 3/4"x 7 1/4"	5 1/8"x 6"	(3) 2x12	1/4"x 11"	(3) 1 3/4"x 9 1/2"	3 1/8"x 10 1/2"	[1] (3) 2"x 14"	1/2"x 11"	(2) 1 3/4"x 14"	3 1/8"x 13 1/2"
C-1	(3) 2x12	1/4"x 11"	(2) 1 3/4"x 9 1/2"	5 1/8"x 7 1/2"	—	1/2"x 11"	(2) 1 3/4"x 14"	6 3/4"x 10 1/2"	—	(2) 1/2"x 11"	(2) 1 3/4"x 16"	5 1/8"x 15"
D-1	(3) 2x12	1/4"x 11"	(2) 1 3/4"x 9 1/2"	5 1/8"x 7 1/2"	—	(2) 1/2"x 9"	(2) 1 3/4"x 14"	6 3/4"x 10 1/2"	—	(2) 1/2"x 11"	(3) 1 3/4"x 18"	5 1/8"x 15"
E-1	(3) 2x12	1/4"x 11"	(2) 1 3/4"x 9 1/2"	5 1/8"x 7 1/2"	—	(2) 1/2"x 9"	(3) 1 3/4"x 11 7/8"	6 3/4"x 10 1/2"	—	(2) 1/2"x 11"	(2) 1 3/4"x 16"	5 1/8"x 15"
F-1	—	1/2"x 11"	(3) 1 3/4"x 9 1/2"	6 3/4"x 9"	—	(2) 1/2"x 11"	(3) 1 3/4"x 14"	6 3/4"x 12"	—	t	—	6 3/4"x 16 1/2"
A-2	(3) 2x12	1/4"x 9"	(2) 1 3/4"x 9 1/2"	5 1/8"x 7 1/2"	—	1/2"x 11"	(2) 1 3/4"x 14"	5 1/8"x 10 1/2"	—	(2) 1/2"x 11"	—	5 1/8"x 15"
B-2	(2) 2x10	1/4"x 7"	(2) 1 3/4"x 7 1/4"	5 1/8"x 6"	(3) 2x12	1/4"x 11"	(2) 1 3/4"x 11 7/8"	3 1/8"x 10 1/2"	[1] (3) 2 x 14	3/4"x 9"	(2) 1 3/4"x 14"	3 1/8"x 13 1/2"
C-2	—	1/2"x 9"	(2) 1 3/4"x 11 7/8"	6 3/4"x 7 1/2"	—	3/4"x 11"	(2) 1 3/4"x 16"	5 1/8"x 12"	—	t	—	5 1/8"x 16"
D-2	—	1/2"x 9"	(3) 1 3/4"x 9 1/2"	6 3/4"x 7 1/2"	—	3/4"x 11"	(3) 1 3/4"x 14"	6 3/4"x 12"	—	t	—	5 1/8"x 16"
E-2	—	1/2"x 9"	(2) 1 3/4"x 11 7/8"	6 3/4"x 7 1/2"	—	3/4"x 11"	(3) 1 3/4"x 14"	6 3/4"x 12"	—	t	(3) 1 3/4"x 18"	6 3/4"x 15"
F-2	—	1/2"x 11"	(2) 1 3/4"x 11 7/8"	5 1/8"x 12"	—	(2) 1/2"x 11"	(3) 1 3/4"x 16"	6 3/4"x 15"	—	t	—	6 3/4"x 19 1/2"

t Indicates use of steel "I" beam

1 Indicates use of higher grade lumber

60 SNOW LOAD

	8 FT. OPENING				12 FT. OPENING				16 FT. OPENING			
	WOOD HEADER	STEEL * FLITCH PL	LAM-PLYWOOD	GLU-LAM	WOOD HEADER	STEEL * FLITCH PL	LAM-PLYWOOD	GLU-LAM	WOOD HEADER	STEEL * FLITCH PL	LAM-PLYWOOD	GLU-LAM
A-1	(3) 2x12 [1]	1/4"x 11"	(2) 1 3/4"x 9 1/2"	5 1/8"x 7 1/2"	—	1/2"x 9"	(2) 1 3/4"x 14"	6 3/4"x 10 1/2"	—	(2) 1/2"x 11"	(3) 1 3/4"x 16"	5 1/8"x 15"
B-1	(2) 2x10	—	(2) 1 3/4"x 7 1/4"	5 1/8"x 6"	(2) 2x14	1/4"x 11"	(2) 1 3/4"x 11 7/8"	3 1/8"x 10 1/2"	[1] (3) 2x14	1/2"x 11"	(2) 1 3/4"x 14"	3 1/8"x 13 1/2"
C-1	—	1/2"x 11"	(2) 1 3/4"x 11 7/8"	6 3/4"x 9"	—	(2) 1/2"x 11"	(2) 1 3/4"x 16"	6 3/4"x 12"	—	t	—	6 3/4"x 16 1/2"
D-1	(3) 2x14	1/2"x 9"	(2) 1 3/4"x 11 7/8"	5 1/8"x 9"	—	(2) 1/2"x 11"	(3) 1 3/4"x 14"	6 3/4"x 12"	—	t	(3) 1 3/4"x 18"	6 3/4"x 15"
E-1	(3) 2x12 [1]	1/4"x 11"	(2) 1 3/4"x 9 1/2"	5 1/8"x 7 1/2"	—	(2) 1/2"x 9"	(2) 1 3/4"x 14"	5 1/8"x 12"	—	(2) 1/2"x 11"	(3) 1 3/4"x 16"	5 1/8"x 15"
F-1	—	1/2"x 11"	(3) 1 3/4"x 11 7/8"	6 3/4"x 10 1/2"	—	t	(3) 1 3/4"x 16"	6 3/4"x 15"	—	t	—	—
A-2	(3) 2x12 [1]	1/2"x 9"	(2) 1 3/4"x 11 7/8"	6 3/4"x 7 1/2"	—	3/4"x 11"	(2) 1 3/4"x 16"	5 1/8"x 13 1/2"	—	t	(3) 1 3/4"x 18"	5 1/8"x 16 1/2"
B-2	(2) 2x10	1/4"x 7"	(2) 1 3/4"x 7 1/4"	5 1/8"x 6"	[1] (2) 2x14	1/4"x 11"	(2) 1 3/4"x 11 7/8"	3 1/8"x 10 1/2"	[1] (3) 2x14	3/4"x 9"	(2) 1 3/4"x 14"	3 1/8"x 13 1/2"
C-2	—	1/2"x 11"	(2) 1 3/4"x 11 7/8"	5 1/8"x 12"	—	(2) 1/2"x 11"	(2) 1 3/4"x 18"	6 3/4"x 15"	—	t	—	—
D-2	—	1/2"x 11"	(3) 1 3/4"x 9 1/2"	6 3/4"x 9"	—	(2) 1/2"x 11"	(3) 1 3/4"x 16"	6 3/4"x 13 1/2"	—	t	(4) 1/3/4"x 18"	6 3/4"x 16 1/2"
E-2	(3) 2x14	1/2"x 9"	(3) 1 3/4"x 9 1/2"	6 3/4"x 7 1/2"	—	3/4"x 11"	(3) 1 3/4"x 14"	5 1/8"x 13 1/2"	—	t	(3) 1 3/4"x 18"	6 3/4"x 15"
F-2	—	3/4"x 11"	(3) 1 3/4"x 11 7/8"	6 3/4"x 12"	—	t	(3) 1 3/4"x 18"	6 3/4"x 18"	—	t	—	—

t Indicates use of steel "I" beam

1 Indicates use of higher grade lumber

---

1 Uniform loading condition assumed for all girders. Deflection limit = L/360.

2 Lumber species - Hem Fir, #2 or better, f=1400 psi for #1, 1150 psi for #2. E=1.5 psi for #1, 1.4 psi for #2. Higher stress lumber such as D.Fir, S.Y. Pine may be substituted. (National Forest Products Association.)

3 Steel - ASTM A36, Fy (min. yield stress) = 36 ksi, allowable stress = .60 Fy = 22 ksi. (American Institute of Steel Construction, Inc.) Steel Flitch Plates to be assembled with min. 3/8" bolts staggered at 2'0" o.c., top and bottom.

4 Laminated Plywood - Fb-2800 psi, Fv=285 psi, E=2.0 psi. Nailing pattern for assembly of multiple units to be min. (2) rows of 16d nails @ 12" o.c. (3) rows of 16d nails @ 12" o.c. for 14", 16", 18" beams (American National Standards Institute, Inc.).

5 Glu-Lam Timber - Western Species - 1 1/2" thick laminations, Fb=2400 psi, Fv=165 psi, E=1.8 psi (American Institute of Timber Construction).

6 Floor loads: 40 psf live, 10 psf dead—Ceiling loads: 20 psf live, 10 psf dead

ROOF BREAK DETAILS

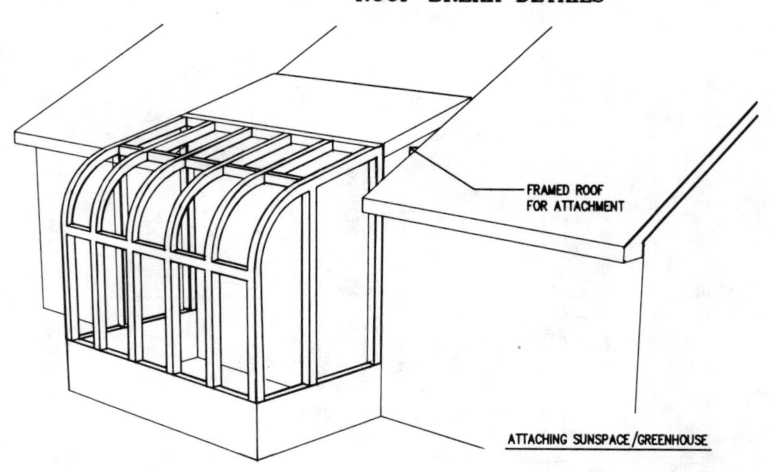

Additions frequently cause a break to occur in a roofline; this is often a desired effect that is inherent to the design. At other times, particularly when the addition has not had proper architectural preplanning, it could be an unfortunate occurrence. The upper figure shows a roof break that is required to attach a greenhouse structure to a one- or one-and-one-half-story home. This little detail is often not planned for by the untrained person. If properly treated with matching materials, it can look just fine, but be aware of its necessity.

When a jog in a plan occurs, a roof break is likely. The size of this break is directly related to the size of the jog in plan and the pitch of the roof. A one-foot setback on an addition with a 6:12 roof pitch with produce a 6-inch roof break. That's fine, but a 4-inch setback on the same addition will produce a 2-inch roof break, which, in most instances, is unacceptable. The only exception would be a hip-roof offset. A two-inch break against a wall is difficult to flash well and will be too insignificant—it might even look like an error. So be careful when adding additions. Either match the rooflines, or make certain that there is ample offset.

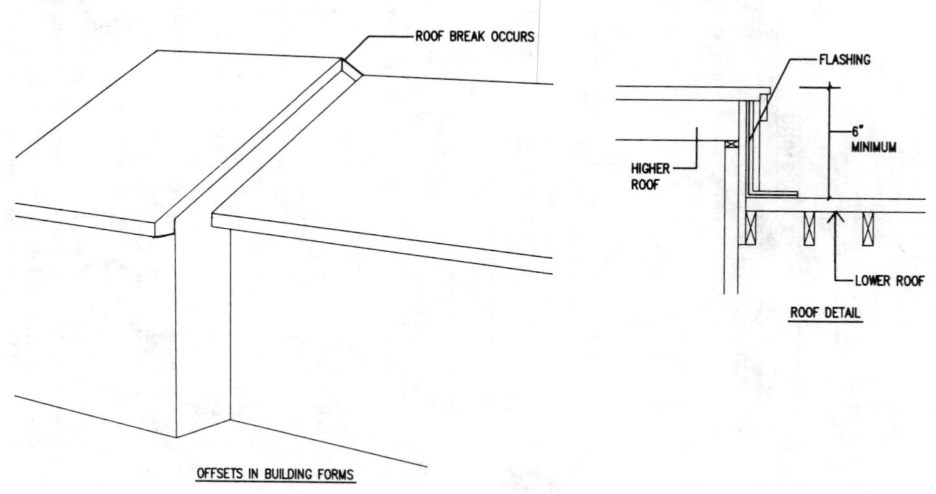

DOUBLE ROOF CONSTRUCTION

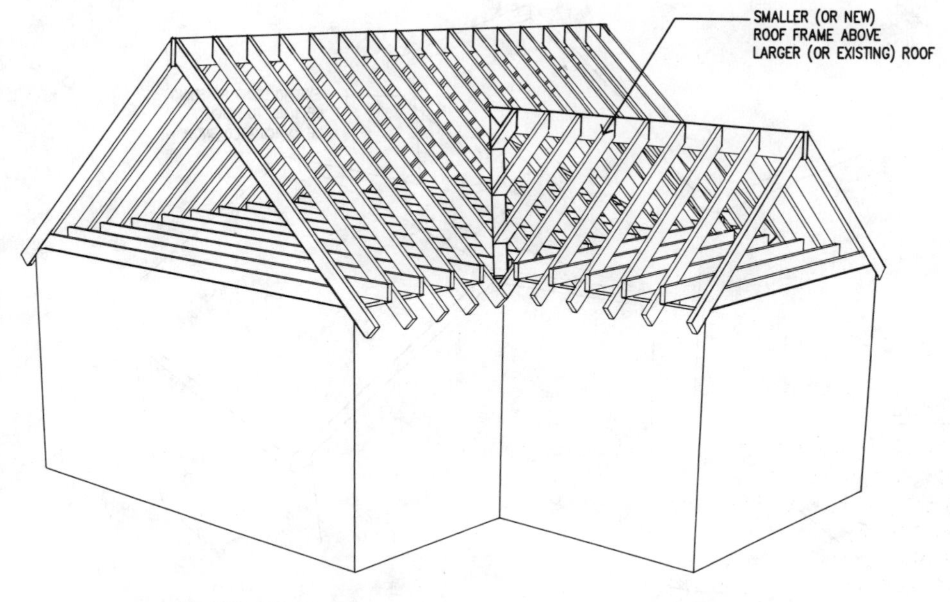

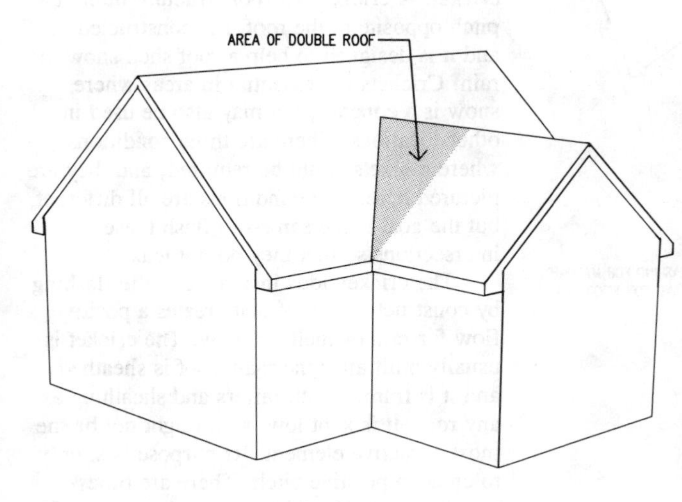

Whenever possible, most contractors and carpenters will use double-roof construction today when framing intersecting rooflines. Many do-it-yourself contractors do not understand it. You have read about structural valleys and headers and are ready to cut apart an old roof, or even frame two new ones that way. It is not necessary. The practice pictured builds one roof, usually the smaller, above the other roof. The rafters of the upper roof are attached to nailing strips that run along the lower roof; these nailing strips replace the structural valleys of the old practice.

This method is easier and quicker. It does require extra lumber for rafters, but eliminates the need for structural valleys. It also is easier to frame a room with a vaulted ceiling that runs past an intersecting roofline. In such a case, you might frame the smaller (vaulted ceiling) roofline first and frame the main roof over, but make sure you ensure against the outward spreading of the lower roof.

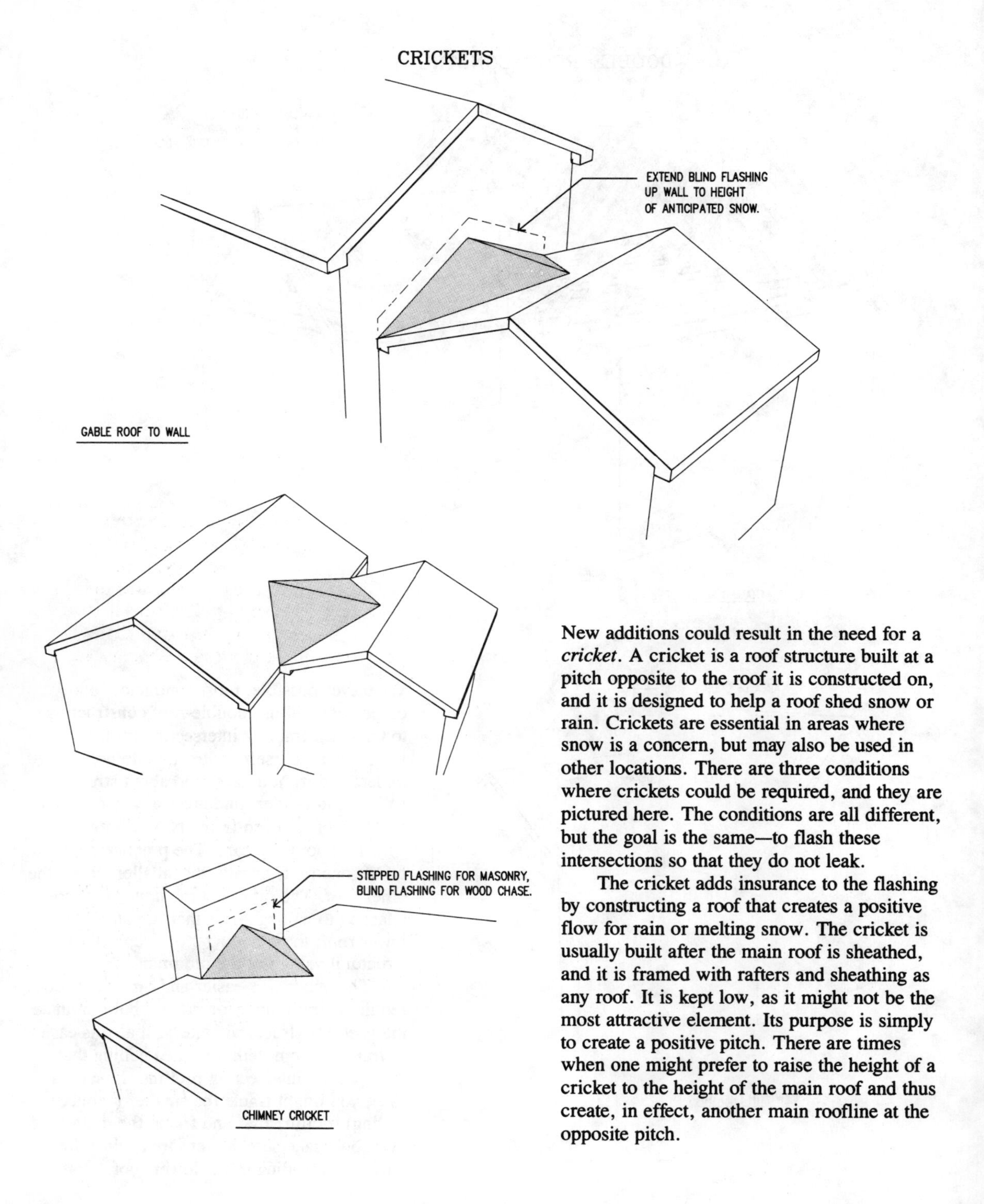

New additions could result in the need for a *cricket*. A cricket is a roof structure built at a pitch opposite to the roof it is constructed on, and it is designed to help a roof shed snow or rain. Crickets are essential in areas where snow is a concern, but may also be used in other locations. There are three conditions where crickets could be required, and they are pictured here. The conditions are all different, but the goal is the same—to flash these intersections so that they do not leak.

The cricket adds insurance to the flashing by constructing a roof that creates a positive flow for rain or melting snow. The cricket is usually built after the main roof is sheathed, and it is framed with rafters and sheathing as any roof. It is kept low, as it might not be the most attractive element. Its purpose is simply to create a positive pitch. There are times when one might prefer to raise the height of a cricket to the height of the main roof and thus create, in effect, another main roofline at the opposite pitch.

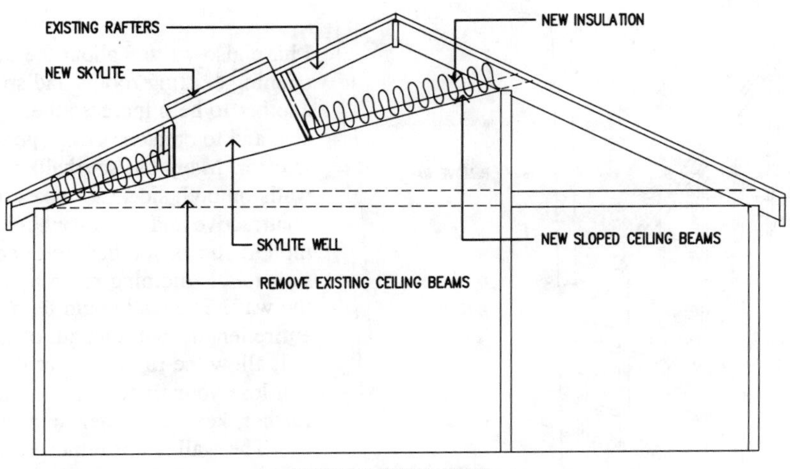

VAULTING AN EXISTING CEILING

I have often written about the desirability of adding volume to spatially enhance a room. This detail shows you how to do it over an existing room covered by conventional ceiling beams and rafters. A trussed roof requires rebuilding the trusses, but a skylight could be installed as shown here. New ceiling beams at a lower slope, as pictured, will provide the depth for good roof insulation and also will prevent against the possible bowing of the outside wall. The skylight well should be splayed as shown to spread the light. The new ceiling beams should only be eliminated if the existing roof is supported against spreading out, and if the rafters can provide ample space for insulation.

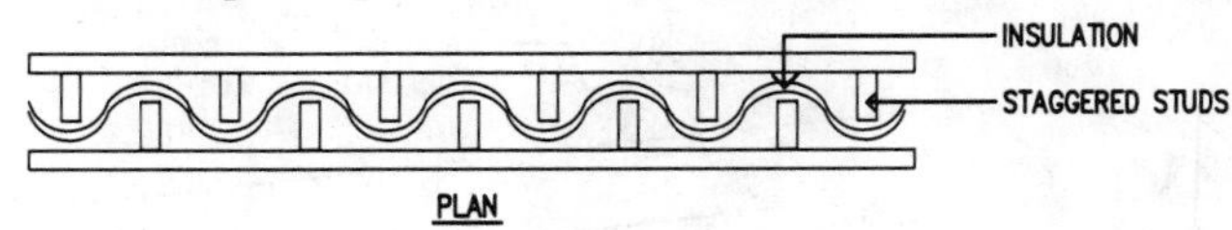

PLAN

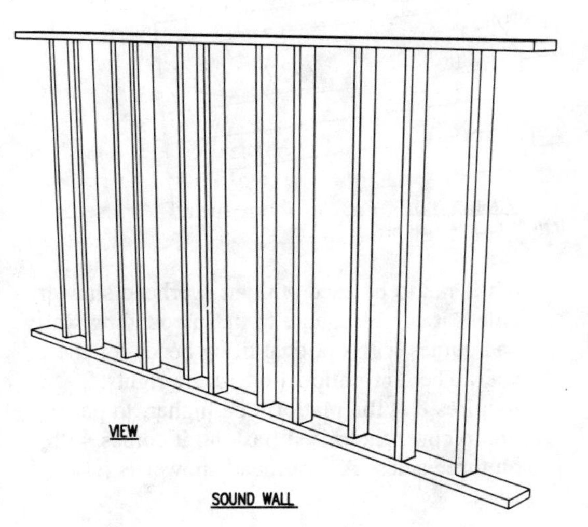

SOUND WALL

Sometimes the remodeling of a home may place a bedroom near an activity room and sound insulation would be a nice luxury. Forget about just insulating the wall; it helps, but it is not enough. The best solution is two separate, fully independent walls with an air space between, but this could take up too much area. The next best is the staggered stud wall illustrated here. Other than at the top and bottom, there are no surfaces that will carry sound from one side to the other. The insulation helps by weaving it through the cavity. Also remember to keep any ducts or pipes out of this wall, and don't place electrical outlets back to back.

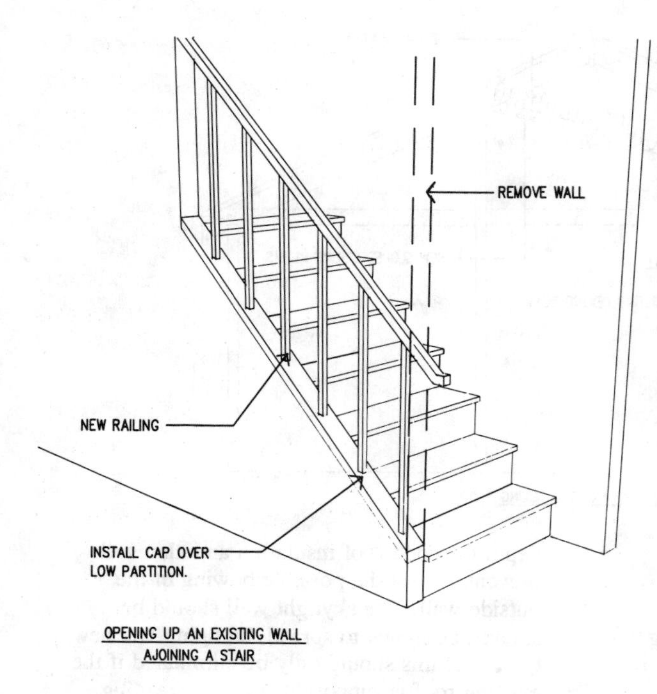

OPENING UP AN EXISTING WALL
AJOINING A STAIR

I have also written about the desirability of opening existing rooms and spaces to one another to help increase their impression of size and to create a more open flowing interior. Many old stairhalls are hemmed by walls on both sides; it makes them unattractive and claustrophobic. The detail at the left shows you how to open the side of a stair to an adjoining room by removing part of the wall. The wall could be removed for its entire length, but a length of six to seven steps will allow the rail to end at the wall before you lose your fingers. If you do extend further, keep this latter item in mind.

The wall is not removed to the floor but will stop at the top of the stringer alongside the steps and will be capped there with an oak piece. This will allow the stair to remain untouched. If the wall extends lower, the side of the stair may have to be refinished. You could open both sides of a stair the same way.

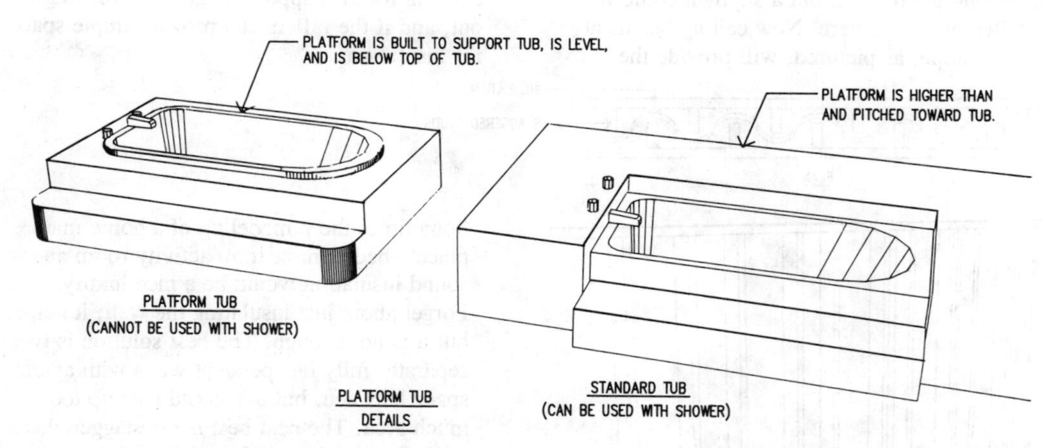

PLATFORM TUB
DETAILS

If you plan on having a platform tub the carpenter will have to know how to frame the platform for it. There is confusion on the subject, as there are two distinct types of platforms, which depend on the type of tub you choose. The platform at the left is for a tub that does not have a skirt and that is intended to sit on top of a platform. Such a tub is not to be used with an overhead shower, unless it can be square to its surrounding walls and comes with optional tiling beads for the wall. The conventional tub at the right requires that the platform be higher, to permit tile to cover the raised flashing it comes with on three sides. An overhead shower is just fine.

I have referred to staging many times. What is it? *Staging* refers to the construction of a remodeling project in stages or parts. It results in every project as a consequence of the fact that certain tasks cannot begin until others are complete. For example, interior floors cannot be installed until the roof is made watertight. Staging can also result from a distinct preplanning effort that focuses on completing various parts or tasks separately and in a predetermined order. It is this latter aspect of staging that I am focusing on.

There are two reasons to plan for staging. The first is to enable you to maintain sanity while living through an extensive interior remodeling project. The goal is to plan the construction in such a fashion that you can live as comfortably as possible while the seemingly endless mess is around you. To do this properly requires that this subject be part of the design program, and that as a design is being developed, staging becomes one of the issues factored into the design equation.

The plan is usually to start with the exterior shell of the new additions while not opening up to the current rooms. You might have to stay out of an existing room (or two) during the day, as a safety precaution, but you retain its evening functionality. Frequently, the next step involves the finishing of the addition, so that it can be made livable. The doors or connections can wait or be installed now, depending on the project. You then move into the newly finished rooms and vacate some parts of the old house so that the interior can be changed. You screen off the parts being worked on as best as possible. The number of moves and steps required will be dependent on the complexity of the interior renovation, the speed of your construction capability, and your pocketbook.

Kitchens and bathrooms present unique problems if they are to be renovated. The easiest solution is to create new rooms at first so they can be used while the project continues. That is all right if you are building a new kitchen. If it will be remodeled in place, it has to be carefully coordinated so that "down time" of the kitchen is minimized. The same required coordination is absolutely necessary when remodeling existing bathrooms if there are no other baths to utilize.

There is a second reason to plan for staging, and I have already alluded to it. It is the cost factor. It could be that your wish list far exceeds your capacity to pay, but you are unwilling or unable to pare it down. Staging could present a viable solution so that you ultimately achieve all the items your needs demand or your home requires. The major difference between this type of staging and the prior one is that much longer gaps of time enter the planning process, and you have to consider the potential living condition of your home at each step. But it might be worth it.

The plans on the next two pages present some graphic examples of the subject.

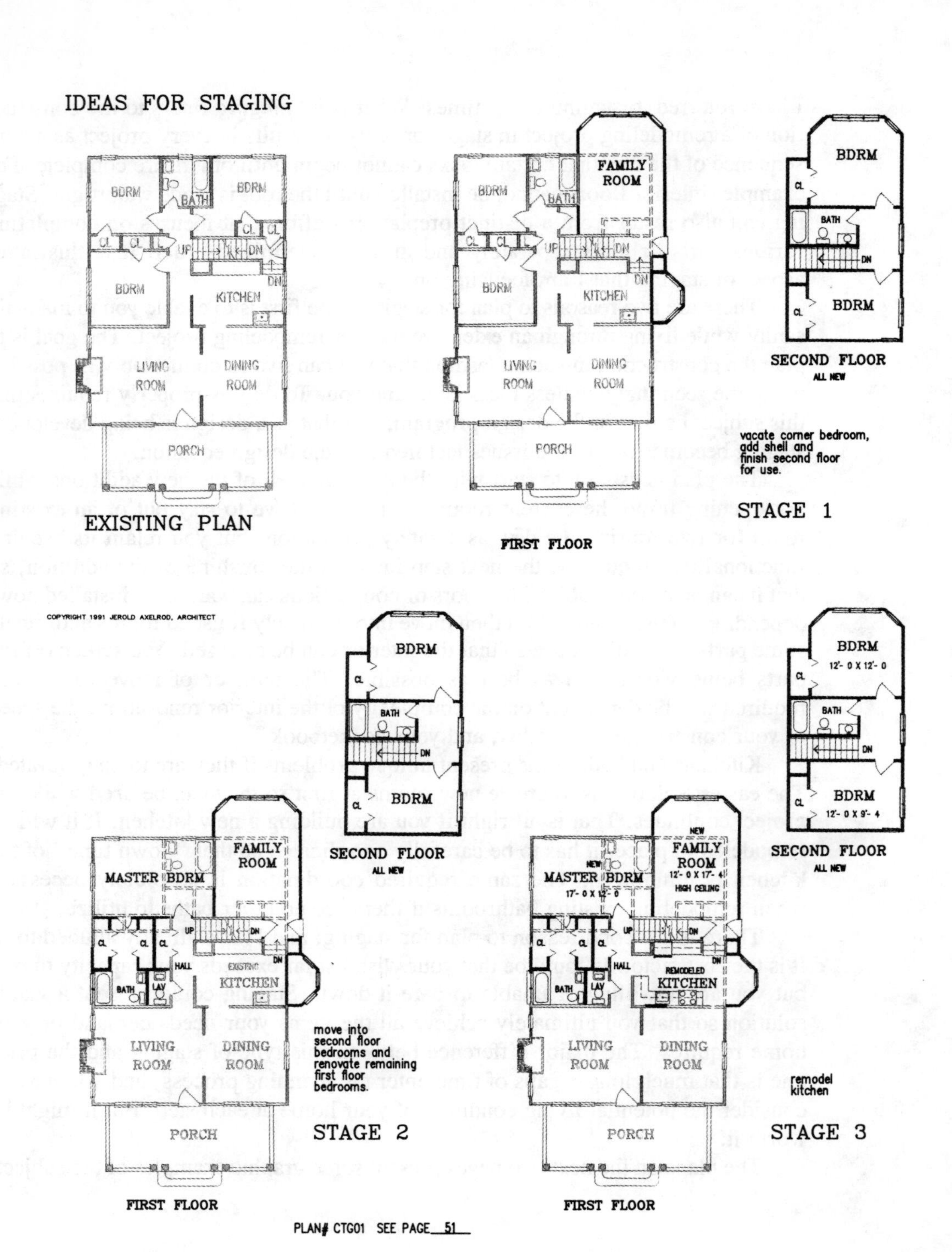
IDEAS FOR STAGING
EXISTING PLAN
BDRM
BATH
BDRM
CL
UP
DN
BDRM
KITCHEN
LIVING ROOM
DINING ROOM
PORCH
COPYRIGHT 1991 JEROLD AXELROD, ARCHITECT
FAMILY ROOM
FIRST FLOOR
SECOND FLOOR
ALL NEW
vacate corner bedroom, add shell and finish second floor for use.
STAGE 1
MASTER BDRM
EXISTING KITCHEN
HALL
LAV
move into second floor bedrooms and renovate remaining first floor bedrooms.
STAGE 2
NEW
12'- 0 X 17'- 4
HIGH CEILING
17'- 0 X 12'- 0
REMODELED KITCHEN
12'- 0 X 12'- 0
12'- 0 X 9'- 4
remodel kitchen
STAGE 3
PLAN# CTG01 SEE PAGE 51

IDEAS FOR STAGING

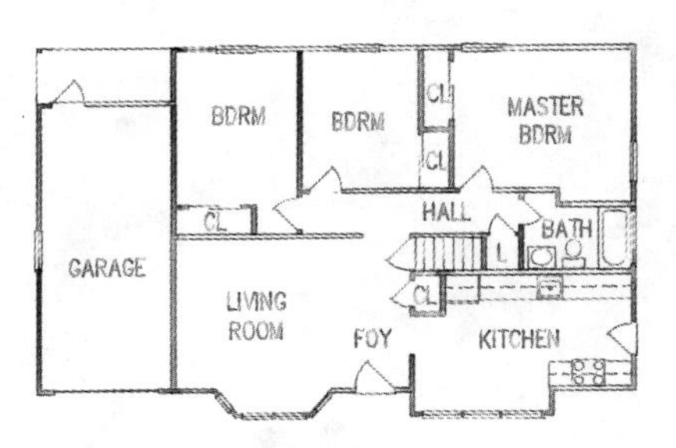

EXISTING PLAN

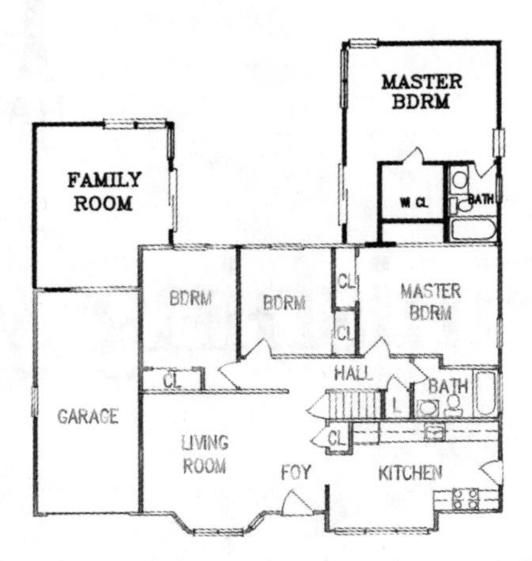

add shell for additions
and finish master suite

STAGE 1

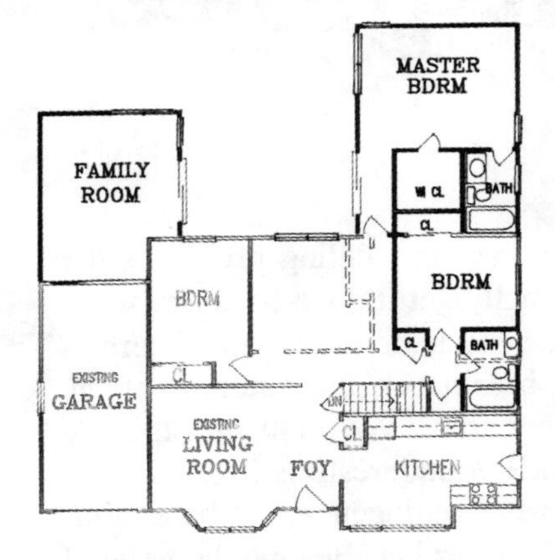

move into new master bedroom,
vacate center and old master bedrooms.
finish new middle bedroom and bath.

STAGE 2

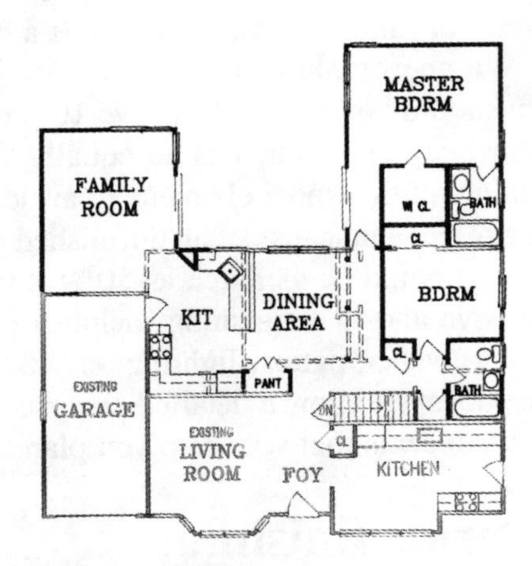

finish family room, new
kitchen and dining area.

STAGE 3

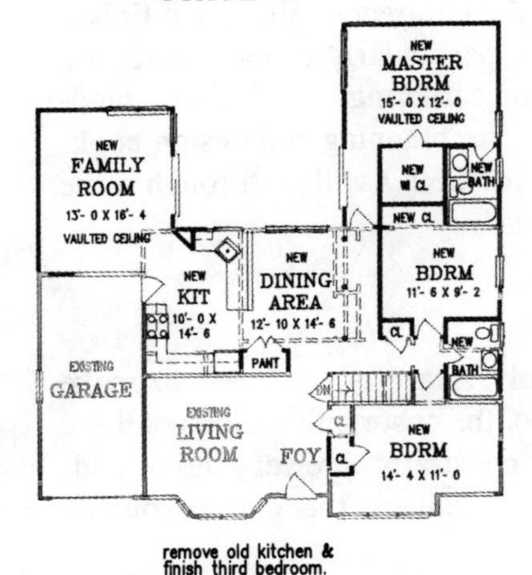

remove old kitchen &
finish third bedroom.

STAGE 4

PLAN# RAN05 SEE PAGE 286

12
CHAPTER

Finishing your project

One of the guaranteed, natural occurrences of every remodeling project is the "finishing squeeze." There are the pressures to finally get it done because, without doubt, it will take longer than you or your contractor estimated. There is another squeeze, though, that has more crucial bearing on the outcome of the project, and it, too, mounts in tension as the project draws to conclusion. It is the scarcity of funds that can develop as a consequence of underestimating.

Whomever you consult will advise you to leave a contingency; the bare minimum should be 5%, but I believe 10% or more is a safer bet. Beyond the issue of contingency, however, lies an equally significant concern—that of properly preplanning all the finish elements to avoid having your project end in an unfinished state. A frequent cause of an unfinished project is the failure to realize, that finishing costs could be as much as 50% of the project, particularly if one's tastes are expensive and that finishing includes planning for carpeting, furnishings, landscaping, walks, patios, lighting, etc. Because this is a planning and design book, space does not allow a detailed analysis of the subject, but I will go through some broad concepts that will help you plan to get it done.

Exterior finishing

If you are building an addition, you are certainly aware of the need to make choices about windows and siding materials, and of the cost ramifications of these choices. I am an advocate of suggesting, if budget constraints are very tight—and, they frequently are—that you postpone those finish elements that can reasonably be accomplished later.

However, I do not place siding or windows in that category. There are times when someone will want stone veneer but can only afford aluminum siding, so they request that a stone ledge be prepared in the foundation to accommodate future stone. Do it if you want, but the likelihood is that you will never add the stone, as the siding of your choice will grow on you.

So choose your siding carefully; give thought to its texture and its color. Make sure that the color harmonizes with the color of your windows and roof. Obtain samples of all three—plus the exterior fascia or trim—and put them side by side, just as you would match a shirt, tie, and jacket. This is at times a difficult decision for many people. Photographs in magazines can help enormously, as can a trip around the neighborhood or to a new subdivision.

Window choices can be equally perplexing, especially because of the proliferation of styles, types, and finishes readily available today. Again, this is not the place for a window choice dissertation, but for some design concepts. If your tastes are very traditional, double-hung windows are a likely choice, and they are comparatively inexpensive. That should not rule out casements, awning, or sliding windows, though, particularly if you install them with divided lights. Sliding windows tend to be the least expensive, and casements are the most expensive because of the hardware. Give thought to using fixed windows, which can be less costly.

White is the preferred traditional window color; use darker trimmed windows for a contemporary exterior or for a rustic, woodsy look. But large areas of white framed glass can also be appropriate to a very modern look when used with white stained siding.

Do give thought to the use of those special shapes. They are expensive, but just one can become the focal point of a room. They can be used in any style of home. In a more contemporary home, use them without divided lights. For an eclectic look, mix them with divided lights over large, undivided windows.

Roof choices are somewhat easier. With the exception of those areas where wood shingles or clay tile are the preferred choices, fiberglass roof shingles are the norm. There is a wide choice of textures and colors with these. Pick the texture first, as it could be a significant cost factor. The heavy, layered look, which imitates wood shingles, comes in several styles, each with its own price and selection of colors.

There are a number of other final decisions that are necessary for you to make to totally preplan and budget for the exterior finish. An addition will require some new landscaping. It might call for a new patio, the replacement of walks and entrances, or maybe even a driveway. These must be considered and budgeted. You might be able to defer a deck or some landscaping for a year or so, but it could be disheartening if you didn't plan for it.

Finally, don't forget about exterior lighting; you could defer some fixtures for a while, if necessary, but remember to do the wiring with your project.

Interior finishing

The same philosophy about postponing certain elements, if necessary, applies to finishing the interior of your remodeling project. As with exterior finishes, however, there are certain areas where you should not plan to use temporary finishes. Items that will be built-in should be permanent; study your choices well. Compromise as necessary, but be confident with your compromise. You do not want to replace kitchen cabinets, plumbing fixtures, or other built-ins again.

Carefully examine the cost of appliances for your proposed new kitchen. Top-of-the-line appliances can cause sticker shock, yet you cannot postpone that choice. What is the point of a new kitchen with old, dated appliances? So budget carefully for these. Your choice of cabinets is a significant cost factor; an imported mica cabinet could be ten times the cost of a simple domestic factory line.

There are similar cost variances in bathroom fixtures and fittings. Every manufacturer produces top-of-the-line fixtures and basic fixtures plus a whole range in between. Obtaining prices for these is not difficult. If you are working with a contractor, he will obtain them for you. If you are doing it yourself, your local plumbing supply company can provide you cost comparisons. Please don't forget about bathroom fittings. There is such an exciting variety of choices available today, and none are necessarily budget-shattering.

Plan for built-in lighting with your project. It is another item that cannot be readily postponed. These are all items that have a significant effect on your budget, and which you need to reconcile before starting. Other items—such as floor finishes, wall paneling, wallpaper, and hanging light fixtures—could be deferred, if necessary. The only exception might be certain floorings, such as ceramic tile or wood, particularly if the floor is in a kitchen or bathroom; those are tough to defer. On the other hand, a ceramic tile family room or an oak living room could be done later, if necessary. If you are carpeting initially, you could pick a less expensive carpet with the intention of replacing it in a few years.

Finally let's talk about furnishings. A completely finished project would include all the elements necessary for you to enjoy the new or remodeled space, and that includes the furnishings. I strongly suggest budgeting for furnishing, even if it is only for a temporary purchase. If you plan for it in advance, whatever choices you make, you will be avoiding that "finishing squeeze."

Personal nuances

Whether or not you will be adding on, or remodeling from a design in this book, you will be embarking upon a project that will enable you to establish a stronger personal identity with your home. In earlier chapters, I spoke about exerting your personal preferences and establishing a program that satisfies your personal needs. I also spoke about the likelihood of making changes from these plans, and the special circumstances of your lot, which you should take advantage of. All of these deal with the same issue—that of creating a personally satisfying remodeling project.

The same holds true for this discussion on finishing. Exert your tastes and your personality in your choice of finishes. If you do so, you will likely enjoy a completed remodeling that will satisfy your fondest wishes and be a rewarding achievement for you and your family.

Glossary

accessible A term used to describe rooms or facilities designed for persons confined to a wheelchair.

angle bay An extension from a wall usually containing three sections of windows, the end two of which are at an acute angle to the wall.

as-of-right As used in zoning discussions, a property owner's rights as permitted by the local zoning ordinance.

barrier-free Rooms or facilities that are free of impediments to handicapped persons; *see* accessible.

borrowed light Natural light entering one room but originating in another, adjacent room.

box bay An extension from a wall comprised primarily of windows, the ends of which are perpendicular to the wall.

cantilever A part of a floor that extends out over the wall below; also used as a verb (*to cantilever*).

cape A type of house originating on Cape Cod; it is a one-and-one-half-story design containing steep rooflines with a smaller second floor footprint; also known as a *farm ranch, cottage* or *chalet*.

circulation The paths of travel used for travel from room to room within a home.

colonial A colloquial name used in certain parts of the country for a two-story type of home; is often confused with "colonial" styling.

compartmented bath A bathroom that contains at least two separate spaces, where certain fixtures (usually a water closet) are separated from others.

conical Pertaining to a multisided hip roof shaped in the form of a cone.

contemporary style A difficult term to define because of varying regional interpretations; contemporary in the midwest and east usually refers to a modern design that does not follow any classical or traditional styles; in the south and west it can frequently refer to the prevalent "in" style, which might be an eclectic style that borrows from traditional styles.

country kitchen A large multifunction space, usually containing a kitchen, a defined eating area, and a sitting area used for informal entertaining and family activities.

country style An informal exterior design and interior decorating style that provides a simple, comfortable, unpretentious quality with frequent references to the past; can be eclectic and very individual.

cricket A small, roofed structure set above a main roof to aid the flow of snow and rain from the roof.

dormer A small, roofed extension protruding through a roof from a second floor room, with window(s) at the end.

easement As used in property rights, an area of a lot that might be reserved for utilities or some other third party's use, on which the lot owner's rights of use may be restricted.

eaves The low end of a gable or hip roof.

eclectic style A design styling that draws from various sources; a blending of styles.

elevation (front, rear, etc.) The architectural term for the individual views of each side of a house drawn straight-on without perspective; also used for interior walls, cabinets, fixtures.

elevation (height) The height of individual elements of your home or property above some established datum.

elliptical Shaped in the form of an ellipse; a curved shape that is shallower than a circular form; used in windows.

farm ranch *See* cape.

fenestration An architectural term referring to the placement of windows in a building or wall.

frieze An element of exterior wall trim; usually a horizontal band directly below a roof or above an entrance; also known as a *cornice*.

gable end The vertical wall at the end of a gable roof, usually in the shape of a triangle.

gable roof A roof that slopes down in two directions from its top (or ridge) in two planes.

galley kitchen A narrow kitchen usually no more than 9 ft. wide with cabinets on two opposite sides.

grade The ground surrounding your house; also the slope of the ground.

great room A large room that combines several functions such as living/dining, kitchen/family, living/family or living/dining/kitchen, etc.

header A horizontal structural member carrying loads from other members; in a wall, it is the member over windows, doors, and other openings; the word is sometimes interchanged with *girder*, the difference being that a girder carries larger loads and is frequently exposed.

high ranch A type of house that has a one-story layout raised out of the ground above an exposed basement (also known as a *split-foyer* or *bilevel*).

hip roof A roof that slopes down from its peak in all directions, usually having four planes.

knee wall A low wall under a roof that forms the sides of second floor rooms with partially sloping ceilings.

modern style Interchanged with and like *contemporary*, it is not always a clearly agreed-upon term; more often than not it is a design that is devoid of any references to anything old, but that definition is changing.

peninsula kitchen A kitchen with cabinets that return from a wall into the center of a room.

pitch roof The pitch of a roof is the slope of a roof usually expressed in inches of height per 12 inches of width (i.e., 5 in 12).

post-modern style A current architectural style that mixes modern designs with older, classical elements; an eclectic style.

quoins A decorative treatment of the exterior corners of a house utilizing alternating raised blocks, usually done in masonry (brick, stone) or stucco.

ranch A colloquial name used in many parts of the country for a one-story type of home; also known as a *rambler*.

reverse gable A gable roof turned perpendicular to another gable roof, forming intersecting valleys with the main roof and resembling a T in plan.

ridge line The high point of a gable or hip roof; the top member is referred to as a ridge beam.

shed roof A roof that slopes in one direction only from its top to bottom, in one plane.

split-foyer *See* high ranch.

style The style of a house refers to exterior design and its related components (i.e., country, contemporary, tudor).

sunporch A term used for an enclosed porch with an abundance of glass in its walls and roof; also sometimes a *greenhouse*.

swale A low point or depression in the grade adjoining a home to carry water away from the structure.

transom A glazed light above a standard-height window or door, usually square-topped; curves and half-rounds are not usually referred to as transoms, although elliptical units are; it's another area where terminology is not uniform.

trombe wall A masonry wall usually located with a sunporch or greenhouse, its purpose being to store heat obtained by passive solar gain.

type The type of a house refers to the nature of the floor plan (i.e., one-story, two-story, split-level, etc.).

valley roof The line formed by two intersecting, downward-sloping rooflines.

variance In zoning, a waiver of the zoning ordinance that is obtained to permit construction otherwise not allowed by the ordinance.

water closet Technical jargon for a toilet.

Plans Index

General Index

Note to readers: Page numbers appearing in italics refer to illustrations in the book.

M

O

P

R

S

T